Lecture Notes in Computer Science 14465

The series Lecture Notes in Computer Science (LNCS), including its subseries Lecture Notes in Artificial Intelligence (LNAI) and Lecture Notes in Bioinformatics (LNBI), has established itself as a medium for the publication of new developments in computer science and information technology research, teaching, and education.

LNCS enjoys close cooperation with the computer science R & D community, the series counts many renowned academics among its volume editors and paper authors, and collaborates with prestigious societies. Its mission is to serve this international community by providing an invaluable service, mainly focused on the publication of conference and workshop proceedings and postproceedings. LNCS commenced publication in 1973.

Michael A. Bekos · Markus Chimani
Editors

Graph Drawing and Network Visualization

31st International Symposium, GD 2023
Isola delle Femmine, Palermo, Italy, September 20–22, 2023
Revised Selected Papers, Part I

Editors
Michael A. Bekos
University of Ioannina
Ioannina, Greece

Markus Chimani
Osnabrück University
Osnabrück, Germany

ISSN 0302-9743 ISSN 1611-3349 (electronic)
Lecture Notes in Computer Science
ISBN 978-3-031-49271-6 ISBN 978-3-031-49272-3 (eBook)
https://doi.org/10.1007/978-3-031-49272-3

This Springer imprint is published by the registered company Springer Nature Switzerland AG
The registered company address is: Gewerbestrasse 11, 6330 Cham, Switzerland

Paper in this product is recyclable.

Preface

This volume contains the papers presented at GD 2023, the 31st International Symposium on Graph Drawing and Network Visualization, held on September 20–22, 2023 in Isola delle Femmine (Palermo), Italy. Graph drawing is concerned with the geometric representation of graphs and constitutes the algorithmic core of network visualization. Graph drawing and network visualization are motivated by applications where it is crucial to visually analyze and interact with relational datasets. Information about the conference series and past symposia is maintained at http://www.graphdrawing.org.

A total of 122 participants from 18 different countries attended the conference. With regards to the program itself, regular papers could be submitted to one of two distinct tracks: Track 1 for papers on combinatorial and algorithmic aspects of graph drawing and Track 2 for papers on experimental, applied, and network visualization aspects. Short papers were given a separate category, which welcomed both theoretical and applied contributions. An additional track was devoted to poster submissions. All the tracks were handled by a single Program Committee. As in previous editions of GD, the papers in the different tracks did not compete with each other, but all program committee members were invited to review papers from either track in a "light-weight double-blind" process.

In response to the call for papers, the Program Committee received a total of 114 submissions, consisting of 100 papers (52 in Track 1, 23 in Track 2, and 25 in the short paper category) and 14 posters. More than 300 reviews were provided, about a third having been contributed by external sub-reviewers. After extensive electronic discussions by the Program Committee via EasyChair, interspersed with virtual meetings of the Program Chairs producing incremental accept/reject proposals, 31 long papers, 7 short papers, and 11 posters were selected for inclusion in the scientific program of GD 2023. This resulted in an overall paper acceptance rate (not considering posters) of 38% (46% in Track 1, 30% in Track 2, and 28% in the short paper category). As is common in GD, some hard choices had to be made in particular during the final acceptance/rejection round, where several papers that clearly had merit still did not make the cut. However, the number of submitted high-quality papers speaks for the community. Authors published an electronic version of their accepted papers on the arXiv e-print repository; a conference index with links to these contributions was made available before the conference.

There were two invited lectures at GD 2023. Monique Teillaud from INRIA Nancy - Grand Est, LORIA (France) described the intrinsics of "*The CGAL Project*", while Michael Kaufmann from Universität Tübingen (Germany) focused "*On Orthogonal Drawings of Plane and Not So Plane Graphs*". Abstracts of both invited lectures are included in these proceedings.

The conference gave out best paper awards in Track 1 and Track 2, as well as a best presentation award and a best poster award. Based on a majority vote of the Program Committee, "On the Biplanarity of Blowups" by David Eppstein was chosen as the best paper in Track 1. In Track 2, the best paper was chosen to be "Celtic-Graph: Drawing Graphs as Celtic Knots and Links" by Peter Eades, Niklas Gröne, Karsten Klein, Patrick Eades, Leo Schreiber, Ulf Hailer, and Falk Schreiber. Based on a majority vote of conference participants, the best presentation award was given to Julia Katheder for her presentation of the paper "Weakly and Strongly Fan-Planar Graphs". Also chosen by the conference audience, the best poster award was given to "What Happens at Dagstuhl? Uncovering Patterns through Visualization" by Felix Klesen, Jacob Miller, Fabrizio Montecchiani, Martin Nöllenburg, and Markus Wallinger. Many thanks to Springer whose sponsorship funded the prize money for these awards.

A PhD School on graph databases was held on the day prior to the conference. The lectures were led by Nikolay Yakovets (TU Eindhoven), Andreas Kollegger (Neo4J), and Fouli Argyriou (yWorks). This year, we had the special treat of celebrating Giuseppe Liotta's 60th birthday on the day after the conference – an opportunity most of the conference participants happily took. Congratulations, Beppe, for so many successful and graph drawing inspiring years!

As is traditional, the 31st Annual Graph Drawing Contest was held during the conference. The contest was divided into two parts, creative topics and the live challenge. The creative topics task featured a single graph: a board-game recommendations network (a data set containing the top-100 games found on www.boardgamegeek.com). The live challenge focused on minimizing the number of crossings on point set embeddings. There were two categories: manual and automatic. We thank the Contest Committee, chaired by Wouter Meulemans, for preparing interesting and challenging contest problems. A report on the contest is included in these proceedings.

Many people and organizations contributed to the success of GD 2023. We would like to thank all members of the Program Committee and the external reviewers for carefully reviewing and discussing the submitted papers and posters; this was crucial for putting together a strong and interesting program. Also thanks to all authors who chose GD 2023 as the publication venue for their research.

We are grateful for the support of our "Gold" sponsors Tom Sawyer Software and yWorks, our "Bronze" sponsors Springer and Neo4J, and our "Contributor" Bazuel. Their generous support helped to ensure the continued success of this conference.

Last but not least, the organizing committee did a wonderful job in ensuring a smooth and joyful conference experience both for the scientific and the non-scientific parts. The credit for this remarkable achievement goes wholly to the organizing co-chairs, Emilio Di Giacomo, Fabrizio Montecchiani, and Alessandra Tappini. They in turn would like to express their thanks to the other local organizers and volunteers, including Carla Binucci, Luca Grilli, Giacomo Ortali, and Tommaso Piselli.

The 32nd International Symposium on Graph Drawing and Network Visualization (GD 2024) will take place September 18–20, 2024, in Vienna, Austria. Stefan Felsner and Karsten Klein will co-chair the Program Committee, and Robert Ganian and Martin Nöllenburg will co-chair the Organizing Committee.

October 2023

Michael A. Bekos
Markus Chimani

Organization

Steering Committee

Patrizio Angelini	John Cabot University, Italy
Michael A. Bekos	University of Ioannina, Greece
Markus Chimani	Osnabrück University, Germany
Giuseppe Di Battista	Roma Tre University, Italy
Emilio Di Giacomo	University of Perugia, Italy
Stefan Felsner	Technische Universität Berlin, Germany
Reinhard von Hanxleden	University of Kiel, Germany
Karsten Klein	University of Konstanz, Germany
Stephen G. Kobourov (Chair)	University of Arizona, USA
Anna Lubiw	University of Waterloo, Canada
Roberto Tamassia	Brown University, USA
Ioannis G. Tollis	ICS-FORTH and University of Crete, Greece
Alexander Wolff	University of Würzburg, Germany

Program Committee

Md. Jawaherul Alam	Amazon, USA
Daniel Archambault	Swansea University, UK
Martin Balko	Charles University in Prague, Czech Republic
Michael A. Bekos (Co-chair)	University of Ioannina, Greece
Steven Chaplick	Maastricht University, The Netherlands
Markus Chimani (Co-chair)	Osnabrück University, Germany
Sabine Cornelsen	University of Konstanz, Germany
Eva Czabarka	University of South Carolina, USA
Emilio Di Giacomo	University of Perugia, Italy
Christian Duncan	Quinnipiac University, USA
Stefan Felsner	TU Berlin, Germany
Fabrizio Frati	Roma Tre University, Italy
Petr Hliněný	Masaryk University in Brno, Czech Republic
Andreas Kerren	Linköping University, Sweden
Fabian Klute	Polytechnic University of Catalonia, Spain
Anna Lubiw	University of Waterloo, Canada
Tamara Mchedlidze	Utrecht University, The Netherlands
Debajyoti Mondal	University of Saskatchewan, Canada
Fabrizio Montecchiani	University of Perugia, Italy
Martin Nöllenburg	Technische Universität Wien, Austria

Yoshio Okamoto	University of Electro-Communications, Japan
Chrysanthi Raftopoulou	National Technical University of Athens, Greece
Lena Schlipf	Universität Tübingen, Germany
Jens M. Schmidt	University of Rostock, Germany
Matthias Stallmann	North Carolina State University, USA
Csaba Toth	California State University Northridge, USA
Torsten Ueckerdt	Karlsruhe Institute of Technology, Germany
Johannes Zink	Universität Würzburg, Germany

Organizing Committee

Carla Binucci	University of Perugia, Italy
Emilio Di Giacomo (Co-chair)	University of Perugia, Italy
Luca Grilli	University of Perugia, Italy
Fabrizio Montecchiani (Co-chair)	University of Perugia, Italy
Giacomo Ortali	University of Perugia, Italy
Tommaso Piselli	University of Perugia, Italy
Alessandra Tappini (Co-chair)	University of Perugia, Italy

Contest Committee

Philipp Kindermann	University of Trier, Germany
Fabian Klute	Polytechnic University of Catalonia, Spain
Tamara Mchedlidze	Utrecht University, The Netherlands
Debajyoti Mondal	University of Saskatchewan, Canada
Wouter Meulemans (Chair)	TU Eindhoven, The Netherlands

External Reviewers

Agarwal, Shivam
Aichholzer, Oswin
Akitaya, Hugo
Angelini, Patrizio
Arleo, Alessio
Aronov, Boris
Arseneva, Elena
Barát, János
Behrisch, Michael
Bergold, Helena
Bhore, Sujoy
Biniaz, Ahmad
Binucci, Carla
Blažej, Václav
Blum, Johannes
Bonichon, Nicolas
Brandenburg, Franz
Cardinal, Jean
Čermák, Filip
Chakraborty, Dibyayan
Cooper, Joshua
D'Elia, Marco
Da Lozzo, Giordano
Di Bartolomeo, Sara
Diatzko, Gregor
Didimo, Walter
Dijk, Thomas C. Van

Dobler, Alexander
Dumitrescu, Adrian
Dunne, Cody
Dutle, Aaron
Eiben, Eduard
Eppstein, David
Firman, Oksana
Fujiwara, Takanori
Fulek, Radoslav
Förster, Henry
Gethner, Ellen
Goaoc, Xavier
Gonçalves, Daniel
Grilli, Luca
Gronemann, Martin
Grosso, Fabrizio
Gupta, Siddharth
Guspiel, Grzegorz
Hamalainen, Rimma
Hamm, Thekla
Hegemann, Tim
Hoffmann, Michael
Hušek, Radek
Joret, Gwenaël
Jungeblut, Paul
Katheder, Julia
Katsanou, Eleni
Khazaliya, Liana
Kindermann, Philipp
Klawitter, Jonathan
Klemz, Boris
Klesen, Felix
Kobourov, Stephen
Kratochvil, Jan
Kucher, Kostiantyn
Kypridemou, Elektra
Lauff, Robert
Linhares, Claudio
Liu, Kevin
Mann, Ryan
Martínez Sandoval, Leonardo Ignacio
Masařík, Tomáš
McGee, Fintan
Meuwese, Ruben
Mueller, Tobias
Opler, Michal
Ortali, Giacomo
Ortlieb, Christian
Parada, Irene
Patrignani, Maurizio
Piselli, Tommaso
Pokrývka, Filip
Pupyrev, Sergey
Purchase, Helen
Reddy, Meghana M.
Roch, Sandro
Rollin, Jonathan
Rosenke, Christian
Rutter, Ignaz
Samal, Robert
Schaefer, Marcus
Scheibner, Mark
Scheucher, Manfred
Schnider, Patrick
Schröder, Felix
Schulz, André
Shahrokhi, Farhad
Sieper, Marie Diana
Staals, Frank
Steiner, Raphael
Storandt, Sabine
Stumpf, Peter
Szekely, Laszlo
Tappini, Alessandra
Tkadlec, Josef
Toth, Geza
van Goethem, Arthur
van Wageningen, Simon
Vogtenhuber, Birgit
Wallinger, Markus
Wang, Yong
Wang, Zhiyu
Wolff, Alexander
Wood, David R.
Wu, Hsiang-Yun
Wulms, Jules
Xia, Ge
Yip, Chi Hoi
Zeng, Ji

Sponsors

Gold Sponsors

Bronze Sponsors

Contributors

Invited Talks

The CGAL Project – cgal.org

Monique Teillaud

Inria Nancy – Grand Est, LORIA, France
monique.teillaud@inria.fr

Abstract. CGAL, The Computational Geometry Algorithms Library, provides easy access to efficient and reliable geometric algorithms in the form of a C++ library. It is used both in academia and industry, in various areas needing geometric computation, such as geographic information systems, computer aided design, molecular biology, medical imaging, computer graphics, and robotics.

The CGAL Project is a joint effort of several research groups in academia and their industrial partner GeometryFactory. It started in 1996 as a consortium of seven academic sites in Europe and Israel. In 2003, it has become an Open Source project, inviting developers from around the world to join.

This talk is definitely not going to be a CGAL course. Instead I will try to give insight about the history of the project and the multiple ingredients that have made it successful over the years: technical choices, development tools, project organization, license choice, etc, without forgetting the most important: the CGAL people.

On Orthogonal Drawings of Plane and Not So Plane Graphs

Michael Kaufmann

Wilhelm-Schickard-Institut für Informatik,
Universität Tübingen, Tübingen, Germany
michael.kaufmann@uni-tuebingen.de

Abstract. The orthogonal drawing model is one of the basic models in Graph Drawing. Algorithms and variants have developed from the very beginning of the field, over the years and it is still very present in recent works.

In my talk, I will make a personal walk through the last 30 years of orthogonal graph drawing. I will highlight some topics that I like and which you possibly don't know, still containing remarkable techniques. At the end, I will give some open questions and directions from the past that might be remarkable.

Contents – Part I

Linear Layouts

Geometric Aspects

Visualization Challenges

Graph Representations

Contents – Part II

Frameworks

Algorithmics

Posters

Best Papers

On the Biplanarity of Blowups

David Eppstein(✉)

Computer Science Department, University of California, Irvine, Irvine, USA
eppstein@uci.edu

Abstract. The 2-blowup of a graph is obtained by replacing each vertex with two non-adjacent copies; a graph is biplanar if it is the union of two planar graphs. We disprove a conjecture of Gethner that 2-blowups of planar graphs are biplanar: iterated Kleetopes are counterexamples. Additionally, we construct biplanar drawings of 2-blowups of planar graphs whose duals have two-path induced path partitions, and drawings with split thickness two of 2-blowups of 3-chromatic planar graphs, and of graphs that can be decomposed into a Hamiltonian path and a dual Hamiltonian path.

Keywords: Graph thickness · Split thickness · Graph blowups · Kleetopes

1 Introduction

In a 2018 survey on the Earth–Moon problem, Ellen Gethner conjectured that 2-blowups of planar graphs are always biplanar [11]. In this paper we refute this conjecture by showing that 2-blowups of iterated Kleetopes are non-biplanar, and more strongly do not have split thickness two.

1.1 Definitions and Preliminaries

Before detailing our results, let us unpack this terminology: what are Kleetopes, biplanarity and split thickness, and blowups?

Polyhedral graphs are the graphs of convex polyhedra. By Steinitz's theorem, these are exactly the 3-vertex-connected planar graphs [23]. Polyhedral graphs have unique planar embeddings [19], whose faces are exactly the *peripheral cycles*, cycles such that every two edges not in the cycle are part of a path with interior vertices disjoint from the cycle [25]. Every *maximal planar graph* with ≥ 4 vertices, one to which no edges can be added while preserving planarity, is polyhedral.

The *Kleetope* of a polyhedral graph (named by Branko Grünbaum for Victor Klee [13]) is a maximal planar graph obtained by adding a new vertex within every face, adjacent to all the vertices of the face. Geometrically, it can be formed by attaching a pyramid to every face, simultaneously. An *iterated Kleetope* is the

M. A. Bekos and M. Chimani (Eds.): GD 2023, LNCS 14465, pp. 3–17, 2023.
https://doi.org/10.1007/978-3-031-49272-3_1

result of repeatedly applying this operation a given number of times. Following notation from our previous work [8], we let KG denote the Kleetope of a graph G and K^iG denote the result of applying the Kleetope operation i times to G.

Thickness is the minimum number of planar subgraphs needed to cover all edges of a given graph. Equivalently, it is the minimum number of edge colors needed to draw the graph in the plane with colored edges so each crossing has edges of two different colors. A graph is *biplanar* if its thickness is at most two. Thus, a biplanar drawing of a graph can be interpreted as a pair of planar drawings of two subgraphs of the given graph that, together, include all of the graph edges. Repeated edges are never necessary and for technical reasons we forbid them.

Little was known about the structure of biplanar graphs. They include all graphs of maximum degree four [5,14], and therefore cannot have more structure than degree-four graphs. An NP-completeness reduction for biplanarity by Mansfield [20] can be used to construct infinitely many non-biplanar graphs. Additionally, as Sýkora et al. observed, 5-regular graphs of girth ≥10 are too dense for their girth to be biplanar [24].

Split thickness is a generalization of thickness in which we form a single planar drawing with multiple copies of each vertex, which are not required to be near each other in the drawing. Each edge of the graph appears once, connecting an arbitrary pair of copies of its endpoints. A drawing has split thickness k if each vertex has at most k copies, and the split thickness of a graph G is the minimum number k such that G has a drawing with split thickness k [9]. Split thickness is less than or equal to thickness, but they can diverge, even for complete graphs: K_{12} has split thickness two [15] but K_9 already has thickness three [2,26]. Thickness has its origin in the Earth–Moon problem, posed by Gerhard Ringel in 1959 [22], which in graph-theoretic terms asks for the maximum chromatic number of biplanar graphs. In the same way, split thickness corresponds to the older m-pire coloring problem [10].

Blowups of graphs are formed by duplicating their vertices a given number of times. More specifically, the (open) k-blowup of a graph G, which we denote as kG,[1] is obtained by making k copies of each vertex of G, and by connecting two vertices in kG whenever they are copies of adjacent vertices in G. Two copies of the same vertex are not adjacent. In a *closed blowup*, the copies are adjacent.

Albertson, Boutin, and Gethner [1] proved that the closed 2-blowup of any tree or forest is planar and therefore that the closed 2-blowup of any graph of arboricity a has thickness at most a. Here, *arboricity* is the minimum number of forests that cover all edges of a graph. Adding a leaf to a forest corresponds, in the closed 2-blowup, to gluing a K_4 subgraph onto an edge, which preserves planarity, and the a forests that cover a graph of arboricity a can be blown up by induction in this way, separately from each other (Fig. 1).

[1] Blowups are a standard concept but their notation varies significantly. Other choices from the literature include $G(k)$, $G[k]$, G^k, and $G^{(k)}$.

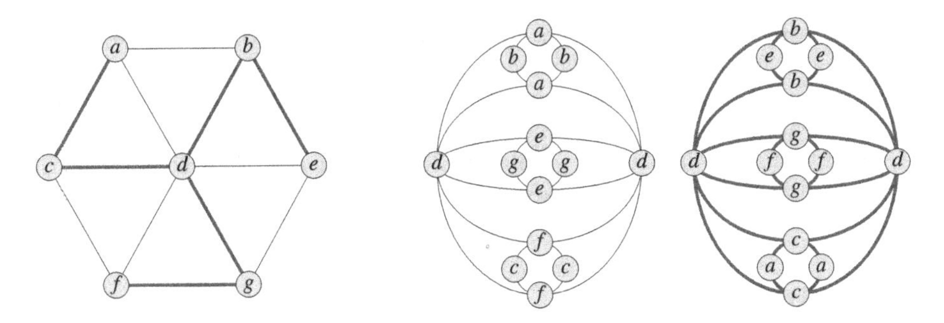

Fig. 1. Decomposition of a seven-vertex wheel graph into two trees (left) and the corresponding biplanar drawing of its 2-blowup (right)

Planar graphs have arboricity at most three [21], and this is tight for planar graphs with more than $2n-2$ edges, including most maximal planar graphs. Therefore, their 2-blowups have thickness at most three. Gethner's conjecture asks whether there are planar graphs for which the resulting thickness-three drawing is optimal or whether smaller thickness, two, can always be achieved.

1.2 New Results

Our main result is that for all sufficiently large maximal planar graphs G, $2K^3G$ does not have thickness two and does not have split thickness two. This gives a proof of non-biplanarity for a natural and sparse class of graphs that (unlike previous methods for proving non-biplanarity) allows short cycles.

To complement this result, we provide biplanar or split thickness two drawings for the blowups of three natural classes of planar graphs:

- We construct a biplanar drawing of the 2-blowup of any planar graph whose faces can be decomposed into two *outerpaths*, strips of polygons connected edge-to-edge with the topology of a path (see Definition 3). Equivalently, this structure is a partition of the dual graph into two induced paths.
- When G can be decomposed into two outerpaths, we construct a split thickness two drawing of its Kleetope KG. In the special case of the tetrahedral graph K_4, this construction can be used for the iterated Kleetope K^2K_4.
- When G and its dual have disjoint Hamiltonian paths, we construct a split thickness two drawing of the 2-blowup of G.
- We construct a split thickness two drawing of the 2-blowup of any 3-chromatic planar graph, and more generally a drawing with split thickness k of the k-blowup of these graphs.

These drawing algorithms motivate our use of iterated Kleetopes in constructing non-biplanar blowups of planar graphs, because Kleetopes are far from having the properties needed to make these algorithms work. Kleetopes of maximal planar graphs are far from 3-chromatic: each added vertex forms a K_4 subgraph, an obstacle to 3-coloring. And iterated Kleetopes are far from being decomposable into outerpaths, as their dual graphs have no long induced paths. The underlying planar graphs for each of our drawing algorithms include infinitely many maximal planar graphs, showing that it is not merely the large size and maximality of iterated Kleetopes that prevents their blowups from having drawings. Additionally, because triangle-free planar graphs are 3-chromatic [12], these constructions suggest that the connection of Sýkora et al. between girth and non-biplanarity is unlikely to help construct planar graphs with non-biplanar blowups.

2 Iterated Kleetopes

In this section we show that some 2-blowups of iterated Kleetopes are not biplanar and do not have split thickness two or less. As in our previous work on the geometric realization of iterated Kleetopes [8], our approach uses the observation that any realization or drawing of an iterated Kleetope must be based on a realization or drawing of a graph with one fewer iteration. This simpler drawing can be recovered from the final drawing by removing the vertices added in the Kleetope process. Using this observation, we build up a sequence of stronger properties for the biplanar and split thickness two embeddings of these graphs, as the number of Kleetope iterations increases. Eventually, these properties will become so strong that they lead to an impossibility.

Definition 1. *For a vertex v of graph G, it is convenient to denote the two copies of v in $2G$ by v_0 and v_1. We distinguish these from the two* images *of v_0 and the two images of v_1 in a biplanar or split thickness two drawing of $2G$. In such a drawing, v itself has four images, two from v_0 and two from v_1.*

We need the following definitions in the proof of our first lemma.

Definition 2. *Define the* excess *of a face in a planar, biplanar, or split thickness two drawing of a graph to be the number of edges in the face, minus three, so triangles have excess zero and all other faces have positive excess. Define the total excess of the drawing to be the sum of all face excesses.*

The total excess of a drawing equals the amount by which the number of edges in the graph falls short of the maximum possible number of edges in a drawing of its type, and so can be calculated only from a graph and the type of its drawing, independent of how it is drawn:

- A planar drawing of an n-vertex graph can have at most $3n-6$ edges, and if there are m edges then the total excess is $(3n-6)-m$.
- A biplanar drawing of an n-vertex graph can have at most $6n-12$ edges, and if there are m edges then the total excess is $(6n-12)-m$.
- A split thickness two drawing of an n-vertex graph can have at most $6n-6$ edges, and if there are m edges then the total excess is $(6n-6)-m$.

Lemma 1. *Let G be a maximal planar graph with n vertices. If $n \geq 49$, then in any biplanar drawing D of $2G$ some vertex v of G has images that are only incident to triangles. If $n \geq 73$, then for any split thickness two drawing some vertex v has the same property. We say that v has* triangulated neighborhoods.

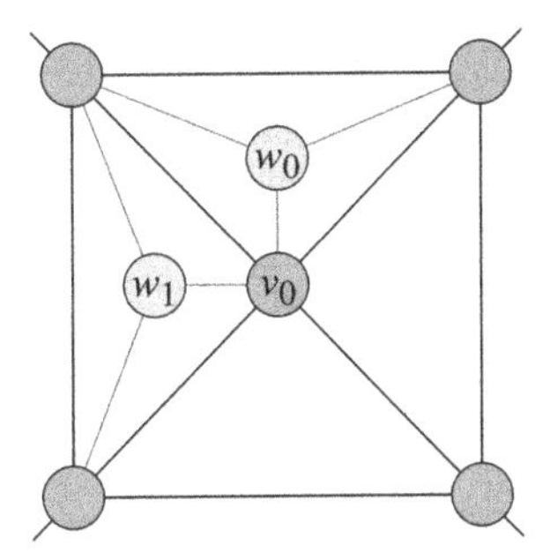

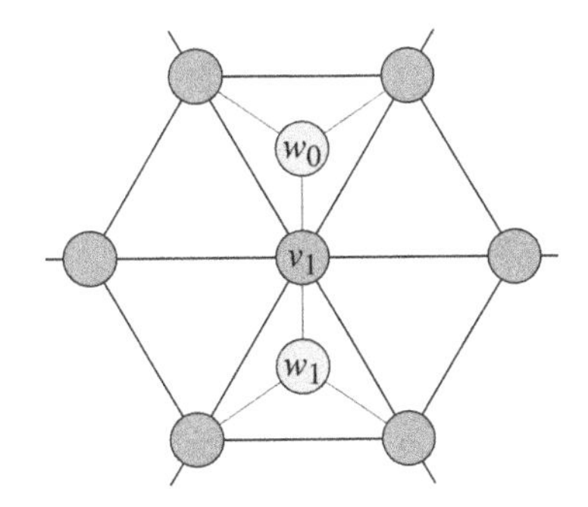

Fig. 2. Left: Illustration for Lemma 2: If v in G has triangulated neighborhoods, and w is any neighbor of v added in KG, then the images of w must lie in two triangles incident to images of v, connected to all six triangle vertices. In this example, the four images of w are neighbors of only two images of v, but they may instead be neighbors of three or four images of v.

Proof. The number of vertices that belong to non-triangular faces in a planar, biplanar, or split thickness two drawing is maximized when the total excess is distributed among disjoint quadrilateral faces. In this case, the number of such vertices is four times the excess. Any other distribution of the total excess, or non-disjointness among the non-triangular faces, produces fewer such vertices.

Because G is maximal planar, it has excess zero and $3n-6$ edges. Its blowup $2G$ has $2n$ vertices and $12n-24$ edges, four copies of each edge in G. Therefore, any biplanar drawing of G has excess 12, and any split thickness two drawing of G has excess 18. The number of vertices that can belong to a non-triangular face in these drawings is, respectively, 48 and 72. For larger values of n, some vertex has triangulated neighborhoods. □

Lemma 2. *Let G be a maximal planar graph, and consider any biplanar drawing or split thickness two drawing D of $2KG$, and the restriction of the same drawing to G. If some vertex v of G has triangulated neighborhoods in the restriction to G, then any neighbor w of v in $KG \setminus G$ has its four images each drawn surrounded by exactly three triangular faces. We say that w has* triangular neighborhoods. *If G has n vertices with $n \geq 49$, then in any biplanar drawing D of $2KG$ some vertex w of KG has triangular neighborhoods. If $n \geq 73$, then for any split thickness two drawing of $2KG$ some vertex w has triangular neighborhoods.*

Proof. As an added vertex in KG, w has degree three, so its copy w_0 in $2KG$ has degree six. To be adjacent to both v_0 and v_1, the two images of w_0 must lie in two triangular faces of the restricted drawing containing v_0 and v_1, as shown in Fig. 2. (No face contains both v_0 and v_1, because they have triangular neighborhoods and are not adjacent.) This placement limits w_0 to having as neighbors only the six vertices of these two triangles, matching its degree, so it must be connected to all six of these vertices. The same argument applies to w_1.

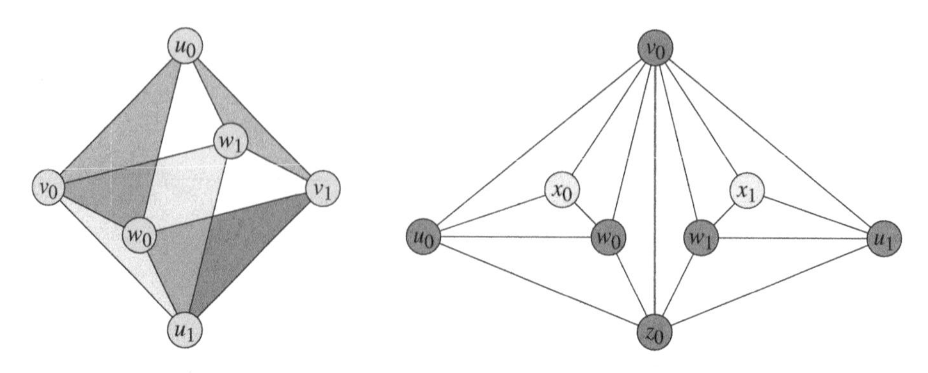

Fig. 3. Left: Illustration for Lemma 3: partition of 2Δ into four triangles, for a triangle $\Delta = uvw$. Right: Illustration for Theorem 1: For the restriction of a given drawing to $2KG$, and for w in KG, images w_0 and w_1 have triangular neighborhoods sharing edge v_0z_0 (red). The third vertices of these triangular neighborhoods, u_0 and u_1, are distinct images of a neighbor u of w. For vertex x in K^2G adjacent to w and to u, two images have triangular neighborhoods without shared edges. (Color figure online)

The existence of w in biplanar or split thickness drawings of $2KG$ for graphs with many vertices follows by applying this argument to the vertex v with triangulated neighborhoods given by Lemma 1. □

In Lemma 2, the two triangular neighborhoods of w_0 must be disjoint, so they cover all six distinct neighbors of w_0 in $2KG$. Similarly, the two neighborhoods of w_1 must be disjoint. However, a neighborhood of w_0 may share a vertex or an edge with a neighborhood of w_1. (It cannot share edges with two neighborhoods because then those two neighborhoods would not be disjoint.) In fact, some sharing is necessary:

Lemma 3. *Let t be a vertex of a planar graph H having three neighbors, all adjacent. Suppose that t has triangular neighborhoods in a biplanar or split thickness two drawing of $2H$. Then it is impossible for all four images of t to have edge-disjoint neighborhoods in this drawing.*

Proof. Let Δ be the triangle of neighbors of t in H. 2Δ is isomorphic to $K_{2,2,2}$, the graph of a regular octahedron. We are assuming our drawings have no repeated edges, so two images of t with edge-disjoint neighborhoods in the drawing must come from edge-disjoint triangles of 2Δ. Thus, if the four images of t could be drawn with edge-disjoint triangular neighborhoods, these neighborhoods would form four edge-disjoint triangles of 2Δ. But in any subdivision of 2Δ into four edge-disjoint triangles (Fig. 3, left), each two triangles share a vertex, and together cover only five vertices of 2Δ. If the four images of t were placed in images of these four triangles, the two triangular neighborhoods of t_0 would miss one of the six neighbors of t_0 in $2H$, as would the triangular neighborhoods of t_1, preventing the drawing from being valid. Therefore, no such drawing is possible. □

Theorem 1. *Let G be a maximal planar graph with n vertices. If $n \geq 49$, then $2K^3G$ has no biplanar drawing, and if $n \geq 73$, then $2K^3G$ has no split thickness two drawing.*

Proof. Suppose for a contradiction that such a drawing existed, and consider the drawings within it of $2K^2G$ and of $2KG$. We will find a vertex with triangular neighborhoods in each of these drawings so that the four neighborhoods of the four images of the chosen vertex are nearly disjoint: these neighborhoods can share at most two edges total in $2KG$, at most one edge in $2K^2G$, and no edges in $2K^3G$. The existence of a vertex in K^3G whose triangular neighborhoods share no edges will contradict Lemma 3, showing that no such drawing can exist. To do this, we consider each level of iteration successively, as follows:

- In the drawing of $2KG$, Lemma 2 gives us a vertex w of KG with triangular neighborhoods. The neighborhood of each image of w shares at most one edge with other neighborhoods of images of w. For biplanar drawings this is immediate (only one other image is in the same planar subgraph and can share an edge with it). For split thickness two drawings, a neighborhood of w_0 that shares edges with the neighborhoods of both images of w_1 is impossible, because then the two images of w_1 would share a vertex, preventing them from covering all six neighbors of w_1. Therefore, among the neighborhoods of all four images of w, there are at most two shared edges.
- To find a vertex x in K^2G with triangular neighborhoods in the drawing of $2K^2G$, with at most one edge shared among the neighborhoods of its four images, consider the vertex w found above in KG. If the neighborhoods of w have at most one shared edge in the drawing of $2KG$, let x be any neighbor of w added in forming K^2G from KG. Then x must again have triangular neighborhoods by Lemma 2. Because these triangular neighborhoods must be interior to the triangular neighborhoods of w, they can have at most one shared edge (the same edge as the one shared by the neighborhoods of w).
Suppose, on the other hand, that the triangular neighborhoods of w share exactly two edges. Let Δ_0, Δ_0', Δ_1, and Δ_1' be the four triangles in $2KG$ neighboring the two images of w_0 and w_1, respectively, with an edge shared by Δ_0 and Δ_1 and another edge shared by Δ_0' and Δ_1'. Because these triangles can only share one edge, and each image of w must be adjacent to images of all three neighbors of w, the vertices of Δ_0 and Δ_1 that are not on the shared edge must be distinct images of the same vertex u. Choose a vertex x of $K^2G \setminus KG$, adjacent to w and to u.
Because w has triangular neighborhoods in KG, its four images in the drawing of $2KG$ are each surrounded by three triangles, formed by images of w and two of its neighbors. For each image, only one of these triangles consists of the three neighbors of x. Thus, the four images of x must be placed in these four triangles. Two of these four triangles are subdivisions of Δ_0 and Δ_1, containing the non-shared images of u, and therefore do not share any edge with each other. The only possible shared edge among the triangular neighborhoods of x is the edge shared by Δ_0', and Δ_1', within which lie the

other two images of x. Thus, by choosing x in K^2G we have eliminated one shared edge between triangular neighborhoods. See Fig. 3, right.
- If the four triangular neighborhoods of x in the drawing of $2K^2G$ are edge-disjoint, we already have a contradiction with Lemma 3. Otherwise, we must find a vertex t in K^3G whose triangular neighborhoods are edge-disjoint, giving us the desired contradiction. To do so, consider the four triangular neighborhoods of x in the drawing of $2K^2G$, only two of which share one edge. For the two triangles that share an edge, the two non-shared vertices of these triangles must be distinct images of the same vertex y in K^2G. Choose a vertex t of $K^3G \setminus K^2G$, adjacent to x and to y. Then the four triangles in which z must be placed lie within the four triangular neighborhoods of x, away from the shared edge of these triangular neighborhoods, so they cannot share any edges with each other. By Lemma 3, this is an impossibility. □

When G is not maximal planar or is too small for the theorem to apply directly, more iterations of the Kleetope operation can be used before applying the same argument. Thus, there exists an i such that for every plane graph G, $2K^iG$ is not biplanar and has no split thickness two drawing.

3 Drawings from Triangle Strips

Definition 3. *An* outerpath *is an outerplanar graph whose weak dual (the adjacency graph of its bounded faces) is a path.*

Suppose that the dual vertices of a planar graph G can be partitioned into two subsets that each induce a path in the dual graph. Then the dual cut edges between these two subsets correspond, in G itself, to a Hamiltonian cycle that partitions G into two outerpaths. Figure 4 depicts an example of this sort of *two-outerpath decomposition* for the graph of the regular icosahedron.

Theorem 2. *If a planar graph G has a two-outerpath decomposition, then $2G$ has a biplanar drawing.*

Proof. We may assume without loss of generality, by triangulating each outerplanar graph if necessary, that both outerpaths are maximal outerplanar: each of their faces is a triangle. If this triangulation step adds two copies of the same edge to the graph, it is not a problem, because the edges added in this triangulation step will be removed from the final drawing.

In each outerpath, the triangular faces form a linear sequence, separated by the internal edges of the outerpath, which are also linearly ordered. In the blowup $2G$, number the two copies of each vertex v as v_0 and v_1. If uv is one of the diagonals of one of the outerpaths (that is, one of its internal edges), then $2uv$ is a four-vertex cycle $u_0v_0u_1v_1$.

We will construct a biplanar drawing of $2G$ with each plane containing all copies of interior edges of one of the two outerpaths, and two out of the four copies of each boundary edge. We draw copies of the interior edges as nested

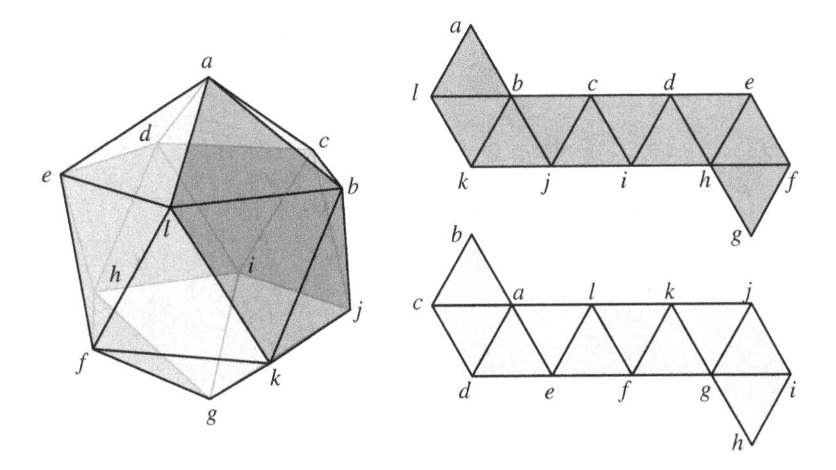

Fig. 4. Decomposition of an icosahedron into two outerpaths.

quadrilaterals, one for each diagonal of its outerpath, in the same order that these diagonals appear within the outerpath. If we draw these quadrilaterals one at a time, from the innermost to the outermost, then each two consecutive quadrilaterals share two opposite vertices, corresponding to the single shared endpoint of the two diagonals.

In each pair of consecutive quadrilaterals, the outer quadrilateral has two potential orientations with respect to the inner one: if the two consecutive diagonals are uv and vw, with quadrilateral $u_0v_0u_1v_1$ drawn inside quadrilateral $v_0w_0v_1w_1$, then these quadrilaterals may be drawn so that pairs u_0w_0 and u_1w_1 are adjacent, or so that pairs u_0w_1 and u_1w_0 are adjacent. In one plane we always choose the orientation with u_0w_0 and u_1w_1 adjacent, and we connect these pairs of vertices by an edge. In the other plane, we always choose the orientation with u_0w_1 and u_1w_0 adjacent, and we connect these pairs of vertices by an edge. In this way, we draw all four copies of each boundary edge, except the edges incident to the two ears (triangles with two boundary edges).

It remains to draw the ears. Each has two boundary edges sharing a vertex, which we call the *ear vertex*. These two boundary edges have not yet been drawn in the plane of their outerpath. The third edge of the ear is a diagonal whose images form the innermost or outermost quadrilateral in its plane. We place both copies of the ear vertex inside or outside this quadrilateral (respectively as it is innermost or outermost), connected to its two neighbors in the ear with the same numbering convention: in the plane where the quadrilaterals are oriented with u_0w_0 and u_1w_1 adjacent, we connect each ear vertex to neighbors with the same subscript, and in the other plane we connect each ear vertex to neighbors with the opposite subscript.

Thus, all copies of the diagonals of one strip and all copies of boundary edges that connect copies having the same index are drawn in one plane. All copies of

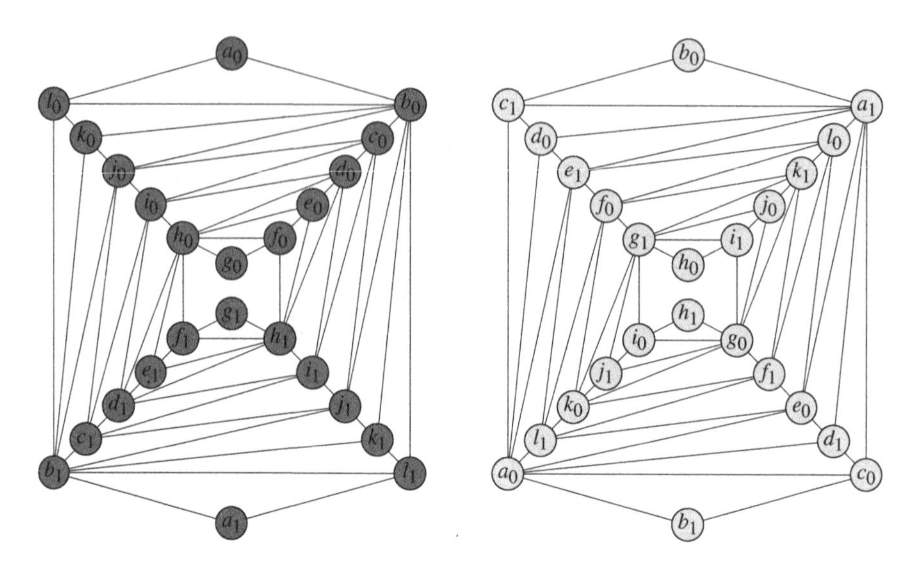

Fig. 5. Biplanar drawing of the 2-blowup of an icosahedron corresponding to the outerpath decomposition of Fig. 4.

the diagonals of the other strip and all copies of boundary edges that connect copies having different indices are drawn in the other plane. The result is a biplanar drawing of the entire blowup $2G$. □

Figure 5 depicts the drawing obtained by applying Theorem 2 to the outerpath decomposition of the icosahedron depicted in Fig. 4. The following extension of this result is noteworthy in connection with our iterated Kleetope counterexample:

Theorem 3. *If a maximal planar graph G has a decomposition into two outerplanar graphs, coming from an induced path partition of its dual graph into two paths, then $2KG$ has a drawing with split thickness two.*

Proof. From the drawing of Theorem 2 for $2G$, add two more edges from the ear vertices to the other vertices on the hexagonal faces they belong to, so that each of the two planes of the drawing contains four images of each triangular face of its outerplanar subgraph of G, in two disjoint pairs. These four triangles can be used to place the four images of each added vertex of KG. □

For instance, the triakis icosahedron, the Kleetope of the icosahedron, is a maximal planar graph whose edges all have total degree ≥ 13. (This is the maximum possible for the minimum total degree of an edge, by Kotzig's theorem [17].) Because the icosahedron has a two-outerpath decomposition, we can apply Theorem 3 to its Kleetope, producing a drawing with split thickness two of the 2-blowup of the triakis icosahedron.

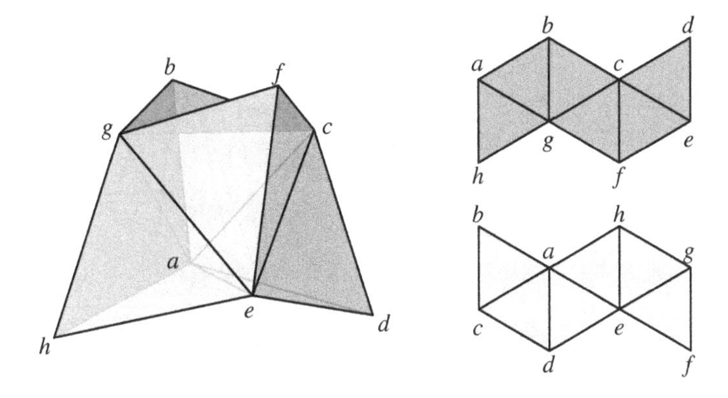

Fig. 6. Outerpath decomposition of KK_4.

In general, we cannot extend this construction to higher-order Kleetopes. When a graph G has a two-outerpath decomposition, the drawings of $2KG$ produced from G by Theorem 3 again have four images of each triangular face of KG, but some of these quadruples of images cannot be grouped into disjoint pairs. However, in the method of Theorem 3 each added vertex of K^2G corresponds to two vertices in $2K^2G$, and the two images of each of these two vertices must be placed in disjoint triangles, in order to provide all six of its adjacencies. Therefore, this method does not provide drawings of $2K^2G$. However, in one special case, for $G = K_4$ (the graph of a tetrahedron), a different method works. In this case, the Kleetope KK_4 has an outerpath decomposition, shown in Fig. 6. Therefore, applying Theorem 3 we can obtain a split thickness two drawing of $2K^2K_4$.

A very similar drawing algorithm to the one in Theorem 2 can be used for graphs with a different form of decomposition into triangle strips.

Definition 4. *Let G be a planar graph having both a Hamiltonian path P and a dual Hamiltonian path P^*, with no edge and its dual edge belonging to both paths. Then we call (P, P^*) a* path–copath decomposition. *It is a special case of the* tree–cotree decomposition *formed from any spanning tree of a planar graph and the dual spanning tree formed by the duals of the complementary set of edges [6].*

Theorem 4. *Let planar graph G have a path–copath decomposition. Then $2G$ has a drawing with split thickness two.*

Proof. We may triangulate the faces of G, if necessary, preserving the existence of a path–copath decomposition by choosing added diagonals that split each face into an outerpath. As a result, the dual path P^* of the path–copath decomposition (P, P^*) becomes a triangulated outerpath. Each vertex of G may appear multiple times on the boundary of this outerpath, with multiplicity equal to its degree in P (at most two, because P is a path). The outerpath can be formed from G by cutting the plane along each edge of P; as a result, each edge of P appears exactly twice on the boundary of the outerpath.

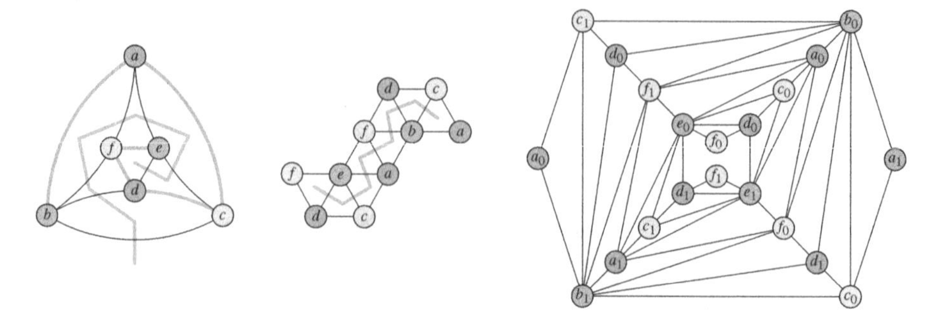

Fig. 7. Path–copath decomposition of the octahedral graph $K_{2,2,2}$ (left), the outerpath obtained by cutting the path (center), and the split thickness 2 drawing of $2K_{2,2,2} = K_{4,4,4}$ obtained from Theorem 4 (right).

We apply the method of Theorem 2 to this single outerpath, producing a drawing in which each appearance of a vertex v of G on the boundary of the outerpath produces images of both copies of v. If v appears once on the boundary of the outerpath, its two copies appear once; if v appears twice, its two copies appear twice, giving this drawing split thickness two. This drawing style automatically produces all four images of each edge interior to the outerpath. For an edge uv of path P, appearing twice on the boundary of the outerpath, we choose arbitrarily which appearance of uv on the boundary of the outerpath is used to draw edges u_0v_0 and u_1v_1, and which is used to draw edges u_0v_1 and u_1v_0. In this way, all four images of uv are drawn correctly. □

Figure 7 depicts an example.

4 Drawings from Colorings

We show in this section that the blowup $2G$ of a 3-colored planar graph G has split thickness at most two. We do not know whether all such blowups are biplanar; our construction does not produce a biplanar drawing. More generally, we show that kG has split thickness at most k; the result for $2G$ is a special case.

Theorem 5. *Let G be planar and 3-chromatic; then kG has a drawing with split thickness k.*

Proof. Color the vertices of G red, blue, and yellow, and number the copies of each vertex in kG from 0 to $k-1$. Draw kG as k^2 disjoint copies of G, where for (i, j) with $0 \leq i, j < k$ we draw a copy of G consisting of the copies of red vertices numbered i, the copies of blue vertices numbered j, and the copies of yellow vertices numbered $-(i+j) \bmod k$. As the disjoint union of k^2 planar drawings, the result is planar. Each edge of kG appears in one copy of G, and each vertex in kG has images in k copies of G. As a planar drawing with k images of each vertex, it is a drawing with split thickness k. □

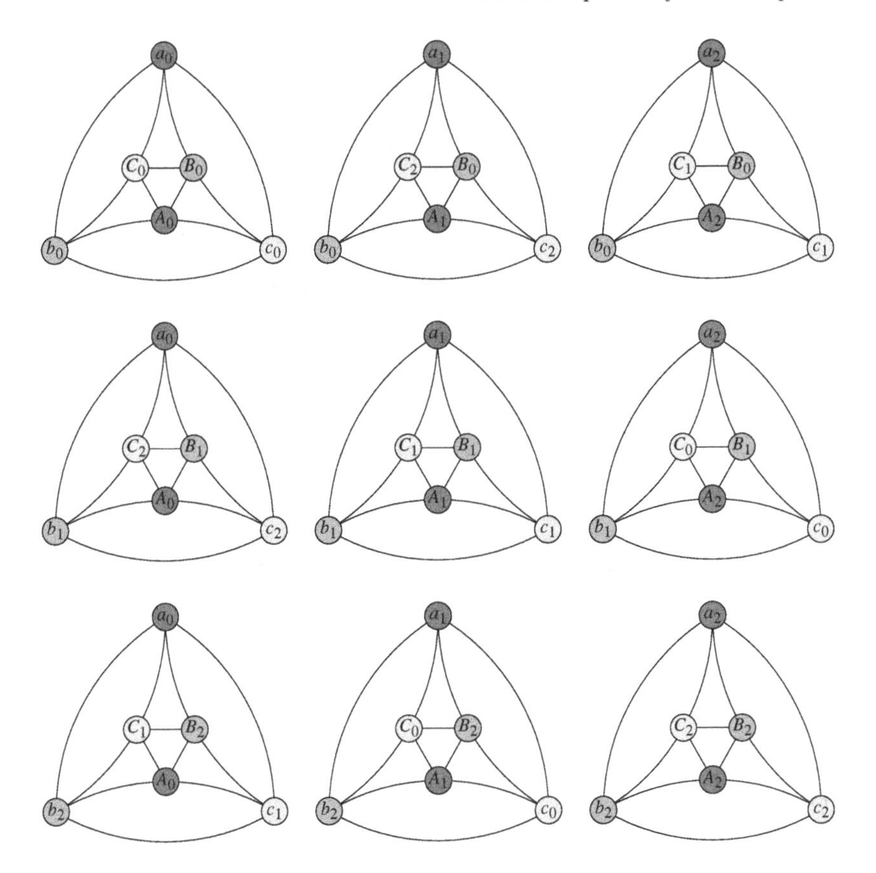

Fig. 8. Theorem 5 applied to the graph $3K_{2,2,2} = K_{6,6,6}$. Each vertex is labeled with a letter (its position in $K_{2,2,2}$), a number (its index as a copy in the blowup), and a color in the coloring of $K_{2,2,2}$. Each letter-number combination has three images, so this is a drawing with split thickness three. $K_{6,6,6}$ has 108 edges, but drawings of 18-vertex graphs with split thickness two can have at most 102 edges, so this drawing is optimal.

Figure 8 shows a drawing of $K_{6,6,6}$ with split thickness three, obtained by applying this construction to the triple blowup of the graph of the octahedron. We remark that, when applied to planar bipartite graphs, the same construction yields a thickness k drawing of the k-blowup: if we group together the copies of G that would have index i for the vertices of the missing color, then each copy of each vertex appears once in each group. In the case $k = 2$, it is also possible to find biplanar drawings of the 2-blowups of planar bipartite graphs in a different way, using the fact that these graphs have arboricity at most two.

5 Conclusions

We have shown that 2-blowups of iterated Kleetopes are not biplanar, but that 2-blowups of planar graphs with outerpath decompositions are biplanar. Additionally, we have shown that 2-blowups of graphs with path–copath decompositions have split thickness at most 2, and k-blowups of planar graphs with chromatic number at most three have split thickness at most k.

Several natural questions remain open for future research:

- Is it ever possible for the 2-blowup of a 3-chromatic planar graph to be non-biplanar? Is it ever possible for the 2-blowup of a 4-vertex-connected planar graph to be non-biplanar?
- What is the computational complexity of finding biplanar drawings of 2-blowups of planar graphs? Biplanarity is NP-complete in general [20], but the proof does not apply to this special case.
- What is the computational complexity of finding a two-outerpath decomposition? Partition into two induced paths is NP-complete for general graphs [18], but although its planar case is closely related to Hamiltonicity of the dual graph, we are unaware of complexity results for this case.
- Can Theorem 2, on drawing 2-blowups of graphs with a two-outerpath decomposition, be extended from thickness to geometric thickness? Geometric thickness (also called real linear thickness) is similar to thickness, but requires vertices to have the same geometric placement in each planar subgraph and requires edges to be drawn as non-crossing line segments [3–5,7,16].

Acknowledgements. This research was supported in part by NSF grant CCF-2212129.

References

1. Albertson, M.O., Boutin, D.L., Gethner, E.: The thickness and chromatic number of r-inflated graphs. Discret. Math. **310**(20), 2725–2734 (2010). https://doi.org/10.1016/j.disc.2010.04.019
2. Battle, J., Harary, F., Kodama, Y.: Every planar graph with nine points has a nonplanar complement. Bull. Am. Math. Soc. **68**, 569–571 (1962). https://doi.org/10.1090/S0002-9904-1962-10850-7
3. Dillencourt, M.B., Eppstein, D., Hirschberg, D.S.: Geometric thickness of complete graphs. J. Graph Algorithms Appl. **4**(3), 5–17 (2000). https://doi.org/10.7155/jgaa.00023
4. Dujmović, V., Wood, D.R.: Graph treewidth and geometric thickness parameters. Discret. Comput. Geom. **37**(4), 641–670 (2007). https://doi.org/10.1007/s00454-007-1318-7
5. Duncan, C.A., Eppstein, D., Kobourov, S.: The geometric thickness of low degree graphs. In: Snoeyink, J., Boissonnat, J.D. (eds.) Proceedings of the 20th ACM Symposium on Computational Geometry, Brooklyn, New York, USA, 8–11 June 2004, pp. 340–346. ACM (2004). https://doi.org/10.1145/997817.997868
6. Eppstein, D.: Dynamic generators of topologically embedded graphs. In: Proceedings of the Fourteenth Annual ACM–SIAM Symposium on Discrete Algorithms, Baltimore, Maryland, USA, 12–14 January 2003, pp. 599–608. Association for Computing Machinery and Society for Industrial and Applied Mathematics (2003)
7. Eppstein, D.: Separating thickness from geometric thickness. In: Pach, J. (ed.) Towards a Theory of Geometric Graphs, Contemporary Mathematics, vol. 342, pp. 75–86. American Mathematical Society (2004)

8. Eppstein, D.: On polyhedral realization with isosceles triangles. Graphs Comb. **37**(4), 1247–1269 (2021). https://doi.org/10.1007/s00373-021-02314-9
9. Eppstein, D., et al.: On the planar split thickness of graphs. Algorithmica **80**(3), 977–994 (2018). https://doi.org/10.1007/s00453-017-0328-y
10. Gardner, M.: Mathematical Games: the coloring of unusual maps leads into uncharted territory. Sci. Am. **242**(2), 14–23 (1980). https://doi.org/10.1038/scientificamerican0280-14
11. Gethner, E.: To the moon and beyond. In: Gera, R., Haynes, T.W., Hedetniemi, S.T. (eds.) Graph Theory. PBM, pp. 115–133. Springer, Cham (2018). https://doi.org/10.1007/978-3-319-97686-0_11
12. Grötzsch, H.: Zur Theorie der diskreten Gebilde, VII: Ein Dreifarbensatz für dreikreisfreie Netze auf der Kugel. Wiss. Z. Martin-Luther-U., Halle-Wittenberg, Math.-Nat. Reihe **8**, 109–120 (1959)
13. Grünbaum, B.: Unambiguous polyhedral graphs. Israel J. Math. **1**(4), 235–238 (1963). https://doi.org/10.1007/BF02759726
14. Halton, J.H.: On the thickness of graphs of given degree. Inf. Sci. **54**(3), 219–238 (1991). https://doi.org/10.1016/0020-0255(91)90052-V
15. Heawood, P.J.: Map colour theorem. Q. J. Math. **24**, 332–338 (1890)
16. Kainen, P.C.: Thickness and coarseness of graphs. Abh. Math. Semin. Univ. Hambg. **39**, 88–95 (1973). https://doi.org/10.1007/BF02992822
17. Kotzig, A.: Contribution to the theory of Eulerian polyhedra. Matematicko-Fyzikálny Časopis **5**, 101–113 (1955)
18. Le, H.O., Le, V.B., Müller, H.: Splitting a graph into disjoint induced paths or cycles. Discret. Appl. Math. **131**(1), 199–212 (2003). https://doi.org/10.1016/S0166-218X(02)00425-0
19. Mac Lane, S.: A structural characterization of planar combinatorial graphs. Duke Math. J. **3**(3), 460–472 (1937). https://doi.org/10.1215/S0012-7094-37-00336-3
20. Mansfield, A.: Determining the thickness of graphs is NP-hard. Math. Proc. Cambridge Philos. Soc. **93**(1), 9–23 (1983). https://doi.org/10.1017/S030500410006028X
21. Nash-Williams, C.S.J.A.: Edge-disjoint spanning trees of finite graphs. J. Lond. Math. Soc. **36**, 445–450 (1961). https://doi.org/10.1112/jlms/s1-36.1.445
22. Ringel, G.: Färbungsprobleme auf Flächen und Graphen, Mathematische Monographien, vol. 2. VEB Deutscher Verlag der Wissenschaften, Berlin (1959)
23. Steinitz, E.: Polyeder und Raumeinteilungen. In: Encyclopädie der mathematischen Wissenschaften, vol. IIIAB12, pp. 1–139 (1922)
24. Sýkora, O., Székely, L.A., Vrt'o, I.: A note on Halton's conjecture. Inf. Sci. **164**(1–4), 61–64 (2004). https://doi.org/10.1016/j.ins.2003.06.008
25. Tutte, W.T.: How to draw a graph. Proc. Lond. Math. Soc. (Third Series) **13**, 743–767 (1963). https://doi.org/10.1112/plms/s3-13.1.743
26. Tutte, W.T.: The non-biplanar character of the complete 9-graph. Can. Math. Bull. **6**, 319–330 (1963). https://doi.org/10.4153/CMB-1963-026-x

CelticGraph: Drawing Graphs as Celtic Knots and Links

Peter Eades[1], Niklas Gröne[2(✉)], Karsten Klein[2], Patrick Eades[3], Leo Schreiber[4], Ulf Hailer[5], and Falk Schreiber[2,6]

[1] School of Computer Science, University of Sydney, Sydney, Australia
peter.eades@sydney.edu.au
[2] Department of Computer and Information Science, University of Konstanz, Konstanz, Germany
{niklas.groene,karsten.klein,falk.schreiber}@uni-konstanz.de
[3] School of Computing and Information Systems, Melbourne University, Melbourne, Australia
patrick.eades@unimelb.edu.au
[4] Uhldingen, Germany
leo.schreiber@posteo.de
[5] Department of History, University of Konstanz, Konstanz, Germany
ulf.hailer@uni-konstanz.de
[6] Faculty of Information Technology, Monash University, Melbourne, Australia

Abstract. Celtic knots are an ancient art form often attributed to Celtic cultures, used to decorate monuments and manuscripts, and to symbolise eternity and interconnectedness. This paper describes the framework CelticGraph to draw graphs as Celtic knots and links. The drawing process raises interesting combinatorial concepts in the theory of circuits in planar graphs. Further, CelticGraph uses a novel algorithm to represent edges as Bézier curves, aiming to show each link as a smooth curve with limited curvature.

Keywords: Celtic Art · Knot Theory · Interactive Interfaces

1 Introduction

Celtic knots are an ancient art form often attributed to Celtic cultures. These elaborate designs (also called "endless knots") were used to decorate monuments and manuscripts, and they were often used to symbolise eternity and interconnectedness. Celtic knots are a well-known visual representation made up of a variety of interlaced knots, lines, and stylised graphical representations. The patterns often form continuous loops with no beginning or end (knot) or a set of such loops (links). In this paper we will use the Celtic knot visualisation metaphor to represent specific graphs in the form of "knot diagrams".

L. Schreiber—Independent Scholar.

M. A. Bekos and M. Chimani (Eds.): GD 2023, LNCS 14465, pp. 18–35, 2023.
https://doi.org/10.1007/978-3-031-49272-3_2

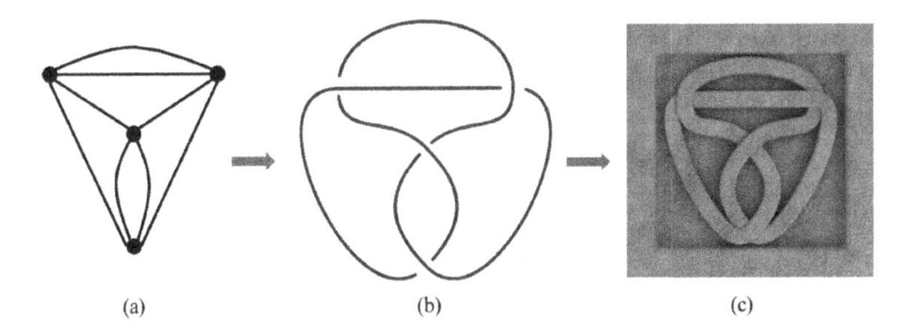

Fig. 1. Using the `CelticGraph` framework, we take the graph K_4' (a) (K_4 with some duplicate edges), and create a knot drawing of K_4' (b); from this we render the graph as a Celtic Knot with a sandstone texture (c).

We show how to draw a 4-regular planar graph[1] as a knot/link diagram. This involves constructing certain circuits[2] in the 4-regular planar graph. Further, we show how to route graph edges so that the underlying links are aesthetically pleasing. This involves some optimisation problems for cubic Bézier curves. We also provide an implementation of the presented methods as an add-on for `Vanted` [25]. This system allows to transform a graph into a knot (link) representation and to interactively change the layout of both, the graph as well as the knot. In addition, knots can be exported to the 3D renderer Blender [4] to allow for artistic 3D renderings of the knot. Figure 1 shows a 4-regular planar graph, its knot representation and a rendering of the knot.

2 Background

This paper has its roots in three disciplines: Mathematical knot theory, Celtic cultural history, and graph drawing. We briefly review the relevant parts of these diverse fields in Sect. 2.1, 2.2, and 2.3. Further, in Sect. 2.4 we review relevant properties of Bézier curves, which are a key ingredient to `CelticGraph`.

2.1 Knot Theory

The mathematical theory of knots and links investigates interlacing curves in three dimensions; this theory has a long and distinguished history in Mathematics [24]. The motivating problem of Knot Theory is *equivalence*: whether two knots can be deformed into each other. A common technique involves projecting the given curves from three dimensions into the plane; the resulting "knot diagram" is a 4-regular planar graph, with vertices at the points where the curve crosses itself (in the projection). For example, a picture of the *trefoil knot* and its knot diagram are in Fig. 2. Properties of a knot or link may be deduced from

[1] Graphs used in this paper can contain multiple edges and loops (also called pseudographs or multigraphs).

[2] For the formal definition of a circuit see Sect. 4.

the knot diagram, and the equivalence problem can sometimes be solved using knot diagrams.

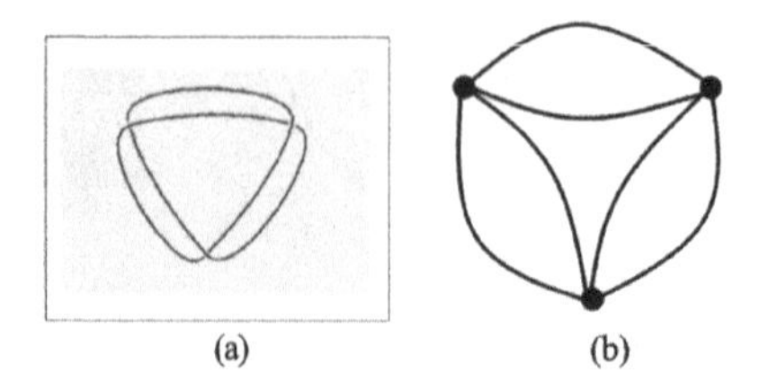

Fig. 2. (a) The trefoil knot and (b) its resulting "knot diagram".

2.2 Celtic Art

Knot patterns ("Celtic knots") are often described as a characteristic ornament of so-called "Celtic art". In fact, since the epoch of the Waldalgesheim style ($4^{th}/3^{rd}$ century BC), Celtic art (resp. ornamentation) is characterised by complex, often geometric patterns of interlinked, opposing or interwoven discs, loops and spirals. The floral models originate from Mediterranean art; in the Celtic context, they were deconstructed, abstracted, arranged paratactically or intertwined [31,38]. It is still unclear whether the import of Mediterranean ornamental models was accompanied by the adoption of their meaning. However, the selective reception of only certain motifs suggests rather an adaptation based on specific Celtic ideas, which we cannot reconstruct exactly due to a lack of written sources.

In today's popular understanding, a special role in the transmission of actual or supposed "Celtic" art is attributed to the early medieval art of Ireland [29,36]. However, such a restriction of Irish or insular art to exclusively Celtic origins would ignore the historical development of the insular-Celtic context in Ireland and the British Isles. The early medieval art of Ireland is partially rooted in indigenous Celtic traditions, but was also shaped by Late Antique Roman, Germanic and Anglo-Saxon, Viking and Mediterranean-Oriental models [18]. The knot and tendril patterns of the $7^{th}/8^{th}$ century can also be traced back to Mediterranean-Oriental manuscripts. Such patterns were subsequently used in Anglo-Saxon art, transmitted by braided ribbon ornaments and other patterns in the Germanic "Tierstil" (e. g. on Late Antique soldiers' belts). For example, the famous Tara Brooch created in Ireland in the late 7th or early 8th century features both native and Germanic a combination of corresponding motifs [37]. Also the knot patterns and braided/spiral ornaments described as typical "Celtic" such as in the Book of Kells [33] and other manuscripts can be linked to Germanic/Anglo-Saxon and late Roman traditions. The ornamentation today often perceived as "Celtic" is therefore less exclusive or typical "Celtic", but rather a result of diverse influences that reflect an equally complex historical-political development [30]. So it is not surprising that the so-called Celtic motifs often presented in tattoo studios of the 21st century, such as braided bands, are not of Celtic but Germanic origin [32].

Note that while "Celtic knots" are related to the mathematical theory of knots, the prime motivation of the two topics is different. For example, the *Bowen knot* [7], a commonly used decorative knot that appears in Celtic cultures, is uninteresting in the mathematical sense (it is clearly an "unknot").

2.3 Graph Drawing as Art

Note that the purpose of `CelticGraph` is different from most Graph Drawing systems. Our aim is to produce decorative and artistically pleasing pictures of graphs, not to make pictures of graphs that effectively convey information and insight into data sets. Other examples of such kind of graph drawing approaches include the system by Devroye and Kruszewski to render images of botanical trees based on random binary trees [8], a system `GDot-i` for drawing graphs as dot paintings inspired by the dot painting style of Central Australia [12,22], and research on bobbin lacework [23]. Also related to our work are Lombardi graph drawings, artistic representations of graphs that contain edges represented as circular arcs and vertices represented with perfect angular resolution [10].

2.4 Bézier Curves

The Gestalt law of continuity [28] implies that humans are more likely to follow continuous and smooth lines rather than broken or jagged lines. To draw graphs as Celtic knots, certain circuits in the graph need to be drawn as smooth curves.

Computer Graphics has developed many models for smooth curves; one of the simplest is a *Bézier curve*. A *cubic Bézier curve* with control points p_0, p_1, p_2, p_3 is defined parametrically by:

$$p(t) = (1-t)^3 p_0 + 3(1-t)^2 t p_1 + 3(1-t)t^2 p_2 + t^3 p_3, \tag{1}$$

for $0 \le t \le 1$. The following properties of cubic Bézier curves are well-known [17]:

- The endpoints of the curve are the first and last control points, that is, $p(0) = p_0$ and $p(1) = p_3$.
- Every point on the curve lies within the convex hull of its control points.
- The line segments (p_0, p_1) and (p_3, p_2) are tangent to the curve at p_0 and p_3 respectively. We say (p_0, p_1) and (p_3, p_2) are *control tangents* of the curve.
- The curve is C^k *smooth* for all $k > 0$, that is, all the derivatives are continuous.

Drawing each edge of a graph as a cubic Bézier curve ensures smoothness in the edges, and can improve readability [40]. However, for `CelticGraph` we need certain *circuits* in the graph to be smooth curves, so we need the curves representing certain incident edges to be *joined smoothly*. Suppose that $p(t)$ and $q(t)$ are two cubic Bézier curves that meet at a common endpoint. Then the curve formed by joining $p(t)$ and $q(t)$ is C^1 smooth as long as the control tangents to each curve at the common endpoint form a straight line; see Fig. 3.

Mathematically, C^1 smoothness is adequate. However, the infinitesimality of Mathematics sometimes does not model human perception well. For example, the curve in Fig. 4 is mathematically smooth, but given a fixed-resolution screen and the limits of human perception, it appears to have a non-differentiable "kink".

For this reason, it is desirable that the *curvature* [14] of each edge is not too large. Informally, the curvature $\kappa(t)$ is the "sharpness" of the curve. More formally, $\kappa(t)$ is the inverse of the radius of the largest circle that can sit on the curve at $p(t)$ without crossing the curve. For a cubic Bézier curve $p(t) = (x(t), y(t))$, the curvature at $p(t)$ is given by [17]:

$$\kappa(t) = \frac{|\dot{x}\ddot{y} - \ddot{x}\dot{y}|}{(\dot{x}^2 + \dot{y}^2)^{1.5}}, \tag{2}$$

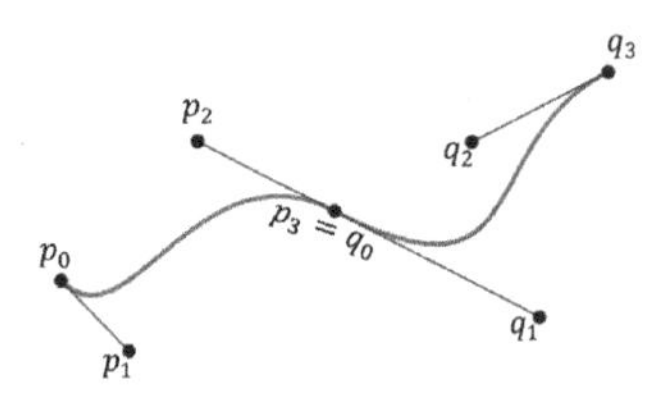

Fig. 3. Two cubic Bézier curves with control points p_0, p_1, p_2, p_3 and q_0, q_1, q_2, q_3, meeting at the point $p_3 = q_0$. The control tangents are shown in black; note that the points p_2, p_3, q_1 lie on a straight line and the join is C_1 and visually smooth.

Fig. 4. A cubic Bézier curve with a "kink", i. e. a point of large curvature, near the middle. The curve is C^1 smooth; but the kink, together with the limits of human perception and screen resolution, mean that the curve does not look smooth.

where $\dot{f}$ denotes the derivative of f with respect to t. Note that $\kappa(t)$ is continuous except for values of t where both $\dot{x}(t)$ and $\dot{y}(t)$ are zero. For `CelticGraph`, we need C^1 smooth curves with reasonably small curvature.

2.5 Related Work

Knot Diagrams as Lombardi Graph Drawings. Closely related are Lombardi graph drawings which are graph drawings with circular-arc edges and perfect angular resolution [10]. Previous studies have demonstrated that a significant group of 4-regular planar graphs can be represented as plane Lombardi graph drawings [26]. However, there are certain restrictions. Notably, if a planar graph contains a loop, it cannot be depicted as a Lombardi drawing. In our approach, every 4-regular planar graph can be transformed in a knot (links).

Celtic Knots by Tiling and Algorithmic Design Methods. Celtic knots can be created using tiling and algorithmic design methods. George Bain introduced a formal method for creating Celtic knot patterns [2], which subsequently has been simplified to a three-grid system by Iain Bain [3,19]. Klempien-Hinrichs and von Totth study the generation of Celtic knots using collage grammars [27]. And Even-Zohar et al. investigate sets of planar curves which yield diagrams for all knots [13]. None of those methods use graphs or are graph drawing approaches.

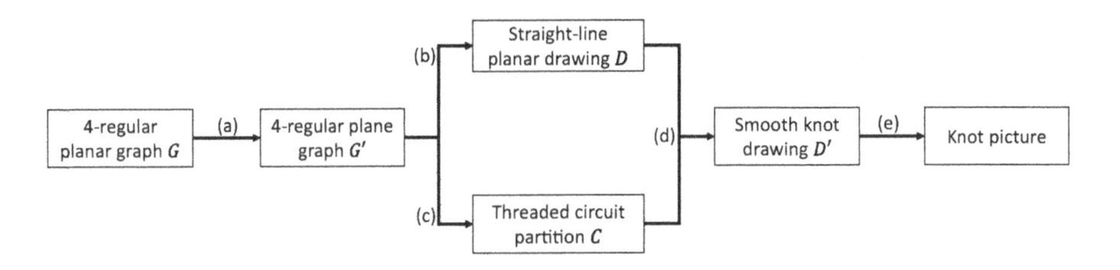

Fig. 5. The *CelticGraph* process

Drawing Graphs with Bézier Curve Edges. A number of network visualisation systems use Bézier curves as edges. These include `yWorks` [41], `GraphViz` [21], `Vanted` [25], `Vizaj` [39], and the framework proposed in [20]. In many cases, such systems allow the user to route the curves by adjusting control points, but few provide automatic computation of the curves. However, there are some exceptions. For example, in the `GraphViz` system, Bézier curve edges are routed within polygons to avoid edge crossings [1]. Force-directed methods are also popular for computing control points of Bézier curve edges [6,15,16]. Brandes et al. present a similar method to the "cross" method in Sect. 5, applied to transport networks [5]. However, only [15] considers smoothness in more than one edge. None of those systems or approaches consider Celtic knots.

3 Overview of the `CelticGraph` process

In this Section we outline `CelticGraph`, our framework for creating aesthetically pleasing pictures of 4-regular planar graphs as knots. The `CelticGraph` procedure is shown in Fig. 5; it has 5 steps:

(a) Create a topological embedding G' of the input 4-regular planar graph G.
(b) Create a planar straight-line drawing D of the plane graph G'.
(c) Create a special circuit partition of G', called a "threaded circuit partition".
(d) Using the straight-line drawing D and the threaded circuit partition C, create a drawing D' of G with cubic Bézier curves as edges.
(e) Render the drawing D' as a knot, on the screen or with a 3D printer.

The first two steps can be done using standard Graph Drawing methods [9]. Steps (c) and (d) are described in the following Sections, step (e) can be done using standard rendering methods.

4 Step (c): Finding the Threaded Circuit Partition

Here we define *threaded circuit partition*, a special kind of circuit partition of a plane graph, and show how to find it in linear time.

A *circuit* in a graph G is a list of distinct edges $(e_0, e_1, \ldots, e_{k-1})$ such that e_i and e_{i+1} share a vertex $i = 0, 1, \ldots, k-1$ (here, and in the remainder of this paper, indices in a circuit of length k are taken modulo k.) We can write the circuit as a list of vertices $(u_0, u_1, \ldots, u_{k-1})$ where $e_i = (u_i, u_{i+1})$. Note that a vertex can appear more than once in a circuit, but an edge cannot. A set $C = \{c_0, c_1, \ldots, c_{h-1}\}$ of circuits in a graph G such that every edge of G is in exactly one c_j is a *circuit partition* for G. Given a circuit partition, we can regard G as a directed graph by directing each edge so that each c_i is a directed circuit.

A path (α, β, γ) of length two (that is, two edges (α, β) and (β, γ)) in a 4-regular plane graph G is a *thread* if edges (α, β) and (β, γ) are not contiguous in the cyclic order of edges around β. This means that there is an edge between (α, β) and (β, γ) in both counterclockwise and clockwise directions in the circular order of edges around β. We say that β is the *midpoint* of the thread (α, β, γ). Note that each vertex in G is the midpoint of two threads; see Fig. 6(a). For every edge (α, β) in G, there is a unique thread (α, β, γ); we say that the edge (β, γ) is the *next edge after* (α, β). For each vertex u_j on a circuit $c = (u_0, u_1, \ldots, u_{k-1})$ with $k > 1$ there is a path $p_j = (u_{j-1}, u_j, u_{j+1})$ of length two such that u_j is the midpoint of p_j. In fact we can consider that the circuit c consists of k paths of length two. We say that the circuit c is *threaded* if for each j, the path $p_j = (u_{j-1}, u_j, u_{j+1})$ is a thread. Note that in such a circuit, the edge (u_j, u_{j+1}) is the (unique) next edge after (u_{j-1}, u_j) for each j. A circuit partition $C = \{c_0, c_1, \ldots, c_{h-1}\}$ is *threaded* if each circuit c_j is threaded. In the case that $h = 1$, a threaded circuit partition defines a *threaded Euler circuit*; see Fig. 6(b).

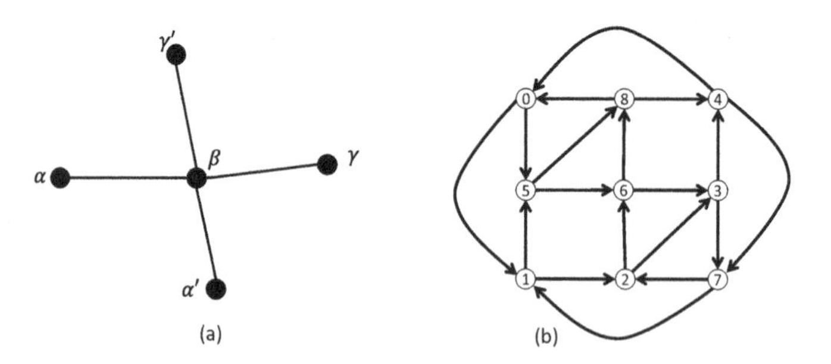

Fig. 6. (a) Two threads, each with midpoint β. (b) Plane 4-regular graph with a threaded Euler circuit $(0, 1, 2, 3, 4, 0, 5, 6, 3, 7, 1, 5, 8, 4, 7, 2, 6, 8)$.

An assignment $v(p) \in \{-1, +1\}$ of an integer -1 or $+1$ to each thread p of a 4-regular plane graph G is an *under-over assignment*. Note that for each vertex β of G, there are two threads p_β and p'_β with midpoint β. We say that an under-over assignment υ is *consistent* if $\upsilon(p_\beta) = -\upsilon(p'_\beta)$ for each vertex β.

An under-over assignment υ is *alternating* on the circuit $(p_0, p_1, \ldots, p_{k-1})$ if $\upsilon(p_i) = -\upsilon(p_{i+1})$ for each i. An under-over assignment for a graph with a threaded circuit partition C is *alternating* if it is alternating on each circuit in C.

Intuitively, a consistent under-over assignment designates which thread passes under or over which thread, and an alternating under-over assignment corresponds to an alternating knot or link [34].

The following theorem gives the essential properties of threaded circuit partitions that are essential for CelticGraph.

Theorem 1. *Every 4-regular plane graph has a unique threaded circuit partition, and this threaded circuit partition has a consistent alternating under-over assignment. Further, this threaded circuit partition can be found in linear time.*

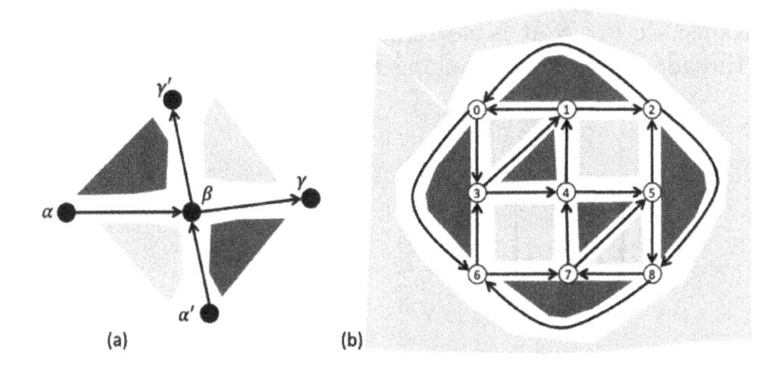

Fig. 7. (a) Two threads: (α, β, γ) has under-over assignment $+1$ (since the face to the left of (α, β) is green), and $(\alpha', \beta, \gamma')$ has under-over assignment -1 (since the face to the left of (α', β) is blue). (b) The faces of the graph are coloured according to its bipartition; note that each vertex has two incoming edges: one has a blue face to the left, the other has a green face to the left, and that the faces on the left of the threaded Euler circuit alternate in colour.

Proof. The existence and uniqueness of the threaded circuit partition follows from the fact that every edge has a unique next edge. A simple linear-time algorithm to find the threaded circuit partition is to repeatedly choose an edge e that is not currently in a circuit, then repeatedly choose the next edge after e until we return to e. We can direct every edge of a 4-regular planar graph G so that each circuit in a given threaded circuit partition C is a directed circuit. This means that we can sensibly define the "left" and "right" faces of an edge. Since a 4-regular plane graph is bridgeless [35], no face is both "left" and "right".

Since the planar dual graph of a 4-regular planar graph is bipartite [35], the faces can be coloured *green* and *blue*, such that no two faces of the same colour share an edge, see Fig. 7. An immediate consequence is that the sequence of left faces to (directed) edges in a threaded circuit *alternate* in colour. Now consider a thread (α, β, γ) in a (directed) threaded circuit in a threaded circuit partition. If the face to the left of (α, β) is green, then assign $+1$ to the path (α, β, γ); otherwise assign -1 to (α, β, γ). Note that the face to the left of (β, γ) is the opposite colour of the face to the left of (α, β), and so the under-over assignment is alternating. Further it is consistent, since at each vertex there is precisely one incoming arc with a green face on the left, and precisely one incoming arc with a blue face on the left.

Threaded Euler Circuits. Celtic knots are sometimes called "endless knots", and can be used to symbolise eternity. For this reason, a threaded *Euler* circuit is desirable; such a circuit gives a drawing of the graph as a knot rather than a link. Using the algorithms in the proof of Theorem 1, one can test whether a given plane graph has a threaded Euler circuit in linear time. Note that different topological embeddings of a given planar graph may have different threaded circuit partitions; see Fig. 8. It is clear that, in some cases, we can increase the length of a threaded circuit by changing the embedding. It is tempting to try

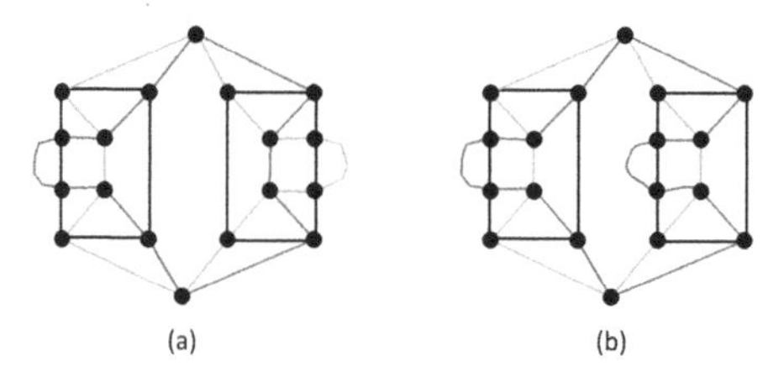

Fig. 8. Two topological embeddings of a planar graph. In (a), the plane graph has a threaded circuit partition of 4 circuits, with two circuits of size 6 (in black) and two circuits of size 12 (in blue and orange). In (b), the plane graph still has 4 threaded circuits: the two of size 6 are unchanged, but the lengths of the blue and orange circuits are 14 and 10 respectively. (Color figure online)

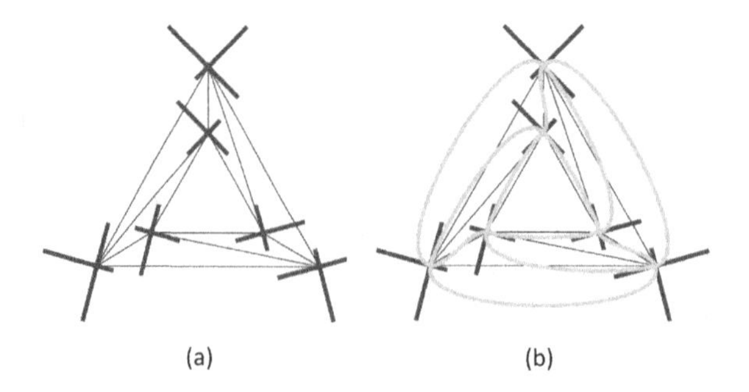

Fig. 9. (a) The 3-prism with a cross at each vertex. (b) Edges drawn as Bézier curves, using the arms of the crosses as control tangents.

to find a method to adjust the embedding to get a threaded Euler circuit. However, it can be shown that changing the embedding cannot change the *number* of threaded circuits in a threaded circuit partition; see Appendix in the arxiv version. [11].

5 Step (d): Smooth Knot Drawing with Bézier Curves

Step (d) takes a straight-line drawing D of the input graph G, and replaces the straight-line edges by cubic Bézier curves in a way that ensures that each circuit in the threaded circuit partition found in step (c) is smooth.

A central concept for the smooth drawing method is a "cross" χ_u at each vertex u. For each u, χ_u consists of 4 line segments called "arms". The four arms are all at right angles to each other, leading to a perfect angular resolution. Each arm of χ_u has an endpoint at u. This is illustrated in Fig. 9(a). Each edge (u, v) then is drawn as a cubic Bézier curve with endpoints u and v, and the control tangents of the curve are arms of the crosses χ_u and χ_v (illustrated in Fig. 9(b)).

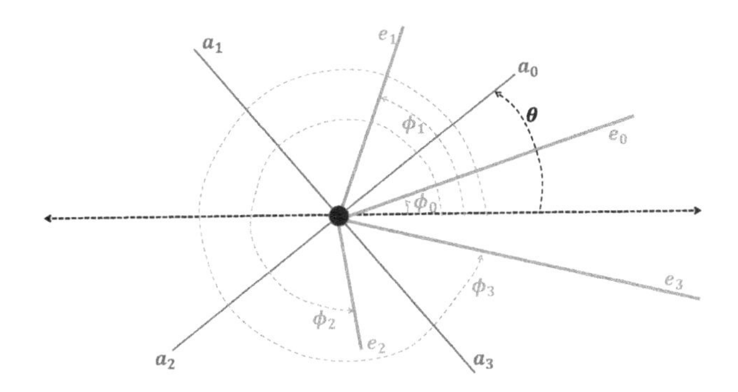

Fig. 10. A cross rotated by an angle of θ. Here the cross is in blue, the edges of the graph are in orange.

For this approach, we need to choose three parameters for each cross χ_u:

1. The mapping between the four arms of χ_u and the four edges incident to u.
2. The angle of orientation of the cross.
3. The length of each arm of the cross.

These parameters are discussed in the next subsections. The methods described in Sect. 5.1 and 5.2 are analogous to the methods in [5]; Sect. 5.3 is not.

5.1 The Edge-Arm Mapping

Suppose that u is a vertex in the straight-line drawing D of the input plane graph. We want to choose the mapping between the arms of the cross χ_u and the edges incident with u so that the arms are approximately in line with the edges.

Now suppose that the edges incident with u are e_0, e_1, e_2, e_3 in counterclockwise order around u. For each $i = 0, 1, 2, 3$ we choose an arm α_i of the cross χ_u corresponding to e_i so that the counterclockwise order of arms around u is the same as the order of edges around u; that is, the counterclockwise order of arms is $\alpha_0, \alpha_1, \alpha_2, \alpha_3$. Note that this method separates multi-edges.

5.2 The Orientation of the Cross

To improve the alignment of the arms of the crosses with the edges, we rotate each cross. Suppose that the counterclockwise angle that edge e_i makes with the horizontal direction is ϕ_i. We want to rotate the cross by an angle θ to align with the edges, as best as possible. This is illustrated in Fig. 10.

Consider the sum of squares error in rotating by θ; this is:

$$f(\theta) = \sum_{i=0}^{i=3} \left(\theta + \frac{i\pi}{2} - \phi_i\right)^2. \tag{3}$$

To minimise $f(\theta)$, we solve $f'(\theta) = 0$ and choose the optimum value:

$$\theta^* = \frac{1}{4}\left(\sum_{i=0}^{i=3} \phi_i\right) - \frac{3\pi}{4}. \tag{4}$$

In Fig. 11, we show a graph with crosses oriented by this method.

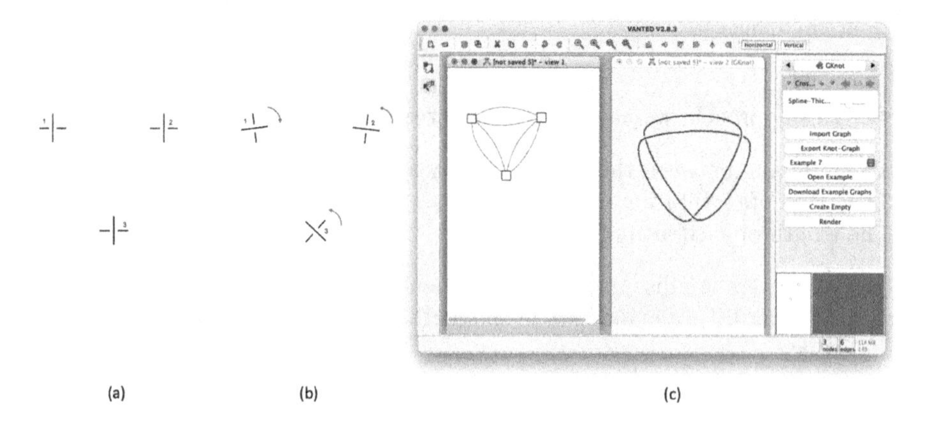

Fig. 11. A graph with crosses oriented to align with edges as much as possible before (a) and after (b) applying the algorithm, and shown in *Vanted* (c).

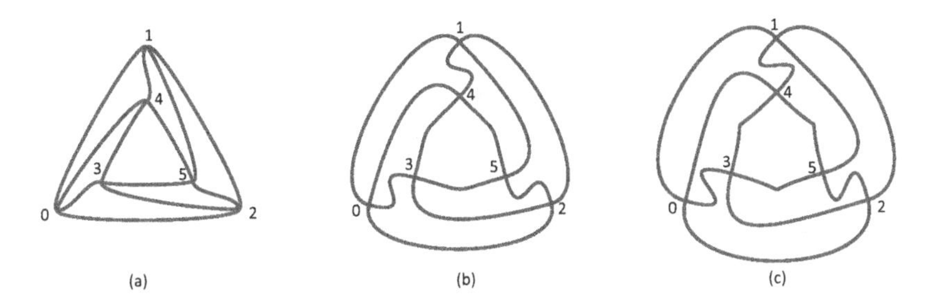

Fig. 12. Three drawings of the 3-prism, differing in edge curvature.

5.3 Arm Length

Note that the "apparent smoothness" of an edge depends on its curvature. We illustrate this with Fig. 12, which shows three Bézier curve drawings of the 3-prism. This graph has 3 threaded circuits, and we want to draw it so that each one of these threaded circuits appears as a smooth curve with limited curvature. In Fig. 12(a), the arms of the crosses are all very short, resulting in a Bézier curve drawing which is very close to a straight-line drawing. Each edge has low curvature in the middle and high curvature around the endpoints. The high curvature near their endpoints results in a lack of apparent smoothness where two Bézier curves join (at the vertices); it is difficult to discern the three threaded circuits. The arms of the crosses are longer in Fig. 12(b), resulting in better curvature at the endpoints. However, here each of the edges $(0,3),(1,4)$, and $(2,5)$ have two points of large curvature; this is undesirable. In Fig. 12(c), the arms of the crosses are longer still. Each of the edges $(0,3),(1,4)$, and $(2,5)$ again have two points of large curvature, but the edges $(3,4),(4,5)$, and $(5,3)$ are worse: each has a "kink" (a point of very high curvature, despite being C^1-smooth).

Next we describe three approaches to choosing the lengths of the arms of the crosses, aiming to give sufficiently small curvature.

Uniform Arm Lengths. The curvature of the edge varies with lengths of the arms, and we want to ensure that the maximum curvature in each edge is not too

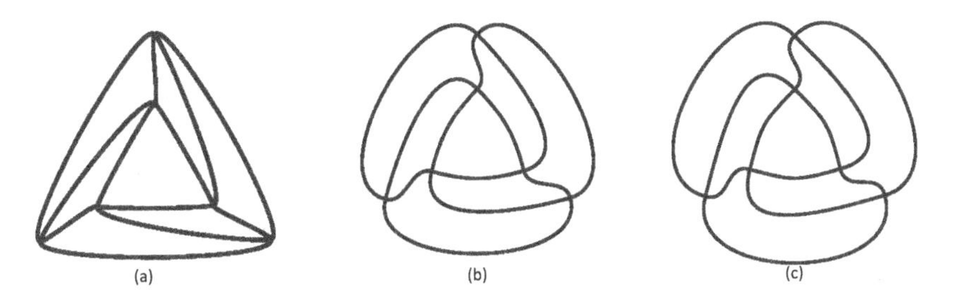

Fig. 13. The uniformly proportional approach: (a) $\alpha = 0.2$; (b) $\alpha = 0.4$; (c) $\alpha = 0.6$.

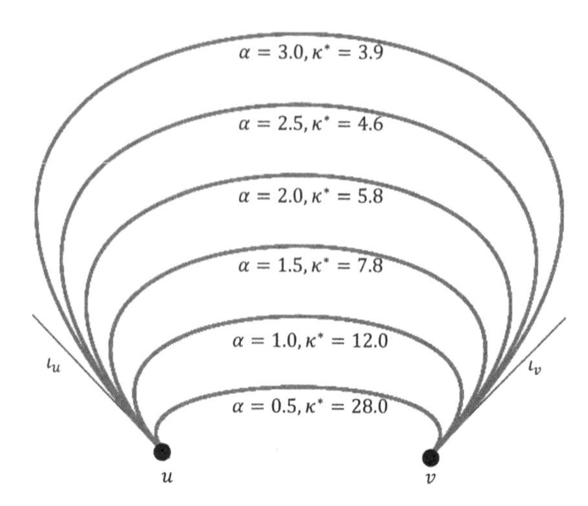

Fig. 14. "Ballooning" curves: as λ_u and λ_v increase, curvature falls but the curve becomes very long.

large. The simplest approach is to use *uniform arm lengths*, that is, judiciously choose a global value λ and set the length of every arm length to λ. The drawings of the 3-prism in Fig. 12 have uniform arm lengths: λ in Fig. 12(a) is quite small, in Fig. 12(c) it is relatively large, and (b) is in between. In fact, the problem with the uniform arm length approach is typified in Fig. 12: if λ is small, the curvature is high near the endpoints for all edges, and increasing λ increases the curvature away from the endpoints, especially in the shorter edges. There is no uniform value of λ that gives good curvature in both short and long edges.

Uniformly Proportional Arm Lengths. An approach that aims to overcome the problems of uniform arm length is to use *uniformly proportional arm lengths*: we judiciously choose a global value α, and then set the lengths of the two arms for edge (u, v) to $\alpha d(u, v)$, where $d(u, v)$ is the Euclidean distance between u and v. Figure 13 shows typical results for the uniformly proportional approach. For $\alpha = 0.2$ the drawing is similar to Fig. 13(a), and has similar problems. But for values of α near 0.5 (Fig. 13(b) and (c)), we have acceptable results; in particular, the shorter edges have acceptable curvature.

Optimal Arm Lengths. A third approach is to choose the arm lengths at each end of an edge (u, v) to minimise maximum curvature, as follows. Suppose $\kappa(t, \lambda_u, \lambda_v)$ is the curvature of the edge (u, v) at point t on the curve, when the arm lengths are λ_u and λ_v at u and v respectively. From Equation (2), we note that

$$\frac{\partial}{\partial t}\kappa(t, \lambda_u, \lambda_v) = \left| \frac{\dddot{x}\dot{y} - \dddot{y}\dot{x}}{(\dot{x}^2 + \dot{y}^2)^{1.5}} - \frac{\ddot{x}\dot{y} - \ddot{y}\dot{x}}{3(\ddot{x} + \ddot{y})^{2.5}} \right| \tag{5}$$

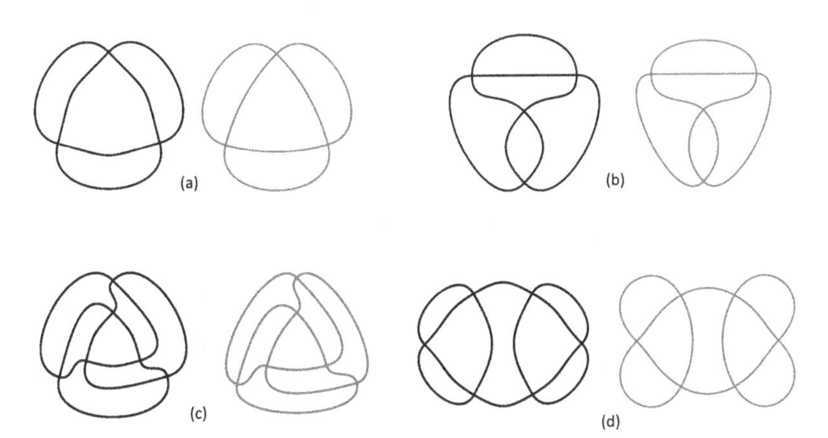

Fig. 15. Comparison of proportional arm length (blue) and optimal arm length (magenta): (a) Trefoil; (b) K_4 knot; (c) 3-prism; (d) Love knot. (Color figure online)

as long as $(\dot{x}^2 + \dot{y}^2) \neq 0$ and $\ddot{x}\dot{y} \neq \ddot{y}\dot{x}$. Since both x and y are cubic functions of t, Eq. (5) is not as complex as it seems, and it is straightforward (but tedious, because of the edge cases) to maximise $\kappa(t, \lambda_u, \lambda_v)$ over t; that is, to find the maximum curvature $\kappa^*(\lambda_u, \lambda_v)$:

$$\kappa^*(\lambda_u, \lambda_v) = \max_{0 \leq t \leq 1} \kappa(t, \lambda_u, \lambda_v).$$

Now we want to choose the arm lengths λ_u and λ_v to minimise $\kappa^*(\lambda_u, \lambda_v)$. Suppose that the unit vectors in the directions of the appropriate arms of χ_u and χ_v are ι_u and ι_v respectively. Note that we can express the internal control points p_1 and p_2 of the Bézier curve in terms of λ_u and λ_v:

$$p_1 = (1 - \lambda_u)u + \lambda_u \iota_u, \quad p_2 = (1 - \lambda_v)v + \lambda_v \iota_v.$$

In this way, $\kappa^*(\lambda_u, \lambda_v)$ is linear in both λ_u and λ_v and finding a minimum point for $\kappa^*(\lambda_u, \lambda_v)$ is straightforward. However, in some cases, an edge with globally minimum maximum curvature may not be desirable. In Fig. 14, for example, the curvature decreases as λ_u and λ_v increase; for large values of λ_u and λ_v the curvature is quite low. The problem is that these large values make the curve very long (it "balloons" out), which might also cause unintended edge crossings.

For this reason, we choose an upper bounds ϵ_u and ϵ_v and take a minimum constrained by $\lambda_u \leq \epsilon_u$ and $\lambda_v \leq \epsilon_v$:

$$\kappa^*_{\min} = \min_{0 \leq \lambda_u \leq \epsilon_u, 0 \leq \lambda_v \leq \epsilon_v} \kappa^*(\lambda_u, \lambda_v).$$

We have found that $\epsilon_u = \epsilon_v = 0.75d(u, v)$ gives good results, where $d(u, v)$ is the distance between u and v. Values of λ_u and λ_v that achieve the (constrained) minimum $\kappa^*_{\min}$ are then used by the Bézier curves. In practice, using such optimal arm lengths gives better results than using uniformly proportional arm lengths. In some cases the difference is not significant, but in others the optimal edges appear to be much smoother. See Fig. 15 for examples.

Fig. 16. Examples of Celtic knot renderings in different media including a 3D printed version (mid left).

6 CelticGraph Implementation as a Vanted Add-on and Rendering

CelticGraph has been implemented as an add-on of Vanted, a tool for interactive visualisation and analysis of networks. Figure 1 shows an example workflow; the first step is implemented as Vanted [25] add-on, the second is done by Blender [4].

Vanted allows a user to load or create 4-regular graphs, either by importing from files (e. g. a .gml file), by selecting from examples, or by creating a new graph by hand. The individual vertices of the graph are then mapped into the data structure of a cross, containing position and the rotation and control points of the to-be-generated Bézier curves. The graph is translated into a knot (link) using the methods for optimal cross rotation and arm length computation described in the previous sections. Vertex positions can be interactively changed, either by interacting with the underlying graph, or by interacting directly with the visualisation of the knot. Once a visualisation satisfies the expectation of the user, the Bézier curves can be exported for further use in Blender.

We implemented a Python script and a geometry node tree in Blender which allows importing the information into Blender and rendering the knot (links), either using a set of predefined media or interactively; the script can also run as batch process with selected parameters and media. Figure 16 shows examples of Celtic knot renderings in different media such as in metal, in stone, with additional decoration and so on; knots can be also printed in 3D. More examples can be found in the gallery of our web page http://celticknots.online which also provides the Vanted add-on, Blender file and a short manual.

Acknowledgements. Partly funded by the Deutsche Forschungsgemeinschaft (DFG, German Research Foundation) - Project-ID 251654672 - TRR 161.

References

1. Abello, J., Gansner, E.: Short and smooth polygonal paths. In: Lucchesi, C.L., Moura, A.V. (eds.) LATIN 1998. LNCS, vol. 1380, pp. 151–162. Springer, Heidelberg (1998). https://doi.org/10.1007/BFb0054318
2. Bain, G.: Celtic Art: The Methods of Construction. Dover Publications, Mineola (1973)
3. Bain, I.: Celtic Knotwork. Sterling Publishing Co., New York (1986)
4. Blender Online Community: Blender - a 3D modelling and rendering package. Blender Foundation (2018). http://www.blender.org
5. Brandes, U., Shubina, G., Tamassia, R.: Improving angular resolution in visualizations of geographic networks. In: de Leeuw, W.C., van Liere, R. (eds.) Data Visualization 2000, pp. 23–32. Springer, Eurographics (2000). https://doi.org/10.1007/978-3-7091-6783-0_3
6. Brandes, U., Wagner, D.: Using graph layout to visualize train interconnection data. J. Graph Algor. Appl. **4**, 135–155 (2000)
7. Clark, H.: A Short and Easy Introduction to Heraldry. Printed for H. Washbourn, London (1827)

8. Devroye, L., Kruszewski, P.: The botanical beauty of random binary trees. In: Brandenburg, F.J. (ed.) GD 1995. LNCS, vol. 1027, pp. 166–177. Springer, Heidelberg (1996). https://doi.org/10.1007/BFb0021801
9. Di Battista, G., Eades, P., Tamassia, R., Tollis, I.G.: Graph Drawing: Algorithms for the Visualization of Graphs. Prentice Hall, Upper Saddle River (1999)
10. Duncan, C.A., Eppstein, D., Goodrich, M.T., Kobourov, S.G., Nöllenburg, M.: Lombardi drawings of graphs. In: Brandes, U., Cornelsen, S. (eds.) GD 2010. LNCS, vol. 6502, pp. 195–207. Springer, Heidelberg (2011). https://doi.org/10.1007/978-3-642-18469-7_18
11. Eades, P., Gröne, N., Klein, K., Eades, P., Schreiber, L., Schreiber, F.: Celticgraph: drawing graphs as celtic knots and links (2023)
12. Eades, P., Hong, S.H., McGrane, M., Meidiana, A.: GDot-i: interactive system for dot paintings of graphs. In: Krone, M., Lenti, S., Schmidt, J. (eds.) EuroVis 2022 - Posters. The Eurographics Association (2022)
13. Even-Zohar, C., Hass, J., Linial, N., Nowik, T.: Universal knot diagrams. J. Knot Theory Ramifications 28(07), 1950031 (2019)
14. Ferguson, R.: An easier derivation of the curvature formula from first principles. Aust. Senior Math. J. **32**, 16–22 (2018)
15. Fink, M., Haverkort, H., Nöllenburg, M., Roberts, M., Schuhmann, J., Wolff, A.: Drawing metro maps using Bézier curves. In: Didimo, W., Patrignani, M. (eds.) GD 2012. LNCS, vol. 7704, pp. 463–474. Springer, Heidelberg (2013). https://doi.org/10.1007/978-3-642-36763-2_41
16. Finkel, B., Tamassia, R.: Curvilinear graph drawing using the force-directed method. In: Pach, J. (ed.) GD 2004. LNCS, vol. 3383, pp. 448–453. Springer, Heidelberg (2005). https://doi.org/10.1007/978-3-540-31843-9_46
17. Foley, J.D., van Dam, A., Feiner, S., Hughes, J.F.: Computer Graphics - Principles and Practice, 2nd edn. Addison-Wesley, Boston (1990)
18. Fries-Knoblach, J.: Die Kelten, pp. 138–142. Kohlhammer-Urban (2012)
19. Glassner, A.: Celtic knotwork, part 1. IEEE Comput. Graph. Appl. **19**(5), 78–84 (1999)
20. Goodrich, M.T., Wagner, C.G.: A framework for drawing planar graphs with curves and polylines. In: Whitesides, S.H. (ed.) GD 1998. LNCS, vol. 1547, pp. 153–166. Springer, Heidelberg (1998). https://doi.org/10.1007/3-540-37623-2_12
21. Graphviz. https://graphviz.org/
22. Hong, S.H., Eades, P., Torkel, M.: Gdot: drawing graphs with dots and circles. In: 2021 IEEE 14th Pacific Visualization Symposium (PacificVis), pp. 156–165 (2021). https://doi.org/10.1109/PacificVis52677.2021.00029
23. Irvine, V., Biedl, T., Kaplan, C.S.: Quasiperiodic bobbin lace patterns. J. Math. Arts **14**(3), 177–198 (2020)
24. James, I.M.: History of Topology. North Holland (1999)
25. Junker, B.H., Klukas, C., Schreiber, F.: VANTED: a system for advanced data analysis and visualization in the context of biological networks. BMC Bioinf. **7**(1), 1–13 (2006)
26. Kindermann, P., Kobourov, S., Löffler, M., Nöllenburg, M., Schulz, A., Vogtenhuber, B.: Lombardi drawings of knots and links. In: Frati, F., Ma, K.-L. (eds.) GD 2017. LNCS, vol. 10692, pp. 113–126. Springer, Cham (2018). https://doi.org/10.1007/978-3-319-73915-1_10
27. Klempien-Hinrichs, R., von Totth, C.: Generation of celtic key patterns with tree-based collage grammars. Electron. Commun. Eur. Assoc. Softw. Sci. Technol. **26**, 205–222 (2010)

28. Koffka, K.: Principles of Gestalt Psychology. Harcort Brace and Co., San Diego (1935)
29. Maier, B.: Die Geschichte, Kultur, Sprache, pp. 158–159. utb (2015)
30. Maier, B.: Die Geschichte, Kultur, Sprache, p. 159. utb (2015)
31. Müller, F.: Die Kunst der Kelten. C.H. Beck Wissen (2012)
32. Müller, F.: Die Kunst der Kelten, p. 119. C.H. Beck Wissen (2012)
33. Monks of St. Columba's order of Iona: The Book of Kells. at the Old Library in Trinity College Dublin (9th century)
34. Murasugi, K.: Knot theory and its applications (1993)
35. Nishizeki, T., Rahman, M.S.: Planar graph drawing, Lecture Notes Series on Computing, vol. 12. World Scientific (2004)
36. Röber, R.: Exhibition catalogue Archäologisches Landesmuseum Stuttgart: Die Welt der Kelten. Zentren der Macht - Kostbarkeiten der Kunst, pp. 460–521. Jan Thorbecke Verlag (2012)
37. Röber, R.: Exhibition catalogue Archäologisches Landesmuseum Stuttgart: Die Welt der Kelten. Zentren der Macht - Kostbarkeiten der Kunst, pp. 512–515. Jan Thorbecke Verlag (2012)
38. Rieckhoff, S., Biel, J.: Die Kelten in Deutschland, pp. 197–206. Konrad Theiss Verlag (2001)
39. Rolland, T., De Vico Fallani, F.: Vizaj-a free online interactive software for visualizing spatial networks. PLoS ONE **18**(3), e0282181 (2023)
40. Xu, K., Rooney, C., Passmore, P., Ham, D.H., Nguyen, P.H.: A user study on curved edges in graph visualization. IEEE Trans. Visual Comput. Graphics **18**(12), 2449–2456 (2012)
41. yworks. https://yWorks.org/

Beyond Planarity

Min-k-planar Drawings of Graphs

Carla Binucci[1(✉)], Aaron Büngener[2], Giuseppe Di Battista[3], Walter Didimo[1], Vida Dujmović[4], Seok-Hee Hong[5], Michael Kaufmann[2], Giuseppe Liotta[1], Pat Morin[6], and Alessandra Tappini[1]

[1] Università degli Studi di Perugia, Perugia, Italy
{carla.binucci,walter.didimo,giuseppe.liotta,alessandra.tappini}@unipg.it
[2] University of Tübingen, Tübingen, Germany
aaron.buengener@student.uni-tuebingen.de,
michael.kaufmann@uni-tuebingen.de
[3] Università Roma Tre, Rome, Italy
giuseppe.dibattista@uniroma3.it
[4] University of Ottawa, Ottawa, Canada
vdujmovi@uOttawa.ca
[5] University of Sydney, Sydney, Australia
seokhee.hong@sydney.edu.au
[6] Carleton University, Ottawa, Canada
morin@scs.carleton.ca

Abstract. The study of nonplanar drawings of graphs with restricted crossing configurations is a well-established topic in graph drawing, often referred to as *beyond-planar graph drawing*. One of the most studied types of drawings in this area are the *k-planar drawings* ($k \geq 1$), where each edge cannot cross more than k times. We generalize k-planar drawings, by introducing the new family of *min-k-planar drawings*. In a min-k-planar drawing edges can cross an arbitrary number of times, but for any two crossing edges, one of the two must have no more than k crossings. We prove a general upper bound on the number of edges of min-k-planar drawings, a finer upper bound for $k = 3$, and tight upper bounds for $k = 1, 2$. Also, we study the inclusion relations between min-k-planar graphs (i.e., graphs admitting min-k-planar drawings) and k-planar graphs.

Keywords: Beyond planarity · k-planarity · edge density

1 Introduction

Beyond planarity [21,26] is a recent area of focus in graph drawing and topological graph theory, having its foundations established in the 1970s and 1980s.

Research started at the Summer Workshop on Graph Drawing (SWGD) 2022, and partially supported by: (i) University of Perugia, Ricerca Base, Proj. AIDMIX (2021) and RICBA22CB; (ii) MUR PRIN Proj. 2022TS4Y3N - "EXPAND: scalable algorithms for EXPloratory Analyses of heterogeneous and dynamic Networked Data"; (iii) MUR PRIN Proj. 2022ME9Z78 - "NextGRAAL: Next-generation algorithms for constrained GRAph visuALization".

M. A. Bekos and M. Chimani (Eds.): GD 2023, LNCS 14465, pp. 39–52, 2023.
https://doi.org/10.1007/978-3-031-49272-3_3

It comprises works on graphs that go beyond planar graphs in the sense that several, mostly local, crossing configurations are forbidden. The simplest are *1-planar* graphs, where at most one crossing per edge is allowed [28,32], and their generalization k*-planar* graphs, where at most $k \geq 1$ crossings per edge are tolerated [13,21,25,30,31]. Other prominent examples of graph classes are *fan-planar* graphs [11,12,16,17,27], where several edges might cross the same edge but they should be adjacent to the same vertex, and k*-gap-planar* graphs ($k \geq 1$) [8–10], where for each pair of crossing edges one of the two edges contains a small gap through which the other edge can pass, and at most k gaps per edge are allowed. Another popular family is the one of k*-quasiplanar* graphs, which forbids k mutually crossing edges [2–5,24]. Mostly, edge density and inclusion relations of different beyond-planar graph classes have been studied [5,21,26].

In this paper we introduce a new graph family that generalizes k-planar graphs by permitting certain edges to have more than k crossings. Namely, for each two crossing edges we require that at least one of them contains at most k crossings. Formally, this graph family is defined as follows:

Definition 1. *A graph G is* min-k-planar ($k \geq 1$) *if it admits a drawing on the plane, called* min-k-planar *drawing, such that for any two crossing edges e and e' it holds* $\min\{\mathrm{cr}(e), \mathrm{cr}(e')\} \leq k$, *where* $\mathrm{cr}(e)$ *and* $\mathrm{cr}(e')$ *are the number of crossings of e and e', respectively.*

Clearly, every k-planar drawing Γ is also min-k-planar, but not vice versa. A *crossing edge* in Γ with more than k crossings is *heavy*, otherwise it is *light*. There are two main motivations behind the study of min-k-planar graphs:

(i) From a theoretical perspective, when a graph is not k-planar we may want to draw it by allowing some heavy edges, whose removal yields a k-planar drawing. In this respect, if m is the total number of edges in the graph, we will prove that the number of heavy edges in a min-k-planar drawing is at most $\frac{k}{2k+1} \cdot m$, which varies in the interval $[\frac{m}{3}, \frac{m}{2})$.

(ii) From a practical perspective, even if a graph is k-planar, allowing (few) pairwise-independent heavy edges may reduce the visual complexity of the layout, even when the total number of crossings grows. For example, Fig. 1 shows two drawings of the same portion of a graph. Despite the drawing in Fig. 1(a) is 2-planar and has fewer crossings in total, the one in Fig. 1(b) appears more readable; it is not 2-planar, but it is min-2-planar.

Min-k-planar graphs are also implicitly studied in [33,34], proving that the underlying graph of a convex min-k-planar drawing has treewidth $3k + 11$.

Contribution. We study the edge density of min-k-planar graphs (Sect. 3) and their inclusion relations with k-planar graphs (Sect. 4). After giving general bounds on edge and crossing numbers, we focus on $k \in \{1, 2, 3\}$:

- We provide tight upper bounds on the maximum number of edges of min-1-planar and min-2-planar graphs. Namely, we prove that n-vertex min-1-planar graphs and min-2-planar graphs have at most $4n - 8$ edges and at

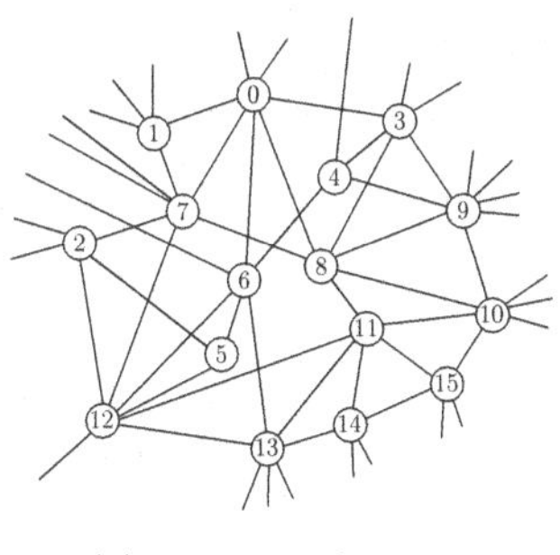

(a) 2-planar drawing

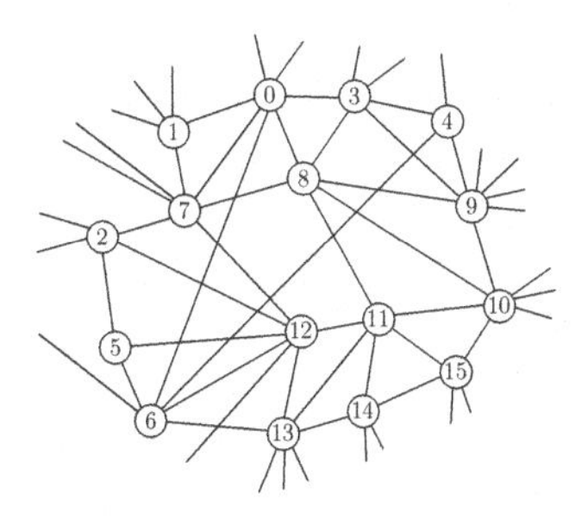

(b) min-2-planar drawing

Fig. 1. Two drawings of the same portion of a graph: (a) is 2-planar and has 10 crossings; (b) is min-2-planar, is not 2-planar, and has 12 crossings; it contains two "heavy" edges incident to vertex 6, each with several crossings.

most $5n - 10$ edges, respectively, as for 1-planar and 2-planar graphs. For min-3-planar graphs we give an upper bound of $6n - 12$ and show min-3-planar graphs with $5.6n - O(1)$ edges, hence having density higher than the one of every 3-planar graph.

- Despite the maximum density of min-k-planar graphs for $k = 1, 2$ equals the one of k-planar graphs, we show that 1-planar and 2-planar graphs are proper sub-classes of min-1-planar and min-2-planar graphs (as for $k = 3$). However, the min-1-planar graphs that can reach the maximum density of $4n - 8$ are also 1-planar (i.e., the two classes coincide), while this is not true for $k = 2, 3$.

Section 2 introduces notation and terminology; final remarks and open problems are in Sect. 5. All full proofs are in [14].

2 Basic Definitions

We only deal with connected graphs. A graph is *simple* if it does not contain multiple edges and self-loops. A graph with multiple edges but not self-loops is also called a *multi-graph*. Let G be any (not necessarily simple) graph. We denote by $V(G)$ and $E(G)$ the set of vertices and the set of edges of G, respectively. A *drawing* Γ of G maps each vertex $v \in V(G)$ to a distinct point in the plane and each edge $uv \in E(G)$ to a simple Jordan arc between the points corresponding to u and v. We always assume that Γ is a *simple* drawing, that is: (*i*) two *adjacent* edges (i.e., edges that share a vertex) do not intersect, except at their common endpoint (in particular, no edge is self-crossing); (*ii*) two *independent* (i.e. non-adjacent) edges intersect at most in one of their interior points, called a *crossing point*; (*iii*) no three edges intersect at a common crossing point.

Let Γ be a drawing of G. A *vertex of* Γ is either a point corresponding to a vertex of G, called a *real-vertex*, or a point corresponding to a crossing point, called a *crossing-vertex* or simply a *crossing*. We remark that in the literature a plane graph obtained by replacing crossing points with dummy vertices is often referred to as a *planarization* [20]. We denote by $V(\Gamma)$ the set of vertices of Γ. An *edge* of Γ is a curve connecting two vertices of Γ. We denote by $E(\Gamma)$ the set of edges of Γ. An edge $e \in E(\Gamma)$ is a portion of an edge in $E(G)$, which we denote by $\overline{e}$; if both the endpoints of e are real-vertices, then e and $\overline{e}$ coincide.

Drawing Γ subdivides the plane into topologically connected regions, called *faces*. The boundary of a face consists of a cyclical sequence of vertices (real- or crossing-vertices) and edges of Γ. We denote by $F(\Gamma)$ the set of faces of Γ. Exactly one face in $F(\Gamma)$ corresponds to an infinite region of the plane, called the *external face* of Γ; the other faces are the *internal faces* of Γ. If the boundary of a face f of Γ contains a vertex v (or an edge e), we say that f *contains* v (or e).

In the following, if not specified, we denote by $n = |V(G)|$ and $m = |E(G)|$ the number of vertices and the number of edges of G, respectively.

Degree of Vertices and Faces. For a vertex $v \in V(G)$, denote by $\deg_G(v)$ the *degree of v in G*, i.e., the number of edges incident to v. Analogously, for a vertex $v \in V(\Gamma)$, denote by $\deg_\Gamma(v)$ the *degree of v in Γ*. Note that, if $v \in V(G)$ then $\deg_\Gamma(v) = \deg_G(v)$, while if v is a crossing-vertex then $\deg_\Gamma(v) = 4$. For a face $f \in F(\Gamma)$, denote by $\deg_\Gamma(f)$ the *degree of f*, i.e., the number of times we traverse vertices (either real- or crossing-vertices) while walking on the boundary of f clockwise. Each vertex contributes to $\deg_\Gamma(f)$ the number of times we traverse it (possibly more than once if the boundary of f is not a simple cycle). Also, denote by $\deg^r_\Gamma(f)$ the *real-vertex degree of f*, i.e., the number of times we traverse a real-vertex of Γ while walking on the boundary of f clockwise. Again, each real-vertex contributes to $\deg^r_\Gamma(f)$ the number of times we traverse it. Finally, $\deg^c_\Gamma(f)$ denotes the number of times we traverse a crossing-vertex of Γ while walking on the boundary of f clockwise. Clearly, $\deg_\Gamma(f) = \deg^r_\Gamma(f) + \deg^c_\Gamma(f)$.

We say that a face $f \in F(\Gamma)$ is an *h-real face*, for $h \geq 0$, if $\deg^r_\Gamma(f) = h$. An h-real face of degree d is called an *h-real d-gon*. For $k = 2, 3, 4, 5, 6$, a face that is an h-real k-gon, is also called an *h-real bigon* ($k = 2$), an *h-real triangle* ($k = 3$), an *h-real quadrilateral* ($k = 4$), an *h-real pentagon* ($k = 5$), and an *h-real hexagon* ($k = 6$), respectively. An edge $e = uv \in E(\Gamma)$ is an *h-real edge* ($h \in \{0, 1, 2\}$) if $|\{u, v\} \cap V(G)| = h$, i.e., e contains h real-vertices.

Beyond-Planar Graphs. A family $\mathcal{F}$ of *beyond-planar graphs* is a set of (non-planar) graphs that admit drawings with desired or forbidden edge-crossing configurations [21]. The *edge density* of a graph $G \in \mathcal{F}$ is the ratio between its number m of edges and its number n of vertices. Graph G is *maximally dense* if it has the maximum edge density over all graphs of $\mathcal{F}$ with n vertices. Graph G is *optimal* if it has the maximum edge density over all graphs in $\mathcal{F}$. Note that $\mathcal{F}$ might not contain optimal graphs for all values of n (see, e.g., [21]).

3 Edge Density of Min-k-planar Graphs

We start by proving some general bounds on the number of crossings in a min-k-planar drawing and on the number of edges of min-k-planar graphs.

Property 1. Any min-k-planar drawing Γ of a graph G (with $k \geq 1$) has at most $k \cdot \ell$ crossings, where ℓ is the number of light edges of G in Γ.

Proof. Two heavy edges cannot cross, thus each crossing in Γ belongs to at least one light edge. Since each light edge has at most k crossings, the bound follows. □

Property 2. Let Γ be a min-k-planar drawing of an m-edge graph G (with $k \geq 1$). The number of heavy edges of G in Γ is at most $\frac{k}{2k+1} \cdot m$.

Proof. Let h and ℓ be the number of heavy edges and the number of light edges of G in Γ, respectively. Observe that $m \geq h + \ell$. By definition, each heavy edge contains at least $(k+1)$ crossings, and two heavy edges do not cross. Hence, the number of crossings in Γ is at least $h \cdot (k+1)$. By Property 1, we have $h \cdot (k+1) \leq k \cdot \ell \leq k \cdot m - k \cdot h$, which implies $h \leq \frac{k}{2k+1} \cdot m$. □

We now give a general bound on the edge density of min-k-planar simple graphs, for any $k \geq 2$. Finer bounds for $k = 1, 2, 3$ are given in the next sections.

Theorem 1. *For any min-k-planar simple graph G with n vertices and m edges it holds $m \leq \min\{5.39\sqrt{k} \cdot n, (3.81\sqrt{k} + 3) \cdot n\}$ when $k \geq 2$.*

Proof (Sketch). Let $\mu = \min\{5.39\sqrt{k} \cdot n, (3.81\sqrt{k}+3) \cdot n\}$. Note that $\mu = 5.39\sqrt{k} \cdot n$ when $2 \leq k \leq 3$, while $\mu = (3.81\sqrt{k} + 3) \cdot n$ when $k \geq 4$.

Suppose first that $2 \leq k \leq 3$. If $m < 6.95n$, the relation $m \leq 5.39\sqrt{k} \cdot n$ trivially holds. If $m \geq 6.95n$, let $\mathrm{cr}(G)$ be the minimum number of crossings required by any min-k-planar drawing Γ of G. The improved version by Ackerman of the popular Crossing Lemma (Theorem 6 in [1]) implies that $\mathrm{cr}(G) \geq \frac{1}{29}\frac{m^3}{n^2}$. If ℓ is the number of light edges of G in Γ, by Property 1 we have $\mathrm{cr}(G) \leq k \cdot \ell \leq k \cdot m$. Hence $\frac{1}{29}\frac{m^3}{n^2} \leq k \cdot m$, which yields $m \leq 5.39\sqrt{k} \cdot n$.

Suppose now that $k \geq 4$ and let Γ be any min-k-planar drawing of G with ℓ light edges. Since no two heavy edges cross, the subgraph of G consisting of all heavy and crossing-free edges in Γ has at most $3n - 6$ edges, hence $m \leq \ell + 3n - 6$. Let G' be the subgraph of G consisting of the ℓ light edges of G only. By applying Ackerman's version of the Crossing Lemma to G', one can prove that $\ell \leq 3.81\sqrt{k} \cdot n$. Therefore, $m \leq \ell + 3n - 6 \leq \ell + 3n \leq (3.81\sqrt{k} + 3) \cdot n$. □

3.1 Density of Min-1-planar Graphs

Let Γ be a min-1-planar drawing of a graph G. We color each edge of $E(G)$ either red or green with the following rule: (i) edges that are crossing-free in Γ are colored red; (ii) if $\{e_1, e_2\} \in E(G)$ is a pair of edges that cross in Γ, with $\mathrm{cr}(e_1) \geq \mathrm{cr}(e_2)$, we color e_1 as green and e_2 as red (if $\mathrm{cr}(e_1) = \mathrm{cr}(e_2) = 1$, the red edge is chosen arbitrarily). Note that, since Γ is a min-1-planar drawing, each red edge is crossed at most once, hence the above coloring rule is well-defined. In particular, heavy edges are always colored green, while if two light edges cross, one is colored green and the other is colored red. Hence, the subgraph induced by the red edges is a plane graph, called the *red subgraph of G defined by Γ*, or simply the *red subgraph of Γ*. The following lemma is proved in [14].

Lemma 1. *Let G be a simple graph and let Γ be a min-1-planar drawing of G. We can always augment Γ with edges in such a way that the new drawing is still min-1-planar and all faces of its red subgraph have degree three.*

We now prove a tight bound on the edge density of min-1-planar graphs.

Theorem 2. *Any n-vertex min-1-planar simple graph has at most $4n-8$ edges, and this bound is tight.*

Proof. Let Γ be a min-1-planar drawing of a simple graph G with n vertices. By Lemma 1, we can augment Γ (and hence G) with new edges, in such a way that the new drawing Γ' (and the corresponding graph G') is min-1-planar and its red subgraph Γ'_r is a triangulated planar graph. Hence, Γ'_r has exactly $3n-6$ edges and $2n-4$ faces. Every green edge of G' (which is also a green edge of G) traverses at least two faces of Γ'_r. Also, since Γ' is a min-1-planar drawing and the red subgraph has only triangular faces, each face of the red subgraph is crossed by at most one green edge. Hence the number of green edges is at most $\frac{2n-4}{2}=n-2$, and therefore G' has at most $(3n-6)+(n-2)=4n-8$ edges in total. Since G is a subgraph of G', then also G has at most $4n-8$ edges.

About the tightness of the bound, we recall that optimal 1-planar graphs with n vertices (which are also min-1-planar) have $4n-8$ edges [18,31,32]. □

Plugging the bound of Theorem 2 into the bound of Property 2, we immediately get that any min-1-planar drawing has at most $\frac{4}{3}n-\frac{8}{3}$ heavy edges. We considerably improve this bound in the next theorem.

Theorem 3. *Any n-vertex min-1-planar drawing has at most $\frac{2}{3}n-1$ heavy edges. Further, there exist min-1-planar drawings with $\frac{2}{3}n-O(1)$ heavy edges.*

Proof. Let Γ be a min-1-planar drawing of a simple graph G with n vertices. As in the proof of Theorem 2, by Lemma 1 we can augment Γ with new red edges, in such a way that the new drawing Γ' is min-1-planar and its red subgraph Γ'_r has all faces of degree three. Hence, Γ'_r has exactly $3n-6$ edges and $2n-4$ faces. Clearly, the number of heavy edges of Γ' is not smaller than the one of Γ. By definition, every heavy edge of Γ' is crossed at least twice, hence it traverses at least three faces of Γ'_r. As before, each face of the red subgraph is crossed by at most one heavy edge. Hence the number of heavy edges is at most $\frac{2n-4}{3}\leq\frac{2}{3}n-1$. For the lower bound, we refer to the left part of Fig. 11 in [14]. □

3.2 Density of Min-2-planar Graphs

Proving a tight bound on the edge density of min-2-planar graphs is more challenging than for min-1-planar graphs. Observe that there are min-2-planar simple graphs with $5n-10$ edges, namely the optimal 2-planar graphs [13]. Each optimal 2-planar drawing consists of a subset of planar edges forming faces of size five (i.e., pentagons), and each face is filled up with five more edges that cross each other twice. In the following we prove that $5n - 10$ is also an upper bound to the number of edges of min-2-planar graphs. To this aim, for any $k \geq 1$, we introduce a class of multi-graphs that generalize min-k-planar simple graphs.

Let G be a (multi-)graph (without self-loops) and let Γ be a (simple) drawing of G. A set of parallel edges of G between the same pair of vertices is called a *bundle* of G. We say that Γ is *bundle-proper* if for every bundle in G: (*i*) at most one of the edges of the bundle is involved in a crossing; and (*ii*) Γ has no face bounded only by two edges of the bundle (i.e., no face of Γ is a 2-real bigon). We remark that, in the literature, two parallel edges that form a face of degree two are called *homotopic*. Hence, property (*ii*) is equivalent to saying that a *bundle-proper* drawing does not contain homotopic parallel edges.

Graph G is *bundle-proper min-k-planar* if it admits a (simple) drawing Γ that is both min-k-planar and bundle-proper. If G has n vertices and has the maximum number of edges over all bundle-proper min-k-planar n-vertex graphs, then we say that G is a *maximally-dense* bundle-proper min-k-planar graph. Consider a pair (G, Γ), where G is an n-vertex bundle-proper min-k-planar graph and Γ is a bundle-proper min-k-planar drawing of G. We say that (G, Γ) is a *maximally-dense crossing-minimal bundle-proper min-k-planar pair* if G is maximally-dense and Γ has the minimum number of crossings over all bundle-proper min-k-planar drawings of maximally-dense bundle-proper min-k-planar n-vertex graphs. The proof of the next lemma is in [14].

Lemma 2. *Let (G, Γ) be a maximally-dense crossing-minimal bundle-proper min-k-planar pair. These properties hold: (a) If a face f of Γ contains two distinct real-vertices u and v, then f contains an edge uv. (b) For each face f of Γ, $\deg_\Gamma(f) \geq 3$. (c) A face f of Γ with $\deg^r_\Gamma(f) \geq 3$ is a 3-real triangle.*

To prove the upper bound we use *discharging* techniques. See [1,2,15,22] for previous works that use this tool. Define a *charging function* $\text{ch} : F(\Gamma) \to \mathbb{R}$ such that, for each $f \in F(\Gamma)$:

$$\text{ch}(f) = \deg_\Gamma(f) + \deg^r_\Gamma(f) - 4 = 2\deg^r_\Gamma(f) + \deg^c_\Gamma(f) - 4 \tag{1}$$

The value $\text{ch}(f)$ is called the *initial charge* of f. Using Euler's formula, it is not difficult to see that the following equality holds (refer to [2] for details):

$$\sum_{f \in F(\Gamma)} \text{ch}(f) = 4n - 8 \tag{2}$$

The goal of a discharging technique is to derive from the initial charging function $\text{ch}(\cdot)$ a new function $\text{ch}'(\cdot)$ that satisfies two properties: (C1) $\text{ch}'(f) \geq \alpha \deg^r_\Gamma(f)$, for some real number $\alpha > 0$; and (C2) $\sum_{f \in F(\Gamma)} \text{ch}'(f) \leq \sum_{f \in F(\Gamma)} \text{ch}(f)$.

If $\alpha > 0$ is a number for which a function $\text{ch}'(\cdot)$ satisfies (C1) and (C2), by Eq. (2) we get: $4n - 8 = \sum_{f \in F(\Gamma)} \text{ch}(f) \geq \sum_{f \in F(\Gamma)} \text{ch}'(f) \geq \alpha \sum_{f \in F(\Gamma)} \deg^r_\Gamma(f)$. Also, since $\sum_{f \in F(\Gamma)} \deg^r_\Gamma(f) = \sum_{v \in V(G)} \deg_G(v) = 2m$, we get the following:

$$m \leq \frac{2}{\alpha}(n - 2) \tag{3}$$

Thus, Eq. (3) can be exploited to prove upper bounds on the edge density of a graph for specific values of α, whenever we find a charging function $\mathrm{ch}'(\cdot)$ that fulfills (C1) and (C2). We prove the following.

Theorem 4. *Any n-vertex min-2-planar simple graph has at most $5n-10$ edges, and this bound is tight.*

Proof (Sketch). We already observed that there exist min-2-planar simple graphs with $5n-10$ edges (e.g., the optimal 2-planar graphs). We now prove that min-2-planar simple graphs have at most $5n-10$ edges. Since any simple graph is also bundle-proper, we can show that the bound holds more in general for bundle-proper min-2-planar. Also, we can restrict to maximally-dense bundle-proper min-2-planar graphs, and in particular to crossing-minimal drawings. Let (G, Γ) be any maximally-dense crossing-minimal bundle-proper min-2-planar pair, with $|V(G)| = n$. We show the existence of a charging function $\mathrm{ch}'(\cdot)$ that satisfies (C1) and (C2) for $\alpha = \frac{2}{5}$, so the result will follow from Eq. (3).

Consider the initial charging function $\mathrm{ch}(\cdot)$ defined in Eq. (1). For each type of triangle t we analyze the value of $\mathrm{ch}(t)$ and the deficit/excess w.r.t. $\frac{2}{5}\deg^r_\Gamma(t)$. If t is a 0-real triangle, $\mathrm{ch}(t) = -1 < 0 = \frac{2}{5}\deg^r_\Gamma(t)$, thus t has a deficit of 1. If t is a 1-real triangle, $\mathrm{ch}(t) = 0 < \frac{2}{5} = \frac{2}{5}\deg^r_\Gamma(t)$, thus t has a deficit of $\frac{2}{5}$. If t is a 2-real triangle, $\mathrm{ch}(t) = 1 > \frac{4}{5} = \frac{2}{5}\deg^r_\Gamma(t)$, thus t has an excess of $\frac{1}{5}$. If t is a 3-real triangle, $\mathrm{ch}(t) = 2 > \frac{6}{5} = \frac{2}{5}\deg^r_\Gamma(t)$, thus t has an excess of $\frac{4}{5}$.

Also, if f is any face of Γ with $\deg_\Gamma(f) \geq 4$, then $\mathrm{ch}(f) = 2\deg^r_\Gamma(f) + \deg^c_\Gamma(f) - 4 = \deg_\Gamma(f) - 4 + \deg^r_\Gamma(f) \geq \deg^r_\Gamma(f) \geq \frac{2}{5}\deg^r_\Gamma(f)$. Therefore $\mathrm{ch}(\cdot)$ only fails to satisfy (C1) at 0-real and 1-real triangles. We begin by setting $\mathrm{ch}'(f) = \mathrm{ch}(f)$ for each face f of Γ and we explain how to modify $\mathrm{ch}'(\cdot)$ in such a way that $\mathrm{ch}'(f) \geq \frac{2}{5}\deg^r_\Gamma(f)$ for each face $f \in F(\Gamma)$, thus satisfying (C1), and such that the total charge remains the same, thus satisfying (C2).

Fixing 0-Real Triangles. Let t be a 0-real triangle in Γ with edges e_1, e_2, and e_3. Refer to Fig. 2. The edges $\overline{e}_1$, $\overline{e}_2$ and $\overline{e}_3$ are three pairwise crossing edges of G. Since Γ is a simple drawing, $\overline{e}_1$, $\overline{e}_2$ and $\overline{e}_3$ are independent edges of G (i.e., their six end-vertices are all distinct). Also, since Γ is min-2-planar, at least two of these three edges, say $\overline{e}_2$ and $\overline{e}_3$, do not cross other edges of G in Γ. This implies that each of the two end-vertices of $\overline{e}_2$ shares a face with an end-vertex of $\overline{e}_3$. Hence, by Lemma 2(a), the four vertices of $\overline{e}_2$ and $\overline{e}_3$ form a 4-cycle $e'\overline{e}_2e''\overline{e}_3$ in G and Γ contains a 2-real quadrilateral f_1 bounded by portions of e'', $\overline{e}_1$, $\overline{e}_2$, $\overline{e}_3$, and a 2-real triangle f_2 bounded by portions of e', $\overline{e}_2$, $\overline{e}_3$.

The charge of f_1 is $\mathrm{ch}'(f_1) = 2$, with an excess of $\frac{6}{5}$ w.r.t. $\frac{2}{5}\deg^r_\Gamma(f_1) = \frac{4}{5}$. The charge of f_2 is $\mathrm{ch}'(f_2) = 1$, with an excess of $\frac{1}{5}$ w.r.t. $\frac{2}{5}\deg^r_\Gamma(f_2) = \frac{4}{5}$. We reduce $\mathrm{ch}'(f_1)$ by $\frac{4}{5}$, reduce $\mathrm{ch}'(f_2)$ by $\frac{1}{5}$, and increase $\mathrm{ch}'(t)$ by 1. After that, the total charge is unchanged and all the three faces t, f_1, and f_2 satisfy (C1). Namely, $\mathrm{ch}'(t) = 0$ (it has no deficit/excess), $\mathrm{ch}'(f_1) = \frac{6}{5}$ (it has an excess of $\frac{2}{5}$), and $\mathrm{ch}'(f_2) = \frac{4}{5}$ (it has no deficit/excess). In the remainder of the proof, we call each of the faces f_1 and f_2 a *0-real triangle-neighboring face*. Each 0-real triangle-neighboring face that is a 2-real triangle (as f_2) shares its unique

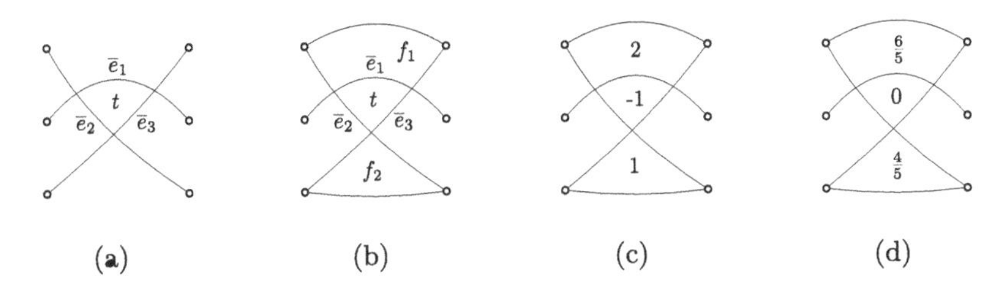

Fig. 2. (a) A 0-real triangle t. (b) A 2-real quadrilateral f_1 and a 2-real triangle f_2 neighboring t. (c) The initial charges. (d) The charges after a redistribution.

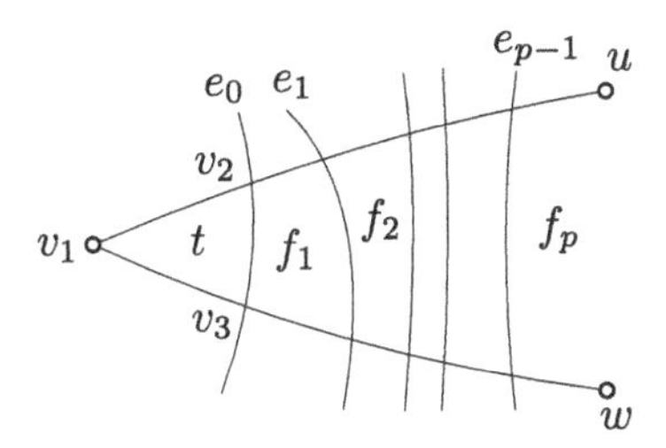

Fig. 3. The demand path of a 1-real triangle t, ending at a face f_p.

crossing vertex with a 0-real triangle; each 0-real triangle-neighboring face that is a 2-real quadrilateral (as f_1) shares its unique 0-real edge with a 0-real triangle.

Fixing 1-Real Triangles. Let t be a 1-real triangle, with real-vertex v_1 and crossing-vertices v_2 and v_3. Refer to Fig. 3 for an illustration. Let $e_0 = v_2v_3$ be the 0-real edge of t, and let f_1 be the face of Γ that shares e_0 with t. If f_1 is a 0-real quadrilateral, denote by e_1 the 0-real edge of f_1 not adjacent to e_0, and by f_2 the face of Γ that shares e_1 with f_1. If f_2 is a 0-real quadrilateral, denote by e_2 the 0-real edge of f_2 not adjacent to e_1, and by f_3 the face of Γ that shares e_2 with f_2. We continue in this way until we encounter a face f_p ($p \geq 1$) that is not a 0-real quadrilateral. This procedure determines a sequence of faces $f_0, f_1, f_2, \dots f_p$, and a sequence of 0-real edges $e_0, e_1, \dots, e_{p-1}$ such that $f_0 = t$, f_i is a 0-real quadrilateral for each $i \in \{1, \dots, p-1\}$, f_p is not a 0-real quadrilateral, and the faces f_i and f_{i-1} share edge e_{i-1} ($i \in \{1, \dots, p\}$).

Note that $\deg_\Gamma(f_p) \geq 4$. Namely, let $e = v_1v_2$ and $e' = v_1v_3$, and let $\overline{e} = v_1u$ and $\overline{e'} = v_1w$ be the edges of G that contain e and e'. Since f_p has at least two crossing-vertices, if f_p were a triangle then it would be either a 0-real triangle or a 1-real triangle. If f_p were a 0-real triangle then $\overline{e}$ and $\overline{e'}$ would cross in Γ, which is impossible as $\overline{e}$ and $\overline{e'}$ are adjacent edges and Γ is a simple drawing. If f_p were a 1-real triangle then $u = w$, i.e., $\overline{e}$ and $\overline{e'}$ would be parallel edges both involved in a crossing, which is impossible as Γ is bundle-proper. Hence, $\deg_\Gamma(f_p) \geq 4$ and, as observed at the beginning of this proof, $\mathrm{ch}'(f_p) \geq \frac{2}{5}\deg^r_\Gamma(f_p)$. Also, the charge excess of f_p is larger than $\frac{2}{5}$. Namely, this excess is $x = 2\deg^r_\Gamma(f_p) + \deg^c_\Gamma(f_p) - 4 - \frac{2}{5}\deg^r_\Gamma(f_p) = \deg_\Gamma(f) + \frac{3}{5}\deg^r_\Gamma(f) - 4$. If f_p has no real-vertices, it must have at least five crossing-vertices (as f_p is not a 0-real quadrilateral), which implies $x \geq 1 > \frac{2}{5}$. If f_p has at least one real-vertex then $x \geq \frac{3}{5} > \frac{2}{5}$.

Therefore, the idea is to fill the $\frac{2}{5}$ charge deficit of t by moving an equivalent amount of charge from f_p to t. We say that t *demands from* f_p *through edge* e_{p-1} a charge amount of $\frac{2}{5}$. We call $f_0, \ldots, f_p$ (which is a path in the dual of Γ) the *demand path* for t. Hence, for each 1-real triangle t of Γ whose demand path ends at a face $f = f_p$, we decrease $\mathrm{ch}'(f)$ by $\frac{2}{5}$ and increase $\mathrm{ch}'(t)$ from 0 to $\frac{2}{5}$. Note that f cannot be a 0-real triangle-neighboring face. Indeed, f is not a triangle, and if f is a 2-real quadrilateral then its 0-real edge is shared either with a 0-real quadrilateral or directly with the 1-real triangle t. It follows that the set of faces whose charge is affected by fixing 1-real triangles does not intersect with the set of faces whose charge is affected by fixing 0-real triangles.

From the reasoning above, after we have fixed all 1-real triangles, we may have problems only if multiple 1-real triangles demanded from the same face f. In this case, f might no longer satisfy (C1). The rest of the proof (see [14]) shows which faces may be in this situation and how to fix their charge. □

Combining Theorem 4 with Property 2 we immediately get that any min-2-planar drawing has at most $2n - 4$ heavy edges. The next theorem considerably improves this bound by exploiting discharging techniques (see [14]).

Theorem 5. *Any n-vertex min-2-planar drawing has at most $\frac{6}{5}(n-2)$ heavy edges. Further, there exist min-2-planar drawings with $n - O(1)$ heavy edges.*

3.3 Density of Min-3-planar Graphs

For the family of min-3-planar graphs we consider graphs that can contain non-homotopic parallel edges. Indeed, it is known that n-vertex 3-planar graphs that are simple have at most $5.5n - 15$ edges [31], but this bound is not tight. On the other hand, a tight upper bound is known for 3-planar graphs that can contain non-homotopic multiple edges, namely $5.5n - 11$ [29]. We give upper bounds on the edge density and on the density of the heavy edges in min-3-planar graphs. The proofs still exploit discharging techniques (see [14]).

Theorem 6. *Any min-3-planar graph with n vertices has at most $6n - 12$ edges.*

Theorem 7. *Any n-vertex min-3-planar drawing has at most $2(n-2)$ heavy edges. Further, there exist min-3-planar drawings with $\frac{6}{5}n - O(1)$ heavy edges.*

4 Relationships with k-planar Graphs

The next theorem shows that while the family of min-1-planar graphs properly contains the family of 1-planar graphs, the two classes coincide when we restrict to optimal graphs, i.e., those with $4n - 8$ edges.

Theorem 8. *1-planar graphs are a proper subset of min-1-planar graphs, while optimal min-1-planar graphs are optimal 1-planar.*

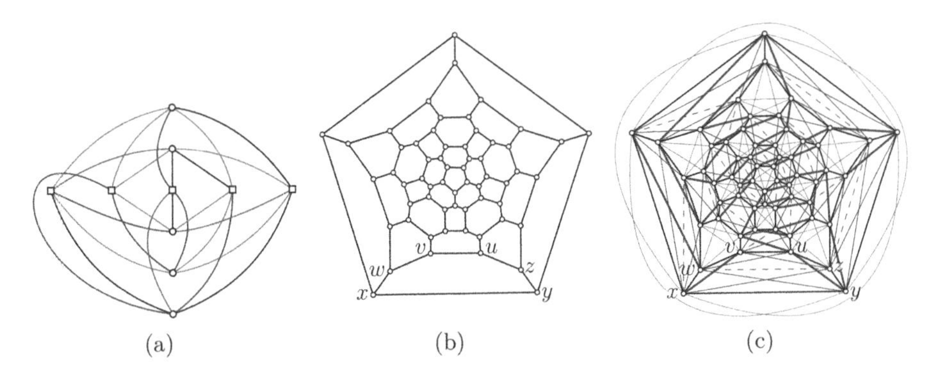

Fig. 4. (a) A min-2-planar drawing of $K_{5,5}$. (b) A planar drawing Γ of the truncated icosahedral graph G. (c) A min-2-planar drawing Γ' of the graph G', obtained by adding 5 edges to each pentagonal face and 7 edges to each hexagonal face of Γ.

Proof. Any 1-planar graph is min-1-planar. By the NP-hardness of testing whether a given planar graph plus a single edge is 1-planar [19], we know that there are such graphs that are not 1-planar, while any planar graph that is extended by a single edge can be drawn min-1-planar. Hence, 1-planar graphs are a proper subset of min-1-planar graphs. Finally, as in the proof of Theorem 2, in every optimal min-1-planar drawing the red subgraph is maximal planar and each green edge traverses exactly two faces of the red subgraph. Hence, each green edge crosses exactly once, i.e., the drawing is also (optimal) 1-planar. □

Unlike min-1-planar graphs, we show that min-2-planar graphs are a proper superset of the 2-planar graphs even when we restrict to optimal graphs.

Theorem 9. *2-planar graphs are a proper subset of min-2-planar graphs, and there are optimal min-2-planar graphs that are not optimal 2-planar.*

Proof (Sketch). First observe that there exist non-optimal min-2-planar graphs that are not 2-planar. For example, $K_{5,5}$ is not 2-planar [6,7], while Fig. 4(a) illustrates a min-2-planar drawing of $K_{5,5}$. To construct an optimal min-2-planar graph that is not 2-planar, start from the truncated icosahedral graph in Fig. 4(b), consisting of 12 pentagonal faces, 20 hexagonal faces, 60 vertices and 90 edges. Then enrich it with 5 edges inside each pentagonal face and 7 edges inside each hexagonal face, as in Fig. 4(c). The new graph is min-2-planar and has $5n - 10$ edges. Conversely, it is not 2-planar as it contains vertices of degree 10 and 11, while it is known that optimal 2-planar graphs must have only vertices with degree multiple of three [23]. See [14] for details. □

In contrast to 1- and 2-planar graphs, the maximum densities of 3-planar and min-3-planar graphs differ.

Theorem 10. *There are min-3-planar (non-simple) graphs denser than optimal 3-planar (non-simple) graphs.*

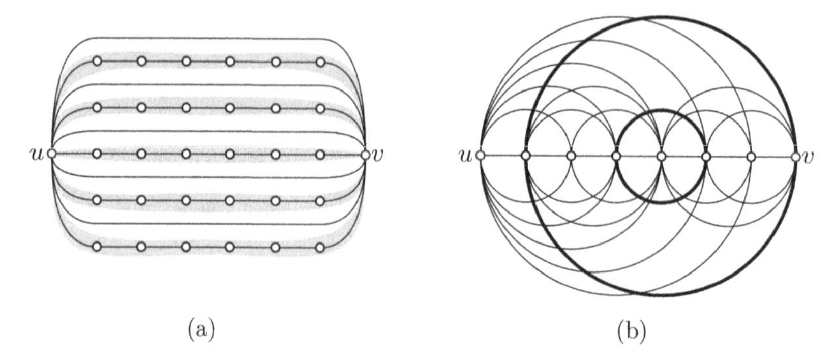

Fig. 5. Illustration of the construction of Theorem 10. If in the graph of figure (a) we replace each shaded chain with a copy of the graph of figure (b), we get a min-3-planar graph that is not 3-planar. The bold edges are heavy edges.

Proof (Sketch). First, consider a planar graph G, and a corresponding drawing Γ, consisting of h parallel chains ($h \geq 1$), each with 8 vertices, sharing the two end-vertices u and v, and interleaved by h copies of edge uv; refer to Fig. 5(a). Then, construct a new graph G', and a corresponding drawing Γ', obtained from G, and from Γ, by replacing each parallel chain with a copy of the graph G'' depicted in Fig. 5(b). In the drawing Γ', each copy of G'' has the same edge crossings as the drawing illustrated in Fig. 5(b). Graph G'' has 8 vertices and 33 edges, and it is min-3-planar. The bold edges are the heavy edges in the drawing of Fig. 5(b)). It can be proved that G'' has $5.\overline{6}n - 11.\overline{3}$ edges, while it is known that every 3-planar graph has at most $5.5n - 11$ edges [29]. □

5 Final Remarks and Open Problems

About edge density, one can ask whether the bound of Theorem 6 for min-3-planar graphs is tight or if it can be further lowered. Providing finer bounds for $k \geq 4$ is also interesting. Another classical research direction is to establish inclusion or incomparability relations between min-k-planar graphs and classes of beyond-planar graphs other than k-planar graphs. The next two lemmas provide initial results in this direction (see [14] for their proofs). In particular, Lemma 3 leaves as open what is the relationship between min-2-planar graphs and 1-gap-planar graphs (which have the same maximum edge density). Lemma 4 implies that min-2-planar graphs and fan-planar graphs are incomparable classes, even if they have the same maximum edge density.

Lemma 3. *Min-k-planar graphs are a subset of k-gap-planar graphs and of $(k+2)$-quasiplanar graphs, for every $k \geq 1$.*

Lemma 4. *For any given $k \geq 2$, fan-planar and min-k-planar graphs are incomparable, i.e., each of the two classes contains graphs that are not in the other.*

References

1. Ackerman, E.: On topological graphs with at most four crossings per edge. Comput. Geom. **85** (2019). https://doi.org/10.1016/j.comgeo.2019.101574
2. Ackerman, E., Tardos, G.: On the maximum number of edges in quasi-planar graphs. J. Comb. Theory Ser. A **114**(3), 563–571 (2007). https://doi.org/10.1016/j.jcta.2006.08.002
3. Agarwal, P.K., Aronov, B., Pach, J., Pollack, R., Sharir, M.: Quasi-planar graphs have a linear number of edges. In: Brandenburg, F. (ed.) GD 1995. LNCS, vol. 1027, pp. 1–7. Springer, Cham (1995). https://doi.org/10.1007/BFb0021784
4. Agarwal, P.K., Aronov, B., Pach, J., Pollack, R., Sharir, M.: Quasi-planar graphs have a linear number of edges. Combinatorica **17**(1), 1–9 (1997). https://doi.org/10.1007/BF01196127
5. Angelini, P., et al.: Simple k-planar graphs are simple (k+1)-quasiplanar. J. Comb. Theory Ser. B **142**, 1–35 (2020). https://doi.org/10.1016/j.jctb.2019.08.006
6. Angelini, P., Bekos, M.A., Kaufmann, M., Schneck, T.: Efficient generation of different topological representations of graphs beyond-planarity. In: Archambault, D., Tóth, C.D. (eds.) GD 2019. LNCS, vol. 11904, pp. 253–267. Springer, Cham (2019). https://doi.org/10.1007/978-3-030-35802-0_20
7. Angelini, P., Bekos, M.A., Kaufmann, M., Schneck, T.: Efficient generation of different topological representations of graphs beyond-planarity. J. Graph Algorithms Appl. **24**(4), 573–601 (2020). https://doi.org/10.7155/jgaa.00531
8. Bachmaier, C., Rutter, I., Stumpf, P.: 1-gap planarity of complete bipartite graphs. In: Biedl, T.C., Kerren, A. (eds.) GD 2018. LNCS, vol. 11282, pp. 646–648. Springer, Cham (2018)
9. Bae, S.W., et al.: Gap-planar graphs. In: Frati, F., Ma, K.-L. (eds.) GD 2017. LNCS, vol. 10692, pp. 531–545. Springer, Cham (2018). https://doi.org/10.1007/978-3-319-73915-1_41
10. Bae, S.W., et al.: Gap-planar graphs. Theor. Comput. Sci. **745**, 36–52 (2018). https://doi.org/10.1016/j.tcs.2018.05.029
11. Bekos, M.A., Cornelsen, S., Grilli, L., Hong, S.-H., Kaufmann, M.: On the recognition of fan-planar and maximal outer-fan-planar graphs. In: Duncan, C., Symvonis, A. (eds.) GD 2014. LNCS, vol. 8871, pp. 198–209. Springer, Heidelberg (2014). https://doi.org/10.1007/978-3-662-45803-7_17
12. Bekos, M.A., Cornelsen, S., Grilli, L., Hong, S., Kaufmann, M.: On the recognition of fan-planar and maximal outer-fan-planar graphs. Algorithmica **79**(2), 401–427 (2017). https://doi.org/10.1007/s00453-016-0200-5
13. Bekos, M.A., Kaufmann, M., Raftopoulou, C.N.: On optimal 2- and 3-planar graphs. In: Aronov, B., Katz, M.J. (eds.) SoCG. LIPIcs, vol. 77, pp. 16:1–16:16. Schloss Dagstuhl (2017). https://doi.org/10.4230/LIPIcs.SoCG.2017.16
14. Binucci, C., et al.: Min-k-planar drawings of graphs. CoRR 2308.13401 (2023). https://arxiv.org/abs/2308.13401
15. Binucci, C., et al.: Nonplanar graph drawings with k vertices per face. In: Paulusma, D., Ries, B. (eds.) WG 2023. LNCS, vol. 14093, pp. 86–100. Springer, Cham (2023). https://doi.org/10.1007/978-3-031-43380-1_7
16. Binucci, C., et al.: Fan-planarity: properties and complexity. Theor. Comput. Sci. **589**, 76–86 (2015). https://doi.org/10.1016/j.tcs.2015.04.020
17. Binucci, C., Di Giacomo, E., Didimo, W., Montecchiani, F., Patrignani, M., Tollis, I.G.: Fan-planar graphs: combinatorial properties and complexity results. In: Duncan, C., Symvonis, A. (eds.) GD 2014. LNCS, vol. 8871, pp. 186–197. Springer, Heidelberg (2014). https://doi.org/10.1007/978-3-662-45803-7_16

18. Bodendiek, R., Schumacher, H., Wagner, K.: Über 1-optimale graphen. Math. Nachr. **117**, 323–339 (1984)
19. Cabello, S., Mohar, B.: Adding one edge to planar graphs makes crossing number and 1-planarity hard. SIAM J. Comput. **42**(5), 1803–1829 (2013). https://doi.org/10.1137/120872310
20. Di Battista, G., Eades, P., Tamassia, R., Tollis, I.G.: Graph Drawing: Algorithms for the Visualization of Graphs. Prentice-Hall (1999)
21. Didimo, W., Liotta, G., Montecchiani, F.: A survey on graph drawing beyond planarity. ACM Comput. Surv. **52**(1), 4:1–4:37 (2019). https://doi.org/10.1145/3301281
22. Dujmovic, V., Gudmundsson, J., Morin, P., Wolle, T.: Notes on large angle crossing graphs. Chicago J. Theor. Comput. Sci. **2011** (2011)
23. Förster, H., Kaufmann, M., Raftopoulou, C.N.: Recognizing and embedding simple optimal 2-planar graphs. In: Purchase, H.C., Rutter, I. (eds.) GD 2021. LNCS, vol. 12868, pp. 87–100. Springer, Cham (2021). https://doi.org/10.1007/978-3-030-92931-2_6
24. Fox, J., Pach, J., Suk, A.: The number of edges in k-quasi-planar graphs. SIAM J. Discret. Math. **27**(1), 550–561 (2013). https://doi.org/10.1137/110858586
25. Grigoriev, A., Bodlaender, H.L.: Algorithms for graphs embeddable with few crossings per edge. Algorithmica **49**(1), 1–11 (2007). https://doi.org/10.1007/s00453-007-0010-x
26. Hong, S., Tokuyama, T. (eds.): Beyond Planar Graphs. Springer, Cham (2020). https://doi.org/10.1007/978-981-15-6533-5
27. Kaufmann, M., Ueckerdt, T.: The density of fan-planar graphs. Electron. J. Comb. **29**(1) (2022). https://doi.org/10.37236/10521
28. Kobourov, S.G., Liotta, G., Montecchiani, F.: An annotated bibliography on 1-planarity. Comput. Sci. Rev. **25**, 49–67 (2017). https://doi.org/10.1016/j.cosrev.2017.06.002
29. Pach, J., Radoicic, R., Tardos, G., Tóth, G.: Improving the crossing lemma by finding more crossings in sparse graphs. Discret. Computat. Geom. **36**(4), 527–552 (2006). https://doi.org/10.1007/s00454-006-1264-9
30. Pach, J., Tóth, G.: Graphs drawn with few crossings per edge. In: North, S. (ed.) GD 1996. LNCS, vol. 1190, pp. 345–354. Springer, Heidelberg (1997). https://doi.org/10.1007/3-540-62495-3_59
31. Pach, J., Tóth, G.: Graphs drawn with few crossings per edge. Combinatorica **17**(3), 427–439 (1997). https://doi.org/10.1007/BF01215922
32. Ringel, G.: Ein Sechsfarbenproblem auf der Kugel. Abh. Math. Sem. Univ. Hamb. **29**, 107–117 (1965)
33. Wood, D.R., Telle, J.A.: Planar decompositions and the crossing number of graphs with an excluded minor. In: Kaufmann, M., Wagner, D. (eds.) GD 2006. LNCS, vol. 4372, pp. 150–161. Springer, Heidelberg (2007). https://doi.org/10.1007/978-3-540-70904-6_16
34. Wood, D.R., Telle, J.A.: Planar decompositions and the crossing number of graphs with an excluded minor. New York J. Math. **13**, 117–146 (2007). https://nyjm.albany.edu/j/2007/13-8.pdf

Weakly and Strongly Fan-Planar Graphs

Otfried Cheong[1], Henry Förster[2], Julia Katheder[2(✉)], Maximilian Pfister[2], and Lena Schlipf[2]

[1] SCALGO, Aarhus, Denmark
otfried@scalgo.com
[2] Wilhelm-Schickard-Institut für Informatik, Universität Tübingen, Tübingen, Germany
{henry.foerster,julia.katheder,maximilian.pfister, lena.schlipf}@uni-tuebingen.de

Abstract. We study two notions of fan-planarity introduced by (Cheong et al., GD22), called weak and strong fan-planarity, which separate two non-equivalent definitions of fan-planarity in the literature. We prove that not every weakly fan-planar graph is strongly fan-planar, while the upper bound on the edge density is the same for both families.

Keywords: fan-planarity · density · weak vs. strong

1 Introduction

Crossings in graph drawings are known to heavily impede readability [17,18]. Unfortunately, however, minimizing the number of crossings is NP-complete [9], while many real-world networks turn out to be non-planar. Fortunately, readable drawings of non-planar graphs can be obtained by limiting the *topology* [16] or the *geometry* of crossings [11,12]. Based on these experimental findings, the research direction of *graph drawing beyond planarity* has emerged. This line of research is dedicated to the study of so-called *beyond planar* graph classes that are defined by forbidden edge-crossing patterns. More precisely, a graph belonging to such a class admits a drawing in which the forbidden pattern is absent. Important beyond planar graph classes are k-planar graphs, where the forbidden pattern is $k+1$ crossings on the same edge, k-quasiplanar graphs, where k mutually crossing edges are prohibited, and RAC-graphs, where edges are not allowed to cross at non-right angles. We refer the interested reader to the survey by Didimo et al. [8] and a recent book [10] on beyond planarity.

In this paper, we study *fan-planar graphs*, which admit *fan-planar drawings*. In these drawings, each edge e may only be crossed by a *fan* of edges, that is a bundle of edges sharing a common endpoint, called *anchor* of e, that all cross e from the same side. Kaufmann and Ueckerdt introduced this graph class in 2014 [13] and described the aforementioned requirement with two forbidden patterns, Pattern (I) and (II) in Fig. 1: the first forbids two edges crossing e to be non-adjacent whereas the second forbids crossings of e by adjacent edges with the common endpoint on different sides of e. Since their introduction, fan-planar graphs have received a lot of attention in the scientific community; see [2] for an overview.

The work of J. Katheder is supported by DFG grant Ka 812-18/2.

M. A. Bekos and M. Chimani (Eds.): GD 2023, LNCS 14465, pp. 53–68, 2023.
https://doi.org/10.1007/978-3-031-49272-3_4

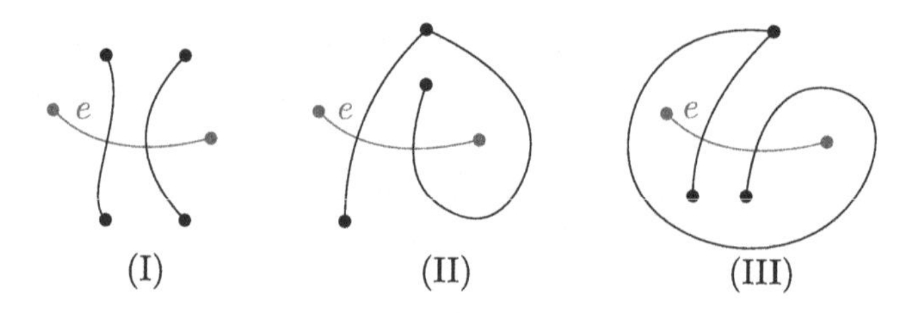

Fig. 1. Forbidden configurations.

Recently, Klemz et al. [15] pointed out a missing case in the proof of the edge density upper-bound in the preprint that introduced fan-planar graphs [13]. This case was consequently fixed in the journal version [14] by the original authors. However, in the process, they introduced a third forbidden pattern, namely Pattern (III) of Fig. 1, which is quite reminiscent of the previously defined Pattern (II). Namely, in both patterns, the drawing restricted to the edge e and the two edges crossing e has two connected regions, called *cells*, where one is bounded and the other one is unbounded. The difference between the two configurations is that in Pattern (II), one endpoint of e lies in the bounded cell, while in Pattern (III), both endpoints are contained there. In the new definition of *fan-planarity*, both endpoints of e must lie in the unbounded cell.

The new forbidden Pattern (III) poses a problem for the existing literature on fan-planarity. Namely, while some previous results still apply when forbidding Pattern (III), other existing results, e.g., the bound of $4n - 12$ on the number of edges of n-vertex bipartite "*fan-planar*" graphs [1], build on the lemmas of the original paper [13] and may be affected by the recent changes to the definition.

Cheong et al. [7] introduced a clear distinction of the two models with the notions of *weak* and *strong fan-planarity*. Namely, *weak fan-planarity* allows Pattern (III) whereas *strong fan-planarity* forbids it. The original definition of fan-planarity [13] coincides with *weak fan-planarity* whereas *strong fan-planarity* matches the definition of the new journal version [14]. Graphs admitting such drawings are called *weakly fan-planar* and *strongly fan-planar*, respectively.

The family of graphs that admits drawings where only Pattern (I) is forbidden, called *adjacency-crossing* graphs, has also been studied by Brandenburg [4]. He showed that there are adjacency-crossing graphs that are *not* weakly fan-planar, so weakly fan-planar graphs form a proper subset of the adjacency-crossing graphs. Moreover, he shows that for any n-vertex adjacency-crossing graph with m edges, one can construct a weakly fan-planar graph with n vertices that also has m edges. Brandenburg concluded from this that an n-vertex adjacency-crossing graph has at most $5n - 10$ edges, since that was the bound claimed by [13] for "fan-planar" graphs. Since that bound holds only under strong fan-planarity, this conclusion contains a gap, which the present paper fills.

Our Contribution. First, we prove that the family of strongly fan-planar graphs is a proper subset of the weakly fan-planar graphs. Together with Brandenburg's result, this implies that the two inclusions of strongly fan-planar graphs inside weakly fan-planar graphs and weakly fan-planar graphs inside adjacency-crossing graphs are both proper. We then continue to show that the known upper bound on the edge-density of strongly fan-planar graphs (namely $5n - 10$ for an n-vertex graph) carries over to weakly fan-planar graphs. This implies that also Brandenburg's bound [4] is in fact correct. We also prove that the known upper bound of $4n - 12$ on the number of edges of an n-vertex *bipartite* strongly fan-planar graph carries over to bipartite weakly fan-planar graphs.

2 Not Every Weakly Fan-Planar Graph is Strongly Fan-Planar

In this section, we will establish that strongly fan-planar graphs form a proper subset of weakly fan-planar graphs by constructing a graph G with a weakly fan-planar drawing Γ, where Pattern (III) cannot be avoided in Γ.

In order to guarantee the existence of at least one Pattern (III) in any valid weakly fan-planar drawing of G, we will use the following key idea. We start with a planar graph with a unique embedding. We will then make every edge of this planar graph "uncrossable" by replacing it with a suitable gadget introduced by Binucci et al. [3]. Afterwards, we insert into every face of the planar graph a small gadget graph (shown in Fig. 2a), which can only be drawn with a quadrangular outer face if we allow (III). Note that this gadget graph itself is in fact strongly fan-planar, as shown in Fig. 2b, and, hence, does not serve itself as an example of a weakly but not strongly fan-planar graph. In order to achieve our goal, we will leverage the following lemma:

Lemma 1 (Binucci et al. [3]). *Let $\mathcal{P}$ be the planarization*[1] *of any weakly fan-planar drawing of K_7. Then, between any pair of vertices of K_7, there exists a path in $\mathcal{P}$ that contains no real edge of the K_7.*

Moreover, we will use the following definition throughout the paper.

Definition 1. *Let Γ be a weakly fan-planar drawing of a graph G. Γ is said to be* minimal, *if, among all weakly fan-planar drawings of G, it contains the smallest possible number of triples of edges that form Pattern (III).*

Theorem 1. *There exists a weakly fan-planar graph that does not admit a strongly fan-planar drawing.*

[1] In a *planarization* $\mathcal{P}$ of a non-planar drawing of a graph G, each crossing is replaced with a dummy-vertex that subdivides both edges involved in the crossing. We call an edge of $\mathcal{P}$ that is not incident to any dummy-vertex a *real edge of G*.

Proof. Let G_0 be a 3-connected planar quadrangulation; e.g., one that is obtained by the construction in [5]. Note that by construction, G_0 is bipartite and has a unique embedding into $\mathbb{R}^2$ up to the choice of the outer face and a mirroring [19]. Next, we insert a copy of our gadget graph H shown in Fig. 2a into every face f of G_0 by identifying the outer cycle of H with the facial cycle of f. Denote by G_1 this supergraph of G_0. We use the color scheme of Fig. 2a to color all edges of G_1 – in particular, the edges of G_0 form a subset of the red edges of G_1. In the next step, we substitute every *red* edge of G_1 by a K_7 and denote the resulting graph by G. We claim that G is weakly fan-planar, but not strongly fan-planar.

For the first statement, observe that K_7 admits a weakly fan-planar drawing, see Fig. 2c, and that our gadget graph H has a weakly fan-planar drawing shown in Fig. 2a. Combining both, we obtain a weakly fan-planar drawing of G.

Consider now the second statement. Let Γ be a weakly fan-planar drawing of G that is minimal and that has the smallest number of crossings among all minimal weakly fan-planar drawings of G. We will prove that Γ contains at least one Pattern (III), which implies by our choice of Γ that *every* weakly fan-planar drawing of G requires at least one Pattern (III).

Consider a red edge $ab \in G_1$ and denote by $a = v_1, \dots, v_7 = b$ the vertices of the K_7 which substitute ab in G. By Lemma 1, in the drawing Γ of this K_7, there exists a sequence S of crossed edges from a to b; see Fig. 2c. By (I), no edge which is not incident to one of $v_1, \dots, v_7$ can intersect S. By construction, the only edges incident to vertices $v_2, \dots, v_6$ are edges of the K_7. Hence, the only edges that can potentially cross S and interact with the remainder of G are incident to either $a = v_1$ or $b = v_7$. Suppose for a contradiction that there exists an edge incident to a or b that crosses S in Γ such that its other endpoint is not one of the vertices of the K_7. But then we can easily reroute the edge such that its crossing with S is avoided, see Fig. 2d, a contradiction to our choice of Γ.

We interpret Γ as a drawing Γ' of G_1, where the red edges are *uncrossed.* Since G_0 has a unique planar embedding into $\mathbb{R}^2$ up to the choice of the outer face and a mirroring and since G_0 consists only of the red edges, in Γ', G_0 is drawn as a planar graph, all faces of which are quadrilaterals. Since the red edges are uncrossed, each quadrilateral face must contain a copy of our gadget H. Indeed, since each vertex of H is connected by a path to both u and u', it must lie in a face that contains both u and u'. But by 3-connectivity of G_0, this face is unique.

Let f be a bounded quadrilateral face of G_0. As argued above, f contains a copy of H in its interior in Γ', i.e. vertices v, v', w, w', z and z', refer to Fig. 2a, lie in the interior of f. Consider the subgraph H' of this H consisting of its red edges. Since the red edges are uncrossed, H' is drawn without crossings in Γ. Since the black edges do not cross H', a small case analysis shows that the vertices u', w, z, and w' must lie in the same face of H', and in fact the embedding of Fig. 2a is unique up to symmetry with respect to the vertical axis. Thus, the blue edges (u, v), (u, v') and (w, w') can only be drawn in the indicated way, but that means that the blue edges form (III), which concludes the proof. □

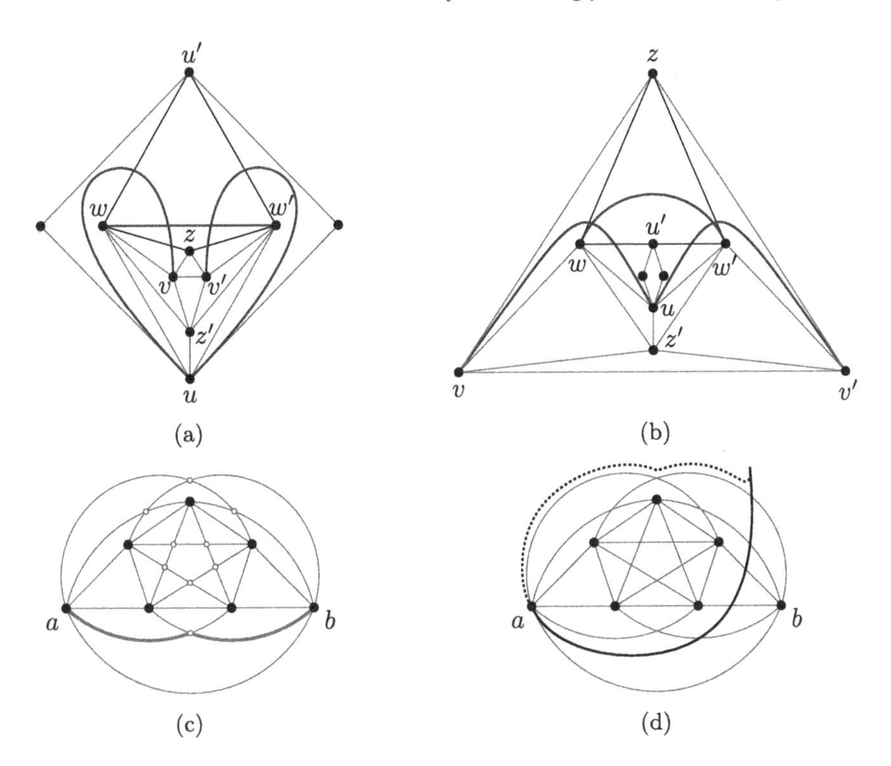

Fig. 2. (a) Gadget graph H. (b) Gadget graph H with (z, v, v') chosen as the outer face. Note that there is no pattern (III). (c) A planarization of a fan-planar drawing of K_7, where the bold edges form a path from a to b that contains no *uncrossed edge* of the K_7. (d) An edge incident to a that crosses the K_7 to avoid this crossing. (Color figure online)

3 Density of Weakly Fan-Planar Graphs

In this section, we show that the density results for strongly fan-planar graphs also transfer to the weakly fan-planar setting. Let us call a triple e, e_ℓ, e_r of edges in a weakly fan-planar drawing a *heart* if e_ℓ and e_r share an endpoint u, both cross e such that they form Pattern (III), and the part of e between the crossings with e_ℓ and e_r is not crossed by any edge of the graph, see Fig. 3. In the remainder of the paper, we call the intersection points of e and e_ℓ (e_r, resp.) x_ℓ (x_r, resp.).

Lemma 2. *Let Γ be a weakly fan-planar drawing that is not strongly fan-planar. Then Γ contains a heart $\mathcal{H}$.*

Proof. By assumption, Γ contains three edges e, e_ℓ, e_r that form Pattern (III), where e_ℓ and e_r share endpoint u and cross e. Let E' be the set of edges that cross e. By (I) and (II), any edge $e' \in E'$ must be incident to u, and by (II) it must cross e from the same side as e_ℓ and e_r. The edges of E' cannot cross each

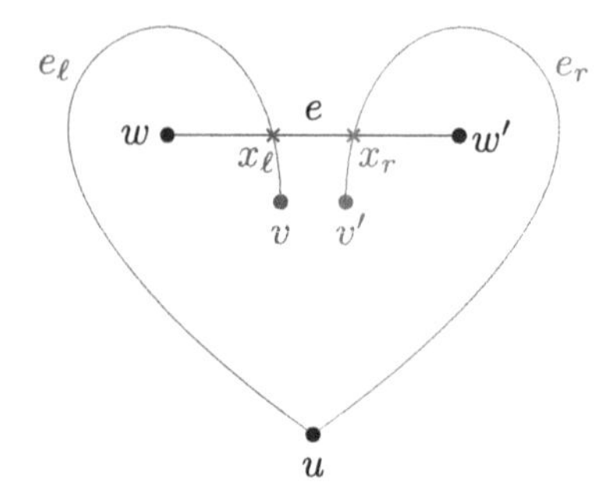

Fig. 3. A heart.

other since they share an endpoint, and each edge $e' \in E'$ forms Pattern (III) either with e and e_r, or with e and e_ℓ. Let $E_\ell \subset E'$ be the set of edges of the first kind, $E_r = E' \setminus E_\ell$ the second kind. If we order E' by their crossing point with e along e, then we first encounter all elements of E_ℓ, then all elements of E_r. The last element of E_ℓ and the first element of E_r form a heart with e. □

We will call the sets E_ℓ and E_r as defined in the previous proof the *left valve* and the *right valve* of the heart $\mathcal{H} = e, e_\ell, e_r$, respectively. We denote by H the edge set containing both valves of $\mathcal{H}$ and the edge e, namely $H = E_\ell \cup E_r \cup e$. In the following, we will define an edge-rerouting operation that will later allow us to reduce the number of hearts in a weakly fan-planar drawing under certain conditions.

Flipping the Valve of a Heart. Consider a heart $\mathcal{H}$ formed by the edges e, e_ℓ, e_r in a weakly fan-planar drawing Γ; refer to Fig. 4a for a visualization. In the following, we will define an operation that we call *flipping* a valve of $\mathcal{H}$ resulting in the drawing Γ'. We describe the flip of E_ℓ, as the other case is symmetric. The general idea is to redraw the edges in E_ℓ "close" to the ones in the other valve E_r, in particular, mainly following the curve of e_r.

Let $e^\ell_1, e^\ell_2, \dots e^\ell_k$ be the edges of E_ℓ in the order that they intersect edge e in Γ starting at w, i.e. $e^\ell_k = e_\ell$. We will draw the curve γ^i of e^ℓ_i in three parts, denoted as $\gamma^i_1, \gamma^i_2, \gamma^i_3$. We consider the edges in reverse order and start with $e^\ell_k = e_\ell$. The first part γ^k_1 of the curve of e^ℓ_k in Γ' follows the curve of e_r slightly outside until x_r, then γ^k_2 follows e until x_ℓ, where the curve intersects e and afterwards it inherits its original curve in Γ as the last part γ^k_3. So, assume that we have drawn e^ℓ_i with $2 < i \leq k$. The curve of e^ℓ_{i-1} follows the curve of e^ℓ_i (slightly outside) until x_r, where it follows e until the intersection point of e^l_{i-1} with e in Γ. Here, the curve intersects e and then again inherits its original curve in Γ until it reaches its endpoint different from u. After this operation, the edges of $E_\ell \cup E_r$ do not cross by simplicity and e does not form Pattern (I) to (III) with $E_\ell \cup E_r$, thus we can make the following observation, illustrated in Fig. 4b.

Observation 1. *Let Γ' be the drawing obtained from flipping a valve of a heart $\mathcal{H}$. Then, $\Gamma'[H]$ is a strongly fan-planar drawing.*

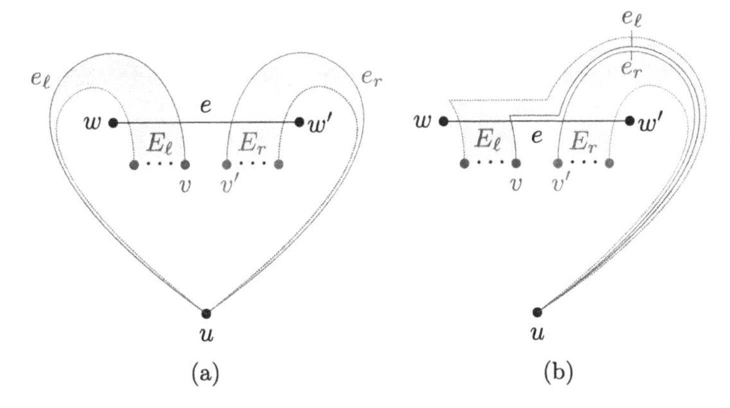

Fig. 4. (a) Illustration of the setting previous to the flip-operation. (b) Transformation from Γ to Γ' by flipping E_ℓ.

Moreover, in the entire drawing Γ' (not limited to H) resulting from a flip, new crossings can arise only in a restricted part of the flipped edges.

Lemma 3. *Let $\mathcal{H}$ be a heart formed by edges e, e_ℓ, e_r in Γ and $E_\ell = \{e_1^\ell, e_2^\ell, \dots e_k^\ell\}$ be the edges of a flipped valve of $\mathcal{H}$ in Γ'. Then any crossing introduced by the flipping operation occurs on the partial curves $\gamma_1^1, \gamma_1^2, \dots \gamma_1^k$.*

Proof. Let us consider a flipped edge $e_i^\ell \in E_\ell$ and its curve γ^i in Γ'. Clearly, any additional crossing that we introduce may only occur on the first part γ_1^i or second part γ_2^i of γ^i, as the last part γ_3^i is inherited from Γ. By construction, γ_2^i is crossing free, as the segment of e between x_ℓ and x_r in Γ is crossing-free since e_ℓ, e_r and e form a heart and by considering the edges in the reverse order that they intersect e (starting at w), they do not intersect each other. □

For our later proofs of Theorems 2 and 3 on the edge density of (bipartite) weakly fan-planar drawings, we need to show that a certain configuration, as described in the following lemma, cannot occur in a *minimal* weakly fan-planar drawing.

Lemma 4. *Let e, e_ℓ, e_r be a heart in a minimal weakly fan-planar drawing Γ. Then there is no edge $e' \neq e$ in Γ that crosses both e_ℓ and e_r.*

Proof. Assume for a contradiction that there exists an edge $e' \neq e$ that intersects both e_ℓ and e_r. By (I), e and e' share an endpoint, say w, see Fig. 5a. This implies by (I) and (II) that *every* edge which crosses e_ℓ or e_r is incident to w. W.l.o.g. assume that x_ℓ is encountered before x_r when traversing e starting at w. First, notice that e, e' and e_r can in turn form (III) themselves, refer to Fig. 5b.

Based on this observation, we will define four sets of edges that will be helpful in the remainder. In particular, we construct these sets based on their drawing in Γ—while some of the edges may be redrawn at a later time, they will always belong to their corresponding sets. Let E_ℓ and E_r be the sets of edges

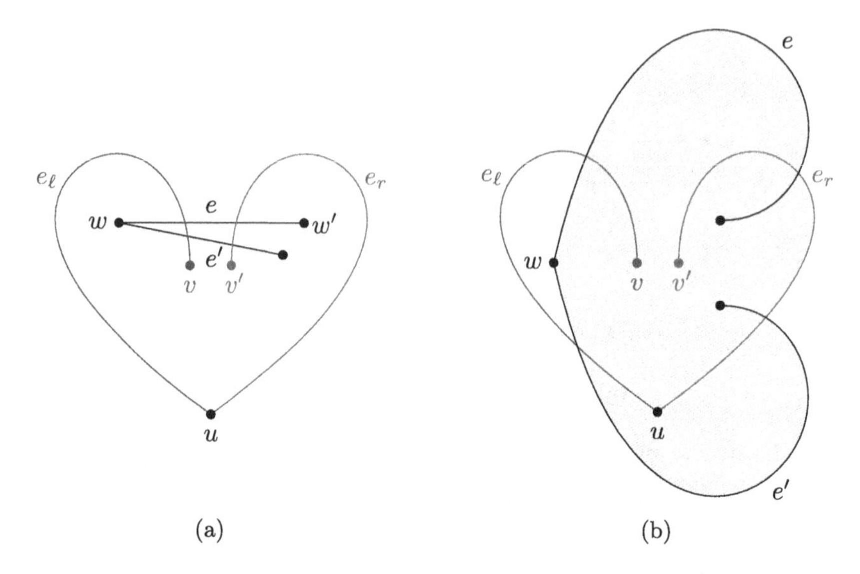

Fig. 5. (a) A single heart $\mathcal{H} = e, e_\ell, e_r$. (b) A double heart formed by $\mathcal{H} = e, e_\ell, e_r$ and $\mathcal{H}' = e_r, e, e'$, the area bounded by $\mathcal{H}'$ is indicated in gray.

that correspond to the left valve and the right valve of the heart $\mathcal{H}$ induced by e, e_ℓ, e_r—in particular, $e_\ell \in E_\ell$ and $e_r \in E_r$. Further, let E_t be the set of edges that cross both e_ℓ and e_r and do not form (III) with e_r and e when traversing e_ℓ starting at v. Complementary, let E_b be the set of edges that also cross e_ℓ and e_r and are not contained in E_t. Further, let e be the first edge in E_t that is encountered when traversing e_ℓ (e_r) starting at u. In the following, we will distinguish between two cases. First, if $E_b = \emptyset$, we call $\mathcal{H}$ a *single heart.* Otherwise, that is when $E_b \neq \emptyset$, we assume that e' is the last edge in E_b that is encountered when traversing e_ℓ starting at u (and therefore also when traversing e_r). We say that $\mathcal{H}$ forms a *double heart* with $\mathcal{H}'$ defined by the edges e_r, e, e'; see Fig. 6.

The main idea is to apply the flip operation to reduce the number of hearts in the resulting (weakly fan-planar, as to be shown) drawing Γ' to obtain a contradiction to the minimality of Γ. For a single heart, one flip will suffice while two flips will be necessary for a double heart. In both cases, we start by flipping the valve E_ℓ as previously defined to obtain the drawing Γ'. We first assert:

Γ' is weakly fan-planar. By Observation 1, any new edge triple forming Pattern (I) or (II) in Γ' must involve at least one edge in the flipped valve E_ℓ and at least one edge belonging to neither E_ℓ nor E_r. Consider a flipped edge $e_i^\ell \in E_\ell$. By Lemma 3, any new crossing on the curve γ^i of e_i^ℓ may only occur on its first part γ_1^i in Γ'. Any edge that intersects γ_1^i in Γ' also intersects e_r and is therefore incident to w by Pattern (I). To show that we do not introduce new edge-triples

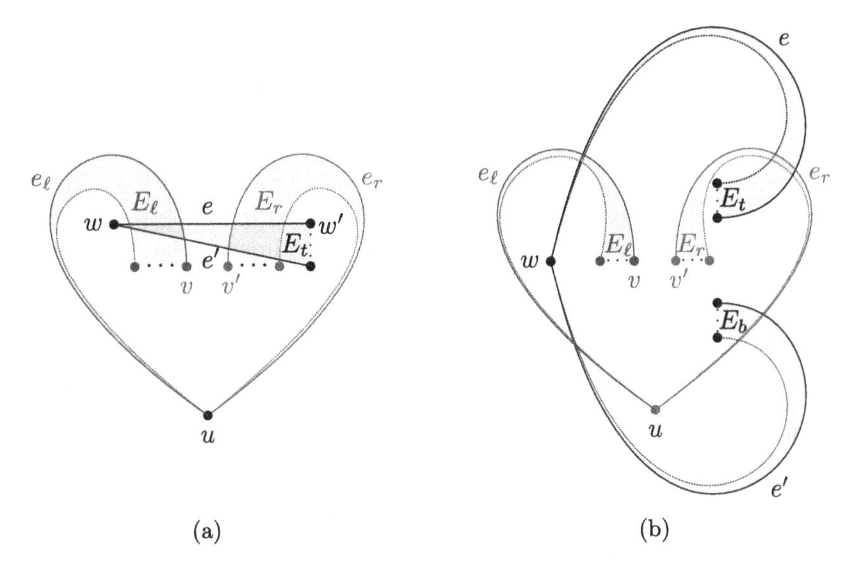

Fig. 6. Illustration of edge sets used in (a) a single heart and (b) a double heart.

forming Pattern (I), it remains to show that all edges in E_ℓ have vertex w as their anchor, even if they only cross e and no other edge in E_t.

We consider two cases. If $\mathcal{H}$ forms a double heart with $\mathcal{H}'$, consider any edge $e_b \in E_b$. Since $e \in E_t$, we observe that each edge $e_\ell \in E_\ell$ is crossed both by e and e_b; see Fig. 7a. Thus, its anchor is w. Otherwise, $\mathcal{H}$ is a single heart. Consider the region $\mathcal{R}$ defined by w, x_ℓ and the intersection of e_ℓ and another edge $e_t \in E_t$ in the original drawing Γ, see Fig. 7b for an illustration of this case.

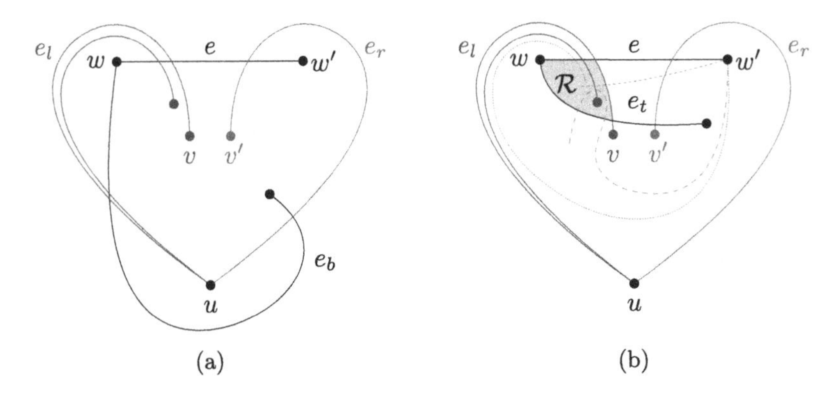

Fig. 7. (a) Every edge in E_ℓ has anchor w if $\mathcal{H}$ is part of a double heart. (b) Any edge of E_ℓ that enters region $\mathcal{R}$ has w as its anchor, if $\mathcal{H}$ forms a single heart.

By definition, every edge in E_ℓ besides e_ℓ enters $\mathcal{R}$ over e. If such an edge is leaving $\mathcal{R}$, then it has to also cross e_t by simplicity, but then its anchor is w and by (I) it can only be crossed by edges that are incident to w. Suppose now an edge of E_ℓ ends in $\mathcal{R}$, but its anchor is not w. Since the edge intersects e, its anchor is therefore w'. But no edge incident to w' can enter $\mathcal{R}$, since it would either cross e_ℓ, whose anchor is w and hence it would coincide with $e = (w, w')$, it cannot cross e by simplicity and if it crosses e_t, then it has to be incident to the anchor of e_t, which is u, but then the curve has to intersect the boundary of $\mathcal{R}$ twice which is impossible.

It remains to show that we do not introduce Pattern (II) by the flip-operation. For the sake of contradiction, assume that there is a new edge triple T forming Pattern (II) after the flip. Note that T involves at least one flipped edge $e_i^\ell \in E_\ell$. We will first establish that the anchor of T is either w or u. Suppose that e_i^ℓ is not incident to the anchor of T, then the anchor of T is w (since all edges in E_ℓ have w as their anchor). Otherwise, the edge crossing e_i^ℓ in T also crosses e_r, hence u is the anchor of T. In the case that w is the anchor of T, observe that all edges in $E_\ell \cup E_r$ are crossed from the same side by edges incident to w when directed from u to their other endvertex. Hence, T cannot form Pattern (II). If u is the anchor of T, it would require a non-empty region between γ_1^i and another partial curve γ_1^j with $j \neq i$ to form Pattern (II), which is a contradiction to our construction.

Thus, we have established that Γ' is indeed weakly fan-planar and we can now consider the number of Patterns (III). We distinguish two cases.

Case 1: $\mathcal{H}$ forms a single heart. By Observation 1, the drawing $\Gamma'[H]$ does not contain any edge triple forming Pattern (III). Hence, any new Pattern (III) in Γ' involves either one or two flipped edges of E_ℓ. In the latter case, since edges in E_ℓ do not cross by simplicity, this would require a non-empty region between the first partial curves of two edges in E_ℓ, which is a contradiction to our construction.

Suppose now that one edge of E_ℓ and two edges e_1, e_2 incident to w form a new Pattern (III). Since the edges in E_ℓ follow the curve of e_r after the flip, e_r is necessarily crossed by one of the two edges incident to w, say e_1, in order to contain u in a closed region. But then e_1 would be already present in Γ and belong to the set E_b, as e, e_ℓ, e_r form Pattern (III), a contradiction to $E_b = \emptyset$.

We conclude that no new Pattern (III) is introduced by Observation 1 while all triples in H forming (III) are eliminated, i.e., the overall number is reduced.

Case 2: $\mathcal{H}$ and $\mathcal{H}'$ form a double heart. First, observe that all edges in the valves $E_\ell \cup E_r$ of $\mathcal{H}$ are anchored by w and all edges of the valves $E_t \cup E_b$ of $\mathcal{H}'$ are anchored by u. In particular, every edge *forming* $\mathcal{H}$ and $\mathcal{H}'$ is contained in one of the four edge sets E_ℓ, E_r, E_t, E_b. Suppose now that there is a third heart $\mathcal{H}''$ distinct from $\mathcal{H}$ that forms a double heart with $\mathcal{H}'$. Hence, there are crossings between the valves of $\mathcal{H}'$ and $\mathcal{H}''$. It follows, that the edges in the valves of $\mathcal{H}''$ are necessarily incident to u by (I), but then all crossings between edges of H' and H'' occur from the same side as between edges of H and H', as otherwise Pattern (II) would be present in Γ. It follows that the hearts $\mathcal{H}$ and $\mathcal{H}''$ are

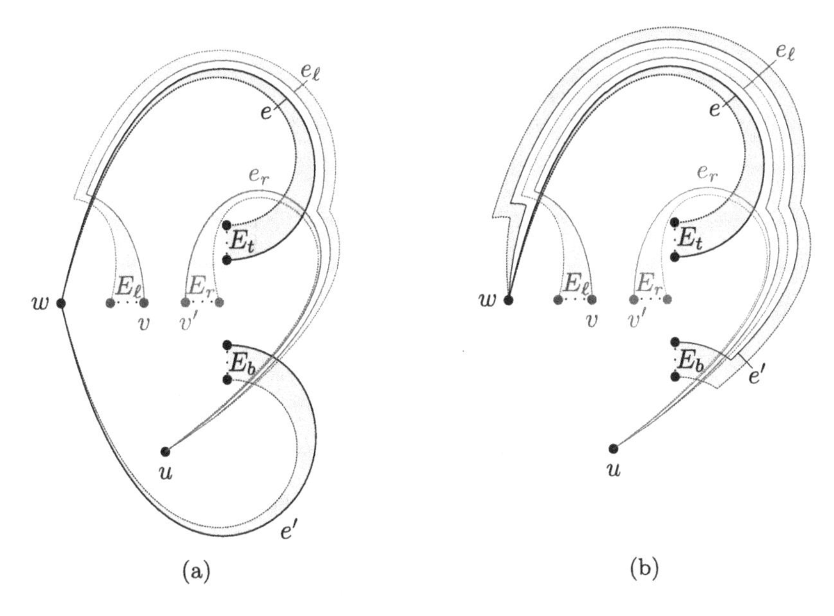

Fig. 8. Illustration of second flip operation in the double heart case in Lemma 4.

identical. Thus, we can consider $\mathcal{H}$ and $\mathcal{H}'$ symmetrically, since in a double heart one always has the other as its counterpart. In this case, the flip of E_ℓ can indeed form new Patterns (III). However, as discussed in the previous case, this can only occur when a flipped edge $e_\ell \in E_\ell$ forms a heart with an edge $e_t \in E_t$ and $e_b \in E_b$. Thus, all new Patterns (III) are contained in the single heart $\mathcal{H}'$ where E_r takes the role of E_t in the discussion of the previous case; see Fig. 8a. We can now proceed as in the previous case by flipping E_b which eliminates all triples forming Pattern (III) as $E_\ell = \emptyset$; see Fig. 8b. □

So far, we investigated minimal weakly fan-planar drawings and their properties. One last ingredient is needed to prove Theorem 2. Given a graph G with a minimal weakly fan-planar drawing, we will sometimes reduce the number of Patterns (III) by modifying G into a different, weakly fan-planar graph G' with the same number of vertices and with the same number of edges. The following lemma states when this modification is possible; see the full version [6] for the proof:

Lemma 5. *Let e_ℓ, e_r and $e = (w, w')$ be a heart $\mathcal{H}$ in a minimal weakly fan-planar drawing Γ of a graph $G = (V, E)$, let $\mathcal{L}$ be the closed curve that consists of the partial curves of the edges e, e_ℓ, e_r up to the crossing points x_ℓ and x_r and let p_1, p_2 be two common neighbors of w and w' outside of $\mathcal{L}$. Then, there is a graph $G' = (V', E')$ with $|V'| = |V|$ and $|E'| = |E|$ that admits a fan-planar drawing Γ' with fewer edge triples forming Pattern (III) than Γ contains.*

We are now ready to prove our main theorem:

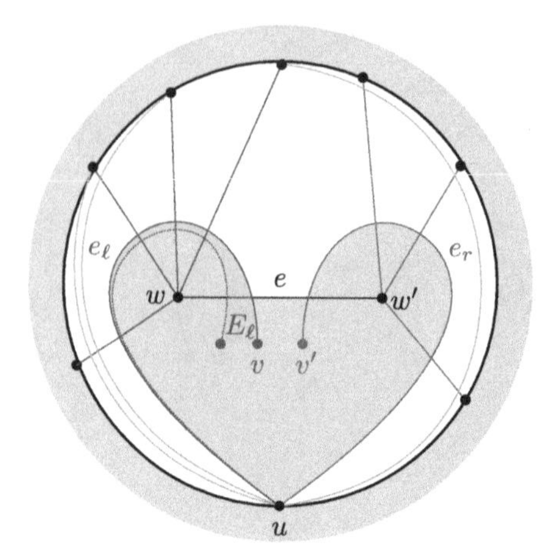

Fig. 9. Illustrations for the proof of Theorem 2.

Theorem 2. *A weakly fan-planar graph G with n vertices has at most $5n - 10$ edges.*

Proof. We proceed by induction on the number of edge triples forming Pattern (III) in a *minimal* weakly fan-planar drawing of G. In the base case, this number is zero, so the drawing is strongly fan-planar. Then G is strongly fan-planar, and has at most $5n - 10$ edges by [14]. For the induction step, consider a graph G and let Γ be a minimal weakly fan-planar drawing of G.

By Lemma 2, Γ contains a heart $e = (w, w'), e_\ell, e_r$, see Fig. 9. If w and w' have at least two common neighbors outside of $\mathcal{L}$, we obtain a graph G' with the same number of vertices and edges with fewer edge triples forming (III) as stated by Lemma 5 and proceed with G'.

Otherwise, denote by G_1 the subgraph of G consisting of those vertices and edges of G that lie (entirely) in the *bounded closed* region bounded by $\mathcal{L}$. In particular, the vertices u, w, v, v', and w', and the edges e_ℓ,e_r, and e all belong to G_1. Similarly, let G_2 be the subgraph of G consisting of those vertices and edges of G that lie entirely in the *unbounded closed* region bounded by $\mathcal{L}$. In particular, vertex $u \in G_2$, but none of the edges e, e_ℓ, e_r is in G_2. Let $|V[G_2]| = r$ and thus $|V[G_1]| = n - (r - 1)$, as u is part of both G_1 and G_2.

Note that the graph G contains edges that are neither in G_1 nor in G_2. We will show how to augment G_2 to G_2' so that it contains an equal number of extra edges. We will create new weakly fan-planar drawings Γ_1' and Γ_2' for G_1 and G_2' that have at least one fewer edge triple forming Pattern (III) each.

We start with G_1. Let E_ℓ and E_r be the left valve and the right valve of our heart, respectively, such that $e_\ell \in E_\ell$ and $e_r \in E_r$ holds. Recall that G_1 lies entirely in the bounded region $\mathcal{L}$ defined by the partial curves of e_ℓ, e_r and e up to their respective intersection points. In particular, this implies that the partial curve of e_ℓ between u and x_ℓ is crossing free. Moreover, the partial curve of e

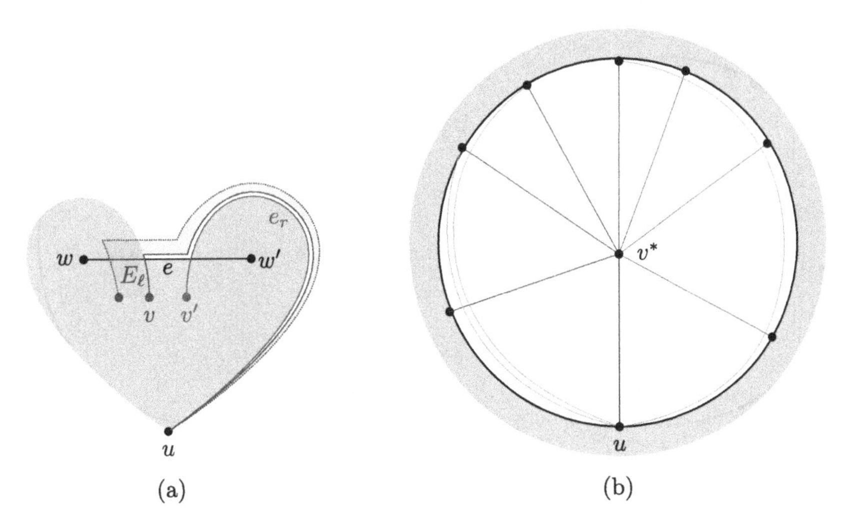

Fig. 10. Illustrations for the proof of Theorem 2.

between x_ℓ and x_r is also crossing free as e_ℓ, e_r and e form a heart. Based on this observation, we flip the left valve E_ℓ, obtaining Γ_1'; see Fig. 10a.

Γ_1' is weakly fan-planar and contains fewer Pattern (III). To show that Γ_1' is weakly fan-planar and contains at least one Pattern (III) less than the drawing of G_1 in Γ, fix an edge $e_i^\ell \in E_\ell$ and let γ^i be its curve in Γ_1'. Recall that by Lemma 3, any new crossings of e_i^ℓ can only occur on the partial curve $\gamma_1^i \subset \gamma^i$. Since no two edges of E_ℓ intersect in Γ_1' by construction and any other edge that would intersect γ^i would also cross e_r in Γ and thus be contained both inside and outside $\mathcal{L}$ by Lemma 4, γ_1^i is uncrossed in Γ_1'. Hence, all crossings of $e_i^\ell \in E_\ell$ are on the part of γ^i which was inherited from Γ; it follows that no new Pattern (I), (II), and (III) is introduced in Γ_1'.

Hence, our new weakly fan-planar drawing Γ_1' has $|E_\ell| \times |E_r| \geq 1$ Pattern (III) less than Γ, and we can apply the inductive assumption to get

$$|E(G_1)| \leq 5|V(G_1)| - 10.$$

Now we consider G_2 and the edges that are neither in G_1 nor in G_2.

Edges that are neither in G_1 nor in G_2. Consider such an edge e'. Clearly, e' crosses $\mathcal{L}$. This crossing cannot be on e by the heart property, so it must be on e_ℓ or e_r. The edge e' cannot cross *both* e_ℓ and e_r by Lemma 4, so e' either crosses e_ℓ and must be incident to w or crosses e_r and must then be incident to w' by (I) and (II). We claim that e' can only cross edges incident to u outside of $\mathcal{L}$; see the light gray edges in Fig. 9. To see this, suppose for a contradiction that e' is crossed by an edge e'' outside of $\mathcal{L}$ which is not incident to u. W.l.o.g. assume that e' is incident to w and crosses e_ℓ, the other case is symmetric. By (I), e'' is then necessarily incident to v, which is contained inside $\mathcal{L}$. However,

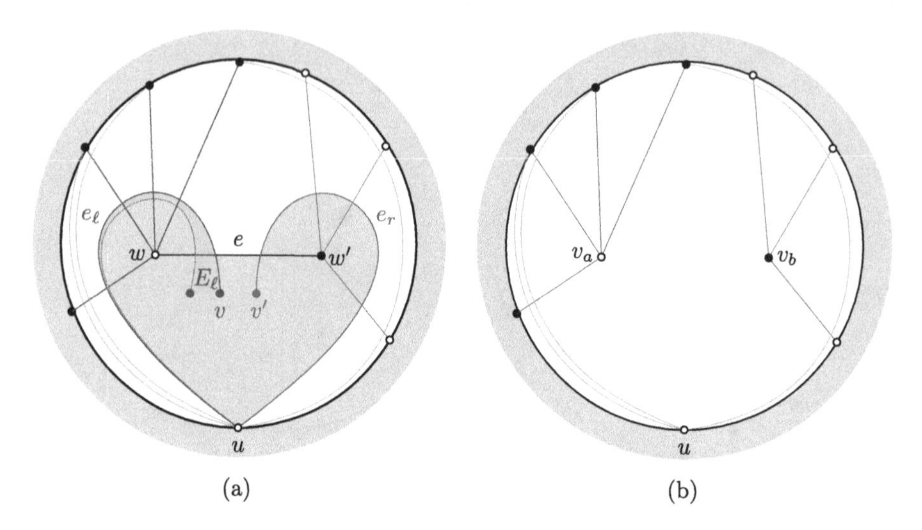

Fig. 11. Illustrations for the proof of Theorem 3.

we already established that only edges such as e' which are incident to w or w' can leave $\mathcal{L}$, a contradiction to e'' crossing e' outside of $\mathcal{L}$.

Augmenting G_2. Let k be the number of edges of G neither in G_1 nor in G_2. Note that w and w' have at most one common neighbor outside of $\mathcal{L}$ (as otherwise we have already applied Lemma 5). Then, we construct a new graph G_2' from G_2 by adding edges as follows. Recall that the weakly fan-planar drawing $\Gamma[G_2]$ derived from Γ contains an empty region inside $\mathcal{L}$ and contains fewer triples forming Pattern (III), as we removed the heart formed by e, e_ℓ, e_r.

We insert a single vertex v^* inside this region, and connect it to all neighbors of w and w' in G_2, and to u; see Fig. 10b. By assumption, w and w' share at most one such vertex and hence there are at least $k-1$ neighbors. Recall that any such edge crosses only edges incident to u outside of $\mathcal{L}$ and therefore fan-planarity is maintained. Hence, we augmented G_2 to the weakly fan-planar graph G_2' that contains $r+1$ vertices and onto which we can apply the induction hypothesis.

We can now bound E(G) as follows:

$$\begin{aligned}|E(G)| &= |E(G_1)| + |E(G_2)| + k = |E(G_1)| + |E(G_2')| \\ &\leq 5(n-r+1) - 10 + 5(r+1) - 10 = 5n - 10\end{aligned}$$

which concludes the proof. □

For bipartite graphs, we proceed in a similar way.

Theorem 3. *An n-vertex bipartite weakly fan-planar graph has at most $4n-12$ edges.*

Proof. Let Γ be a minimal weakly fan-planar drawing of G. We again proceed by induction on the number of edge triples forming Pattern (III). In the base case,

Γ is strongly fan-planar and hence G has at most $4n - 12$ edges by [1]. In the induction step, we proceed as in the previous proof with two differences: First, since G is bipartite, w and w' cannot have a common neighbor, so we do not need to apply Lemma 5. Second, when constructing G_2', instead of inserting a single vertex v^*, we insert two vertices v_a and v_b, and connect v_a with the neighbors of w in G_2, v_b with the neighbors of w' in G_2; thus maintaining bipartiteness; see Fig. 11b. In total, we get

$$\begin{aligned}|E(G)| &= |E(G_1)| + |E(G_2)| + k = |E(G_1)| + |E(G_2')| \\ &\leq 4(n - r + 1) - 12 + 4(r + 2) - 12 = 4n - 12,\end{aligned}$$

which concludes the proof. □

References

1. Angelini, P., Bekos, M.A., Kaufmann, M., Pfister, M., Ueckerdt, T.: Beyond-planarity: Turán-type results for non-planar bipartite graphs. In: Hsu, W., Lee, D., Liao, C. (eds.) 29th International Symposium on Algorithms and Computation, ISAAC 2018, Jiaoxi, Yilan, Taiwan, 16–19 December 2018. LIPIcs, vol. 123, pp. 28:1–28:13. Schloss Dagstuhl - Leibniz-Zentrum für Informatik (2018). https://doi.org/10.4230/LIPIcs.ISAAC.2018.28
2. Bekos, M.A., Grilli, L.: Fan-planar graphs. In: Hong and Tokuyama [10], chap. 8, pp. 131–148. https://doi.org/10.1007/978-981-15-6533-5_8
3. Binucci, C., et al.: Fan-planarity: properties and complexity. Theor. Comput. Sci. **589**, 76–86 (2015). https://doi.org/10.1016/j.tcs.2015.04.020
4. Brandenburg, F.J.: On fan-crossing graphs. Theor. Comput. Sci. **841**, 39–49 (2020). https://doi.org/10.1016/j.tcs.2020.07.002
5. Brinkmann, G., Greenberg, S., Greenhill, C.S., McKay, B.D., Thomas, R., Wollan, P.: Generation of simple quadrangulations of the sphere. Discret. Math. **305**(1–3), 33–54 (2005). https://doi.org/10.1016/j.disc.2005.10.005
6. Cheong, O., Förster, H., Katheder, J., Pfister, M., Schlipf, L.: Weakly and strongly fan-planar graphs. CoRR abs/2308.08966v2 (2023). https://arxiv.org/abs/2308.08966v2
7. Cheong, O., Pfister, M., Schlipf, L.: The thickness of fan-planar graphs is at most three. In: Angelini, P., von Hanxleden, R. (eds.) GD 2022. LNCS, vol. 13764, pp. 247–260. Springer, Cham (2023). https://doi.org/10.1007/978-3-031-22203-0_18
8. Didimo, W., Liotta, G., Montecchiani, F.: A survey on graph drawing beyond planarity. ACM Comput. Surv. **52**(1) (2019). https://doi.org/10.1145/3301281
9. Garey, M.R., Johnson, D.S.: Crossing number is NP-complete. SIAM J. Algebraic Discret. Methods **4**(3), 312–316 (1983). https://doi.org/10.1137/0604033
10. Hong, S., Tokuyama, T.: Beyond Planar Graphs, Communications of NII Shonan Meetings. Springer, Cham (2020). https://doi.org/10.1007/978-981-15-6533-5
11. Huang, W.: Using eye tracking to investigate graph layout effects. In: Hong, S., Ma, K. (eds.) 6th International Asia-Pacific Symposium on Visualization, APVIS 2007, Sydney, Australia, 5–7 February 2007, pp. 97–100. IEEE Computer Society (2007). https://doi.org/10.1109/APVIS.2007.329282
12. Huang, W., Eades, P., Hong, S.: Larger crossing angles make graphs easier to read. J. Vis. Lang. Comput. **25**(4), 452–465 (2014). https://doi.org/10.1016/j.jvlc.2014.03.001

13. Kaufmann, M., Ueckerdt, T.: The density of fan-planar graphs. CoRR abs/1403.6184v1 (2014). http://arxiv.org/abs/1403.6184v1
14. Kaufmann, M., Ueckerdt, T.: The density of fan-planar graphs. Electron. J. Comb. **29**(1) (2022). https://doi.org/10.37236/10521
15. Klemz, B., Knorr, K., Reddy, M.M., Schröder, F.: Simplifying non-simple fan-planar drawings. J. Graph Algorithms Appl. **27**(2), 147–172 (2023). https://doi.org/10.7155/jgaa.00618
16. Mutzel, P.: An alternative method to crossing minimization on hierarchical graphs. In: North, S. (ed.) GD 1996. LNCS, vol. 1190, pp. 318–333. Springer, Heidelberg (1997). https://doi.org/10.1007/3-540-62495-3_57
17. Purchase, H.: Which aesthetic has the greatest effect on human understanding? In: DiBattista, G. (ed.) GD 1997. LNCS, vol. 1353, pp. 248–261. Springer, Heidelberg (1997). https://doi.org/10.1007/3-540-63938-1_67
18. Ware, C., Purchase, H.C., Colpoys, L., McGill, M.: Cognitive measurements of graph aesthetics. Inf. Vis. **1**(2), 103–110 (2002). https://doi.org/10.1057/palgrave.ivs.9500013
19. Whitney, H.: Congruent graphs and the connectivity of graphs. In: Eells, J., Toledo, D. (eds.) Hassler Whitney Collected Papers Contemporary Mathematicians, pp. 61–79. Birkhäuser, Boston (1992). https://doi.org/10.1007/978-1-4612-2972-8_4

On RAC Drawings of Graphs with Two Bends per Edge

Csaba D. Tóth[1,2](✉)

[1] California State University Northridge, Los Angeles, CA, USA
csaba.toth@csun.edu
[2] Tufts University, Medford, MA, USA

Abstract. It is shown that every n-vertex graph that admits a 2-bend RAC drawing in the plane, where the edges are polylines with two bends per edge and any pair of edges can only cross at a right angle, has at most $24n - 26$ edges for $n \geq 3$. This improves upon the previous upper bound of $74.2n$; this is the first improvement in more than 12 years. A crucial ingredient of the proof is an upper bound on the size of plane multigraphs with polyline edges in which the first and last segments are either parallel or orthogonal.

Keywords: right angle crossing · two-bend drawing · density

1 Introduction

Right-angle-crossing drawings (for short, **RAC drawings**) were introduced by Didimo et al. [8]. In a RAC drawing of a graph $G = (V, E)$, the vertices are distinct points in the plane, edges are polylines, each composed of finitely many line segments, and any two edges can cross only at a $90°$ angle. For an integer $b \geq 0$, a **RAC$_b$ drawing** is a RAC drawing in which every edge is a polyline with at most b *bends*; and a **RAC$_b$ graph** is an abstract graph that admits such a drawing. Didimo et al. [7] proved that every RAC$_0$ graph on $n \geq 4$ vertices has at most $4n - 10$ edges, and this bound is tight when $n = 3h - 5$ for all $h \geq 3$; see also [10]. They also showed that every graph is a RAC$_3$ graph.

Angelini et al. [2] proved that every RAC$_1$ graph on n vertices has at most $5.5n - O(1)$ edges, and this bound is the best possible up to an additive constant. Arikushi et al. [5] showed that every RAC$_2$ graph on n vertices has at most $74.2n$ edges, and constructed RAC$_2$ graphs with $\frac{47}{6}n - O(\sqrt{n}) > 7.83n - O(\sqrt{n})$ edges. Recently, Angelini et al. [4] constructed an infinite family of RAC$_2$ graphs with $10n - O(1)$ edges; and conjectured that this lower bound is the best possible.

See recent surveys [7,9] and results [3,11–13] for other aspects of RAC drawings. The concept of RAC drawings was also generalized to angles other than $90°$, and to combinatorial constraints on the crossing patterns in a drawing [1].

The main result of this note is the following theorem.

Theorem 1. *Every RAC_2 graph with $n \geq 3$ vertices has at most $24n - 26$ edges.*

M. A. Bekos and M. Chimani (Eds.): GD 2023, LNCS 14465, pp. 69–77, 2023.
https://doi.org/10.1007/978-3-031-49272-3_5

2 Multigraphs with Angle-Constrained End Segments

A plane multigraph $G = (V, E)$ is a multigraph embedded in the plane such that the vertices are distinct points, and the edges are Jordan arcs between the corresponding vertices (not passing through any other vertex), and any pair of edges may intersect only at vertices. The *multiplicity* of an edge between vertices u and v is the total number of edges in E between u and v.

We define a **plane ortho-fin multigraph** as a plane multigraph $G = (V, E)$ such that every edge $e \in E$ is a polygonal path $e = (p_0, p_1, \ldots, p_k)$ where the first and last edge segments are either parallel or orthogonal, that is, $p_0p_1 \| p_{k-1}p_k$ or $p_0p_1 \perp p_{k-1}p_k$. See Fig. 1 for examples.

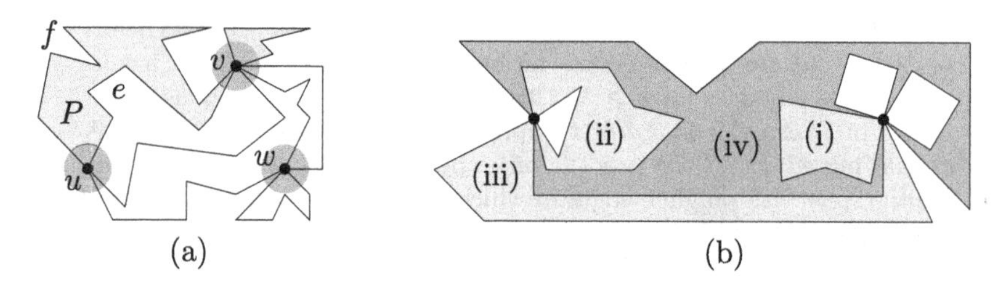

Fig. 1. (a) A plane ortho-fin multigraph with 3 vertices and 7 edges. The parallel edges e and f form a simple polygon P with potential $\Phi(P) = \pi/2$. (b) A plane ortho-fin multigraph with 2 vertices and 7 edges.

It is not difficult to see that in a plane ortho-fin multigraph, every vertex is incident to at most three loops and the multiplicity of any edge between two distinct vertices is at most eight (and both bounds can be attained). Combined with Euler's formula, this would already give an upper bound of $3n + 8(3n - 6) = 27n - 48$ for the size of a plane ortho-fin multigraph with $n \geq 3$ vertices. In this section, we prove a tighter bound of $7n - 3$ (Theorem 2). The key observation is an angle property for loops and double edges (Lemma 1). For a face P of a plane multigraph $G = (V, E)$, let the **potential** $\Phi(P)$ be the sum of interior angles of P over all vertices in V incident to P. A bounded face of a plane multigraph $G = (V, E)$ is called **degenerate** if it is incident to at most 2 vertices in V.

Lemma 1. *Let $G = (V, E)$ be a plane ortho-fin multigraph, and P a face bounded by a loop or by two parallel edges. Then $\Phi(P)$ is a multiple of $\pi/2$.*

Proof. Let P be bounded by a counterclockwise loop $e = (p_0, p_1, \ldots, p_k)$ incident to vertex $v = p_0 = p_k$. Since $p_0p_1 \| p_{k-1}p_k$ or $p_0p_1 \perp p_{k-1}p_k$, then the interior angle of P at v is $\pi/2$, π, or $3\pi/2$. In all three cases, $\Phi(P)$ is a multiple of $\pi/2$.

Let P be bounded by parallel edges $e = (p_0, p_1, \ldots, p_k)$ and $f = (q_0, q_1, \ldots, q_\ell)$ between u and v. Assume w.l.o.g. that e is oriented counterclockwise along P; consequently, f is oriented clockwise. The interior angles of P at u and v are $\angle p_1uq_1$ and $\angle q_\ell vp_k$. Note, in particular, that $\Phi(P) = \angle p_1uq_1 + \angle q_\ell vp_k$, and $\Phi(P)$ depends only on the directions of the vectors $\overrightarrow{up_1}$, $\overrightarrow{uq_1}$, $\overrightarrow{vp_k}$, and $\overrightarrow{vq_\ell}$. If $\overrightarrow{up_1}$ and $\overrightarrow{vp_k}$ have the same direction, and so do $\overrightarrow{uq_1}$ and $\overrightarrow{vq_\ell}$, then $\angle p_1uq_1 + \angle q_\ell vp_k = 2\pi$. In general, the directions of $\overrightarrow{up_1}$ and $\overrightarrow{vp_k}$ (resp., $\overrightarrow{uq_1}$ and $\overrightarrow{vq_\ell}$) differ by a multiple of $\pi/2$. Consequently, $\Phi(P) = \angle p_1uq_1 + \angle q_\ell vp_k$ is also a multiple of $\pi/2$. □

Lemma 2. *Let $G = (V, E)$ be a plane ortho-fin multigraph. Then the potential of every degenerate face is a multiple of $\pi/2$.*

Proof. Let P be a degenerate face. The boundary of P can be classified into the following four types; see Fig. 1(b): (i) a single loop, (ii) two or more loops incident to the same vertex, (iii) two parallel edges, (iv) two parallel edges and $k \geq 1$ loops incident to some endpoints of these edges.

For degenerate faces of Types (i) and (iii), the claim follows directly from Lemma 1. For Type (ii), let P be a degenerate face bounded by $k \geq 2$ loops incident to vertex v. Denote by $Q_1, \ldots, Q_k$ the simple polygons bounded by these loops. By Lemma 1, $\Phi(Q_1), \ldots, \Phi(Q_k)$ are each multiples of $\pi/2$. We may assume w.l.o.g. that P lies in the interior of Q_1, and the exterior of $Q_2, \ldots, Q_k$. Then we have $P = Q_1 \setminus (\bigcup_{i=2}^k Q_i)$, and $\Phi(P) = \Phi(Q_1) - \sum_{i=2}^k \Phi(Q_i)$, which is a multiple of $\pi/2$.

For Type (iv), let P be a degenerate face bounded by two parallel edges between $u, v \in V$ and by $k \geq 1$ loops incident to u or v. Let Q_0 be the polygon bounded by the two parallel edges, and $Q_1, \ldots, Q_k$ the polygons bounded by the loops. Then Q_0 is of Type (iii) and $Q_1, \ldots, Q_k$ are each of Type (i). By Lemma 1, $\Phi(Q_0), \Phi(Q_1), \ldots, \Phi(Q_k)$ are multiples of $\pi/2$. If the outer boundary of P is formed by the two parallel edges, then $P = Q_0 \setminus (\bigcup_{i=1}^k Q_i)$, and $\Phi(P) = \Phi(Q_0) - \sum_{i=1}^k \Phi(Q_i)$. Otherwise, the outer boundary of P is one of the loops, say the boundary of Q_1. In this case, $P = Q_1 \setminus (Q_0 \cup \bigcup_{i=2}^k Q_i)$, and $\Phi(P) = \Phi(Q_1) - \Phi(Q_0) - \sum_{i=3}^k \Phi(Q_i)$. In both cases, $\Phi(P)$ is a multiple of $\pi/2$. □

It turns out that the property in Lemma 2 is sufficient for the analysis. We define a **weak ortho-fin multigraph** as a plane multigraph $G = (V, E)$ such that every edge is a polygonal path, and the potential of every degenerate face is a multiple of $\pi/2$. By Lemma 2, every plane ortho-fin multigraph is a weak ortho-fin multigraph.

A cycle C in a plane multigraph $G = (V, E)$ is a **separating cycle** if both the interior and the exterior of C contains some vertex in V. We also need an elementary lemma to handle subgraphs with one or two vertices. By Euler's formula, a simple planar graph on n vertices has at most euler(n) edges, where euler$(1) = 0$, euler$(2) = 2$, and euler$(n) = 3n - 6$ for $n \geq 3$.

Lemma 3. *Let $G = (V, E)$ be a weak ortho-fin multigraph, where the outer face is bounded by a loop or two parallel edges. If G has n vertices, n_{out} of which are incident to the outer face, and $n_{\text{int}} = n - n_{\text{out}}$, then* euler$(n) \leq 3n_{\text{int}} + n_{\text{out}} - 1$.

Proof. There are two cases to consider: If $n_{\text{out}} = 1$, then $3n_{\text{int}} + n_{\text{out}} - 1 = 3(n-1)+1-1 = 3n-3 \geq$ euler(n) for all $n \geq 1$. If $n_{\text{out}} = 2$, then $3n_{\text{int}}+n_{\text{out}}-1 = 3(n-2) + 2 - 1 = 3n - 5 \geq$ euler(n) for all $n \geq 2$. □

Lemma 4. *Let $G = (V, E)$ be a weak ortho-fin multigraph with n vertices, where the outer face is bounded by a loop or two parallel edges, and the boundary of the outer face is a simple polygon P with n_{int} vertices in its interior. Then G has at most $7n_{\mathrm{int}} + \frac{2}{\pi} \cdot \Phi(P) - 1$ edges in the interior of P.*

Proof. We proceed by induction on the number of separating loops and separating double edges. Let n_{out} denote the number of vertices of the outer face.

Base Case. Assume that G has no separating loops and no separating double edges. Let $G^- = (V, E^-)$ be a simple subgraph of G obtained by deleting loops and keeping only one (arbitrary) copy of parallel edges. Every bounded face in G^- is incident to 3 or more vertices. If we incrementally add the edges of $E \setminus E^-$ to G^-, each new edge subdivides an existing face. Since there are no separating loops or double edges, the insertion of each edge creates a new face incident to at most two vertices, or subdivides such a face into two such faces. Denote by k^- the number of degenerate faces of G. Then $k = |E| - |E^-| \geq |E| - \mathrm{euler}(n)$.

Next, we augment G with new edges to a weak ortho-fin multigraph $G^+ = (V, E^+)$ as follows. Let $F = \{f_1, \ldots, f_k\}$ be the set of degenerate faces of G. In each face $f_i \in F$ with $\Phi(f_i) > \pi/2$, we insert $\frac{2}{\pi}\Phi(f_i) - 1$ edges that subdivide f_i into subfaces each of which have a potential equal $\pi/2$. Specifically, we can subdivide a face f_i with $\Phi(f_i) > \pi/2$ as follows. While f_i has is an interior angle $\theta > \pi/2$ at some vertex $v \in V$, insert a sufficiently small square-shaped loop incident to v, which subdivides f_i into a square Q of potential $\Phi(Q) = \pi/2$, and a face $f_i' = f_i \setminus Q$ of potential $\Phi(f_i') = \Phi(f_i) - \pi/2$. Now we can assume that all interior angles at vertices of V are less than or equal to $\pi/2$. Consider the sequence $\theta_1, \ldots, \theta_m$ of interior angles at consecutive vertices in V along the boundary of f_i. Let $m' \in \{2, \ldots, m\}$ be the minimum integer such that $\sum_{i=1}^{m'} \theta_i > \pi/2$. Subdivide the angle θ_1 (resp., $\theta_{m'}$) by a sufficiently small segment in the interior of f_i into $\theta_1 = \theta_1^- + \theta_1^+ > 0$ (resp., $\theta_{m'} = \theta_{m'}^- + \theta_{m'}^+$) such that $\theta_1^+ + (\sum_{1<i<m'} \theta_i) + \theta_{m'}^- = \pi/2$. Since the face f_i is piecewise linear and path-connected, the two small segments can be augmented into a polygonal path in the interior of f_i that subdivides f_i into $f_i = f_i' \cup f_i''$ with $\Phi(f_i') = \theta_1^+ + (\sum_{1<i<m'} \theta_i) + \theta_{m'}^- = \pi/2$, and $\Phi(f_i'') = \Phi(f_i) - \pi/2$.

Clearly, we have augmented G with $\frac{2}{\pi}\left(\sum_{i=1}^{k} \Phi(f_i)\right) - k$ new edges. The number k^+ of degenerate faces in G^+ equals

$$k^+ = k + \frac{2}{\pi}\left(\sum_{i=1}^{k} \Phi(f_i)\right) - k \geq |E| - \mathrm{euler}(n) + \frac{2}{\pi}\left(\sum_{i=1}^{k} \Phi(f_i)\right) - k. \quad (1)$$

Since the faces of G^+ are interior-disjoint, the sum of interior angles is at most $2\pi n_{\mathrm{int}} + \Phi(P)$. Comparing the lower and upper bounds for the sum of interior angles at vertices in V, we obtain

$$\begin{aligned} k^+ \cdot \frac{\pi}{2} &\leq 2\pi n_{\mathrm{int}} + \Phi(P) \\ k^+ &\leq 4n_{\mathrm{int}} + \frac{2}{\pi}\Phi(P). \end{aligned} \quad (2)$$

The combination of upper bound (1) and the lower bound (2) yields

$$
\begin{aligned}
|E| &\le 4n_{\text{int}} + \frac{2}{\pi}\Phi(P) + \text{euler}(n) + k - \frac{2}{\pi}\sum_{i=1}^{k}\Phi(f_i) && (3)\\
&\le 4n_{\text{int}} + \frac{2}{\pi}\Phi(P) + \text{euler}(n)\\
&= 7n_{\text{int}} + \frac{2}{\pi}\Phi(P) + \big(\text{euler}(n) - 3n_{\text{int}}\big). && (4)
\end{aligned}
$$

The number of edges in the interior of P equals $|E| - n_{\text{out}} \le 7n_{\text{int}} + \frac{2}{\pi}\Phi(P) + \big(\text{euler}(n) - 3n_{\text{int}} - n_{\text{out}}\big) \le 7n_{\text{int}} + \frac{2}{\pi}\Phi(P) - 1$, by Lemma 3, as required.

Induction Step. Assume that G has separating loops or separating double edges. Let $G' = (V', E')$ be the sub-multigraph obtained from G by deleting all vertices that lie in the interior of any loop or in the interior of any cycle formed by parallel edges, and let $n' = |V'|$. Note that the vertices and edges of the outer face have not been deleted.

Let $F' = \{f_1, \ldots, f_k\}$ be the set of degenerate faces of G'. By construction, the vertices in $V \setminus V'$ lie in some faces in F'. Suppose that face f_i contains n_i interior vertices for $i = 1, \ldots, k$. Using (3) for $G' = (V', E')$, we obtain

$$
\begin{aligned}
|E| &\le |E'| + \sum_{i=1}^{k}\left(7n_i + \frac{2}{\pi}\Phi(f_i) - 1\right)\\
&\le \left(4n'_{\text{int}} + \frac{2}{\pi}\Phi(P) + \text{euler}(n') + k - \frac{2}{\pi}\sum_{i=1}^{k}\Phi(f_i)\right) + \sum_{i=1}^{k}\left(7n_i + \frac{2}{\pi}\Phi(f_i) - 1\right)\\
&= 4n'_{\text{int}} + \frac{2}{\pi}\Phi(P) + \text{euler}(n') + \sum_{i=1}^{k} 7n_i\\
&= 7n_{\text{int}} + \frac{2}{\pi}\Phi(P) + \big(\text{euler}(n') - 3n'_{\text{int}}\big).
\end{aligned}
$$

The number of edges in the interior of P is $|E| - n_{\text{out}} \le 7n_{\text{int}} + \frac{2}{\pi}\Phi(P) + \big(\text{euler}(n') - 3n'_{\text{int}} - n'_{\text{out}}\big) \le 7n_{\text{int}} + \frac{2}{\pi}\Phi(P) - 1$, by Lemma 3, as required. This completes the induction step, hence the entire proof. □

Theorem 2. *A weak ortho-fin multigraph on $n \ge 1$ vertices has $\le 7n - 3$ edges.*

Proof. Let n_{out} be the number of outer vertices of a weak ortho-fin multigraph $G = (V, E)$. First assume that $n_{\text{out}} \in \{1, 2\}$. Assume further that the outer face is the exterior of a simple polygon P. Then G has n_{out} edges on the outer face, and Lemma 4 gives a bound for all remaining edges. The sum of interior angles of P at the n_{out} outer vertices is $\Phi(P) < 2\pi n_{\text{out}}$. Overall, we obtain $|E| < n_{\text{out}} + 7(n - n_{\text{out}}) + \frac{2}{\pi}(2\pi)n_{\text{out}} - 1 \le 7n - 2n_{\text{out}} - 1 \le 7n - 3$.

Next assume that $n_{\text{out}} \in \{1, 2\}$, but the outer face has $n_{\text{out}} + \ell$ edges (including at least ℓ loops) for some integer $\ell \ge 0$. Then G is contained in $\ell + 1$ interior-disjoint polygons bounded by these outer edges. The sum of interior angles over all polygons is still less than $2\pi n_{\rm out}$. Applying Lemma 4 over all polygons yields $|E| < (n_{\rm out}+\ell)+(7(n-n_{\rm out})+\frac{2}{\pi}(2\pi)n_{\rm out}-(\ell+1)) \leq 7n-2n_{\rm out}-1 \leq 7n-3$.

Consider now the case that $n_{\rm out} \geq 3$. Let P be a square that contains G in its interior, and v_0 a corner of P. Augment G with the vertex v_0, a loop along the boundary of P, and $n_{\rm out}$ arbitrary edges between v_0 and the outer vertices of G. Since the new edges between v_0 and the outer vertices of G do not bound any degenerate face and $\Phi(P) = \pi/2$, we obtain a weak ortho-fin multigraph. Applying Lemma 4 for the augmented graph, we obtain $|E|+n_{\rm out} \leq 7n+\frac{2}{\pi}(\frac{\pi}{2})-1 = 7n$, hence $|E| \leq 7n-n_{\rm out} \leq 7n-3$. □

3 Proof of Theorem 1

Let $G = (V, E)$ be a RAC$_2$ drawing with $n \geq 3$ vertices. Assume w.l.o.g. that every edge has two bends (by subdividing edge segments if necessary), and the middle segment of every edge has positive or negative slope (not 0 or ∞), by rotating the entire drawing by a small angle if necessary. Each edge has two **end segments** and one **middle segment**. We classify crossings as **end-end**, **end-middle**, and **middle-middle** based on the crossing segments.

Arikushi et al. [5] defined a "block" on the set of $3|E|$ edge segments. First define a symmetric relation on the edge segments: $s_1 \sim s_2$ iff s_1 and s_2 cross. The transitive closure of this relation is an equivalence relation. A **block** is a set of segments in an equivalence class. Note that each block consists of segments of exactly two orthogonal directions; see Fig. 2 for an example.

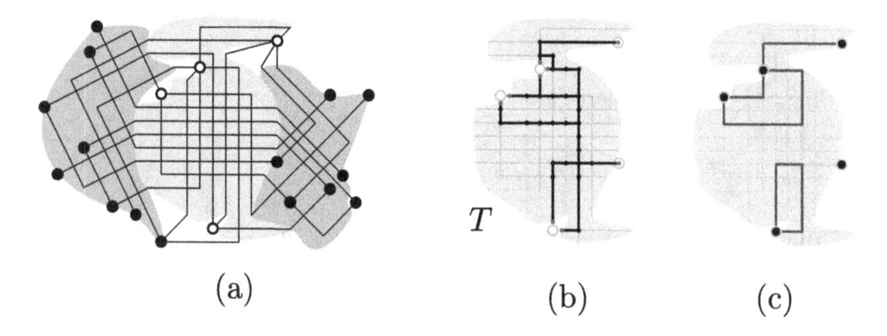

Fig. 2. (a) Three blocks in a RAC$_2$ drawing. (b) A spanning tree T on nine terminals in the middle block. (c) A plane ortho-fin multigraph on the terminals.

3.1 Matching End Segments in Blocks

Let B be a block of $G = (V, E)$. Denote by End(B) the set of end segments in B, and let end(B) = |End(B)|. The segments in B form a connected **arrangement** A, which is a plane straight-line graph: The *vertices* of A are the segment endpoints and all crossings in B, and the edges of A are maximal sub-segments between consecutive vertices of A. We call a vertex of A a **terminal** if it is a vertex in V (that is, an endpoint of some edge in E). If a terminal p is incident to $k > 1$ edges of the arrangement A, we shorten these edges in a sufficiently small ε-neighborhood of p, and split p into k terminals. We may now assume that each terminal has degree 1 in A, hence there are end(B) terminals in A.

Let T be a minimum tree in A that spans all terminals. It is well known that one can find $\lfloor \frac{1}{2}\text{end}(B) \rfloor$ pairs of terminals such that the (unique) paths between these pairs in T are pairwise edge-disjoint (e.g., take a minimum-weight matching of $\lfloor \frac{1}{2}\text{end}(B) \rfloor$ pairs of terminals). If $k > 1$ terminals correspond to the same vertex $p \in V$, then we can extend the paths by $\varepsilon > 0$ to p, and the extended paths are still edge disjoint. Let $\mathcal{E}(B)$ be the set of these paths. We say that an edge $e \in \mathcal{E}(B)$ **represents** an edge $f \in E$ if the first or last edge segment of f contains the first or last edge segment of the path e. By definition, each edge $e \in \mathcal{E}(B)$ represents at most two edges in E.

Let $\mathcal{E}$ be the union of the sets $\mathcal{E}(B)$ over all blocks B; see Fig. 3(left). Note that $H = (V, \mathcal{E})$ is a plane ortho-fin multigraph. Indeed, the edge of H are paths in $\mathcal{E}$. This means that no two edges of H cross. Each edge of H is a path within the same block, and so the first and last segment of each edge of H are either parallel or orthogonal to each other.

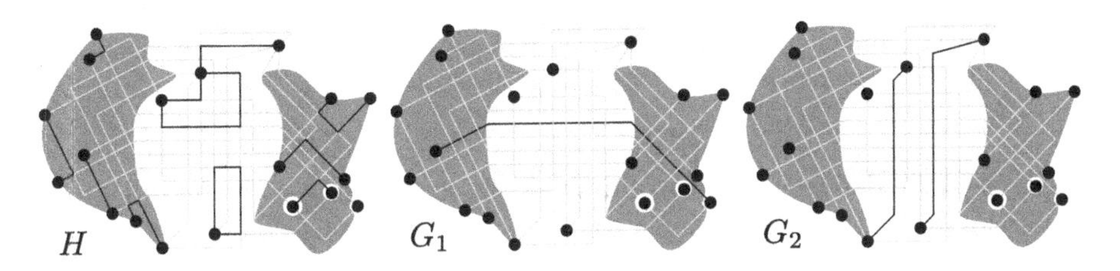

Fig. 3. Graphs H, G_1 and G_2 for the RAC$_2$ drawing in Fig. 2

By Theorem 2, the graph $H = (V, \mathcal{E})$ has at most $7n - 3$ edges, and so it represents at most $2(7n - 3) = 14n - 6$ edges of G.

3.2 Gap Planar Graphs

Let $E_0 \subset E$ be the set of edges in $G = (V, E)$ that are not represented in H, and let $G_0 = (V, E_0)$. Clearly G_0 is a RAC$_2$ drawing with n vertices.

Lemma 5. *In the drawing $G_0 = (V, E_0)$, there is no end-end crossing, and each middle segment is crossed by at most one end segment.*

Proof. A block of G_0 is a subarrangement of a block of G. In every block of G, there is at most one end segment whose edge is not represented by some path in $H = (V, \mathcal{E})$. Consequently, in every block of G_0, there is at most one end segment. Both claims follow. □

Partition $G_0 = (V, E_0)$ into two subgraphs, denoted $G_1 = (V, E_1)$ and $G_2 = (V, E_2)$, such that E_1 contains all edges in E whose middle segments have negative slopes, and $E_2 = E_1 \setminus E_0$; see Fig. 3 for an example.

Lemma 6. *In each of $G_1 = (V, E_1)$ and $G_2 = (V, E_2)$, all crossings are end-middle crossings, and every middle segment has at most one crossing.*

Proof. Since G_1 (resp., G_2) is a RAC_2 drawing, where all middle segments have positive (resp., negative) slopes, then the middle segments do not cross. Combined with Lemma 5, this implies that all crossings are end-middle crossings, and every middle segment crosses at most one end segment. □

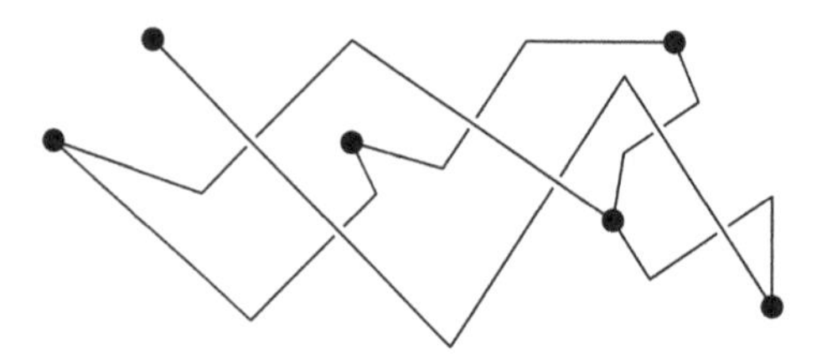

Fig. 4. A RAC_2 drawing: All crossings are between end- and middle-segments, and every middle-segment has positive slope and at most one crossing.

Bae et al. [6] defined a ***k*-gap planar** graph, for an integer $k \geq 0$, as a graph G that can drawn in the plane such that (1) exactly two edges of G cross in any point, (2) each crossing point is *assigned* to one of its two crossing edges, and (3) each edge is assigned with at most k of its crossings.

Lemma 7. *Both $G_1 = (V, E_1)$ and $G_2 = (V, E_2)$ are 1-gap planar.*

Proof. Every crossing is an end-middle crossing by Lemma 6. Assign each crossing to the edge that contains the middle segment involved in the crossing. Then each edge is assigned with at most one crossing by Lemma 6; see Fig. 4. □

Bae et al. [6] proved that every 1-gap planar graph on $n \geq 3$ vertices has at most $5n - 10$ edges, and this bound is the best possible for $n \geq 5$. (They have further proved that a multigraph with $n \geq 3$ vertices that has a 1-gap-planar drawing in which no two parallel edges are homotopic has at most $5n - 10$ edges.) It follows that G_1 and G_2 each have at most $5n - 10$ edges if $n \geq 3$.

Proof. (Proof of Theorem 1.). Let $G = (V, E)$ be a RAC_2 drawing. Graph $H = (V, \mathcal{E})$ represents at most $2(7n - 3) = 14n - 6$ edges of G by Theorem 2. The remaining edges of G are partitioned between G_1 and G_2, each containing at most $5n - 10$ edges for $n \geq 3$; see Fig. 4. Overall, G has at most $24n - 26$ edges if $n \geq 3$. □

Acknowledgements. Work on this paper was initiated at the Tenth Annual Workshop on Geometry and Graphs held at the Bellairs Research Institute, February 3–10, 2023. The author is grateful to Michael Kaufmann and Torsten Ueckerdt for stimulating discussions; and also thanks the GD 2023 reviewers for many helpful comments and suggestions. Research funded in part by NSF awards DMS-1800734 and DMS-2154347.

References

1. Ackerman, E., Fulek, R., Tóth, C.D.: Graphs that admit polyline drawings with few crossing angles. SIAM J. Discret. Math. **26**(1), 305–320 (2012). https://doi.org/10.1137/100819564
2. Angelini, P., Bekos, M.A., Förster, H., Kaufmann, M.: On RAC drawings of graphs with one bend per edge. Theor. Comput. Sci. **828–829**, 42–54 (2020). https://doi.org/10.1016/j.tcs.2020.04.018
3. Angelini, P., Bekos, M.A., Katheder, J., Kaufmann, M., Pfister, M.: RAC drawings of graphs with low degree. In: Proceedings of the 47th Symposium on Mathematical Foundations of Computer Science (MFCS). LIPIcs, vol. 241, pp. 11:1–11:15. Schloss Dagstuhl (2022). https://doi.org/10.4230/LIPIcs.MFCS.2022.11
4. Angelini, P., Bekos, M.A., Katheder, J., Kaufmann, M., Pfister, M., Ueckerdt, T.: Axis-parallel right angle crossing graphs (2023). https://doi.org/10.48550/arXiv.2306.17073, preprint arXiv:2306.17073
5. Arikushi, K., Fulek, R., Keszegh, B., Moric, F., Tóth, C.D.: Graphs that admit right angle crossing drawings. Comput. Geom. **45**(4), 169–177 (2012). https://doi.org/10.1016/j.comgeo.2011.11.008
6. Bae, S.W., et al.: Gap-planar graphs. Theor. Comput. Sci. **745**, 36–52 (2018). https://doi.org/10.1016/j.tcs.2018.05.029
7. Didimo, W.: Right angle crossing drawings of graphs. In: Hong, S.-H., Tokuyama, T. (eds.) Beyond Planar Graphs, pp. 149–169. Springer, Singapore (2020). https://doi.org/10.1007/978-981-15-6533-5_9
8. Didimo, W., Eades, P., Liotta, G.: Drawing graphs with right angle crossings. Theor. Comput. Sci. **412**(39), 5156–5166 (2011). https://doi.org/10.1016/j.tcs.2011.05.025
9. Didimo, W., Liotta, G., Montecchiani, F.: A survey on graph drawing beyond planarity. ACM Comput. Surv. 52(1), 4:1–4:37 (2019). https://doi.org/10.1145/3301281
10. Dujmović, V., Gudmundsson, J., Morin, P., Wolle, T.: Notes on large angle crossing graphs. Chic. J. Theor. Comput. Sci. **2011** (2011). http://cjtcs.cs.uchicago.edu/articles/CATS2010/4/contents.html
11. Förster, H., Kaufmann, M.: On compact RAC drawings. In: Grandoni, F., Herman, G., Sanders, P. (eds.) Proceedings of the 28th European Symposium on Algorithms (ESA). LIPIcs, vol. 173, pp. 53:1–53:21. Schloss Dagstuhl (2020). https://doi.org/10.4230/LIPIcs.ESA.2020.53
12. Rahmati, Z., Emami, F.: RAC drawings in subcubic area. Inf. Process. Lett. **159–160**, 105945 (2020). https://doi.org/10.1016/j.ipl.2020.105945
13. Schaefer, M.: RAC-drawability is $\exists\mathbb{R}$-complete. In: Purchase, H.C., Rutter, I. (eds.) GD 2021. LNCS, vol. 12868, pp. 72–86. Springer, Cham (2021). https://doi.org/10.1007/978-3-030-92931-2_5

String Graphs with Precise Number of Intersections

Petr Chmel[(✉)] and Vít Jelínek

Computer Science Institute, Charles University, Prague, Czech Republic
{chmel,jelinek}@iuuk.mff.cuni.cz

Abstract. A string graph is an intersection graph of curves in the plane. A k-string graph is a graph with a string representation in which every pair of curves intersects in at most k points. We introduce the class of $(=k)$-string graphs as a further restriction of k-string graphs by requiring that every two curves intersect in either zero or precisely k points. We study the hierarchy of these graphs, showing that for any $k \geq 1$, $(=k)$-string graphs are a subclass of $(=k+2)$-string graphs as well as of $(=4k)$-string graphs; however, there are no other inclusions between the classes of $(=k)$-string and $(=\ell)$-string graphs apart from those that are implied by the above rules. In particular, the classes of $(=k)$-string graphs and $(=k+1)$-string graphs are incomparable by inclusion for any k, and the class of $(=2)$-string graphs is not contained in the class of $(=2\ell+1)$-string graphs for any ℓ.

Keywords: intersection graph · string graph · hierarchy

1 Introduction

An *intersection representation* of a graph $G = (V, E)$ is a collection of sets $R = \{R(v) : v \in V\}$ such that for every two distinct vertices $u, v \in V$, the sets $R(u), R(v)$ intersect if and only if u, v form an edge of G. In this paper, we focus on *string representations*, which are intersection representations where the sets $R(v)$ are curves in the plane. Graphs admitting a string representation are known as *string graphs*, and their class is denoted by STRING. They were introduced by Sinden [19].

A common way to further restrict string representations is to bound the number of intersections between a pair of curves – this gives rise to the class of *k-string graphs* (denoted by k-STRING), which are the graphs admitting a string representation in which every two curves intersect in at most k points. This yields a hierarchy of classes parameterized by the number k. Clearly, k-STRING is a subclass of $(k+1)$-STRING for all $k \geq 1$, and Kratochvíl and Matoušek [12]

The first author was supported by the Czech Science Foundation Grant No. 19-27871X. The second author was supported by the Czech Science Foundation Grant No. 23-04949X.

M. A. Bekos and M. Chimani (Eds.): GD 2023, LNCS 14465, pp. 78–92, 2023.
https://doi.org/10.1007/978-3-031-49272-3_6

showed that the inclusions are strict. Moreover, they showed that it is NP-complete to recognize k-string graphs for any fixed $k \geq 1$ [11].

A folklore result shows that any string graph can be represented as an intersection graph of paths on a rectilinear grid; this is known as a VPG representation. Asinowski et al. [1] introduced the classes of B_k-VPG graphs, which are graphs admitting a VPG representation in which every path has at most k bends. Clearly, B_k-VPG is a subclass of B_{k+1}-VPG. Chaplick et al. [4] have shown that these inclusions are strict, and that for any $k \geq 1$, it is NP-complete to recognize the class B_k-VPG; for $k = 0$, NP-completeness was proven by Kratochvíl [10,11].

By way of context, we also mention other hierarchies, such as k-DIR graphs [12]: the intersection graphs of line segments in the plane with at most k different slopes, k-length-segment graphs [2]: the intersection graphs of line segments in the plane with at most k different lengths, k-size-disk graphs [2]: the intersection graphs of disks in the plane with at most k different diameters, and k-interval graphs [20]: the intersection graphs of unions of k intervals on the real line.

We can also find examples of such hierarchies within contact graphs, where intersections are only permitted when the intersection point is an endpoint of at least one of the intersecting curves. We may restrict the number of curves sharing a single point, which yields the class of k-contact graphs [7] – this class can be also restricted further to only line segments and not general strings. Additionally, we can define the class B_k-CPG [6], which is analogous to B_k-VPG, except the graphs must be contact graphs. Recognition of graphs from these classes is also known to be NP-complete [3,8].

In all of the above cases, the graph classes in the hierarchy form a chain of strict inclusions: increasing the parameter k by one yields a strict superclass.

We introduce a new hierarchy of so called $(=k)$-string graphs, which are the intersection graphs of curves in the plane such that every intersection is a crossing and every pair of intersecting curves has *exactly* k crossings. This hierarchy turns out to have a more intricate structure than all the previously mentioned examples. For instance, while $(=k)$-STRING is a strict subclass of $(=k+2)$-STRING for any $k \geq 1$, the two classes $(=k)$-STRING and $(=k+1)$-STRING are incomparable by inclusion. With our results, we will fully characterise all the inclusions between the classes of $(=k)$-string graphs, for $k \in \mathbb{N}$.

Our results show that the parity of k plays a key role in understanding the relations between the classes of $(=k)$-string graphs. This motivates us to introduce *odd-string* graphs, as graphs having a representation where any two intersecting curves have an odd number of intersections. We show that this class is a proper subclass of string graphs; indeed, there are $(=2)$-string graphs which are not odd-string graphs.

We remark that this distinction between curves with odd and even number of intersections is somewhat reminiscent of the distinction between pair-crossing and odd-crossing numbers of graphs, which is relevant in the study of topological graph drawing [14–16].

Let us formally introduce our terminology.

Definition 1 (Proper string representation). *A string representation R is* proper *if every curve in R is a simple piecewise linear curve with finitely many bends, any two curves in R intersect in at most finitely many points, no three curves in R intersect in one point, and no intersection point of two curves coincides with a bend or an endpoint of a curve. Note that this implies that every intersection of two curves is a crossing.*

Definition 2 ($(=k)$-string graphs). *A* $(=k)$-string representation *is a proper representation in which any two curves are either disjoint or intersect in precisely k points. A graph is a* $(=k)$-string graph *if it has a $(=k)$-string representation. The class of such graphs is denoted by $(=k)$-*STRING.

We stress that a $(=k)$-string representation must, by definition, be proper. This prevents us, e.g., from increasing the number of intersection points of two curves by introducing contact points in which the curves touch but do not cross.

2 The Hierarchy

Our main goal in this paper is to understand the inclusion hierarchy of the classes of $(=k)$-string graphs for various $k \in \mathbb{N}$. We begin by two simple results.

Proposition 1. *For all $k \geq 1$, $(=k)$-*STRING *is a subclass of $(=k+2)$-*STRING.

Proposition 2. *For all $k \geq 1$, $(=k)$-*STRING *is a subclass of $(=4k)$-*STRING.

The proofs of these results follow from the operations depicted in Fig. 1, with Proposition 1 using Fig. 1a and Proposition 2 using Fig. 1b. The full proofs appear in the full version [5].

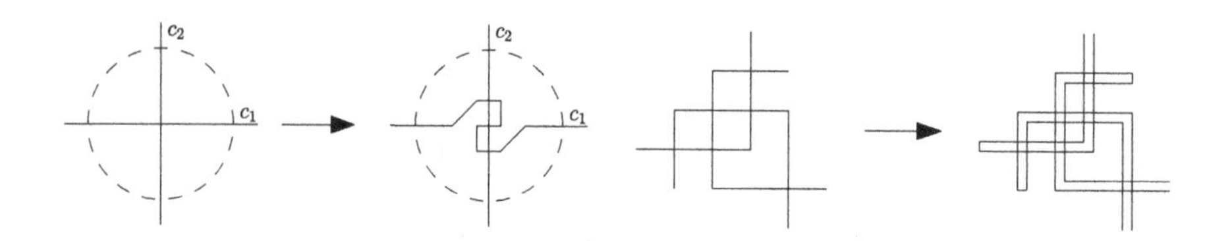

(a) Adding two intersections (b) Quadrupling the intersections

Fig. 1. Two operations for increasing the number of intersection points

We also note, for future reference, that for bipartite graphs, we may only double the representation of just one of the two partition classes.

Proposition 3. *For all $k \geq 1$, every bipartite $(=k)$-string graph is also a $(=2k)$-string graph.*

Note that Proposition 2 implies that any $(=k)$-string representation with k odd can be transformed into a $(=\ell)$-string representation with ℓ even and large enough. However, there is no obvious way to proceed in the opposite direction, towards a representation with an odd number of crossings per pair of curves. This motivates our next definition.

Definition 3 (Odd-string graphs). *An* odd-string representation *of a graph G is a proper string representation R in which any two curves are either disjoint or cross in an odd number of points. A graph G is an* odd-string graph *if it has an odd-string representation. The class of odd-string graphs is denoted by* ODD-STRING.

Of course, for any k odd, $(=k)$-STRING is a subclass of ODD-STRING. Conversely, given any odd-string representation R, we can repeatedly apply the operation from Fig. 1a to transform R into a $(=k)$-string representation for some k odd. This shows that ODD-STRING can equivalently be defined as the union $\bigcup_{\ell \in \mathbb{N}} (=2\ell - 1)$-STRING.

2.1 Hierarchy Non-inclusions

Our main results in this paper are negative results, which essentially show that there are no more inclusions between the classes we consider beyond those implied by Propositions 1 and 2. Specifically, we will prove the following results:

1. for all $1 \leq k < \ell$, the class of $(=\ell)$-string graphs is not a subclass of $(=k)$-string graphs (see Theorem 1),
2. for all $k \geq 1$, the classes of $(=k)$-string graphs is not a subclass of $(=k+1)$-string graphs (see Theorem 2),
3. for any odd k, the class of $(=k)$-string graphs is not a subclass of $(=4k-2)$-string graphs – in particular, the bound $4k$ in Proposition 2 is best possible when it comes to transforming a $(=k)$-string representation with k odd into a $(=\ell)$-string representation with ℓ even (see Theorem 3),
4. there exist $(=2)$-string graphs with no odd-string representation; consequently, ODD-STRING is a proper subclass of STRING (see Theorem 4).

Figure 2 outlines the inclusion between the various $(=k)$-string classes. For reference, the figure also includes the classes k-STRING. We do not have a full characterisation of the inclusions between k-STRING and $(=\ell)$-STRING, although we will later show that k-STRING is a subclass of $(=8k)$-STRING (Theorem 5).

Noodle-Forcing Lemma. Our main tool to prove the non-inclusions in our hierarchy is the Noodle-Forcing Lemma of Chaplick et al. [4], originally devised to prove non-inclusions in the B_k-VPG hierarchy. We will show that the lemma can be adapted to the setting of $(=k)$-string representations. We begin by stating a simplified version of the lemma.

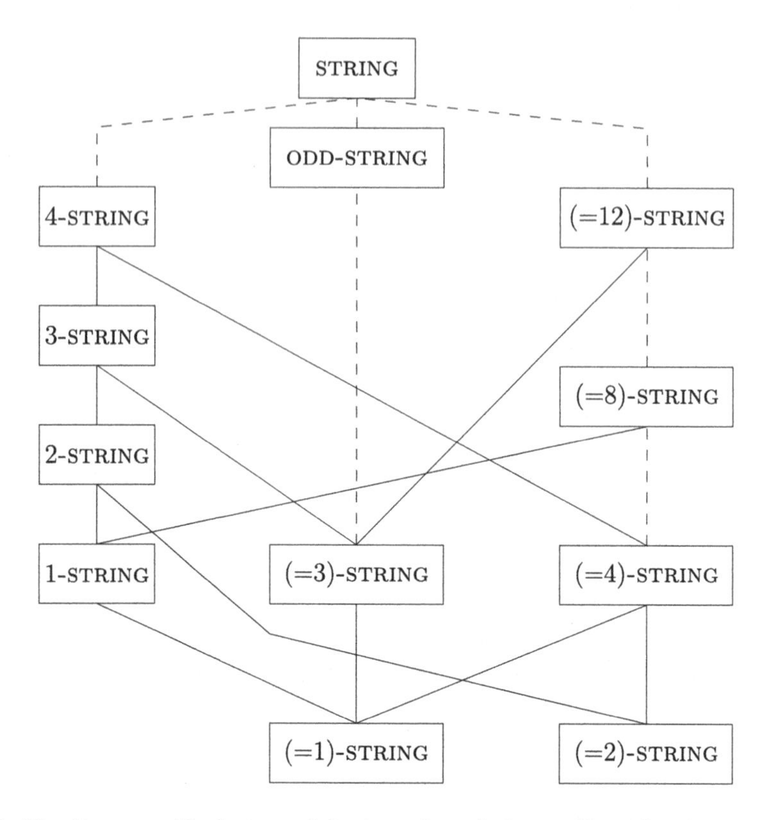

Fig. 2. The diagram of inclusions of the investigated classes. Note that k-string representations are not assumed to be proper.

Lemma 1 (Noodle-Forcing Lemma, Chaplick et al. [4]). *Let $G = (V, E)$ be a graph with a proper string representation $R = \{R(v): v \in V\}$. Then there is a graph $G^{\#}(R) = (V^{\#}, E^{\#})$ containing G as an induced subgraph that has a proper string representation $R^{\#} = \{R^{\#}(v): v \in V^{\#}\}$ such that $R(v) = R^{\#}(v)$ for all $v \in V$ and $R^{\#}(w)$ is a vertical or a horizontal segment for $w \in V^{\#} \setminus V$.*

Moreover, for any $\varepsilon > 0$ any (not necessarily proper) string representation of $G^{\#}$ can be transformed by a homeomorphism of the plane and a circular inversion into a representation $R^{\varepsilon} = \{R^{\varepsilon}(v): v \in V^{\#}\}$ such that for every vertex $v \in V$, the curve $R^{\varepsilon}(v)$ is contained in the $\varepsilon-$neighborhood of $R(v)$ and $R(v)$ is contained in the $\varepsilon-$neighborhood of $R^{\varepsilon}(v)$.

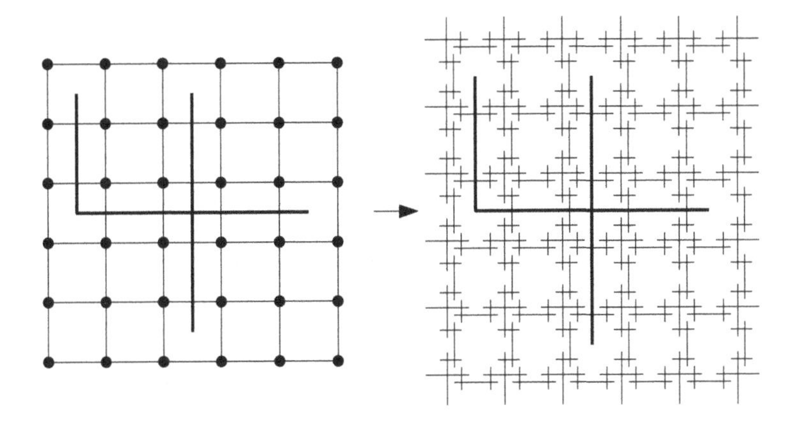

Fig. 3. An example of the construction in the proof of Noodle-Forcing Lemma

The main idea of the proof of the Noodle-Forcing Lemma is to overlay the representation R with a sufficiently fine grid-like mesh of horizontal and vertical segments, as shown in Fig. 3. The relevant details of the construction are presented in the full version [5]. Henceforth, we will use the notation $G^{\#}(R)$ for the specific graph created by the construction illustrated in Fig. 3.

Importantly, we can show that if the Noodle-Forcing Lemma is applied to a $(=k)$-string representation R, then the graph $G^{\#}(R)$ is also in $(=k)$-string.

Lemma 2. *For all $k \geq 1$, given a $(=k)$-string representation R of a graph G, the graph $G^{\#}(R)$ has a $(=k)$-string representation as well.*

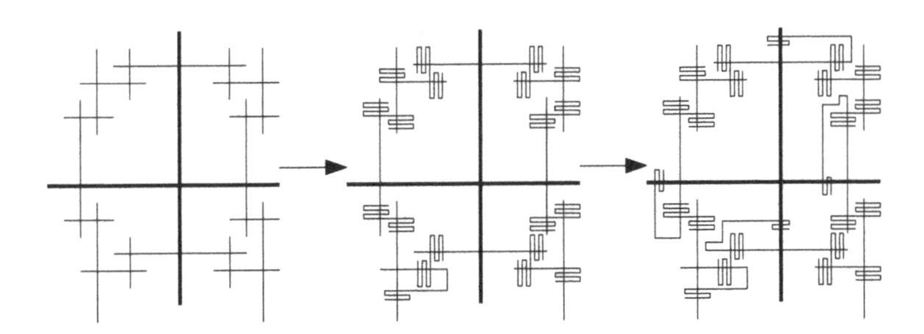

Fig. 4. The extension of the representation $R^{\#}$ into a $(=k)$-string one in two steps

We postpone the proof to the full version [5]; the idea is shown in Fig. 4.

Proofs of the Non-inclusions. Proofs involving the Noodle-Forcing Lemma usually follow the same pattern. The goal is typically to find a graph that belongs to one class, say $(=\ell)$-STRING, but not to another class, say $(=k)$-STRING. We thus begin by choosing a suitable gadget G with an $(=\ell)$-string representation R and apply the lemma to obtain the graph $G^{\#} = G^{\#}(R)$, which by Lemma 2 also belongs to $(=\ell)$-STRING. The main part is then to show that $G^{\#}$ is not in $(=k)$-STRING. This part of the argument will depend on the particular structure of G and R, but there are certain common ideas which we now present.

For contradiction, we assume that a $(=k)$-string representation R' of $G^{\#}$ exists. By the Noodle-Forcing Lemma, we transform R' into a representation R^{ε} of $G^{\#}$ with properties as in the statement of the lemma. Moreover, R^{ε} is also a $(=k)$-string representation as homeomorphisms and circular inversions preserve intersections between curves.

For brevity, we say that we *precook* a $(=k)$-string representation R^{ε} from R' whenever we apply the preceding procedure with $\varepsilon > 0$ small enough so that the properties in the following paragraphs hold.

The curve $R^{\varepsilon}(v)$ is ε-close to $R(v)$, i.e., $R^{\varepsilon}(v)$ is confined into the set $N_{\varepsilon}(v) := \{x\colon \exists y \in R(v)\colon \operatorname{dist}(x,y) < \varepsilon\}$. Following Chaplick et al. [4], we call the set $N_{\varepsilon}(v)$ the *noodle* of v. For small enough ε, each noodle is a simply connected region, and if two curves $R(u)$ and $R(v)$ intersect in ℓ points $p_1, p_2, \dots, p_\ell$, then $N_{\varepsilon}(u)$ and $N_{\varepsilon}(v)$ intersect in ℓ pairwise disjoint parallelograms $Z_1, \dots, Z_\ell$, where Z_i contains p_i (recall that in a proper representation, no crossing point may coincide with a bend of its curve). We call the sets $Z_1, \dots, Z_\ell$ the *zones* of $N_{\varepsilon}(u) \cap N_{\varepsilon}(v)$.

We also assume that ε is small enough so that the distance between any intersection point in R and any endpoint of a curve in R is strictly larger than $\varepsilon > 0$. This is possible since R is proper.

Consider now the intersection of the curve $R^{\varepsilon}(v)$ with a zone Z_i: its connected components shall be called the *fragments* of v in Z_i; see Fig. 5. Each fragment is a curve which either connects two opposite sides of Z_i (we call such fragment a *v-traversal* through Z_i) or has both endpoints on the same side of Z_i (such a fragment is called a *v-reversal*). Choosing ε small enough, we may ensure that each Z_i has at least one v-traversal, since $R^{\varepsilon}(v)$ must reach a point ε-close to each endpoint of $R(v)$, and the endpoints of $R(v)$ can be made at least ε-far from any zone.

Note that any v-traversal partitions its zone into two parts, separating the endpoints of any u-traversal of that zone from each other. Thus, any v-traversal must intersect any u-traversal of the same zone, and in fact they must intersect in an odd number of points. In particular, each zone has at least one intersection between two traversals. Moreover, any v-reversal in Z_i forms a closed loop together with the segment on the boundary of Z_i connecting its endpoints, and any u-fragment has both endpoints outside this loop. It follows that any v-reversal is intersected an even number of times (possibly zero) by any u-fragment.

As an example of this pattern, we show that there exists a $(=k+1)$-string graph that is not a $(=k)$-string graph, in fact it is not even a k-string graph. We note that this implies that the inclusions in Propositions 1 and 2 are strict.

Theorem 1. *For all $k \geq 1$, $(=k+1)$-STRING is not a subclass of k-STRING, and therefore it is not a subclass of $(=\ell)$-STRING for any $1 \leq \ell \leq k$ either.*

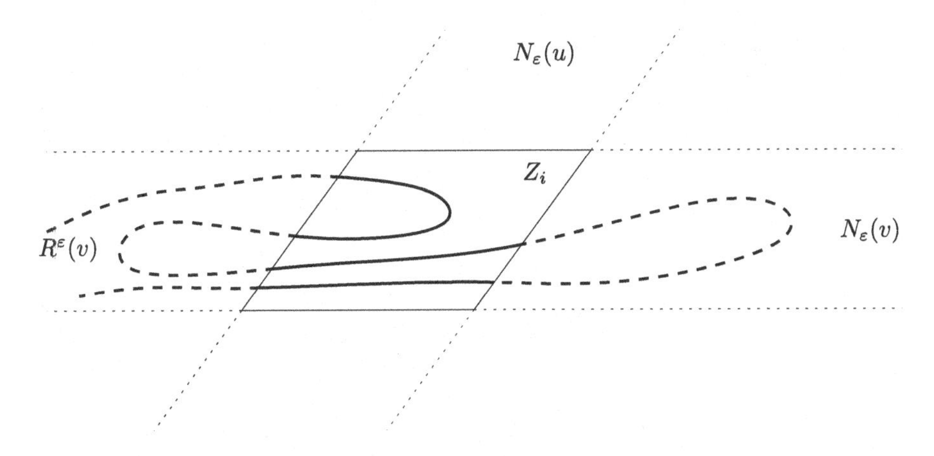

Fig. 5. An example of a zone Z_i at the intersection of two noodles $N_\varepsilon(u)$ and $N_\varepsilon(v)$, in which the curve $R^\varepsilon(v)$ forms one v-reversal and two v-traversals.

Proof. We construct a representation R_{k+1} of a graph $G = K_2$, the complete graph on two vertices. We note that the construction is the same as in the proof of $\mathrm{B}_k-\mathrm{VPG} \subsetneq \mathrm{B}_{k+1}-\mathrm{VPG}$ by Chaplick et al. [4]. As shown in Fig. 6, the representation is $(=k+1)$-string, and we apply the Noodle-Forcing Lemma on the representation to obtain a graph $G^\#$ with representation $R^\#_{k+1}$. By Lemma 2, the graph $G^\#$ has a $(=k+1)$-string representation.

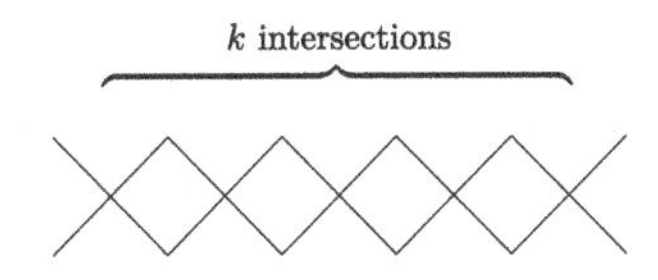

Fig. 6. The sausage construction of representation R_k

We now claim that the graph $G^\#$ has no k-string representation. For contradiction, let R' be a k-string representation of $G^\#$. We precook R' into a k-string representation R^ε.

Let u, v be the two vertices of G. Their two noodles $N(u)$ and $N(v)$ have $k+1$ zones in their intersection, and as observed before, each zone Z_i must have at least one u-traversal and one v-traversal, which must intersect in at least one point. Therefore, there are at least $k+1$ intersection points between $R^\varepsilon(u)$ and $R^\varepsilon(v)$, contradicting the assumption that R^ε is a k-string representation.

The second part of the statement follows immediately, as any $(=\ell)$-string representation of the graph $G^\#$ for $1 \leq \ell \leq k$ is also a k-string representation. □

Our next goal is to show non-inclusion in the opposite direction and construct a $(=k)$-string graph that does not have a $(=k+1)$-string representation. The key role in our argument is played by the concept of *faithful extension* of a string representation R. Informally, a faithful extension of R is obtained by extending each string $R(v)$ of R by attaching two new (possibly empty) parts to it, each starting from an endpoint of the original string $R(v)$ and following it very closely, with the effect of "doubling" some number of intersections which appear along $R(v)$ near its endpoints; see Fig. 7. The fully formal definition of faithful extension is rather technical, and is presented in the full version [5].

To construct a $(=k)$-string graph that does not have a $(=k+1)$-string representation, we proceed in two steps. First, we show that given a $(=k)$-string representation R of a graph G, the $(=k)$-string graph $G^{\#} = G^{\#}(R)$ has a $(=k+1)$-string representation if and only if the representation R can be faithfully extended into a $(=k+1)$-string representation; see Lemma 3. In the second step, we find a graph with a $(=k)$-string representation that cannot be faithfully extended into a $(=k+1)$-string representation; see Lemma 4.

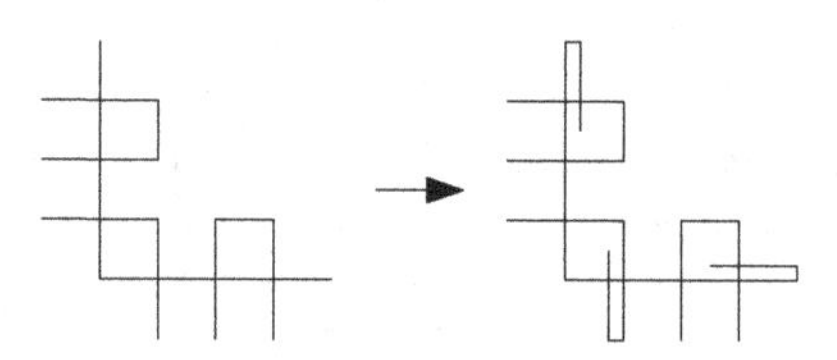

Fig. 7. An example of a faithful extension of a $(=2)$-string representation into a $(=3)$-string representation

Lemma 3. *For all $k \geq 1$, given a $(=k)$-string representation R of a graph G, the graph $G^{\#} = G^{\#}(R)$ has these two properties:*

1. *$G^{\#}$ has a $(=k)$-string representation, and*
2. *$G^{\#}$ has a $(=k+1)$-string representation if and only if the representation R can be faithfully extended into a $(=k+1)$-string representation R^{+} of G.*

The proof of the lemma is postponed to the full version [5].

Lemma 4. *For all $k \geq 1$, there exists a graph G_k with a $(=k)$-string representation R_k that cannot be faithfully extended into a $(=k+1)$-string representation.*

Again, we postpone the proof to the full version [5]. The proof is based on simple case analysis of the representations of graphs in Fig. 8, where the cases are split by possible faithful extensions of some of the strings involved.

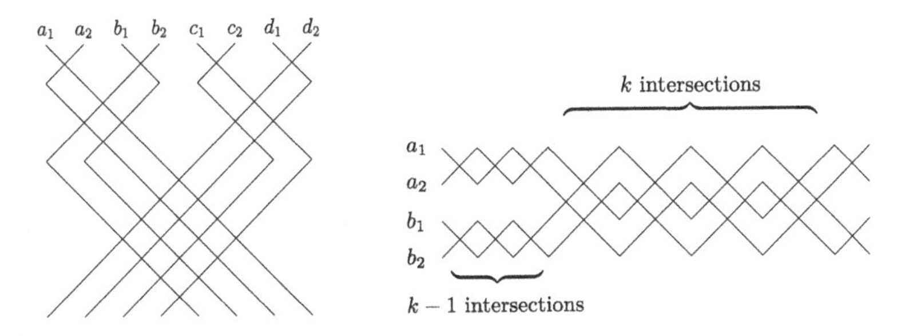

Fig. 8. The graphs G_1 with representation R_1 (left), and G_k with representation R_k for $k \geq 2$ (right)

Theorem 2. *For all $k \geq 1$, $(=k)$-STRING $\not\subseteq$ $(=k+1)$-STRING.*

Proof. This follows immediately: by Lemma 4, there is a graph G_k with its representation R_k that cannot be faithfully extended, and therefore by Lemma 3, the graph $G^{\#}(R_k)$ has a $(=k)$-string representation, but it does not have a $(=k+1)$-string representation, proving the theorem. □

We have seen that a simple doubling argument shows that any $(=k)$-string graph is also a $(=4k)$-string graph. We will now show that this is best possible.

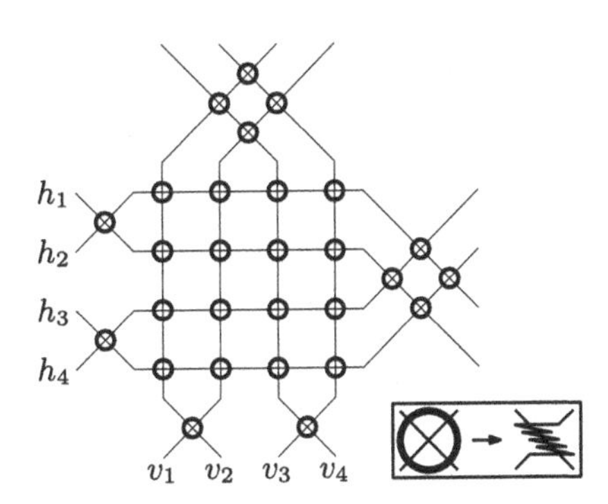

Fig. 9. The representation R used in the proof of Theorem 3. Each circled crossing represents k crossings close to each other, where k is odd.

Theorem 3. *For every odd k, there is a $(=k)$-string graph G_k which is not a $(=4k-2)$-string graph.*

Proof. We use as the starting point the $(=k)$-string representation R of the graph K_8 depicted in Fig. 9. We will call the strings $v_1, \ldots, v_4$ *vertical*, and the remaining four strings *horizontal.* Note that each circled intersection in that figure represents a k-tuple of intersections close to each other.

Applying the Noodle-Forcing Lemma to the representation R, we obtain a $(=k)$-string graph $G_k = G^{\#}(R)$. We claim that G_k has no $(=4k-2)$-string representation. For contradiction, assume that such a representation exists, and precook it into a representation R^{ε}, in which every string $R^{\varepsilon}(u)$ is confined to a sufficiently narrow noodle $N(u)$ surrounding $R(u)$.

We shall use the following terminology: for any two strings u, v of R, their *crossing area*, denoted by $C(u,v)$, is the union of the k zones formed by the intersections of their noodles, together with the parts of the noodles that lie between the zones. Thus, the crossing area forms a contiguous part of each of the two noodles. Removing the crossing area from a noodle $N(u)$ splits $N(u)$ into two connected pieces, which we call the *tips* of $N(u)$ defined by $C(u,v)$. For a vertex u, we say that its intersection with another vertex v is

- *ambiguous*, if at least one endpoint of $R^{\varepsilon}(u)$ is in $C(u,v)$,
- *peripheral*, if both endpoints of $R^{\varepsilon}(u)$ are in the same tip of $N(u)$ defined by $C(u,v)$, and
- *central*, if each endpoint of $R^{\varepsilon}(u)$ is in a different tip of $N(u)$ defined by $C(u,v)$.

Notice that if the intersection with v is peripheral for u, then every zone in $C(u,v)$ has an even number of u-traversals. Similarly, if the intersection is central for u, then every zone in $C(u,v)$ has an odd number of u-traversals. We make the following observations:

1. For two vertices u, v their mutual intersection cannot be peripheral for both of them, since this would mean that each zone in $C(u,v)$ has at least four crossing points of $R^{\varepsilon}(u)$ and $R^{\varepsilon}(v)$, which is impossible in a $(=4k-2)$-string representation.
2. It is also impossible for the intersection of u and v to be central for both vertices, since then every zone would contain an odd number of mutual crossing points.
3. Consider four distinct vertices u, v, v', v'', where the three crossing areas $C(u,v)$, $C(u,v')$ and $C(u,v'')$ appear in this order along the noodle $N(u)$. If neither $C(u,v)$ nor $C(u,v'')$ is peripheral for the vertex u, then $C(u,v')$ is central for u.

Referring to Fig. 9, consider the crossing of v_1 and v_2. Since it cannot be peripheral for both curves by the first observation above, suppose w.l.o.g. that it is non-peripheral for v_1. Likewise, suppose that the crossing of v_3 and v_4 is non-peripheral for v_3. We then look at the intersection of v_1 and v_3, and let $v_i \in \{v_1, v_3\}$ be the vertex for which the intersection is non-peripheral. It follows from the third observation that v_i has central intersections with all the four horizontal vertices $h_1, \ldots, h_4$. By analogous reasoning, a horizontal vertex $h_j \in \{h_1, \ldots, h_4\}$ has central intersections with all the four vertical vertices, including v_i. This however means that $R^{\varepsilon}(v_i)$ and $R^{\varepsilon}(h_j)$ have an odd number of mutual crossings by the second observation, a contradiction. □

We now turn our attention to the class ODD-STRING.

Theorem 4. *There is a* $(=2)$*-string graph which is not an odd-string graph.*

Proof. Let us consider the graph G with representation R as in Fig. 10. We apply the Noodle-Forcing Lemma on R, getting the graph $G^{\#}$. We will show that $G^{\#}$ is not an odd-string graph.

For contradiction, let R' be an odd-string representation of $G^{\#}$, which we precook into a representation R^{ε}. Let $N(u)$ be the noodle of the vertex u. Note that any pair of intersecting noodles $N(u)$, $N(v)$ forms two zones. The *crossing area* of $N(u)$ and $N(v)$, denoted by $C(u,v)$, is then the union of the two zones together with the sections of $N(u)$ and of $N(v)$ that lie between the two zones.

For a pair of intersecting noodles $N(u)$ and $N(v)$, we say that the curve $R^{\varepsilon}(u)$ *covers* the intersection of $N(u)$ and $N(v)$, if $R^{\varepsilon}(u)$ has at least one endpoint in the crossing area $C(u,v)$.

We claim that for any pair of intersecting noodles $N(u)$ and $N(v)$, at least one of the two curves $R^{\varepsilon}(u), R^{\varepsilon}(v)$ must cover their intersection. Indeed, suppose for contradiction that neither of the two curves covers the intersection, and let Z and Z' be the two zones of $N(u) \cap N(v)$. Then either both Z and Z' have an odd number of u-traversals or both Z and Z' have an even number of u-traversals, depending on the positions of the two endpoints of $R^{\varepsilon}(u)$. The same is true for v-traversals as well. Thus, the number of crossings between $R^{\varepsilon}(u)$ and $R^{\varepsilon}(v)$ inside Z has the same parity as the number of such crossings inside Z', showing that the two curves have an even number of crossings overall. This is impossible, since R^{ε} is an odd-string representation. This shows that at least one of the curves covers the intersection.

Now, we note that each of the two bold curves can only cover its crossings with at most two of the thin curves. Therefore, there are two thin curves $R^{\varepsilon}(x), R^{\varepsilon}(y)$ that do not have their crossings with the bold curves covered by the bold curves, and therefore the thin curves have to cover the bold curve crossings with their endpoints. But this implies that neither of the two curves $R^{\varepsilon}(x), R^{\varepsilon}(y)$ can cover the crossing of $N(x)$ and $N(y)$, which is a contradiction. □

This result shows that the class ODD-STRING differs from the class STRING. Moreover, we can summarize our findings by the following proposition.

Proposition 4. *For* $k, \ell \in \mathbb{N}$, $(=k)$-STRING $\subseteq$ $(=\ell)$-STRING *if and only if one of the two conditions holds*

- $k \leq \ell$ *and* k, ℓ *have the same parity, or*
- k *is odd,* ℓ *is even and* $\ell \geq 4k$.

Proof. The backward implication follows from Propositions 1 and 2. The forward implication follows from Theorems 2 (which removes any case with $\ell < k$), 4 (which removes any case where k is even and ℓ is odd), and 3 (which removes any case where k is odd, ℓ is even and $\ell \leq 4k-2$). □

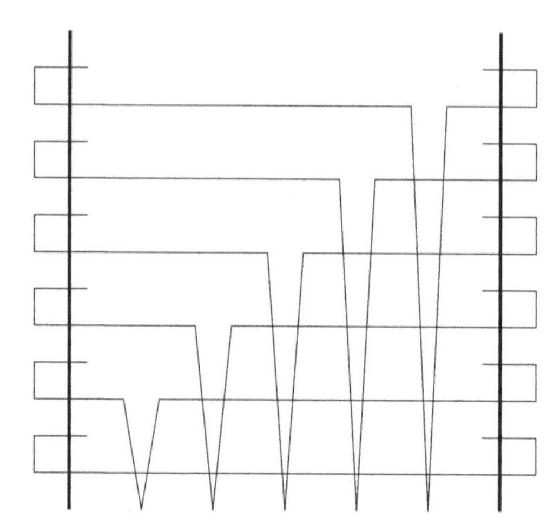

Fig. 10. The graph G with a $(=2)$-string representation for Theorem 4

3 Final Remarks and Open Problems

3.1 $(=\ell)$-String Graphs from k-String Graphs

It would be nice to clarify the inclusions between the classes k-STRING and $(=\ell)$-STRING for various values of ℓ and k. The key problem is to determine, for a given k, the smallest ℓ such that k-STRING is a subclass of $(=\ell)$-STRING.

Immediately from Theorem 2, we can see that (for $k > 1$) ℓ must be at least $k + 2$, since k-string graphs contain both the class of $(=k - 1)$-string graphs and $(=k)$-string graphs, and therefore are not a subclass of either $(=k)$-string or $(=k+1)$-string. On the other hand, our current methods get no better upper bound than $\ell \leq 8k$.

To establish the bound, we first define the class of proper k-string graphs to be the class of all k-string graphs that have a proper k-string representation. Our proof works in two steps: in the first step, we build a proper $2k$-string representation from a k-string representation, and in the second step, we build a $(=4\ell)$-string representation from a proper ℓ-string representation. We obtain the following result, whose proof appears in the full version [5]. We do not know if the bound $8k$ is tight.

Theorem 5. *For all $k \geq 1$, k-STRING is a subclass $(=8k)$-STRING.*

3.2 Complexity

After defining the new classes of odd-string graphs and $(=k)$-string graphs, we may naturally ask about the complexity of their recognition. We can easily see that all mentioned classes contain all grid intersection graphs, i.e., the intersection graphs of horizontal and vertical segments (for $(=2)$-string graphs, we use

Proposition 3) and are contained in string graphs. Therefore, by the sandwiching results of Kratochvíl [9], or Mustaţă and Pergel [13], we see that the recognition problem of ODD-STRING as well as of $(=k)$-STRING is NP-hard.

For $(=k)$-string graphs with fixed k, we can easily see that the problem is in NP (and hence NP-complete). However, when k is a part of the input, it is not clear whether the problem still remains in NP.

For the recognition of ODD-STRING, it is not even clear whether the problem is decidable, as there is no known upper bound on the number of intersections an odd-string graph on n vertices may require in its representation. This situation is reminiscent of a similar issue with the recognition of string graphs, whose decidability used to be an open problem until Schaefer and Štefankovič first showed that the problem is decidable [18], and later with Sedgwick that it belongs to NP [17]. However, we are unable to adapt their methods to odd-string representations.

Another type of complexity questions focuses on the complexity of deciding the existence of a $(=k)$-string representation when we are given a $(=\ell)$-string representation as the input. The following cases seem interesting:

- $k = \ell + 1$ (i.e., can we extend the representation to have one more intersection per pair of curves? It also makes sense to restrict this question to the faithful extensions fully defined in the full version [5]),
- $k = \ell - 2$ (i.e., can we remove two intersections per pair, or is this the "minimal" representation of this graph with respect to the number of intersection points per pair with the given parity?),
- as a generalization of the previous case, $k = \ell - c$ for any $c \in \mathbb{N}$.

We remark that analogous questions have previously been tackled, e.g., for the B_k-VPG hierarchy, where Chaplick et al. [4] showed that recognizing B_k-VPG graphs is NP-complete, even when a B_{k+1}-VPG representation is given as part of the input.

References

1. Asinowski, A., Cohen, E., Golumbic, M.C., Limouzy, V., Lipshteyn, M., Stern, M.: Vertex intersection graphs of paths on a grid. J. Graph Algorithms Appl. **16**(2), 129–150 (2012). https://doi.org/10.7155/jgaa.00253
2. Cabello, S., Jejčič, M.: Refining the hierarchies of classes of geometric intersection graphs. Electron. J. Comb. P1–33 (2017). https://doi.org/10.37236/6040
3. Champseix, N., Galby, E., Munaro, A., Ries, B.: CPG graphs: some structural and hardness results. Discret. Appl. Math. **290**, 17–35 (2021). https://doi.org/10.1016/j.dam.2020.11.018
4. Chaplick, S., Jelínek, V., Kratochvíl, J., Vyskočil, T.: Bend-bounded path intersection graphs: sausages, noodles, and waffles on a grill. In: Golumbic, M.C., Stern, M., Levy, A., Morgenstern, G. (eds.) WG 2012. LNCS, vol. 7551, pp. 274–285. Springer, Heidelberg (2012). https://doi.org/10.1007/978-3-642-34611-8_28
5. Chmel, P., Jelínek, V.: String graphs with precise number of intersections. CoRR abs/2308.15590 (2023). https://arxiv.org/abs/2308.15590

6. Deniz, Z., Galby, E., Munaro, A., Ries, B.: On contact graphs of paths on a grid. In: Biedl, T., Kerren, A. (eds.) GD 2018. LNCS, vol. 11282, pp. 317–330. Springer, Cham (2018). https://doi.org/10.1007/978-3-030-04414-5_22
7. Hliněný, P.: Classes and recognition of curve contact graphs. J. Comb. Theor. Ser. B **74**(1), 87–103 (1998). https://doi.org/10.1006/jctb.1998.1846
8. Hliněný, P.: Contact graphs of line segments are NP-complete. Discret. Math. **235**(1), 95–106 (2001). https://doi.org/10.1016/S0012-365X(00)00263-6
9. Kratochvíl, J.: String graphs. II. Recognizing string graphs is NP-hard. J. Comb. Theor. Ser. B **52**(1), 67–78 (1991). https://doi.org/10.1016/0095-8956(91)90091-W
10. Kratochvíl, J.: A special planar satisfiability problem and a consequence of its NP-completeness. Discret. Appl. Math. **52**(3), 233–252 (1994). https://doi.org/10.1016/0166-218X(94)90143-0
11. Kratochvíl, J., Matoušek, J.: NP-hardness results for intersection graphs. Commentationes Mathematicae Universitatis Carolinae **030**(4), 761–773 (1989). http://eudml.org/doc/17790
12. Kratochvíl, J., Matoušek, J.: Intersection graphs of segments. J. Comb. Theor. Ser. B **62**(2), 289–315 (1994). https://doi.org/10.1006/jctb.1994.1071
13. Mustaţă, I., Pergel, M.: What makes the recognition problem hard for classes related to segment and string graphs? CoRR abs/2201.08498 (2022). https://arxiv.org/abs/2201.08498
14. Pach, J., Tóth, G.: Which crossing number is it anyway? J. Comb. Theor. Ser. B **80**(2), 225–246 (2000). https://doi.org/10.1006/jctb.2000.1978
15. Pelsmajer, M.J., Schaefer, M., Štefankovic, D.: Odd crossing number and crossing number are not the same. Discrete Comput. Geom. **39**(1), 442–454 (2008). https://doi.org/10.1007/s00454-008-9058-x
16. Schaefer, M.: The graph crossing number and its variants: a survey. Electron. J. Comb. DS21-Apr (2012). https://doi.org/10.37236/2713
17. Schaefer, M., Sedgwick, E., Štefankovič, D.: Recognizing string graphs in NP. J. Comput. Syst. Sci. **67**(2), 365–380 (2003). https://doi.org/10.1016/S0022-0000(03)00045-X
18. Schaefer, M., Štefankovič, D.: Decidability of string graphs. J. Comput. Syst. Sci. **68**(2), 319–334 (2004). https://doi.org/10.1016/j.jcss.2003.07.002
19. Sinden, F.W.: Topology of thin film RC circuits. Bell Syst. Tech. J. **45**(9), 1639–1662 (1966). https://doi.org/10.1002/j.1538-7305.1966.tb01713.x
20. Trotter, W.T., Jr., Harary, F.: On double and multiple interval graphs. J. Graph Theory **3**(3), 205–211 (1979). https://doi.org/10.1002/jgt.3190030302

Crossing Numbers

Degenerate Crossing Number and Signed Reversal Distance

Niloufar Fuladi, Alfredo Hubard, and Arnaud de Mesmay(✉)

Univ Gustave Eiffel, CNRS, LIGM, 77454 Marne-la-Vallée, France
niloufar.fuladi@aol.com, alfredo.hubard@univ-eiffel.fr,
arnaud.de-mesmay@univ-eiffel.fr

Abstract. The degenerate crossing number of a graph is the minimum number of transverse crossings among all its drawings, where edges are represented as simple arcs and multiple edges passing through the same point are counted as a single crossing. Interpreting each crossing as a cross-cap induces an embedding into a non-orientable surface. In 2007, Mohar showed that the degenerate crossing number of a graph is at most its non-orientable genus and he conjectured that these quantities are equal for every graph. He also made the stronger conjecture that this also holds for any loopless pseudotriangulation with a fixed embedding scheme.

In this paper, we prove a structure theorem that almost completely classifies the loopless 2-vertex embedding schemes for which the degenerate crossing number equals the non-orientable genus. In particular, we provide a counterexample to Mohar's stronger conjecture, but show that in the vast majority of the 2-vertex cases, the conjecture does hold.

The reversal distance between two signed permutations is the minimum number of reversals that transform one permutation to the other one. If we represent the trajectory of each element of a signed permutation under successive reversals by a simple arc, we obtain a drawing of a 2-vertex embedding scheme with degenerate crossings. Our main result is proved by leveraging this connection and a classical result in genome rearrangement (the Hannenhali-Pevzner algorithm) and can also be understood as an extension of this algorithm when the reversals do not necessarily happen in a monotone order.

Keywords: Degenerate crossing number · Non-orientable surface · Signed reversal distance

1 Introduction

A *cross-cap drawing* of a graph G is a drawing of G on the sphere with g distinct points, called *cross-caps*, such that the drawing is an embedding except at the cross-caps, where multiple edges are allowed to cross transversely, as pictured in Fig. 1.In [14], Pach and Toth introduced the *degenerate crossing number*, denoted by $Cr_{deg}(G)$ which in this language is the minimum number

M. A. Bekos and M. Chimani (Eds.): GD 2023, LNCS 14465, pp. 95–109, 2023.
https://doi.org/10.1007/978-3-031-49272-3_7

of cross-caps for a cross-cap drawing of G to exist, where edges are required to be drawn as simple arcs. In [12], Mohar removed the constraint that edges be simple arcs, leading to the *genus crossing number*, which he proved to be equal to the non-orientable genus of the graph, denoted by $g(G)$. He then made an enticing conjecture claiming that these two crossing numbers are equal. We say that a cross-cap drawing of graph G is *perfect* if there are $g(G)$ cross-caps and every edge intersects each cross-cap at most once. Then this conjecture can be restated as follows:

Fig. 1. A degenerate crossing and a cross-cap placed at this crossing.

Conjecture 1. [12, Conjecture 3.1, Proposition 3.3] Every simple graph has a perfect cross-cap drawing.

Mohar went further and conjectured that for any loopless graph embedding, there exists a perfect cross-cap drawing that is compatible with this embedding, in a sense that we now describe. The terminology that we introduce is equivalent to the PD1S in [12, Section 3]. A *pseudo-triangulation* is a cellularly embedded multi-graph $\phi\colon G \to S$ in which each face has degree three. Denote by $S^2\backslash g\bigotimes$, the sphere minus g tiny disks, and by $(S^2\backslash g\bigotimes)/\sim$ the space obtained by quotienting the boundary of each disk with the antipodal map. Topologically, this amounts gluing a Möbius band on each missing disk, thus yielding the non-orientable surface of genus g, denoted by N_g. We say that an embedded multi-graph $\phi\colon G \to N_g$ *admits a cross-cap drawing* $\phi'\colon G \to (S^2\backslash g\bigotimes)/\sim$, if there is a homeomorphism $f\colon N_g \to (S^2\backslash g\bigotimes)/\sim$ such that $f(\phi(G)) = \phi'(G)$.

Conjecture 2. [12, Conjecture 3.4] For any positive integer g, every loopless pseudo-triangulation of N_g admits a perfect cross-cap drawing.

An even stronger conjecture was hinted at in [12, Paragraph following Conjecture 3.4], suggesting that one could possibly remove the loopless assumption if one forbids separating loops. This strengthening was disproved by Schaefer and Štefankovič [15, Theorem 7].

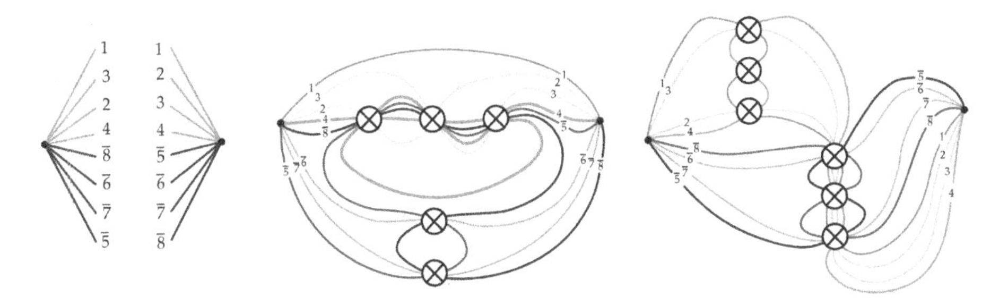

Fig. 2. Left: A loopless 2-vertex scheme made of a *positive block*, in red, consisting of only positive edges, and a *negative block*, in blue, consisting of negative edges. Middle: a cross-cap drawing showing that it has non-orientable genus 5. The bold red edge enters a cross-cap twice. Right: A cross-cap drawing where each edge enters each cross-cap at most once requires 6 cross-caps. (Color figure online)

In addition to their motivation from crossing number theory, these conjectures would also shed light on the difficult task of visualizing high genus embedded graphs, providing an alternate approach to that of Duncan, Goodrich and Kobourov [5], who rely on canonical polygonal schemes [11].

Our Results. A big step towards both these conjectures was achieved by Schaefer and Štefankovič, who proved [15, Theorem 10] that any multi-graph embedded on a non-orientable surface of genus g admits a cross-cap drawing with g cross-caps, in which each edge enters each cross-cap at most *twice*. This theorem applies in particular to one-vertex embedding schemes, and thus suggests a natural approach towards proving Conjectures 1 and 2. First contract a spanning tree to obtain a one-vertex graph and apply this theorem. Then, edges might enter cross-caps twice, but since the initial graph is loopless, one could hope to uncontract some edges so as to spread these two cross-caps on two edges, thus obtaining a perfect cross-cap drawing. Our first result shows that this approach cannot work, as some loopless 2-vertex schemes do not admit perfect cross-cap drawings.

Theorem 1. *A loopless 2-vertex embedding scheme that consists of exactly one non-trivial positive block and one non-trivial negative block admits no perfect cross-cap drawing.*

We refer to Fig. 2 for an example that should provide an intuitive idea of the notion of blocks, and to Sect. 2 for the precise definition. As a corollary, we obtain a counter-example to Conjecture 2:

Corollary 1. *There exist a loopless pseudo-triangulation G that admits no perfect cross-cap drawing.*

Our second contribution and main theorem is a converse to Theorem 1.

Theorem 2. *For any embedding G of a loopless 2-vertex graph on N_g, at least one of the following is true.*

1. *G admits a perfect cross-cap drawing with g cross-caps,*
2. *or the reduced graph of G is one of the two schemes pictured in Fig. 3,*

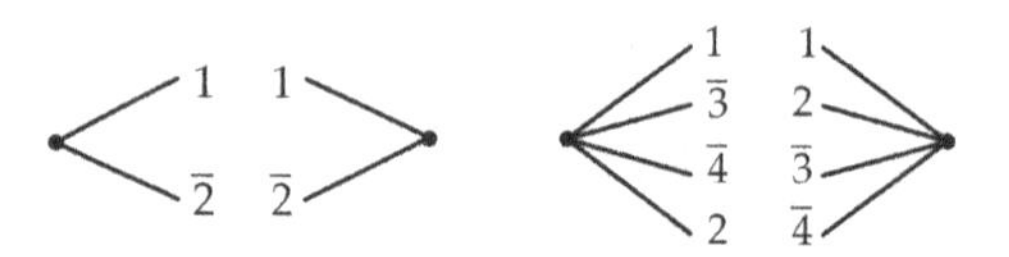

Fig. 3. The only two possible exceptions to perfect cross-cap drawings.

We refer to Sect. 2 for the definition of blocks and reduced graphs. Essentially, Theorem 2 shows that apart from two narrow families of exceptional cases, all the loopless 2-vertex embeddings do satisfy Conjecture 2. As an illustration, Fig. 4 shows that while the example in Fig. 2 does not admit a perfect cross-cap drawing, surprisingly, it does after *adding* two edges to it. It directly follows from Theorem 2 that under standard random models, any loopless 2-vertex embedding scheme admits a perfect cross-cap drawing asymptotically almost surely.

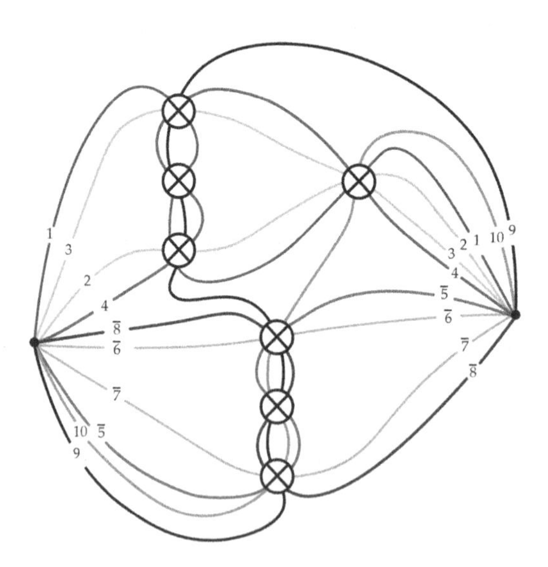

Fig. 4. A perfect cross-cap drawing of Fig. 2 with two additional edges.

Techniques and Connections to Signed Reversal Distance. Our focus on the 2-vertex case in Theorem 2 is further motivated by a connection (introduced in [7]) to computational genomics. An important problem in computational biology is to compute various notions of distance between two genomes (see [4, 6]). Remarkably, one of the most biologically relevant distances is also one of the few that can be calculated efficiently: a one chromosome genome is encoded by a signed permutation (i.e., permutations of integers in which each element has a sign) $\pi = (\pi_1, \ldots, \pi_n)$. The *reversal* of the interval (i, j) acts on π by reversing the order of the elements $\pi_i, \ldots, \pi_j$ as well as their signs, it maps

$$(\pi_1, \pi_2, \ldots, \pi_{i-1}, \pi_i, \pi_{i+1} \ldots, \pi_{j-1}, \pi_j, \pi_{j+1} \ldots \pi_n),$$

to

$$(\pi_1, \pi_2, \ldots, \pi_{i-1}, \overline{\pi_j}, \overline{\pi_{j-1}} \ldots, \overline{\pi_{i+1}}, \overline{\pi_i}, \pi_{j+1} \ldots \pi_n).$$

The *reversal distance* between signed permutations π, π', denoted by $d(\pi, \pi')$ is the minimum number of reversals needed to transform π to π'. A celebrated algorithm of Hannenhalli and Pevzner [9] (see also [2,3]) computes in polynomial time the reversal distance between two signed permutations.

Now, a signed permutation is exactly the combinatorial data in the embedding scheme of a 2-vertex loopless graph, and when tracing the action of each element of a signed permutation under the action of reversals, one obtains a cross-cap drawing of that embedding scheme, where each reversal corresponds to a cross-cap and each edge is an x-monotone curve, and in particular no edge enters twice the same cross-cap (see Fig. 5 for an illustration).

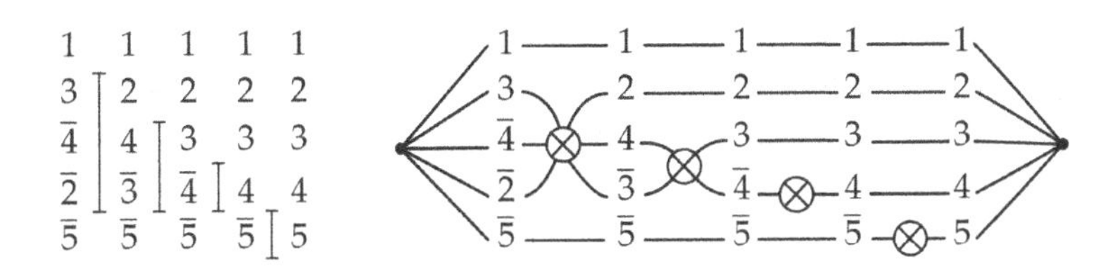

Fig. 5. From sorting permutations with reversals to monotone cross-cap drawings.

This easily implies the inequalities: $g(\pi) \leq Cr_{deg}(\pi) \leq d(\pi, id)$. It turns out that the inequality $g(\pi) \leq d(\pi, id)$ is central to the reversal distance theory, and the cases of equality are well understood. As we explain in Sect. 4 these arguments are natural from the point of view of embedding schemes and our proof of Theorem 2 heavily relies on them. Conversely, Theorem 2 can be reinterpreted in the setting of signed permutations as providing an extension of the Hannenhalli-Pevzner theory.

The proof of Theorem 2 consists of two steps which can readily be made algorithmic: we first *reduce* a signed permutation π to a simpler one $\pi|$ for which we can prove that $g(\pi|) = d(\pi|, id)$ (Lemma 8), then we devise a technique to *blow up* (Lemma 9) the cross-cap drawing of the reduced signed permutation $\pi|$, yielding a perfect cross-cap drawing of the original signed permutation.

2 Preliminaries

Embedding Schemes. In this article, we work with multi-graphs, possibly with loops and multiple edges. An *embedding* of a graph G on a surface S is an injective map $\phi : G \rightarrow S$. We consider two embeddings equivalent if their images are homeomorphic. The *faces* of an embedded graph are the connected components of $S \backslash \phi(G)$. An embedding is *cellular* if its faces are homeomorphic to topological disks. The Euler genus, $eg(G)$, of a cellular embedding of a graph G is the quantity $2-v+e-f$, where v, e and f denote respectively the number of vertices, edges and faces of the embedding of G. If S is orientable, an embedding can be described combinatorially by a *rotation system*, that is, the set of cyclic permutations encoding the order of the edges around each vertex. When S is non-orientable, which will always be the case in this paper, some additional data is required to encode an embedding, which is encompassed in the concept of an *embedding scheme.* We introduce the main definitions and refer to Mohar and Thomassen [13, Section 3.3] for extensive background.

Definition 1 (Embedding scheme). *An* embedding scheme *consists of a triple* (G, ρ, λ) *where*

- G *is a graph,*
- $\rho = \{\rho_v, v \in V(G)\}$, *where each* ρ_v *is a cyclic permutation of the edges incident to* v, *and*
- λ *is a function that assigns a* signature $\{+1, -1\}$ *to each edge of* G.

From an embedding scheme (G, ρ, λ), one can naturally recover the set of *facial cycles* by following the edges and switching sides according to their signatures. Then, pasting a topological disk on each facial cycle yields a cellular embedding of G. Given an embedding scheme (G, ρ, λ), a *flip* at a vertex v yields another embedding scheme of the same graph, in which we reverse the order of the edges incident to v and invert the signature of those edges incident to v that are not loops. We say that two embedding schemes (G, ρ, λ) and (G, ρ', λ') are *equivalent* if one can go from one to the other one by a sequence of flips. Two embedding schemes are equivalent if and only if they induce equivalent cellular embeddings [13, Theorem 3.3.1]. This justifies that equivalence classes of embedding schemes and embedded graphs can be considered as being two representations of the same objects, and we switch freely between the two points of view in this article, sometimes using the shorthand G to denote (G, ρ, λ).

A closed curve in an embedding scheme is one-sided (resp. two-sided) if and only if the signatures of its edges multiply to -1 (resp. $+1$). An embedding scheme is called *orientable* if it only contains two-sided closed curves; otherwise it is called *non-orientable.* A closed curve in a non-orientable embedding scheme is *separating* if cutting along γ yields two connected components. A closed curve in a non-orientable embedding scheme is *orienting* [7,15] if by cutting along γ, we obtain a connected orientable surface.

For two edges a and b in a 2-vertex embedding scheme, we denote by $a \cdot b$ the concatenation of a and b which we will interpret as a cycle. Denoting the vertices of the scheme by v_1 and v_2, we define a *wedge* between a and b, $\omega_{a,b}$:

- If both a and b are negative, then $\omega_{a,b}$ contains all the half-edges in the interval (a, b) in both ρ_{v_1} and ρ_{v_2}.
- If at least one of them is positive, then $\omega_{a,b}$ contains all the half-edges in the interval (a, b) in ρ_{v_1} and (b, a) in ρ_{v_2}.

We say that a wedge *encloses* an edge if it contains both its half-edges or none of them. For example $w_{1,4}$ in Fig. 6 encloses all the edges of the graph. We can recognize orienting and separating curves in an embedding scheme with the following lemmas.

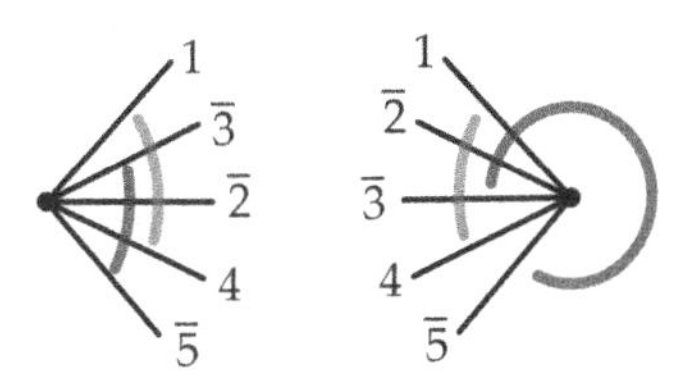

Fig. 6. The wedges $\omega_{1,4}$ and $\omega_{3,5}$ depicted in orange and green respectively. (Color figure online)

Lemma 1. *For two edges a and b in a non-orientable loopless 2-vertex embedding scheme, the cycle a.b is orienting,*

- *if at least one of a and b is positive and $\omega_{a,b}$ contains exactly one end of all negative edges and encloses all the positive edges, or*
- *if both a and b are negative and their wedge contains exactly one end of all positive edges and encloses all the negative edges.*

Lemma 2. *For two edges a and b in a loopless 2-vertex embedding scheme, the cycle a.b is separating if a and b have the same signature and $\omega_{a,b}$ encloses all the edges.*

See Fig. 6 for an example of separating and orienting cycle in a 2-vertex scheme. The proof is obtained by contracting an edge and appealing to similar results for 1-vertex schemes [7, Lemma 2.3], it is included in the full version [8].

2-Vertex Embedding Schemes and Signed Permutations. This paper almost exclusively deals with loopless graphs with two vertices, and in that setting embedding schemes take a particularly simple form. Without loss of generality, one can number the edges so that the cyclic permutation around one of the vertices is the identity. Then the data of the embedding scheme just consists of the cyclic permutation around the other vertex, and the signature of the edges, and thus this amounts to a *signed cyclic permutation*: a cyclic permutation where each number is additionally endowed with a + or − sign. We also use an overline notation $\overline{i}$ to depict negative signs. Therefore, in what follows we freely identify a signed permutation and a 2-vertex embedding scheme. Two edges in a 2-vertex embedding scheme are *homotopic* if either they are both positive and appear consecutively in increasing order (e.g., $(i, i+1)$) or they are both negative and they appear in the reverse order (e.g., $(\overline{i+1}, \overline{i})$).

A *positive block* in a signed permutation is an interval $I = (\pi_i, \dots, \pi_j)$ where all the elements are positive, $\pi_i < \pi_j$, all the integers in $[\pi_i, \pi_j]$ are contained in I. A *negative block* in a signed permutation is an interval $I = (\overline{\pi_i}, \dots \overline{\pi_j})$ where all the elements are negative, $\pi_i > \pi_j$, and all the integers in $[\pi_j, \pi_i]$ are contained in I. A block is *non-trivial* if it is not already sorted, i.e., it is not equal to $(\pi_i, \pi_i+1, \dots \pi_j-1, \pi_j)$ or to $(\overline{\pi_i}, \overline{\pi_i - 1}, \dots, \overline{\pi_j+1}, \overline{\pi_j})$. Our concept of blocks is similar to the notion of hurdles of [9], or more accurately, to the notion of unoriented components in [3]. In both cases, we call π_i and π_j the frames of the block. A block is called *minimal* if it does not contain any block except itself.

We say that a signed permutation is *reduced* if it has no blocks. Given a signed permutation π, its reduced permutation $\pi|$ is a permutation in which we replace every minimal block with a single element of the same sign, and we iterate this process until we arrive at a reduced permutation.

Cross-Cap Drawings. A cross-cap drawing of an embedding scheme (G, ρ, λ) is a cross-cap drawing of (G, ρ, λ) or of an equivalent scheme (under flips), that respects the cyclic permutations and signatures on the edges, i.e., if an edge has signature $+1$ (resp. -1) then it enters an even (resp. odd) number of cross-caps. This definition is equivalent to the cross-cap drawings of embeddings defined in the introduction. For non-orientable embedding schemes, by Euler's formula, the minimum number of cross-caps for a drawing coincides with the Euler genus. For orientable embedding schemes of non-zero Euler genus, one needs exactly one additional cross-cap:

Lemma 3 (**[15, Lemma 6]**). *Let (G, ρ, λ) be an orientable embedding scheme with non-zero Euler genus. Then any cross-cap drawing of G requires $eg(G)+1$ cross-caps, in particular $eg(G)+1$ is odd.*

The next lemma allows us to recognize types of curves in a cross-cap drawing.

Lemma 4. *For any cross-cap drawing of a non-orientable embedding scheme:*

1. *A closed curve is one-sided (two-sided) if and only if it enters an odd (even) number of cross-caps.*
2. *A closed curve is orienting (separating) if and only if it enters each cross-cap an odd (even) number of times.*

We refer to [15, Lemma 3] and [15, Lemma 4] for proofs.

Reversal Distance and Monotone Cross-Cap Drawings. Signed permutations model genomes with a single chromosome in computational biology where they come endowed with the reversal distance. The reversal distance $d(\pi, id)$ is

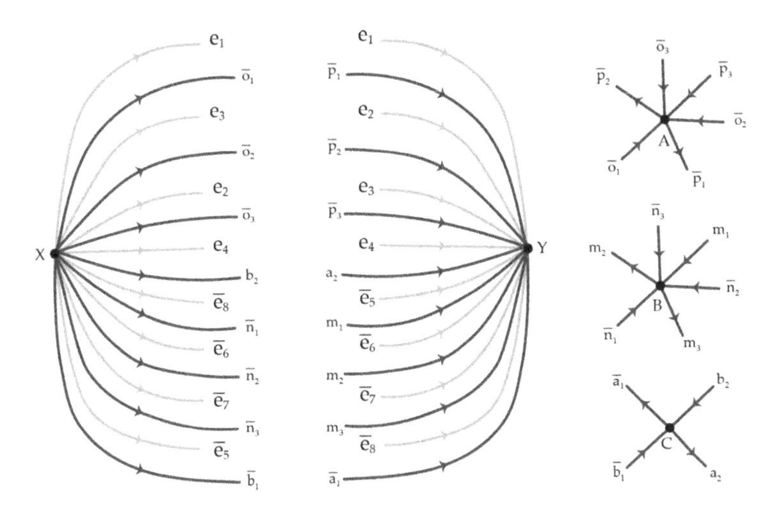

Fig. 7. A pseudo-triangulation of N_5 admitting no perfect cross-cap drawing.

the smallest number d such that there exists a sequence $\{\pi = \pi^1, \pi^2, \dots \pi^d = id\}$ such that (π^i) and (π^{i+1}) differ by a signed reversal. We call such a (not necessarily minimizing) sequence, a *path* of signed permutations.

To any path of signed permutations $\{\pi^1, \pi^2, \dots \pi^d\}$ we can associate a cross-cap drawing. We place a source vertex at $(-1, n/2)$ and a terminal vertex at $(d+1, n/2)$. Edges will be x-monotone piece-wise linear curves between these two vertices, and at each crossing between two such curves we introduce a cross-cap. The edge j emanates from $(-1, n/2)$ to $(0, \pi_j^1)$, for each $k \leq d$ it passes through $(k, \pi_k^j) \in \mathbb{R}^2$, and finally it connects (d, π_j^d) to the terminal vertex at $(d+1, n/2)$. In the remaining of this article, we often forget about the vertices $(-1, n/2)$ and $(d+1, n/2)$ in our illustrations as they play no role, and we always assume that π^d is the identity (see Fig. 5).

3 The Counterexample

In this section, we provide a family of 2-vertex embedding schemes that do not admit a perfect cross-cap drawing. Then we provide an explicit pseudo-triangulation of N_5 (depicted in Fig. 7). , disproving Conjecture 2.

Remark 1. If an embedding scheme G has one positive and one negative block, then so does its flipped version, therefore we do not need to account for the possible flip in the proof of Theorem 1.

In order to prove Theorem 1, we rely on Lemmas 1 and 5.

Lemma 5. *Let G be an embedding scheme that consists of a non-trivial positive block A and a non-trivial negative block B, then $g(G) = g(A) + g(B) - 1$.*

The proof follows directly from the Euler characteristic and is deferred to the full version [8]. We now have all the tools to prove Theorem 1, and refer to Fig. 2 for an example to help follow the proof.

Proof of Theorem 1. Let G be a concatenation of a positive block A with frames a_1 and a_2 and a negative block B with frames b_1 and b_2. Let us assume that ϕ is a perfect cross-cap drawing of G. From Lemma 1 we derive that $a_1 \cdot b_1$ and $a_1 \cdot b_2$ are orienting curves, hence by Lemma 4 each of them enters each cross-cap once. Lemma 2 implies that $b_1 \cdot b_2$ is separating. Therefore by Lemma 4, b_1 and b_2 enter the same cross-caps and do not enter any cross-cap that a_1 enters. Similarly, $a_1 \cdot a_2$ is separating and hence they enter the same cross-caps and no cross-cap that b_1 and b_2 enter. Then A is drawn with g_A cross-caps and B is drawn with g_B cross-caps that are disjoint from the cross-caps that A entered. But by Lemma 5 the non-orientable genus of G is $g_A + g_B - 1$. Therefore, there are not enough cross-caps available to draw both A and B. This concludes. □

Corollary 1 follows at once as we can always add edges and vertices to a scheme to triangulate it without adding loops nor changing its genus, and any perfect cross-cap drawing of the triangulation restricts to a cross-cap drawing of the scheme. We provide in Fig. 7 an example of such a pseudo-triangulation.

4 Topology of the Reversal Distance

In this section, we recall some well-known results from the genomics rearrangements literature[1], which we interpret in the language of embedding schemes (see also [10] for an alternate topological interpretation of these arguments).

4.1 The Bafna-Pevzner Inequality from Euler's Formula

Let π be a signed cyclic permutation, which, as explained in Sect. 2, we think of as a 2-vertex embedding scheme, which requires $g(\pi)$ cross-caps to be drawn. We can compute its number of faces, which we denote by $f(\pi)$ and the number of elements in the permutation corresponds to the number of edges in the scheme, which we denote by $e(\pi)$. Then Euler's formula reads $2 - eg(\pi) = 2 - e(\pi) + f(\pi)$ which simplifies to $eg(\pi) = e(\pi) - f(\pi)$. By Lemma 3, we thus have $d(\pi, id) \geq e(\pi) - f(\pi)$.

A very similar inequality was first discovered by Bafna and Pevzner [1, Theorem 2] without reference to embeddings. Figure 8 shows an example where the inequality is strict: the embedding scheme has Euler genus two, non-orientable genus three and one can show that the signed permutation requires four reversals to be sorted. However, as pictured on the right, it does admit a perfect cross-cap drawing with three cross-caps. Necessarily, in that example, the cross-caps can not be interpreted as reversals: this is apparent here as they do not occur in a monotone order.

[1] The literature primarily deals with sorting standard permutations, while here we are sorting cyclic permutations. In our description, we directly translate their techniques to this cyclic setting.

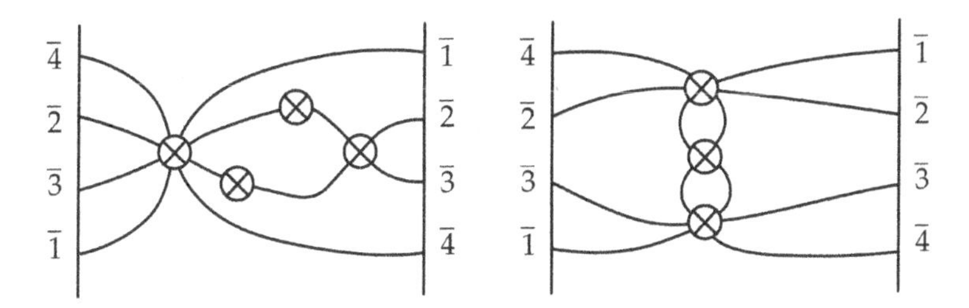

Fig. 8. The embedding scheme depicted in this picture has non-orientable genus 3 but requires four reversals (left). However, it admits a perfect cross-cap drawing with three cross-caps (right).

The starting idea of the Hannenhalli-Pevzner (HP) algorithm is to identify intervals in a signed permutation where applying a reversal is clearly making progress. Given a signed cyclic permutation π, we call a pair of consecutive integers i and $i+1$ *reversible* if they have opposite signs in π (this is called an *oriented pair* in [3]). For a given reversible pair there exist two reversals σ, σ' such that i and $i+1$ are homotopic in $\pi \cdot \sigma$ and in $\pi \cdot \sigma'$. These two reversals are equivalent in the sense that the two permutations that they yield are flipped versions of each other. The following lemma follows from an Euler characteristic argument (see proof in the full version [8]).

Lemma 6. *If $i, i+1$ are a reversible pair in π and σ is a reversal that turns them into homotopic curves in $\pi \cdot \sigma$, then $eg(\pi \cdot \sigma) = eg(\pi) - 1$.*

4.2 The HP Algorithm for Reduced Signed Permutations

It is immediate to see that blocks do not contain reversible pairs of edges, and thus non-trivial blocks form a natural obstruction to applying the reversals described above. Given a reversible pair $(i, i+1)$ in the signed permutation π, let σ be a reversal that turns i and $i+1$ into homotopic edges. The *score* of $(i, i+1)$ is the number of reversible pairs in $\pi \cdot \sigma$.

HP Algorithm: While there is a reversible pair, reverse a pair of maximal score [2].

Theorem 3. *If a signed permutation π is non-orientable and has no non-trivial blocks then $d(\pi, id) = eg(\pi)$, and the HP algorithm gives a sequence of reversals of this optimal length.*

This theorem follows at once from the following lemma.

Lemma 7. *Let π be a non-orientable signed permutation without non-trivial blocks, $(i, i+1)$ be a reversible pair of maximal score, and σ be a reversal that makes i and $i+1$ homotopic such that $\pi \cdot \sigma$ is not the identity. Then $\pi \cdot \sigma$ is non-orientable and has no non-trivial blocks.*

The proof is almost identical to that of [2, Theorem 10], see the full version [8].

Proof of Theorem 3. By the previous lemma, if π is non-orientable has no non-trivial block and σ is a reversal of maximum score, then $\pi \cdot \sigma$ is also non-orientable and also has no non-trivial block. By induction, $d(\pi \cdot \sigma, id) = e(\pi \cdot \sigma) - f(\pi \cdot \sigma) = eg(\pi \cdot \sigma)$. Therefore, by Lemma 6, $eg(\pi) \leq d(\pi, id)$, and $d(\pi, id) \leq d(\pi \cdot \sigma, id) + 1 = eg(\pi \cdot \sigma) + 1 = e(\pi) - f(\pi) = eg(\pi)$. □

5 Perfect Drawings for Most 2-Vertex Graph Embeddings

We say that a cross-cap drawing is *fantastic* if it is perfect and every edge enters at least one cross-cap.

Lemma 8. *Every reduced loopless 2-vertex graph embedding scheme that is different from* (1), $(1, \overline{2})$ *or* $(1, \overline{3}, \overline{4}, 2)$ *admits a fantastic cross-cap drawing.*

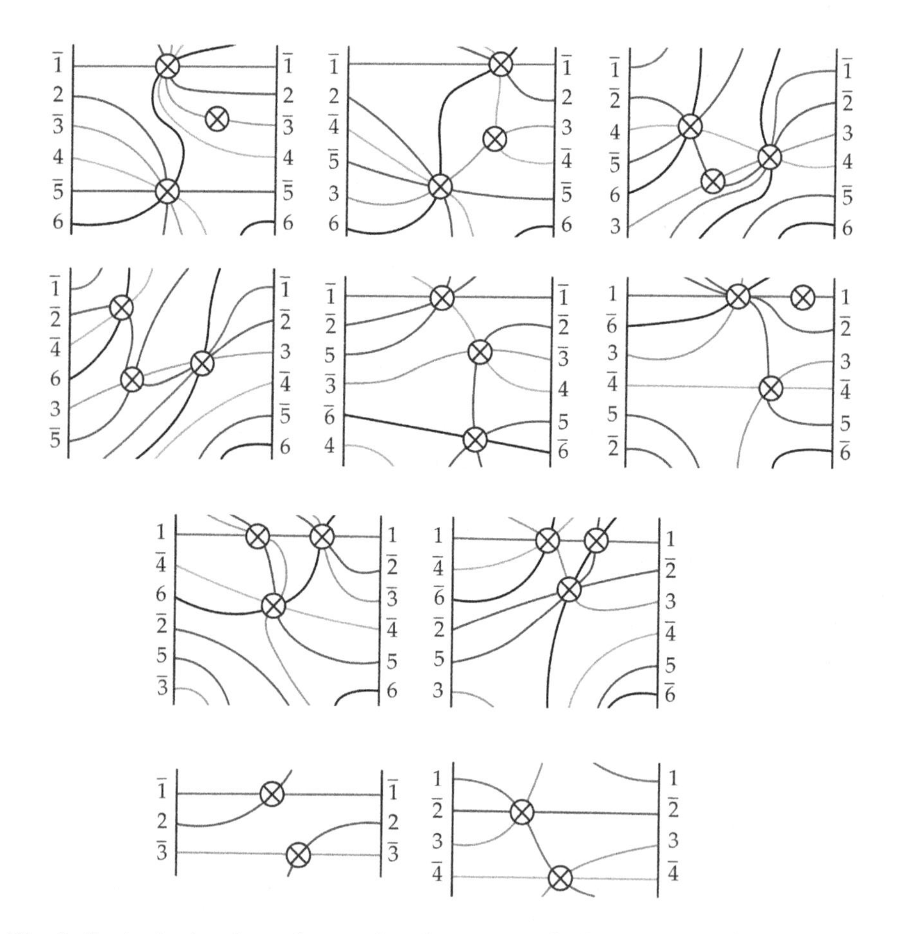

Fig. 9. Fantastic drawings of genus-2 and genus-3 embedding schemes (these are cylindrical drawings: the top is identified to the bottom).

The proof is based on an induction and an exhaustive analysis of all the loopless 2-vertex embedding schemes of genus 2 and 3, as pictured in Fig. 9. It is included in the full version [8].

In order to prove Theorem 2, our strategy is to first look for a fantastic cross-cap drawing for the reduced graph of an embedding scheme. Then, to obtain a drawing for the initial graph, we need to bring back the blocks that we replaced and extend the drawing to the edges of the block. This is achieved via the following blowing-up operation.

Blowing Up a Cross-Cap. Let ϕ be a fantastic drawing for G. By the definition of fantastic drawing, any reduced edge enters at least one cross-cap. Let e be a reduced edge that corresponded to a block X with frames a and b and let $\mathfrak{c}$ be a cross-cap in ϕ that e enters. By Lemma 3, X needs an odd number of cross-caps to be drawn, exactly $eg(X) + 1$. We replace $\mathfrak{c}$ by $eg(X) + 1$ cross-caps and we draw the frames of X, a and b as follows. We draw a following e thoroughly. To draw b we follow e except that we make b enter the new $eg(X) + 1$ cross-caps in the reversed order that a enters. All the other edges outside of X that were entering e are now drawn in the same way as a or b depending on how they were crossing e at $\mathfrak{c}$. Finally we remove e (see Fig. 10).

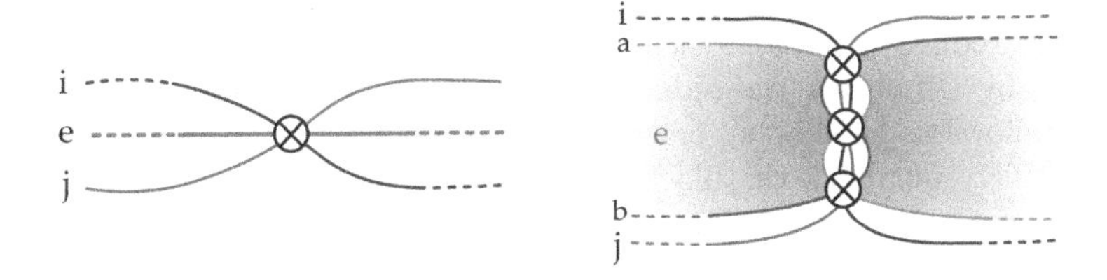

Fig. 10. Blowing up a cross-cap to draw the frames of a block

Repeatedly blowing up a drawing of a reduced permutation $\pi|$ yields a drawing for all edges of π except for the edges inside the blocks. The following lemma shows that these edges can be added in this cross-cap drawing.

Lemma 9. *Let π be a signed permutation on n elements such that all the elements form a non-trivial minimal negative block. The associated embedding scheme admits a perfect cross-cap drawing in which the frames of π, i.e. elements 1 and n, enter all the cross-caps but in opposite order.*

Proof. Let us denote by $g(\pi) = eg(\pi) + 1$ the non-orientable genus of the associated embedding scheme. We know that $\pi_1 = \overline{n}$ and $\pi_n = \overline{1}$. Let us define π' from π by replacing $\overline{1}$ by $n + 1$. The following lemma is proved in the full version [8] using the HP algorithm and Theorem 3.

Lemma 10. *The optimal number of reversals to go from π' to the permutation $(2, 3, \ldots, n + 1)$ is $g(\pi') = g(\pi)$. There exists a sequence of such reversals such that no reversal is applied on the element $n + 1$.*

Now, the sequence of reversals of Lemma 10 gives us a cross-cap drawing for π. By Lemma 1, the edges n and $n+1$ form an orienting cycle, and thus together they have to enter all the $g(\pi')$ cross-caps exactly once. We know that the edge $n+1$ does not enter any cross-cap in this drawing which implies that n enters all the cross-caps exactly once. We can obtain a cross-cap drawing for π from this drawing for π' by drawing 1 entering the cross-caps that n enters with the opposite order as depicted in Fig. 11. □

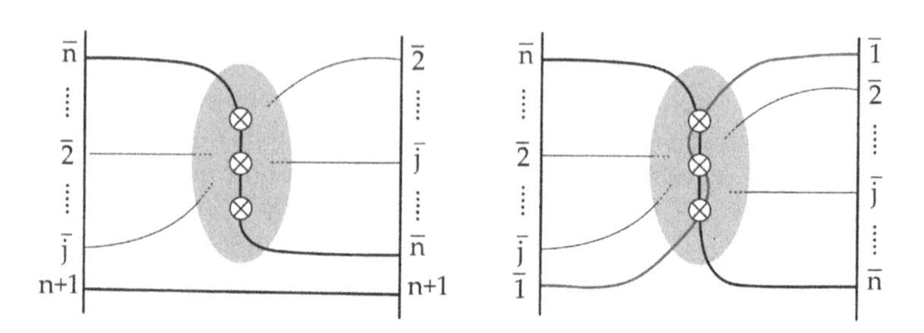

Fig. 11. From a cross-cap drawing of π' to a cross-cap drawing of π.

The idea of the proof of Theorem 2 is as follows. Starting with an embedding scheme, we reduce it. If we reach one of the graphs in Fig. 3, we are done. Otherwise, by Lemma 8, the reduced graph admits a fantastic drawing. Then Lemma 9 allows us to blow-up cross-caps while maintaining perfection, thus we can inductively put back the blocks and obtain a perfect cross-cap drawing of the initial graph. We refer to the full version [8] for the full proof.

Acknowledgements. We are very grateful to the anonymous reviewers for helpful comments. This work was partially supported by the ANR project SoS (ANR-17-CE40-0033).

References

1. Bafna, V., Pevzner, P.A.: Genome rearrangements and sorting by reversals. SIAM J. Comput. **25**(2), 272–289 (1996). https://doi.org/10.1109/SFCS.1993.366872
2. Bergeron, A.: A very elementary presentation of the Hannenhalli-Pevzner theory. In: Amir, A. (ed.) CPM 2001. LNCS, vol. pp, pp. 106–117. Springer, Heidelberg (2001). https://doi.org/10.1016/j.dam.2004.04.010
3. Bergeron, A., Mixtacki, J., Stoye, J.: Reversal distance without hurdles and fortresses. In: Sahinalp, S.C., Muthukrishnan, S., Dogrusoz, U. (eds.) CPM 2004. LNCS, vol. 3109, pp. 388–399. Springer, Heidelberg (2004). https://doi.org/10.1007/978-3-540-27801-6_29
4. Compeau, P., Pevzner, P.: Bioinformatics Algorithms: An Active Learning Approach. Active Learning Publishers La Jolla, California (2015)
5. Duncan, C.A., Goodrich, M.T., Kobourov, S.G.: Planar drawings of higher-genus graphs. J. Graph Algorithms Appl. **15**(1), 7–32 (2011). https://doi.org/10.1007/978-3-642-11805-0_7

6. Fertin, G., Labarre, A., Rusu, I., Vialette, S., Tannier, E.: Combinatorics of Genome Rearrangements. MIT press, Cambridge (2009)
7. Fuladi, N., Hubard, A., de Mesmay, A.: Short topological decompositions of non-orientable surfaces. In: 38th International Symposium on Computational Geometry (SoCG 2022). Schloss Dagstuhl-Leibniz-Zentrum für Informatik (2022). https://doi.org/10.4230/LIPIcs.SoCG.2022.41
8. Fuladi, N., Hubard, A., de Mesmay, A.: Degenerate crossing number and signed reversal distance (2023). https://arxiv.org/pdf/2308.10666v1.pdf
9. Hannenhalli, S., Pevzner, P.A.: Transforming cabbage into turnip: polynomial algorithm for sorting signed permutations by reversals. J. ACM (JACM) **46**(1), 1–27 (1999). https://doi.org/10.1145/300515.300516
10. Huang, F.W., Reidys, C.M.: A topological framework for signed permutations. Discret. Math. **340**(9), 2161–2182 (2017). https://doi.org/10.1016/j.disc.2017.03.019
11. Lazarus, F., Pocchiola, M., Vegter, G., Verroust, A.: Computing a canonical polygonal schema of an orientable triangulated surface. In: Proceedings of the Seventeenth Annual Symposium on Computational Geometry, pp. 80–89 (2001). https://doi.org/10.1145/378583.378630
12. Mohar, B.: The genus crossing number. ARS Math. Contemporanea **2**(2), 157–162 (2009). https://doi.org/10.26493/1855-3974.21.157
13. Mohar, B., Thomassen, C.: Graphs on Surfaces, vol. 10. JHU press, Baltimore (2001)
14. Pach, J., Tóth, G.: Degenerate crossing numbers. Discrete Comput. Geom. **41**(3), 376–384 (2009). https://doi.org/10.1007/s00454-009-9141-y
15. Schaefer, M., Štefankovič, D.: The degenerate crossing number and higher-genus embeddings. J. Graph Algorithms Appl. **26**(1), 35–58 (2022). https://doi.org/10.7155/jgaa.00580

Minimizing an Uncrossed Collection of Drawings

Petr Hliněný[1] and Tomáš Masařík[2(✉)]

[1] Faculty of Informatics of Masaryk University, Brno, Czech Republic
hlineny@fi.muni.cz

[2] Institute of Informatics, Faculty of Mathematics, Informatics and Mechanics, University of Warsaw, Warsaw, Poland
masarik@mimuw.edu.pl

Abstract. In this paper, we introduce the following new concept in graph drawing. Our task is to find a small collection of drawings such that they all together satisfy some property that is useful for graph visualization. We propose investigating a property where each edge is not crossed in at least one drawing in the collection. We call such collection *uncrossed*. This property is motivated by a quintessential problem of the crossing number, where one asks for a drawing where the number of edge crossings is minimum. Indeed, if we are allowed to visualize only one drawing, then the one which minimizes the number of crossings is probably the neatest for the first orientation. However, a collection of drawings where each highlights a different aspect of a graph without any crossings could shed even more light on the graph's structure.

We propose two definitions. First, the *uncrossed number*, minimizes the number of graph drawings in a collection, satisfying the uncrossed property. Second, the *uncrossed crossing number*, minimizes the total number of crossings in the collection that satisfy the uncrossed property. For both definitions, we establish initial results. We prove that the uncrossed crossing number is NP-hard, but there is an FPT algorithm parameterized by the solution size.

Keywords: Crossing Number · Planarity · Thickness · Fixed-parameter Tractability

1 Introduction

Determining the *crossing number* $\mathrm{cr}(G)$ of a graph G, i.e. the smallest possible number $k = \mathrm{cr}(G)$ of pairwise transverse intersections (called *crossings*) of edges in any drawing of G is among the most important theoretical problems in graph drawing. Its computational complexity is well-researched. It is known that graphs with crossing number $k = 0$, i.e. planar graphs, can be easily recognized. Computing the crossing number k of a graph is NP-hard [16], even in very restricted

The full version of this paper is available on arXiv [27].

T. M. was supported by Polish National Science Centre SONATA-17 grant number 2021/43/D/ST6/03312.

M. A. Bekos and M. Chimani (Eds.): GD 2023, LNCS 14465, pp. 110–123, 2023.
https://doi.org/10.1007/978-3-031-49272-3_8

settings of cubic graphs [25], of a fixed rotation system [33] and of near-planar graphs [7], It is also APX-hard [6]. However, there is a fixed-parameter algorithm for the problem [19] with respect to the number of crossings k, and even one that can solve the problem in linear time $\mathcal{O}(f(k)n)$ [28].

More recently, various graph drawing extension problems motivated by graph and network visualization received increased attention. In relation to the traditional crossing number problem, we in particular, mention works on the edge insertion problem [8,9,20] and on the crossing number in a partially predrawn setting [15,24]. Such drawing extension solutions can be useful, for instance, in situations in which a particular region or a feature of a graph should be highlighted, and the rest of the drawing is then completed as nicely as possible while respecting the highlighted part. It is, however, more common that we want to somehow nicely display every part of the given graph. That is, of course, problematic with only one drawing, but what if we are allowed to produce several drawings of the same graph or network at once? On the experimental side, the same idea was actually raised by Biedl, Marks, Ryall, and Whitesides [5] quite long time ago, but does not seem to be actively pursued nowadays.

Motivated by the latter thoughts, this paper proposes and studies the following concept of looking at graph drawings. Instead of finding a single "flawless" drawing (in the plane), we look for a small collection of drawings that fulfill some visualization property(ies) "together", meaning in their union. In that way, each drawing in the collection spotlights a different part of the graph's structure. This seems to be a promising concept as minimizing just one property of a drawing globally, which is traditionally done, might produce a drawing that is nowhere nice locally. Therefore, having to produce only one solution may lead to visualizations that are not as easy to understand.

In the proposed approach, we aim to produce a (smallest) collection of drawings of a graph that all together admit a property useful for visualizing the graph structure. There might be several different sensible candidates for the property. Inspired by the crossing number, we consider a property that requires each edge not to be crossed in at least one drawing in the collection. Therefore, we produce a collection of drawings that emphasize different aspects of a graph in different drawings. Formally, we define the uncrossed number of a graph as follows (see Sect. 2 for the necessary standard definitions).

Definition 1.1 (Uncrossed number). *Let $G = (V, E)$ be a graph. A family of drawings $D_1, \ldots, D_k$ of G, such that each edge $e \in E$ is not crossed in some drawing D_i, $1 \leq i \leq k$, is called an* uncrossed collection of drawings *of G. The* uncrossed number *of a graph G, denoted by* $\mathrm{unc}(G)$, *is defined as the least cardinality k of an uncrossed collection of drawings of G.*

The uncrossed number is related to the traditional *thickness* of a graph (the thickness of G ($\theta(G)$) is the least number of planar subgraphs of G whose union is whole G). Obviously, the thickness of G is at most $\mathrm{unc}(G)$, and G is of thickness 1 (planar) if and only if $\mathrm{unc}(G) = 1$. On the other hand, the uncrossed number is upper bounded by *outerthickness* of a graph (the outerthickness of G ($\theta_o(G)$) is the least number of outerplanar subgraphs of G whose union is whole G). Indeed, we can draw each of the outerplanar graphs without crossings and draw all the other edges of G inside the outerface. It is valuable to know that a planar graph can be edge partitioned into two outerplanar graphs as proven by Gonçalves in 2005 [18]. We summarize all the relations above; For any G:

$$\theta(G) \leq \mathrm{unc}(G) \leq \theta_o(G) \leq 2\theta(G).$$

To give a bit more intuition, we argue in Proposition 3.1 that $\mathrm{unc}(K_7) = 3$ which is the same bound as in the outerthickness, $\theta_o(K_7) = 3$ [21], while indeed $\theta(K_7) = 2$.

Thickness is a very well-studied graph parameter; consult an initial survey [32] or more recent papers [12,13] for further details. It used to be on Garey & Johnson's list of open problems but was proven to be NP-complete already in 1982 by Mansfield [31]. Surprisingly, the complexity of outerthickness has yet to be uncovered. A recent paper by Xu and Zha [42] provides bounds based on the genus of the graph, but in many other aspects, outerthickness still needs to be explored.

Analogously as the uncrossed number is an alternative counterpart of planarity in our approach, the graph crossing number has a natural counterpart in the total crossing number of an uncrossed collection of drawings. Informally, the task is to find a collection of drawings satisfying the uncrossed property such that it has the smallest number of crossings summed over the collection.

Definition 1.2 (Uncrossed crossing number). *Let $G = (V, E)$ be a graph and $k \geq 1$ an integer. The* uncrossed crossing number in k drawings *of a graph G, denoted by $\mathrm{ucr}_k(G)$, equals the least sum $\mathrm{cr}(D_1)+\cdots+\mathrm{cr}(D_{k'})$ over all uncrossed collections $D_1, \ldots, D_{k'}$ of drawings of G where $k' \leq k$, or $\mathrm{ucr}_k(G) = \infty$ if no uncrossed collection exists.*

The uncrossed crossing number *of G is then defined as $\mathrm{ucr}(G) := \min_k \mathrm{ucr}_k(G)$. The least k for which $\mathrm{ucr}(G) = \mathrm{ucr}_k(G)$ is called the* crossing-optimal uncrossed number $\mathrm{ounc}(G)$.

Observe that $\mathrm{ucr}(G)$ is a well-defined integer since there always exists an uncrossed collection of $k = |E(G)|$ drawings of G. Furthermore, $\mathrm{unc}(G) \leq \mathrm{ounc}(G)$, and $\mathrm{ounc}(G) \leq |E(G)|$ since every uncrossed collection of $> |E(G)|$ drawings obviously contains a strict uncrossed subcollection. On the other hand, one can easily construct a family of graphs G such that $\mathrm{unc}(G) = 2$ and $\mathrm{ounc}(G)$ is unbounded; see Proposition 3.2 and Fig. 1. We also have obvious $\mathrm{ucr}(G) \geq \mathrm{ounc}(G) \cdot \mathrm{cr}(G)$, but the gap in this inequality can be arbitrarily large—consider, e.g., the complete graph K_5 with every edge except two disjoint ones blown up to many parallel edges, as in Proposition 3.3.

We also provide some useful estimates on the uncrossed crossing number. We, for instance, give the asymptotics of the uncrossed crossing number for the complete graph in Proposition 3.4, and we give an asymptotically tight analogy of the classical Crossing Lemma for the uncrossed crossing number of graphs with sufficiently many edges in Theorem 3.5.

As discussed, the crossing number is NP-hard even in various restricted settings. We prove that the uncrossed crossing number is NP-complete in Sect. 4 building on ideas of Hliněný and Derňár [26]. We complement this result by presenting an FPT algorithm parameterized by its size in Sect. 5. This is analogous to Grohe's FPT algorithm for the crossing number parameterized by its size [19].

2 Preliminaries

In this paper, we consider multigraphs by default, i.e., our graphs are allowed to have multiple edges (while loops are irrelevant here), with the understanding that we can always subdivide parallel edges without changing the crossing number.

A *drawing* D of a graph G in the Euclidean plane $\mathbb{R}^2$ is a function that maps each vertex $v \in V(G)$ to a distinct point $D(v) \in \mathbb{R}^2$ and each edge $e = uv \in E(G)$ to a simple open curve $D(e) \subset \mathbb{R}^2$ with the ends $D(u)$ and $D(v)$. We require that $D(e)$ is disjoint from $D(w)$ for all $w \in V(G) \setminus \{u, v\}$. Throughout the paper, we will moreover assume that: there are finitely many points that are in an intersection of two edges, no more than two edges intersect in any single point other than a vertex, and whenever two edges intersect in a point, they do so transversally (i.e., not tangentially).

An intersection (a common point) of two edges is called a *crossing* of these edges. A drawing D is *planar* (or a *plane graph*) if D has no crossings, and a graph is *planar* if it has a planar drawing. The number of crossings in a drawing $\mathcal{G}$ is denoted by $\mathrm{cr}(D)$. The (ordinary) *crossing number* $\mathrm{cr}(G)$ *of* G is defined as the minimum of $\mathrm{cr}(D)$ over all drawings D of G.

The following is a useful artifice in crossing numbers research. In a *weighted* graph, each edge is assigned a positive integer (the *weight* of the edge). Now the *weighted crossing number* is defined as the ordinary crossing number, but a crossing between edges e_1 and e_2, say of weights t_1 and t_2, contributes the product $t_1 \cdot t_2$ to the weighted crossing number. For the purpose of computing the crossing number, an edge of integer weight t can be equivalently replaced by a bunch of t parallel edges of weights 1; this is since we can easily redraw every edge of the bunch tightly along the "cheapest" edge of the bunch. The same argument holds also with our new Definition 1.2.

3 Basic Properties

To give the readers a better feeling about the new concept(s) before moving onto the computational-complexity properties, we first present some basic properties of the uncrossed and the uncrossed crossing number.

We start by inspecting the uncrossed number.

Proposition 3.1. *We have* $\mathrm{unc}(K_7) = 3$.

Proof. We say that drawing *realizes* an edge if it is not crossed in that drawing. Suppose that drawings D_1 and D_2 exist. Let D_1 be the drawing that realizes at least as many edges as D_2. As a graph on the uncrossed edges needs to be planar, the number of realized edges in D_1 ranges from 11 (As K_7 has 21 edges in total) to 15 $(3 \cdot 7 - 6)$. Consider faces that are defined by edges that are not crossed in drawing D_i, $i \in \{1, 2\}$. We call them *uncrossed faces* of D_i and we denote by D'_i the subdrawing of D_i consisting of all uncrossed faces.

As we draw K_7, for each pair of vertices, there needs to be a common uncrossed face containing both. We call it the *neighborhood property.*

Claim. * [1] Drawing D'_1 consists of the outerface with at most one vertex inside. We denote such vertex *central* (in drawing D'_1). Moreover, the central vertex (if it exists) is contained in all other faces except for the outerface.

The claim leaves only three possibilities for D'_1;

Case 1: D'_1 form an outerplanar triangulation drawing (case of 11 realized edges), but due to $\theta_o(K_7) = 3$ [21] D'_2 could not be outerplanar in this case,
Case 2: D'_1 has one central vertex contained in 5 uncrossed triangles, one uncrossed 4-cycle, and the outerface has length 6 (case of 11 realized edges),
Case 3: D'_1 has one central vertex contained in 6 uncrossed triangles, and the outerface has length 6 (case of 12 realized edges).

In Cases 2 and 3, we need to realize at least three edges of each vertex except for the central one v_c (of D'_1). Moreover, in D'_2, we need to put v_c to one face, even though we do not need to realize more than one of its edges, say that this will be the outerface. As all the other vertices are on a common face, their realized edges form an outerplanar graph on six vertices. Therefore, there are two vertices of degree 2 in D'_2 which contradicts the option that for each of them, we need to realize three edges.

In Case 1, we already know that D'_2 cannot be outerplanar. As D'_1 is outerplanar there has to be two non-adjacent vertices v_1 and v_2 that each realize 4 edges in D_2. This implies that there are at least three vertices as their common neighbors in D'_2 and thus they create three separate faces. No matter which face we put the remaining vertices in they will be separated from the vertex not on that face. □

We continue with a few easy facts about the uncrossed crossing number.

Proposition 3.2. * *For every integer m there exists a graph G on $2m$ vertices such that* $\mathrm{unc}(G) = 2$ *and* $\mathrm{ounc}(G) \geq m$ *(see Fig. 1).*

Proposition 3.3. * *For every integer $m \geq 3$ there exists a graph G such that* $\mathrm{cr}(G) = 1$ *and* $\mathrm{ucr}(G) = 2m$. *Namely, G can be obtained from the graph K_5 by selecting two disjoint edges and making every other edge "heavy" of weight m.*

[1] Proofs marked by * are available in the appendix of [27].

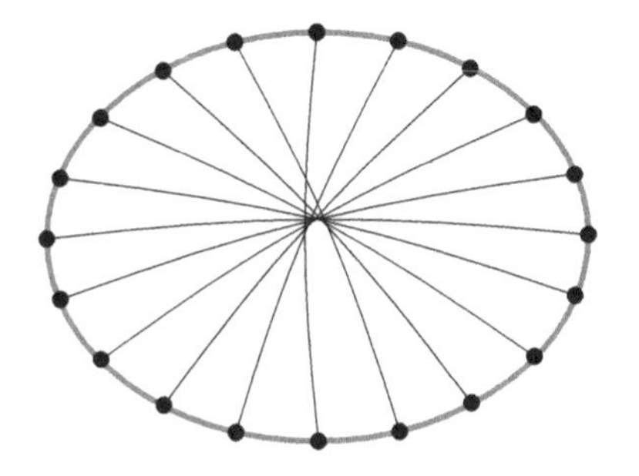
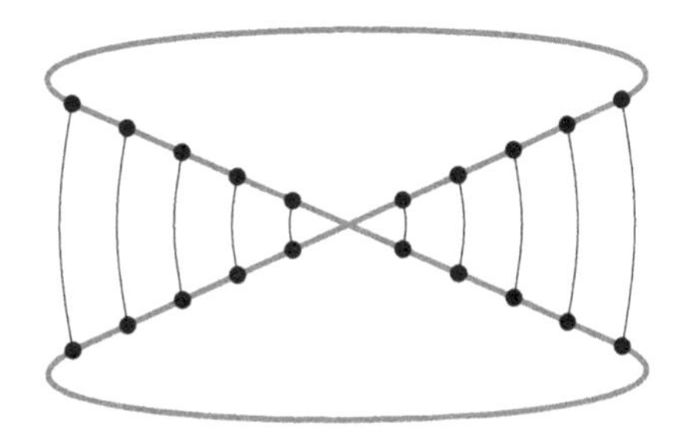

Fig. 1. A illustration of the graph G, for $m = 10$, from the proof of Proposition 3.2; this graph has $\mathrm{unc}(G) = 2$ based on an uncrossed collection of two drawings which both look as depicted on the right.

With respect to the next claim, we remark that the exact value of $\mathrm{cr}(K_n)$ for every n is still an open problem, although the right asymptotic $\mathrm{cr}(K_n) \in \Theta(n^4)$ follows directly from the famous Crossing Lemma [2,30].

Proposition 3.4. * *For every $n \geq 5$, we have $\frac{n}{6} \cdot \mathrm{cr}(K_n) \leq \mathrm{ucr}(K_n) \leq \frac{n^5}{96}$. In particular, $\mathrm{ucr}(K_n) \in \Theta(n^5)$.*

The simple lower-bound proof from Proposition 3.4 can be easily generalized with the help of the aforementioned Crossing Lemma, and the generalization is asymptotically tight by the example of K_n in Proposition 3.4.

Theorem 3.5. *Let G be a simple graph such that $|E(G)| \geq 7|V(G)|$. Then $\mathrm{ucr}(G) \geq |E(G)|^4/(87 \cdot |V(G)|^3)$.*

Proof. Let $D_1, \ldots, D_k$ be any uncrossed collection of drawings of G. By the currently best variant of the Crossing Lemma – in Ackerman [1], we for all $i = 1, \ldots, k$ have $\mathrm{cr}(D_i) \geq |E(G)|^3/(29 \cdot |V(G)|^2)$. Moreover, since a simple graph G can have at most $3|V(G)| - 6$ uncrossed edges in any plane drawing, we get $k \geq |E(G)|/(3|V(G)| - 6) \geq |E(G)|/(3|V(G)|)$, which concludes the universal lower bound $\mathrm{cr}(D_1) + \ldots + \mathrm{cr}(D_k) \geq |E(G)|^4/(87 \cdot |V(G)|^3)$. □

4 Hardness Reduction

Our hardness proof for the uncrossed crossing number builds on the ideas of [26] (Theorem 4.1), which in turn is based on [7].

In the context of [35] we define a *tile* $T = (H, a, b, c, d)$ where H is a graph and a, b, c, d is a sequence of distinct vertices, called here the *corners* of the tile. The *inverted* tile $T^{\updownarrow}$ is the tile (H, a, b, d, c). A tile T is *perfectly connected* if both H and the subgraph $H - \{a, b, c, d\}$ are connected and no edge of H has both ends in $\{a, b, c, d\}$. A *tile drawing* of a tile $T = (H, a, b, c, d)$ is a drawing of the underlying graph H in the unit square such that the vertices a, b, c, d are the upper left, lower left, lower right, and upper right corner, respectively. The

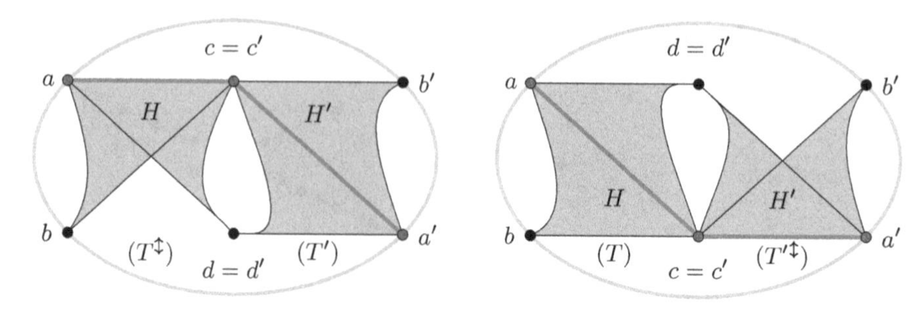

Fig. 2. A sketch of an uncrossed collection of two drawings of the graph G in the proof of Theorem 4.2; the heavy outer cycle C and the many a–c and a'–c' paths are outlined in thick gray. Only edges of H are crossed on the left and only edges of H' on the right. If $\mathrm{tcr}(T^{\updownarrow}) \leq k$, then $\mathrm{ucr}(G) \leq \mathrm{ucr}_2(G) \leq 2k$.

tile crossing number $\mathrm{tcr}(T)$ of a tile T is the minimum number of crossings over all tile drawings of T. A tile T is *planar* if $\mathrm{tcr}(T) = 0$.

Theorem 4.1 ([26, **Definition 9 and Corollary 12**]). *The following problem is* NP*-hard: given an integer k, and a perfectly connected planar tile $T = (H, a, b, c, d)$ such that there exist $2k+1$ edge-disjoint paths between the vertices a and c in H and the vertex d is of degree 1, decide whether* $\mathrm{tcr}(T^{\updownarrow}) \leq k$.

Theorem 4.2. *Given an integer m and a graph G, the problems to decide whether* $\mathrm{ucr}(G) \leq m$ *and whether* $\mathrm{ucr}_2(G) \leq m$ *are both* NP*-complete. This holds even if G contains an edge $f \in E(G)$ such that $G - f$ is planar.*

Proof. Let $T = (H, a, b, c, d)$ be a tile as expected in Theorem 4.1, from which we reduce, and let $T' = (H', a', b', c', d')$ be a copy of T. We take a 4-cycle C on the vertices (a, b, a', b') in this order, such that each edge of C is of weight $2k+1$, or equivalently composed of $2k+1$ parallel edges, and make G as the union $C \cup H \cup H'$ with the identification $c = c'$ and $d = d'$. We may pick f as one of the two edges incident to d and see that $G - f$ is planar. Let $m = 2k$.

If $\mathrm{tcr}(T^{\updownarrow}) \leq k$, then we have a drawing D_1 of G such that; C is uncrossed, H is drawn as in the inverted tile $T^{\updownarrow}$ with $\leq k$ crossings and H' is drawn as the planar tile T' with no crossings. See Fig. 2. We let D_2 be the drawing symmetric to D_1, with H' drawn as in the inverted tile $T'^{\updownarrow}$ and H without crossings. The collection $\{D_1, D_2\}$ certifies that $\mathrm{ucr}(G) \leq \mathrm{ucr}_2(G) \leq k + k = m$.

On the other hand, assume an uncrossed collection of at least two drawings of G with at most m crossings in total. Then one of these drawings, say D_1, has at most $m/2 = k$ crossings. In particular, the cycle C of weight $2k+1$ must be uncrossed (as a crossing "hits" at most two of the parallel edges composing each edge of C), and so C makes the boundary of a valid tile drawing of connected $(H \cup H', a, b, a', b')$. Furthermore, by the assumption of Theorem 4.1 on paths from a to c, there exist $2k+1$ edge-disjoint paths from a to a' in $H \cup H'$, and again there is an a–a' path P without crossings on it in D_1.

Consider now the position of the vertex $d = d'$ with respect to the uncrossed path P in D_1; up to symmetry between H and H', let $d = d'$ lie "below" P (with respect to having C depicted with the vertices a and b' at the top, and b, a' at the bottom, as in Fig. 2 left). Let $P' = P \cap H'$. Let $Q' \subseteq H'$ be a shortest path from $d = d'$ to $V(P')$, which exists in a perfectly connected tile (we do not require Q' to be crossing-free). We now modify the drawing D_1 into D_1' as follows; we delete the subdrawing of H' except $P' \cup Q'$ from D_1, then we contract the crossing-free path P' making $c = a'$, and add a new vertex d'' in a close neighbourhood of a' such that d'' is adjacent only to d by an edge drawn along Q' (and then Q' is deleted). This does not introduce any more crossings than existed in D_1. Since d was of degree 1 in H and is of degree 2 in D_1', we may now simply identify $d = d''$ by prolonging the single edge of d in H. By the assumption of d being "below" P, we get that the cyclic order of the tile vertices in D_1' is now (a, b, d, c), and so D_1' is a valid tile drawing of $(H, a, b, d, c) = T^{\updownarrow}$ with at most $\mathrm{cr}(D_1') \leq \mathrm{cr}(D_1) \leq k$ crossings.

Summarizing the previous, $\mathrm{ucr}(G) \leq m$ or $\mathrm{ucr}_2(G) \leq m$ implies $\mathrm{tcr}(T^{\updownarrow}) \leq \mathrm{cr}(D_1) \leq k$. On the other hand, membership in NP is shown in the standard way; we guess an uncrossed collection of drawings, imagine every crossing point turned into a new vertex subdividing the participating edges, and test planarity. This completes the proof. □

5 FPT Algorithm

We now give the second main result, that computing the uncrossed crossing number is fixed-parameter tractable in the solution value.

Theorem 5.1. *There is a quadratic* FPT*-time algorithm with an integer parameter k that, given a graph G and an integer c, decides whether* $\mathrm{ucr}_c(G) \leq k$.

We first remark that only values of $c \leq k$ in Theorem 5.1 are meaningful since, having an uncrossed collection of $c > k$ drawings with the total number of crossings k, implies that one of the drawings is planar, and so is G, which can be tested beforehand. In particular, $\mathrm{ucr}(G) \leq k \iff \mathrm{ucr}_k(G) \leq k$. Furthermore, if $\mathrm{ucr}(G) \leq k$ in Theorem 5.1, then we can straightforwardly compute $\mathrm{ounc}(G)$.

Theorem 5.1 is analogous to the FPT algorithm of Grohe [19] for the classical crossing number, and our proof is on a high level very similar to [19]. However, the technical details are different, not only because we need a slightly different formulation of the problem itself, but also since the grid-reduction phase of the algorithm [19] uses annotation with uncrossable edges which is not suitable when working with a collection of many drawings.

Let a *hexagonal graph* be a simple plane graph H such that all faces of H are of length 6, all vertices of degree 2 or 3, and the vertices of degree 2 occur only on the outer face of H. A hexagonal graph H is called a *hexagonal r-grid* (also a "wall") if the following holds: there are pairwise-disjoint cycles $C_1, \ldots, C_r \subseteq H$, called the *principal cycles*, such that $V(H) = V(C_1) \cup \cdots \cup V(C_r)$, the cycle C_1 bounds a hexagonal face, and for $i = 2, \ldots, r$ the cycle C_i contains C_{i-1} in its

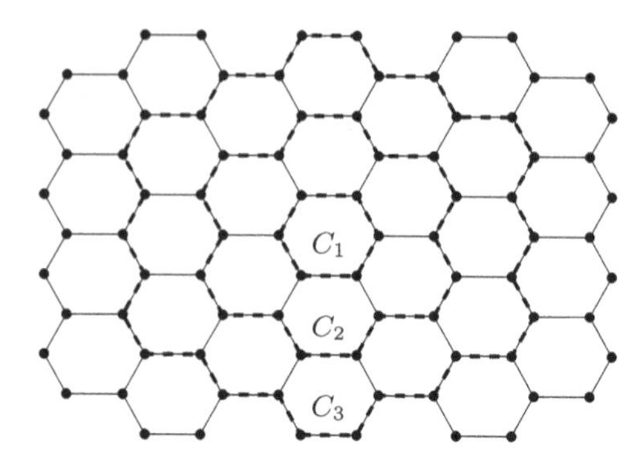

Fig. 3. A fragment of a hexagonal graph, and the principal cycles C_1, C_2, C_3 (thick-dashed) defining a hexagonal 3-grid.

interior (they are nested from outermost C_r to innermost C_1). See Fig. 3, and note that such H is unique up to isomorphism for each r.

More generally, given a graph G, we call the hexagonal r-grid H a *hexagonal r-grid in G*, if there is a subgraph $H_1 \subseteq G$ (also called a *grid in G*) which is isomorphic to a subdivision of H. We canonically extend the meaning of principal cycles to H_1. The interior vertices of H are those not on the outermost principal cycle, and the *interior vertices of H_1* are those which are the interior vertices of H or subdivisions of edges of H incident to interior vertices of H. A grid $H_1 \subseteq G$ is *flat* if the subgraph of G induced on $V(H_1)$ and the vertices of all connected components of $G - V(H_1)$ adjacent to (some) interior vertices of H_1 is planar.

We proceed with the following steps.

Theorem 5.2 ([10,34,37]).* *For any $r \in \mathbb{N}$ there is a linear* FPT*-time algorithm parameterized by r, which given a graph G outputs either a tree decomposition of G of width at most $4^{10}r^{10}$, or a hexagonal r-grid in G.*

Lemma 5.3 ([19,39]). *For $r_1, k \in \mathbb{N}$ and a graph G we can in linear time compute the following: Given a hexagonal r-grid in G, where $r = 400r_1\lceil\sqrt{k}\rceil$, either output a flat hexagonal r_1-grid in G, or answer truthfully that $\mathrm{cr}(G) \geq k+1$.*

Lemma 5.4 ([17]).* *For $k \in \mathbb{N}$, let a graph G contain a flat hexagonal $(4k+4)$-grid $H \subseteq G$, and let $e \in E(H)$ be an edge of the innermost principal cycle of H. If there is a drawing D_1 of $G - e$ such that $\mathrm{cr}(D_1) \leq k$, then there is a drawing D of G such that every crossing of D exists also in D_1. (In particular, $\mathrm{cr}(D) \leq \mathrm{cr}(D_1)$ and e is not crossed in D.)*

For the next lemma, we briefly mention that MSO_2 logic of graphs is logic dealing with multigraphs as two-sorted relational structures with vertices, edges and their incidence relation. In addition to standard propositional logic, MSO_2 quantifies over vertices and edges and their sets. We write $H \models \varphi$ to mean that formula φ evaluates to true on H. For a graph H, we denote by $H^{\bullet s}$ the graph obtained from H by subdividing every edge with s new vertices. Closely following the ideas of Grohe [19], we prove:

Lemma 5.5. * *For $k, c \in \mathbb{N}$ such that $c \leq k$, there is a formula $\varphi_{c,k}$ in the MSO$_2$ logic, such that for every graph H we have $\mathrm{ucr}_c(H) \leq k$ if and only if $H^{\bullet k} \models \varphi_{c,k}$.*

Now, we can finish the main task.

Proof (of Theorem 5.1). Let $r := 400r_1\lceil\sqrt{k}\rceil$ where $r_1 := 4k+4$, and $G_1 := G$. Our algorithm first repeats the following until the listed treewidth condition on G_1 is met.

i. Call the algorithm of Theorem 5.2 to compute a hexagonal r-grid $H_0 \subseteq G_1$, unless the tree-width of G_1 is at most $4^{10}r^{10}$. Then apply Lemma 5.3 onto H_0 to compute a flat hexagonal r_1-grid $H_1 \subseteq G_1$, unless $\mathrm{cr}(G_1) \geq k+1$.
ii. Pick $e \in E(H_1)$ belonging to the innermost principal cycle of H_1 arbitrarily and delete it, $G_1' = G_1 - e$. Trivially, $\mathrm{ucr}_c(G_1') \leq \mathrm{ucr}_c(G_1)$, and from Lemma 5.4 it follows that $\min(\mathrm{ucr}_c(G_1'), k+1) = \min(\mathrm{ucr}_c(G_1), k+1)$: if $\mathrm{ucr}_c(G_1') \leq k$, every drawing of the collection witnessing $\mathrm{ucr}_c(G_1')$ has at most k crossings, and so Lemma 5.4 is applicable and gives an (again) uncrossed collection of drawings of G_1, which implies $\mathrm{ucr}_c(G_1) \leq \mathrm{ucr}_c(G_1')$.
iii. Continue to (i.) with $G_1 := G_1'$, i.e., removing the edge e.

When the first routine is finished, we get $G_1 \subseteq G$ such that;

- $\mathrm{cr}(G_1) \geq k+1$, and hence $\mathrm{cr}(G) \geq k+1$ directly implying $\mathrm{ucr}_c(G) \geq k+1$,
- or the tree-width of G_1 is bounded by $4^{10}r^{10}$, and $\min(\mathrm{ucr}_c(G_1), k+1) = \min(\mathrm{ucr}_c(G), k+1)$.

In the latter case, we apply classical Courcelle's theorem [11] to decide whether $G_1^{\bullet k} \models \varphi_{c,k}$ for the formula of Lemma 5.5, using the tree decomposition of G_1 computed by Theorem 5.2 and made into a tree decomposition of $G_1^{\bullet k}$. This is the sought answer since, by previous, we have $\mathrm{ucr}_c(G) \leq k \iff \mathrm{ucr}_c(G_1) \leq k$.

Regarding runtime, we perform at most $|E(G)|$ iterations in the first stage, and each takes linear FPT-time. In the second stage we spend again linear FPT-time, since testing $G_1^{\bullet k} \models \varphi_{c,k}$ is in linear FPT with respect to the treewidth and the formula size as parameters. □

Remark 5.6. One can easily adapt Theorem 5.2 to actually compute a collection of drawings witnessing $\mathrm{ucr}_c(G) \leq k$. First, the constructive version of Courcelle's theorem, see [29], run on $G_1 \models \varphi_{c,k}$ computes witnesses of the positive answer in the form of a specification of which pairs of edges of G_1 cross and in which order in a collection giving $\mathrm{ucr}_c(G_1) \leq k$. By Lemma 5.4, only these crossings from G_1 are relevant for the corresponding uncrossed collection of drawings of G, and hence we simply turn these crossings of G_1 into new vertices in G, and apply any standard planarity algorithm.

6 Conclusions

Our work brings an interesting connection between two significantly studied concepts in the graph drawing area; the crossing number and the (outer)thickness of graphs. On the side closer to the classical crossing number, we suggest the uncrossed crossing number, and closer to the thickness, we propose the uncrossed number. The initial results show that the uncrossed crossing number behaves quite similarly to the classical crossing number—and hence it seems similarly hard to understand it in full generality. This may spark further interest in combinatorics research, and we hope that particular aspects of the uncrossed crossing number may be found useful in graph drawing.

In the combinatorial direction, for instance, it would be interesting to understand the structure of graphs critical for the uncrossed crossing number. The results of Sect. 5, after an easy modification, also imply that these critical graphs do not contain large grids, and so have bounded tree-width. Hence, one should ask whether critical graphs for the uncrossed crossing number resemble the already known asymptotic structure of classical crossing-critical graphs [14].

On the other hand, our understanding of the uncrossed number is much more fuzzy and difficult to understand. As we discussed, it can be 2-approximated by the outerthickness, but more detailed behavior is yet to be comprehended. Therefore, based on our initial observation, we conjecture several hypotheses regarding this number.

We conjecture that the uncrossed number is NP-hard (probably even para NP-hard). As the NP-hardness of the outerthickness is surprisingly still unknown, exploring a related notion may give us more understanding to attack this problem as well. Probably even more challenging seems the understanding of the complexity of the crossing-optimal uncrossed number.

As the precise bound for thickness [3,4,40] and outerthickness [21,22] of complete and complete bipartite graphs have been determined (almost[2]) completely, we suggest exploring it for the uncrossed number. For simple graphs, we conjecture that except for the planar but not outerplanar cases, $\mathrm{ounc}(K_n) = \mathrm{unc}(K_n) = \theta_o(K_n)$ and $\mathrm{ounc}(K_{m,n}) = \mathrm{unc}(K_{m,n}) = \theta_o(K_{m,n})$.

Another interesting direction is determining precise upper and lower bounds based on the minimum/maximum degree, as was done for outerthickness [36] and thickness [23,38,41].

On the big picture, we proposed a general meta-problem that might highlight various aspects of a graph by providing a collection of drawing such that a particular property is met at least in one drawing in the collection. In this paper, we proposed the first two(three) specific problems that capture this initial motivation, both naturally extending a seminal notion of the crossing number. We would like to see other possible approaches of what property one could impose on a collection of drawings that provide us with some new perspectives on the graph's structure.

---

[2] There are some possible exceptions in case of $\theta(K_{m,n})$ where the exact value is not known; see [36] for further discussion.

References

1. Ackerman, E.: On topological graphs with at most four crossings per edge. Comput. Geom. **85** (2019). https://doi.org/10.1016/j.comgeo.2019.101574
2. Ajtai, M., Chvátal, V., Newborn, M.M., Szemerédi, E.: Crossing-free subgraphs. In: Hammer, P.L., Rosa, A., Sabidussi, G., Turgeon, J. (eds.) Theory and Practice of Combinatorics, North-Holland Mathematics Studies, vol. 60, pp. 9–12. North-Holland (1982). https://doi.org/10.1016/S0304-0208(08)73484-4
3. Alekseev, V.B., Gončakov, V.S.: The thickness of an arbitrary complete graph. Math. USSR-Sbornik **30**(2), 187–202 (1976). https://doi.org/10.1070/sm1976v030n02abeh002267
4. Beineke, L.W., Harary, F., Moon, J.W.: On the thickness of the complete bipartite graph. Math. Proc. Cambridge Philos. Soc. **60**(1), 1–5 (1964). https://doi.org/10.1017/S0305004100037385
5. Biedl, T., Marks, J., Ryall, K., Whitesides, S.: Graph multidrawing: finding nice drawings without defining nice. In: Whitesides, S.H. (ed.) GD 1998. LNCS, vol. 1547, pp. 347–355. Springer, Heidelberg (1998). https://doi.org/10.1007/3-540-37623-2_26
6. Cabello, S.: Hardness of approximation for crossing number. Discrete Comput. Geom. **49**(2), 348–358 (2012). https://doi.org/10.1007/s00454-012-9440-6
7. Cabello, S., Mohar, B.: Adding one edge to planar graphs makes crossing number and 1-planarity hard. SIAM J. Comput. **42**(5), 1803–1829 (2013). https://doi.org/10.1137/120872310
8. Chimani, M., Hliněný, P.: A tighter insertion-based approximation of the crossing number. In: Aceto, L., Henzinger, M., Sgall, J. (eds.) ICALP 2011. LNCS, vol. 6755, pp. 122–134. Springer, Heidelberg (2011). https://doi.org/10.1007/978-3-642-22006-7_11
9. Chimani, M., Hliněný, P.: Inserting multiple edges into a planar graph. In: SoCG. LIPIcs, vol. 51, pp. 30:1–30:15. Schloss Dagstuhl - Leibniz-Zentrum für Informatik (2016). https://doi.org/10.4230/LIPIcs.SoCG.2016.30
10. Chuzhoy, J., Tan, Z.: Towards tight(er) bounds for the excluded grid theorem. J. Comb. Theory, Ser. B **146**, 219–265 (2021). https://doi.org/10.1016/j.jctb.2020.09.010
11. Courcelle, B.: The monadic second-order logic of graphs. I. recognizable sets of finite graphs. Inf. Comput. **85**(1), 12–75 (1990). https://doi.org/10.1016/0890-5401(90)90043-H
12. Dujmović, V., Wood, D.R.: Thickness and antithickness of graphs. J. Comput. Geom. **9**(1) (2018). https://doi.org/10.20382/JOCG.V9I1A12
13. Duncan, C.A.: On graph thickness, geometric thickness, and separator theorems. Comput. Geom. **44**(2), 95–99 (2011). https://doi.org/10.1016/j.comgeo.2010.09.005, special issue of selected papers from the 21st Annual Canadian Conference on Computational Geometry
14. Dvořák, Z., Hliněný, P., Mohar, B.: Structure and generation of crossing-critical graphs. In: SoCG. LIPIcs, vol. 99, pp. 33:1–33:14. Schloss Dagstuhl - Leibniz-Zentrum für Informatik (2018). https://doi.org/10.4230/LIPIcs.SoCG.2018.33
15. Ganian, R., Hamm, T., Klute, F., Parada, I., Vogtenhuber, B.: Crossing-optimal extension of simple drawings. In: Bansal, N., Merelli, E., Worrell, J. (eds.) ICALP 2021. LIPIcs, vol. 198, pp. 72:1–72:17. Schloss Dagstuhl - Leibniz-Zentrum für Informatik (2021). https://doi.org/10.4230/LIPIcs.ICALP.2021.72

16. Garey, M.R., Johnson, D.S.: Crossing number is NP-complete. SIAM J. Algebr. Discrete Meth. **4**(3), 312–316 (1983). https://doi.org/10.1137/060403
17. Geelen, J.F., Richter, R.B., Salazar, G.: Embedding grids in surfaces. Eur. J. Comb. **25**(6), 785–792 (2004). https://doi.org/10.1016/j.ejc.2003.07.007
18. Gonçalves, D.: Edge partition of planar graphs into two outerplanar graphs. In: Proceedings of the Thirty-Seventh Annual ACM Symposium on Theory of Computing, pp. 504–512. STOC 2005, Association for Computing Machinery, New York, NY, USA (2005). https://doi.org/10.1145/1060590.1060666
19. Grohe, M.: Computing crossing numbers in quadratic time. J. Comput. Syst. Sci. **68**(2), 285–302 (2004). https://doi.org/10.1016/j.jcss.2003.07.008
20. Gutwenger, C., Mutzel, P., Weiskircher, R.: Inserting an edge into a planar graph. Algorithmica **41**(4), 289–308 (2005). https://doi.org/10.1007/s00453-004-1128-8
21. Guy, R.K., Nowakowski, R.J.: The outerthickness & outercoarseness of graphs I. The complete graph & the n-Cube, pp. 297–310. Physica-Verlag HD, Heidelberg (1990). https://doi.org/10.1007/978-3-642-46908-4_34
22. Guy, R.K., Nowakowski, R.J.: The outerthickness & outercoarseness of graphs II. the complete bipartite graph. Contemp. Meth. Graph Theory, 313–322 (1990)
23. Halton, J.H.: On the thickness of graphs of given degree. Inf. Sci. **54**(3), 219–238 (1991). https://doi.org/10.1016/0020-0255(91)90052-V
24. Hamm, T., Hliněný, P.: Parameterised partially-predrawn crossing number. In: SoCG. LIPIcs, vol. 224, pp. 46:1–46:15. Schloss Dagstuhl - Leibniz-Zentrum für Informatik (2022). https://doi.org/10.4230/LIPIcs.SoCG.2022.46
25. Hliněný, P.: Crossing number is hard for cubic graphs. J. Comb. Theory, Ser. B **96**(4), 455–471 (2006). https://doi.org/10.1016/j.jctb.2005.09.009
26. Hliněný, P., Derňár, M.: Crossing number is hard for kernelization. In: 32nd International Symposium on Computational Geometry, SoCG 2016, 14–18 June 2016, Boston, MA, USA. LIPIcs, vol. 51, pp. 42:1–42:10. Schloss Dagstuhl - Leibniz-Zentrum fuer Informatik (2016). https://doi.org/10.4230/LIPIcs.SoCG.2016.42
27. Hliněný, P., Masařík, T.: Minimizing an uncrossed collection of drawings. CoRR, pp. 1–17 (2023). https://arxiv.org/abs/2306.09550v2
28. Kawarabayashi, K.I., Reed, B.: Computing crossing number in linear time. In: Proceedings of the Thirty-Ninth Annual ACM Symposium on Theory of Computing, pp. 382–390. STOC 2007, Association for Computing Machinery (2007). https://doi.org/10.1145/1250790.1250848
29. Kneis, J., Langer, A.: A practical approach to Courcelle's theorem. Electron. Notes Theor. Comput. Sci. **251**, 65–81 (2009). https://doi.org/10.1016/j.entcs.2009.08.028
30. Leighton, F.T.: Complexity Issues in VLSI: Optimal Layouts for the Shuffle-exchange Graph and Other Networks. MIT Press, Cambridge, MA, USA (1983)
31. Mansfield, A.: Determining the thickness of graphs is NP-hard. Math. Proc. Cambridge Philos. Soc. **93**(1), 9–23 (1983). https://doi.org/10.1017/S030500410006028X
32. Mutzel, P., Odenthal, T., Scharbrodt, M.: The thickness of graphs: a survey. Graphs Comb. **14**(1), 59–73 (1998). https://doi.org/10.1007/PL00007219
33. Pelsmajer, M.J., Schaefer, M., Štefankovič, D.: Crossing numbers of graphs with rotation systems. Algorithmica **60**(3), 679–702 (2011). https://doi.org/10.1007/s00453-009-9343-y
34. Perković, L., Reed, B.A.: An improved algorithm for finding tree decompositions of small width. Int. J. Found. Comput. Sci. **11**(3), 365–371 (2000). https://doi.org/10.1142/S0129054100000247

35. Pinontoan, B., Richter, R.B.: Crossing numbers of sequences of graphs II: planar tiles. J. Graph Theory **42**(4), 332–341 (2003). https://doi.org/10.1002/jgt.10097
36. Poranen, T., Mäkinen, E.: Remarks on the thickness and outerthickness of a graph. Comput. Math. Appl. **50**(1), 249–254 (2005). https://doi.org/10.1016/j.camwa.2004.10.048
37. Robertson, N., Seymour, P.D.: Graph minors. XIII. the disjoint paths problem. J. Comb. Theory, Ser. B **63**(1), 65–110 (1995). https://doi.org/10.1006/jctb.1995.1006
38. Sýkora, O., Székely, L.A., Vrto, I.: A note on Halton's conjecture. Inf. Sci. **164**(1), 61–64 (2004). https://doi.org/10.1016/j.ins.2003.06.008
39. Thomassen, C.: A simpler proof of the excluded minor theorem for higher surfaces. J. Comb. Theory, Ser. B **70**(2), 306–311 (1997). https://doi.org/10.1006/jctb.1997.1761
40. Vasak, J.M.: The thickness of the complete graph. University of Illinois at Urbana-Champaign (1976). PhD thesis
41. Wessel, W.: Über die abhängigkeit der dicke eines graphen von seinen knotenpunktvalenzen. Geom. Kombinatorik **2**, 235–238 (1984)
42. Xu, B., Zha, X.: Thickness and outerthickness for embedded graphs. Discret. Math. **341**(6), 1688–1695 (2018). https://doi.org/10.1016/j.disc.2018.02.024

Bichromatic Perfect Matchings with Crossings

Oswin Aichholzer[1], Stefan Felsner[2], Rosna Paul[1](✉), Manfred Scheucher[2], and Birgit Vogtenhuber[1]

[1] Institute of Software Technology, Graz University of Technology, Graz, Austria
{oaich,ropaul,bvogt}@ist.tugraz.at
[2] Institute for Mathematics, Technical University of Berlin, Berlin, Germany
{felsner,scheucher}@math.tu-berlin.de

Abstract. We consider bichromatic point sets with n red and n blue points and study straight-line bichromatic perfect matchings on them. We show that every such point set in convex position admits a matching with at least $\frac{3n^2}{8} - \frac{n}{2} + c$ crossings, for some $-\frac{1}{2} \leq c \leq \frac{1}{8}$. This bound is tight since for any $k > \frac{3n^2}{8} - \frac{n}{2} + \frac{1}{8}$ there exist bichromatic point sets that do not admit any perfect matching with k crossings.

Keywords: Perfect matchings · Bichromatic point sets · Crossings

1 Introduction

Let $P = R \cup B$, $|R| = |B| = n$ be a point set in *general position*, that is, no three points of P are collinear. We refer to R and B as the set of red and blue points, respectively. A straight-line matching M on P where every point in R is uniquely matched to a point in B is called a *straight-line bichromatic perfect matching* (all matchings considered in this work are straight-line, so we will mostly omit this term). In this work, we study the existence of bichromatic perfect matchings on P with a fixed number k of crossings, where $0 \leq k \leq \binom{n}{2}$. It is folklore that any P of even size admits a crossing-free perfect matching. Perfect matchings with k crossings in the uncolored setting have been considered in [2]. There it is shown that for every $k \leq \frac{n^2}{16} - O(n\sqrt{n})$, every point set of size $2n$ admits a perfect matching with exactly k crossings and that there exist point sets where every perfect matching has at most $\frac{5n^2}{18}$ crossings. As a direct consequence, there exist bichromatic point sets which do not admit bichromatic perfect matchings with k crossings for $k > \frac{5n^2}{18}$. For $2n$ (uncolored) points in convex position it was shown in [2] that they admit perfect matchings with k crossings for every k in the range from 0 to $\binom{n}{2}$.

O.A. and R.P. supported by the Austrian Science Fund (FWF): W1230. S.F. supported by the DFG Grant FE 340/13-1. M.S. supported by the DFG Grant SCHE 2214/1-1. We thank our anonymous reviewers for the useful suggestions.

M. A. Bekos and M. Chimani (Eds.): GD 2023, LNCS 14465, pp. 124–132, 2023.
https://doi.org/10.1007/978-3-031-49272-3_9

For bichromatic point sets, this situation changes quite significantly. Consider a set P of $2n$ points in convex position (convex point set, for short) with an alternating coloring, that is, every second point along the convex hull is red (and the other points are blue). Moreover, let the number n of red (and blue) points be even. Then the number of crossings in a bichromatic perfect matching M on P is at most $\frac{n(n-2)}{2} = \binom{n}{2} - \frac{n}{2}$. The idea is as follows: Label the points of P as $p_0, p_1, \ldots, p_{2n-1}$ along the boundary of the convex hull. Then p_i cannot be matched to p_{i+n} since both points have the same color. Hence, for any edge e in M, the number of crossings of e is at most $n-2$. As every crossing involves two edges, the number of crossings of M is at most $\frac{n(n-2)}{2} = \binom{n}{2} - \frac{n}{2}$. This bound is tight, since it is possible to construct a bichromatic perfect matching on P with exactly $\binom{n}{2} - \frac{n}{2}$ crossings as follows. For $0 \leq i \leq n-1$, match the point p_i to the point p_{i+n+1}, when i is even. Otherwise, match p_i to p_{i+n-1}. Based on the above observations we state the following question.

Question 1. *For which values of k does every bichromatic convex point set $P = R \cup B$, $|R| = |B| = n$, admit a straight-line bichromatic perfect matching with exactly k crossings?*

The above example implies that if $k > \binom{n}{2} - \frac{n}{2}$, there exist bichromatic point sets with n red and n blue points that do not have any bichromatic perfect matching with k crossings. Thus, the answer to Question 1 can be true only for $k \leq \binom{n}{2} - \frac{n}{2}$. As a main result of this paper, we prove the following theorem.

Theorem 1. *For every n and for every $k > \frac{3n^2}{8} - \frac{n}{2} + \frac{1}{8}$, there exists a bichromatic convex point set with n red and n blue points that does not have a straight-line bichromatic perfect matching with k crossings.*

To show this, we study bichromatic convex point sets and matchings on them with the maximum number of crossings. In Sect. 2, we first determine the maximum number of crossings for certain bichromatic convex point sets, depending on their cardinality modulo 4. Then we prove that this number gives a tight lower bound on the maximum number of crossings in any bichromatic convex point set. We further show some positive results for Question 1 in Sect. 3.

Related Work. A survey by Kano and Urrutia [7] gives an overview of various problems on bichromatic point sets, including matching problems. Crossing-free bichromatic perfect matchings have been studied from various perspectives such as their structure [6,9], linear transformation distance [1], and matchings compatible to each other [4,5]. Sharir and Welzl [10] proved that the number of crossing-free bichromatic perfect matchings on $2n$ points is at most $O(7.61^n)$. However, on bichromatic perfect matchings with crossings much less is known. Pach et al. [8] showed that every straight-line drawing of $K_{n,n}$ contains a crossing family of size at least $n^{1-o(1)}$, where a *crossing family* is a set of pairwise crossing edges. This implies that for any $P = R \cup B$, $|R| = |B| = n$, there exists a bichromatic perfect matching with at least $n^{2-o(1)}$ crossings.

2 Bichromatic Convex Point Sets

Let $\mathcal{C}_{n,n}$ be the collection of all bichromatic convex point sets $P = R \cup B$ with $|R| = |B| = n$. For a point set $P \in \mathcal{C}_{n,n}$, we label the points in P in clockwise direction along the convex hull as $p_0, p_1, \ldots, p_{2n-1}$ and refer to this as the *clockwise ordering*. We will consider all indices modulo $2n$. The number of crossings in a bichromatic perfect matching M_P on P is denoted by $\overline{\text{cr}}(M_P)$. If M_P has the maximum number of crossings among all such matchings on P, then M_P is called a *max-crossing* matching on P. Among all max-crossing matchings for all $P \in \mathcal{C}_{n,n}$, we are interested in matchings with the minimum number of crossings. We call such a matching a *min-max-crossing matching* of $\mathcal{C}_{n,n}$. From now on, we mostly refer to bichromatic perfect matchings just as *matchings*.

A *block* of $P \in \mathcal{C}_{n,n}$ is a maximal set of consecutive points of P of the same color. (If $R_1 = \{p_a, p_{a+1}, \ldots, p_{a+s}\}$ is a red block then p_{a-1} and p_{a+s+1} are blue.) Collecting the blocks of P in clockwise order yields a cyclically ordered partition $(R_1, B_1, R_2, B_2, \ldots, R_s, B_s)$ of P. The coloring of P is called $2s$*–block* if it induces s red and s blue blocks. In particular a $2n$–block coloring is alternating. A $2s$–block coloring where all blocks have the same cardinality is *balanced*; see e.g. Figure 4(a). A balanced $2s$–block coloring only exists if s divides n. To overcome this restriction, we also consider $2s$–block colorings in which block sizes differ by at most 1 as *balanced* $2s$*–block* colorings. Note that for $s = 2$ and any given value of n, there is a unique balanced $2s$–block coloring (up to symmetry).

Theorem 2. *Let $P \in \mathcal{C}_{n,n}$ and $\mathrm{M}_P^{\vee}$ be a max-crossing matching on P. Then*

$$\overline{\text{cr}}(\mathrm{M}_P^{\vee}) \geq \begin{cases} \frac{3n^2}{8} - \frac{n}{2} & \text{if } n \equiv 0 \mod 4 \\ \frac{3n^2}{8} - \frac{n}{2} + \frac{1}{8} & \text{if } n \equiv 1 \mod 4 \\ \frac{3n^2}{8} - \frac{n}{2} - \frac{1}{2} & \text{if } n \equiv 2 \mod 4 \\ \frac{3n^2}{8} - \frac{n}{2} + \frac{1}{8} & \text{if } n \equiv 3 \mod 4 \end{cases}$$

Moreover, equality holds if P has a balanced 4–block coloring. In this case, $\mathrm{M}_P^{\vee}$ is a min-max-crossing matching of $\mathcal{C}_{n,n}$.

Theorem 1 is implied by Theorem 2: consider any point set with balanced 4–block coloring. The bound for the crossings in a max-crossing matching of this point set implies Theorem 1. Theorem 2 will follow directly from Lemma 2 and Lemma 3, which are stated and shown in the next sections.

2.1 Max-Crossing Matching of a Balanced 4–Block Coloring

In this section, we determine the number of crossings in any max-crossing matching of a point set with balanced 4–block coloring. A *crossing family* of a point set P is a set of edges spanned by points from P that pairwise cross.

Lemma 1. *Let $P \in \mathcal{C}_{n,n}$ have a 4–block coloring with blocks R_1, B_1, R_2, B_2 and let $\mathrm{M}_P^{\vee}$ be a max-crossing matching on P. Then for each block $X \in \{R_1, R_2, B_1, B_2\}$, the edges emanating from X form a crossing family.*

Proof. Consider a block X of P and assume w.l.o.g. that $X = \{p_1, p_2, \ldots, p_x\}$. If in a matching M on P, there are two non-crossing edges with endpoints in X, then there also exist two such non-crossing edges with adjacent endpoints in X, i.e., M contains edges (p_i, p_k) and (p_{i+1}, p_j) for some $1 \leq i < x \leq j < k \leq 2n$. Let M' be obtained from M by replacing the two edges by the two crossing edges (p_i, p_j) and (p_{i+1}, p_k) and note that $\overline{\text{cr}}(M') = \overline{\text{cr}}(M) + 1$. □

Consider a max-crossing matching $\text{M}_P^\vee$ on P. Lemma 1 implies that there is some $a \geq 0$ such that in $\text{M}_P^\vee$, the first a points of R_1 are matched to points in B_1 while the last $|R_1| - a$ points of R_1 are matched to points in B_2. Analogously, the first $|B_1| - a$ points of B_1 are matched to points in R_2. Since $|R_2| = n - |R_1|$ and the $|B_1| - a$ points of R_2 are matched to B_1, the first $n - |R_1| - |B_1| + a$ points of R_2 are matched to points in B_2; see Fig. 1. Hence, to get a max-crossing matching on P, it is sufficient to determine the optimal value of a.

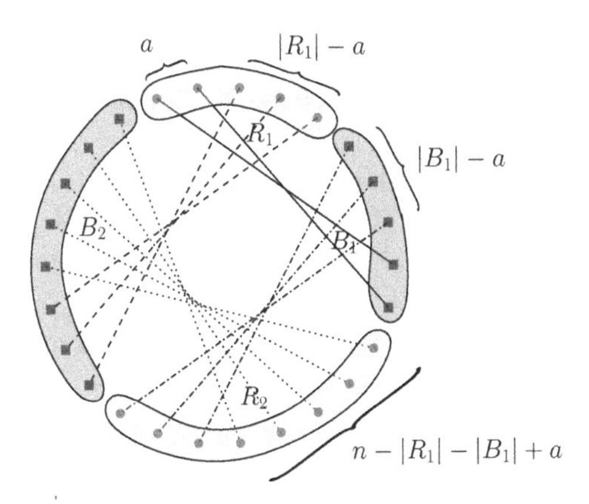

Fig. 1. Structure of a max-crossing matching on a set P with 4–block coloring.

By determining the optimum value of a, we next construct a max-crossing matching on any set P with balanced 4–block coloring and compute its exact crossing number.

Lemma 2. *Let $P \in \mathcal{C}_{n,n}$ have a balanced 4-block coloring and let $\text{M}_P^\vee$ be a max-crossing matching on P. Then*

$$\overline{\text{cr}}(\text{M}_P^\vee) = \begin{cases} \frac{3n^2}{8} - \frac{n}{2} & \textit{if } n \equiv 0 \mod 4 \\ \frac{3n^2}{8} - \frac{n}{2} + \frac{1}{8} & \textit{if } n \equiv 1 \mod 4 \\ \frac{3n^2}{8} - \frac{n}{2} - \frac{1}{2} & \textit{if } n \equiv 2 \mod 4 \\ \frac{3n^2}{8} - \frac{n}{2} + \frac{1}{8} & \textit{if } n \equiv 3 \mod 4 \end{cases}$$

Proof. Let $P \in \mathcal{C}_{n,n}$ have a balanced 4–block coloring with blocks R_1, B_1, R_2, B_2, labeled such that $|R_1|, |B_1| \leq \frac{n}{2}$. Let $r_1 = |R_1| = \lfloor \frac{n}{2} \rfloor$ and $b_1 = |B_1| = \lfloor \frac{n}{2} \rfloor$.

Let M_P be a matching on P such that the first x points of the set R_1 are matched to the last x points of B_1, as a crossing family, for some $x \in \mathbb{N}_0$. The number of pairs of non-crossing edges in M_P is obtained by $(r_1 - x)(b_1 - x) + x(n - r_1 - b_1 + x) = r_1 b_1 - 2xb_1 - 2xr_1 + nx + 2x^2$. To determine the value $x \in \mathbb{N}_0$ that gives the maximum number of crossings, we first calculate the value $x^* \in \mathbb{R}$ for which $f(x) = (n - 2r_1 - 2b_1)x + 2x^2 + r_1 b_1$ attains its minimum. This is achieved by $x^* = \frac{1}{2}(r_1 + b_1 - \frac{n}{2})$. Note that x^* might not be in $\mathbb{N}_0$. Since f is a quadratic function, its minimum over all $x \in \mathbb{N}_0$ is reached for $x = \lfloor x^* \rceil$, where $\lfloor x^* \rceil$ denotes the closest integer of $x^* \in \mathbb{R}$. Then the max-crossing matching $\mathrm{M}_P^\vee$ on P has $\overline{\mathrm{cr}}(\mathrm{M}_P^\vee) = \binom{n}{2} - (n - 2r_1 - 2b_1)x - 2x^2 - r_1 b_1$ many crossings, where $x = \lfloor \frac{r_1+b_1}{2} - \frac{n}{4} \rceil = \lfloor \lfloor \frac{n}{2} \rfloor - \frac{n}{4} \rceil$.

Note that the blocks in P may differ in size by 1 and that also the rounding for obtaining x depends on the value of $n \mod 4$. To account for this, we evaluate each case separately.

Case 1: Let $n \equiv 0 \mod 4$. Then there exists an integer m such that $n = 4m$. Then $r_1 = b_1 = 2m$ and $x = \lfloor \frac{4m}{2} - \frac{4m}{4} \rceil = m$. Hence the number of crossings of $\mathrm{M}_P^\vee$ is $\overline{\mathrm{cr}}(\mathrm{M}_P^\vee) = \binom{4m}{2} - (4m - 4(2m))m - 2m^2 - (2m)^2 = 6m^2 - 2m$. Replacing m by $\frac{n}{4}$ gives $\overline{\mathrm{cr}}(\mathrm{M}_P^\vee) = \frac{3n^2}{8} - \frac{n}{2}$.

Case 2: Let $n \equiv 1 \mod 4$. Then there exists an integer m such that $n = 4m+1$. Then $r_1 = b_1 = 2m$ and $x = \lfloor \frac{4m}{2} - \frac{4m+1}{4} \rceil = m$. Hence the number of crossings of $\mathrm{M}_P^\vee$ is $\overline{\mathrm{cr}}(\mathrm{M}_P^\vee) = \binom{4m+1}{2} - (4m + 1 - 4(2m))m - 2m^2 - (2m)^2 = 6m^2 + m$. Replacing m by $\frac{n-1}{4}$ gives $\overline{\mathrm{cr}}(\mathrm{M}_P^\vee) = \frac{3n^2}{8} - \frac{n}{2} + \frac{1}{8}$.

Case 3: Let $n \equiv 2 \mod 4$. Then there exists an integer m such that $n = 4m+2$. Then $r_1 = b_1 = 2m + 1$ and $x = \lfloor \frac{4m+2}{2} - \frac{4m+2}{4} \rceil = m$. Hence the number of crossings of $\mathrm{M}_P^\vee$ is $\overline{\mathrm{cr}}(\mathrm{M}_P^\vee) = \binom{4m+2}{2} - (4m + 2 - 4(2m+1))m - 2m^2 - (2m+1)^2 = 6m^2 + 4m$. Replacing m by $\frac{n-2}{4}$ gives $\overline{\mathrm{cr}}(\mathrm{M}_P^\vee) = \frac{3n^2}{8} - \frac{n}{2} - \frac{1}{2}$.

Case 4: Let $n \equiv 3 \mod 4$. Then there exists an integer m such that $n = 4m+3$. Then $r_1 = b_1 = 2m + 1$ and $x = \lfloor \frac{4m+2}{2} - \frac{4m+3}{4} \rceil = m$. Hence the number of crossings of $\mathrm{M}_P^\vee$ is $\overline{\mathrm{cr}}(\mathrm{M}_P^\vee) = \binom{4m+3}{2} - (4m + 3 - 4(2m + 1))m - 2m^2 - (2m+1)^2 = 6m^2 + 7m + 2$. Replacing m by $\frac{n-2}{4}$ gives $\overline{\mathrm{cr}}(\mathrm{M}_P^\vee) = \frac{3n^2}{8} - \frac{n}{2} + \frac{1}{8}$.

Altogether this completes the proof of the lemma. □

2.2 Min-Max-Crossing Matching for All Colorings

In the following, we show that the maximum number of crossings of a bichromatic matching on $P \in \mathcal{C}_{n,n}$ is minimized by sets with balanced 4–block coloring. Let $P \in \mathcal{C}_{n,n}$. For any point $v \in P$, the point $w \in P$ is called the antipodal pair of v, if the line through v and w partitions P into two equal sized halves (antipodals exist because the number of points is even). If antipodal pairs v and w are of the same color, then they are *monochromatic antipodal* pairs (in short *m-antipodal* pairs) and if they have different colors then they are *bichromatic antipodal* pairs (in short *b-antipodal* pairs).

Lemma 3. *Let $P, Q \in \mathcal{C}_{n,n}$, where Q has a balanced 4-block coloring and let $\mathrm{M}_P^\vee, \mathrm{M}_Q^\vee$ be max-crossing matchings on P and Q, respectively. Then $\overline{\mathrm{cr}}(\mathrm{M}_P^\vee) \geq \overline{\mathrm{cr}}(\mathrm{M}_Q^\vee)$. That is, $\mathrm{M}_Q^\vee$ is a min-max-crossing matching of the set $\mathcal{C}_{n,n}$.*

To prove Lemma 3, we make use of a variant of the classic ham sandwich theorem [11]. A full proof can be found in Appendix A of the arXiv version [3].

Proof (sketch). We define a matching M_P on P in three steps and then compare its crossings with those of $\mathrm{M}_Q^\vee$, where we will distinguish two cases.

Step 1: Let S be the point set obtained by removing all the b-antipodal pairs from P. If S is empty then all the points in P are b-antipodal pairs. Thus P admits a crossing family of size n (which is a perfect matching) that has more crossings than $\mathrm{M}_Q^\vee$. Hence we may assume that S is non-empty.

Step 2: Partition the set S into four groups as follows. First, we arbitrarily partition S into two consecutive sets S_L and S_R of equal size and note that each part contains half of the blue and half of the red points. Using the ham sandwich theorem, partition S_L into $S_{L,1}$ and $S_{L,2}$ such that each of them has an equal number of red and blue points. Due to the symmetry of S_L and S_R, this partition can be duplicated on $S_{R,1}$. Depending on the ham sandwich cut, $S_{L,1}$, $S_{L,2}$, $S_{R,1}$, and $S_{R,2}$ form four or six bundles of consecutive points along the convex hull. If we have only four bundles, we are done with the partition. So assume that we have six bundles. Then one partition in the part S_R, say $S_{R,2}$, is split into $S_{R,2a}$ and $S_{R,2b}$ by $S_{R,1}$ along the convex hull; see Fig. 2 (left). A similar splitting occurs for S_L. As the points in $S_{R,2b}$ and $S_{L,2a}$, and $S_{R,2a}$ and $S_{L,2b}$ are m-antipodal pairs, the composition of points in $S_{R,2}$ is same as in $S_{L,2a} \cup S_{R,2a}$ and the composition of points in $S_{L,2}$ is same as $S_{L,2b} \cup S_{R,2b}$. That is, if we have four or six bundles, they can always be (re)assembled into four cyclically connected groups $S_{RT}, S_{RB}, S_{LB}, S_{LT}$ of S such that each of these groups has the same number of red and blue points; see again Fig. 2. Moreover, each group is antipodal to another group: S_{RT} is antipodal to S_{LB} and S_{LT} is antipodal to S_{RB}. We call these pairs of groups *matching-pair* groups.

Step 3: Add all the removed b-antipodal pairs back to S to get P. The partition of S induces a partition $P_{RT}, P_{RB}, P_{LB}, P_{LT}$ in P. Note that the number of red (blue) points of P_{RT} and the number of blue (red) points of P_{LB} are equal. The same holds for P_{RB} and P_{LT}. Define the matching M_P on P as follows. For any matching-pair group $(X, Y) \in \{(P_{RT}, P_{LB}), (P_{RB}, P_{LT})\}$, the points in X are matched to points in Y such that any two of the matching edges emanating from the points of the same color on X cross each other.

Assuming that S is non-empty, there are two possible structures for S, depending on whether $|S_L|/2$ is even or odd. We consider them as two separate cases; see Fig. 3. Case 1: All groups in the partition of S have the same number m of red and blue points. Case 2: Each group in one matching-pair group has m red (and blue) points and each group in the other matching-pair has $m + 1$ red (and blue) points.

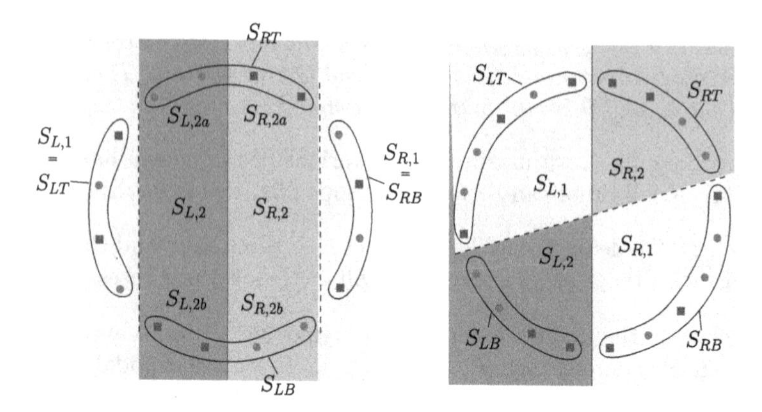

Fig. 2. A bichromatic point set S with 16 points (left) and 20 points (right). In both cases, dotted lines represent the partition of S_L obtained by the ham sandwich theorem, and its mirror on S_R.

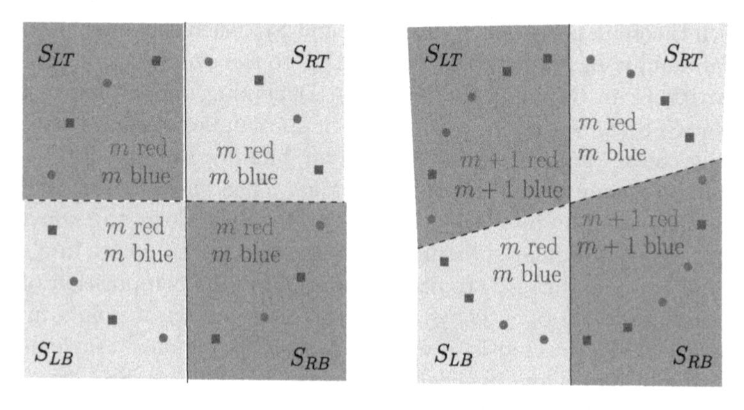

Fig. 3. Possible distribution of red and blue points between the groups of S. (Color figure online)

For Case 1, the number of crossings in M_P is given by $\overline{\text{cr}}(M_P) \geq \binom{m+x_1}{2} + \binom{m+x_2}{2} + \binom{m+y_1}{2} + \binom{m+y_2}{2} + (2m + x_1 + x_2)(2m + y_1 + y_2) + x_1x_2 + y_1y_2$, where x_1, x_2 are the number of b-antipodal points in P_{RT} of the colors red and blue, respectively and y_1, y_2 are the number of b-antipodal points in P_{LT} of the colors red and blue. These b-antipodal pairs always cross in M_P and contribute $x_1x_2 + y_1y_2$ crossings in M_P. Then for a balanced 4–block coloring $Q \in \mathcal{C}_{n,n}$ we have, $\overline{\text{cr}}(\text{M}_Q^\vee) = \frac{3n^2}{8} - \frac{n}{2} + c$, where $n = 4m + x_1 + x_2 + y_1 + y_2$ and $c \in \{\frac{1}{8}, -\frac{1}{2}, 0\}$ by Lemma 2. Comparing the number of crossings in M_P and $\text{M}_Q^\vee$ gives $\overline{\text{cr}}(M_P) - \overline{\text{cr}}(\text{M}_Q^\vee) \geq \frac{(x_1y_1+x_2y_1+x_1y_2+x_2y_2)}{4} + \frac{(x_1^2+x_2^2)}{8} + \frac{(y_1^2+y_2^2)}{8} + \frac{(x_1x_2+y_1y_2)}{4} - c$. As $c \leq \frac{1}{8}$ and since S is a proper subset of P, $x_1 + x_2 + y_1 + y_2 \geq 1$. Thus $\overline{\text{cr}}(M_P) \geq \overline{\text{cr}}(\text{M}_Q^\vee)$.

The reasoning for Case 2 is similar to the one for Case 1 sketched above. □

As mentioned, Theorem 2 follows directly from Lemma 2 and Lemma 3. We remark that balanced 4–block colorings are not the only colorings that admit min-max-crossing matchings. For example, for $n \equiv 0 \mod 4$, a 4-block coloring with block sizes $\frac{n}{2}+1, \frac{n}{2}, \frac{n}{2}-1, \frac{n}{2}$ always induces the same number of crossings as the according balanced 4–block coloring; see Fig. 4.

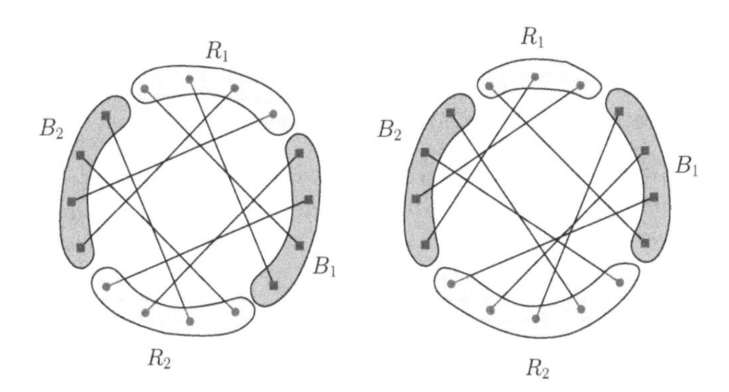

Fig. 4. A balanced 4–block coloring (left) and a slightly unbalanced 4–block coloring (right) on 16 points and max-crossing matchings on them, each with 20 crossings.

3 Further Results

We have shown that for any n and any $k > \frac{3n^2}{8} - \frac{n}{2} + \frac{1}{8}$ there exist point sets $P \in \mathcal{C}_{n,n}$ that do not admit any bichromatic perfect matching with k crossings. It is natural to ask what happens in the range $[1, \frac{3n^2}{8}]$. By straight-forward calculations, any $P \in \mathcal{C}_{n,n}$ with $2n$–block coloring cannot have a bichromatic perfect matching with k crossings for $k \in \{1, 2\}$. Using computers (with SAT framework) we obtained that every $P \in \mathcal{C}_{7,7}$ admits bichromatic perfect matchings with k crossings for any $k \in \{0, 1, \ldots, 15\} \setminus \{1, 2\}$. Based on this, we can show the following proposition. Its proof is deferred to Appendix B of the arXiv version [3].

Proposition 1. *For $n \geq 7$, every $P \in \mathcal{C}_{n,n}$ admits bichromatic perfect matchings with k crossings for any $k \in \{0, 1, \ldots, \frac{15(n-6)}{7}\} \setminus \{1, 2\}$.*

For bichromatic point sets in general (non-convex) position with n red and n blue points, computer assisted results shows that for $n = 6$, every such set admits bichromatic perfect matchings with k crossings for $k \in \{0, 3, 4\}$. With a similar proof as for Proposition 1, it follows that every bichromatic point set with n red and n blue points admits bichromatic perfect matchings with k crossings for $k \in \{0, 1, \ldots, \frac{n}{3}\} \setminus \{1, 2\}$. In ongoing work, we study the range $k \in [\frac{15(n-6)}{7}, \frac{3n^2}{8}]$ for bichromatic convex point sets and the range $k \in [\frac{n}{3}, \frac{5n^2}{18}]$ for bichromatic point sets in general position.

4 Conclusion

We considered max-crossing matchings of bichromatic convex point sets in the plane with n red and n blue points. We gave the exact number of crossings in max-crossing matchings of point sets with balanced 4–block coloring and showed that these matchings are min-max-crossing matchings of bichromatic convex point sets. This result implies a negative answer to Question 1 on the existence of matchings with k crossings in the convex case for $k > \frac{3n^2}{8} - \frac{n}{2} + \frac{1}{8}$. We further answered the question for $k \leq \frac{15(n-6)}{7}$. From a computational point of view, an interesting open question is the following:

Question 2. *Given a bichromatic (convex) point set P and an integer k, what is the computational complexity of deciding whether there is a matching with exactly (or at least) k crossings?*

References

1. Aichholzer, O., Barba, L., Hackl, T., Pilz, A., Vogtenhuber, B.: Linear transformation distance for bichromatic matchings. Comput. Geom. **68**, 77–88 (2018). https://doi.org/10.1016/j.comgeo.2017.05.003
2. Aichholzer, O., et al.: Perfect matchings with crossings. In: Bazgan, C., Fernau, H. (eds.) IWOCA 2022. LNCS, vol. 13270, pp. 46–59. Springer, Cham (2022). https://doi.org/10.1007/978-3-031-06678-8_4
3. Aichholzer, O., Felsner, S., Paul, R., Scheucher, M., Vogtenhuber, B.: Bichromatic perfect matchings with crossings (2023). arXiv:2309.00546v1
4. Aichholzer, O., Hurtado, F., Vogtenhuber, B.: Compatible matchings for bichromatic plane straight-line graphs. In: Proceedings of EuroCG'12, pp. 257–260 (2012). https://www.eurocg.org/2012/booklet.pdf
5. Aloupis, G., Barba, L., Langerman, S., Souvaine, D.L.: Bichromatic compatible matchings. Comput. Geom. **48**(8), 622–633 (2015). https://doi.org/10.1016/j.comgeo.2014.08.009
6. Asinowski, A., Miltzow, T., Rote, G.: Quasi-parallel segments and characterization of unique bichromatic matchings. J. Comput. Geom. **6**(1), 185–219 (2015). https://doi.org/10.20382/jocg.v6i1a8
7. Kano, M., Urrutia, J.: Discrete geometry on colored point sets in the plane – a survey. Graphs Comb. **37**(1), 1–53 (2021). https://doi.org/10.1007/s00373-020-02210-8
8. Pach, J., Rubin, N., Tardos, G.: Planar point sets determine many pairwise crossing segments. Adv. Math. **386**, 107779 (2021). https://doi.org/10.1016/j.aim.2021.107779
9. Savić, M., Stojaković, M.: Structural properties of bichromatic non-crossing matchings. Appl. Math. Comput. **415**, 126695 (2022)
10. Sharir, M., Welzl, E.: On the number of crossing-free matchings, cycles, and partitions. SIAM J. Comput. **36**(3), 695–720 (2006). https://doi.org/10.1137/050636036
11. Tóth, C.D., O'Rourke, J., Goodman, J.E.: Handbook of Discrete and Computational Geometry. 3rd edn. CRC Press, Boca Raton (2017). https://doi.org/10.1201/9781315119601

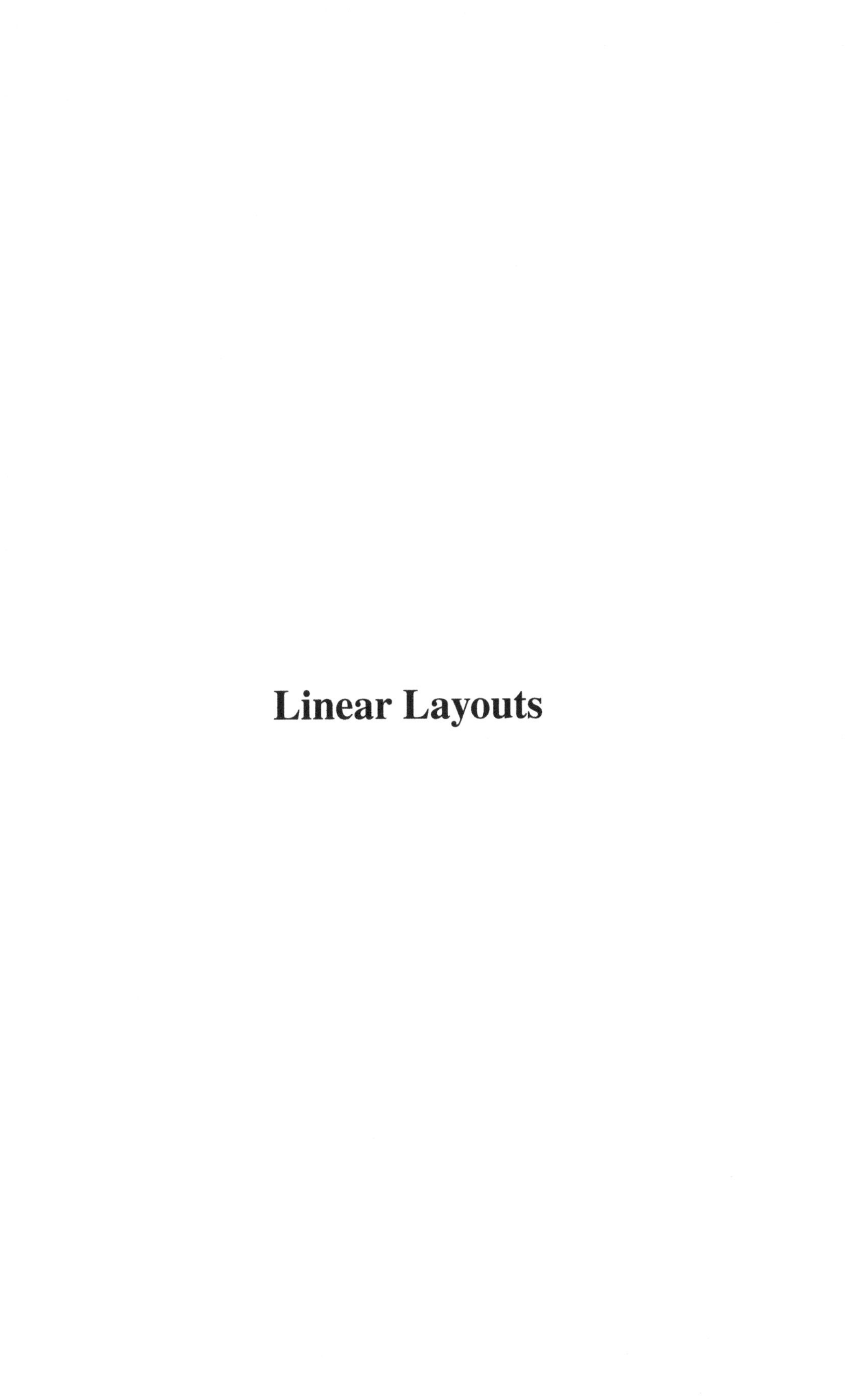

Linear Layouts

On Families of Planar DAGs with Constant Stack Number

Martin Nöllenburg[1] and Sergey Pupyrev[2](✉)

[1] Algorithms and Complexity Group, TU Wien, Vienna, Austria
noellenburg@ac.tuwien.ac.at
[2] Meta, Menlo Park, CA, USA
spupyrev@gmail.com

Abstract. A k-stack layout (or k-page book embedding) of a graph consists of a total order of the vertices, and a partition of the edges into k sets of non-crossing edges with respect to the vertex order. The stack number of a graph is the minimum k such that it admits a k-stack layout. In this paper we study a long-standing problem regarding the stack number of planar directed acyclic graphs (DAGs), for which the vertex order has to respect the orientation of the edges. We investigate upper and lower bounds on the stack number of several families of planar graphs: We improve the constant upper bounds on the stack number of single-source and monotone outerplanar DAGs and of outerpath DAGs, and improve the constant upper bound for upward planar 3-trees. Further, we provide computer-aided lower bounds for upward (outer-) planar DAGs.

Keywords: stack number · twist number · planar directed acyclic graphs

1 Introduction

Let $G = (V, E)$ be a simple graph with n vertices and σ be a total order of the vertex set V. Two edges (u, v) and (w, z) in E with $u <_\sigma w$ *cross* if $u <_\sigma w <_\sigma v <_\sigma z$. A *$k$-stack layout* (*$k$-page book embedding*) of G is a total order of V and a partition of E into k subsets, called *stacks* or *pages*, such that no two edges in the same subset cross. The *stack number* (*page number*, *book thickness*) of G is the minimum k such that G admits a k-stack layout.

Heath et al. [18,19] extended the notion of stack number to directed acyclic graphs (DAGs for short) in a natural way: Given a DAG, $G = (V, E)$, a book embedding of G is defined as for undirected graphs, except that the total order σ of V is now required to be a *linear extension* of the partial order of V induced by E. That is, if G contains a directed edge (u, v) from a vertex u to a vertex v, then $u <_\sigma v$ in any feasible total order σ of V. Heath et al. showed that DAGs with stack number 1 can be characterized and recognized efficiently; however, they proved that, in general, determining the stack number of a DAG is NP-complete.

The main problem raised by Heath et al. [18,19] and studied in several papers [5,11,14,16,17] is whether every upward planar DAG has constant stack

M. A. Bekos and M. Chimani (Eds.): GD 2023, LNCS 14465, pp. 135–151, 2023.
https://doi.org/10.1007/978-3-031-49272-3_10

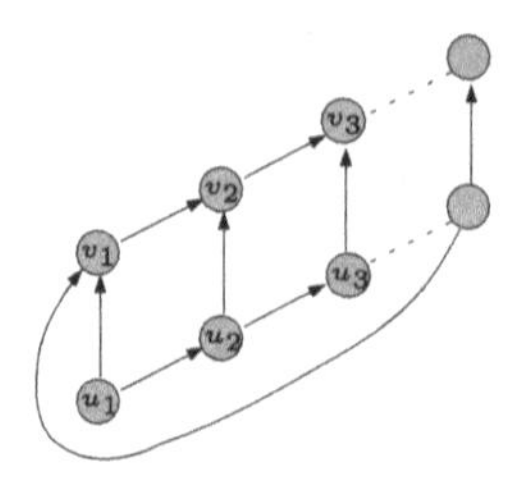

(a) A DAG that requires $n/2$ stacks: edges (u_i, v_i) form an $n/2$-twist

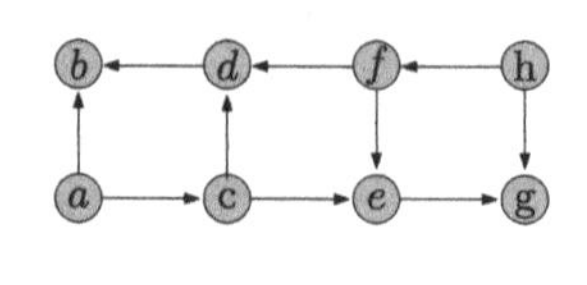

(b) An outerplanar DAG which is not upward planar [29]

Fig. 1. Planar DAGs that (a) need many stacks or (b) are not upward.

number. Recall that an *upward planar* DAG is a DAG that admits a drawing which is simultaneously *upward*, that is, each edge is represented by a curve monotonically increasing in the y-direction, and *planar*, that is, no two edges cross each other.

Open Problem 1. *Is the stack number of every upward planar DAG bounded by a constant?*

Notice that upward planarity is a necessary condition for the question: there exist DAGs which admit a planar non-upward embedding and that require $\Omega(n)$ stacks in any book embedding [19]; see Fig. 1a.

In its general form, Open Problem 1 is still unresolved. Heath et al. [18,19] showed that directed trees and unicyclic DAGs have stack numbers 1 and 2, respectively. Mchedlidze and Symvonis [25] proved that N-free upward planar DAGs, which contain series-parallel digraphs, have stack number 2. Frati et al. [16] gave several conditions under which upward planar triangulations have bounded stack number. In particular, they showed that (i) maximal upward planar 3-trees have a constant stack number, and (ii) planar triangulations with a bounded (directed) diameter have a constant stack number. Notice that the graph in Fig. 1a, that requires $\Omega(n)$ stacks, is a partial planar 3-tree. Thus, it is reasonable to ask whether the stack number is bounded for (non-upward but directed acyclic) 2-trees or their subfamilies, outerplanar graphs, also known as simple 2-trees. This question has been first asked by Heath et al. [19] and recently highlighted by Bekos et al. [7].[1] Bhore et al. [10] gave upper bounds for some upward outerplanar graphs, namely internally-triangulated outerpaths (16 stacks), cacti (6 stacks), and upward outerplanar graphs whose biconnected components are st-outerplanar (8 stacks).

[1] Very recently, Jungeblut et al. [22] resolved the problem by proving that every outerplanar DAG has constant stack number, upper bounded by 24776, while there are directed acyclic (non upward planar) 2-trees with unbounded stack number. Their proof of the upper bound relies on Theorem 1b (see below) as a central tool and their second result solves an open question raised in a preprint version of this paper.

We emphasize that directed acyclic 2-trees are planar but not necessarily upward, and thus, the results of Frati et al. [16] do not apply for this class of graphs. For example, the graph in Fig. 1b is a directed acyclic partial 2-tree (in fact, it is an outerpath DAG) but it cannot be drawn in an upward fashion.

Our Contributions. We investigate upper and lower bounds for the stack number of upward planar DAGs and outerplanar DAGs (oDAGs for short). Throughout the paper, we express the bounds in terms of the maximum size of a *twist* in the vertex order, that is, the maximum number of mutually crossing edges. This parameter, also called the *twist number* of a graph, is tied to the stack number; analyzing the maximum twist size significantly simplifies the arguments at the cost of (slightly) worsened bounds for the stack number. We refer to Sect. 2 for details and formal definitions.

In Sect. 3, we present constant upper bounds for several prominent subclasses of outerplanar DAGs.

Theorem 1.

a. *Every single-source outerplanar DAG has a constant stack number with a vertex order whose twist size is at most* 3.
b. *Every monotone outerplanar DAG has a constant stack number with a vertex order whose twist size is at most* 4.
c. *Every outerpath DAG has a constant stack number with a vertex order whose twist size is at most* 4.

The recent result of Davies [13] implies that every graph with a vertex order whose twist size is at most k, has stack number at most $2k \log_2 k + 2k \log_2 \log_2 k + 10k$ (and for $k = 3$ Davies proves an upper bound of 19). It follows that single-source oDAGs have stack number at most 19, while monotone oDAGs and outerpath DAGs have stack number at most 64. We note that the stack assignment for the provided vertex orders can likely be improved. For example, we show an upper bound of 4 stacks for single-source oDAGs (refer to Lemma 2 in Sect. 3).

Our proof technique utilized for Theorem 1 can be applied to other classes of DAGs. In Sect. 4 we tighten the upper bound on the stack number of upward (maximal) planar 3-trees. Frati et al. [16] bound the stack number of upward planar 3-trees by a function of the size of the maximum twist size without providing an explicit bound. We strengthen their results by presenting an arguably simpler proof that yields an exact (small) bound of 5 on the maximum twist size (which by Davies' result [13] translates into a stack number of at most 85).

Theorem 2. *Every upward planar 3-tree has a constant stack number with a vertex order whose twist size is at most* 5.

The proofs of Theorem 1 and Theorem 2 are constructive and lead to linear-time algorithms for constructing the vertex orders.

Finally, we explore lower bounds on the stack number of planar DAGs in Sect. 5. They rely on computational experiments using a SAT formulation of the book embedding problem.

Theorem 3.

a. *There exists a single-source single-sink upward outerplanar DAG with stack number* 3.
b. *There exists an upward outerplanar DAG with stack number* 4.
c. *There exists an upward planar 3-tree DAG with stack number* 5.

Other Related Work. Book embeddings of undirected graphs received a lot of attention due to their numerous applications. It is known that the graphs with stack number 1 are exactly outerplanar graphs, while graphs with stack number 2 are exactly the subhamiltonian graphs, which implies that it is NP-complete to decide whether a graph admits a 2-stack layout. More generally, every planar graph has stack number at most 4, and the bound is worst-case optimal [9,35].

Stack numbers of directed acyclic graphs have also been extensively studied. Similarly to the undirected case, it is NP-complete to test whether the stack number of a DAG is at most k, even when $k = 2$ [8]. Several works analyzed the stack number of partially ordered sets (posets), which can be viewed as upward planar DAGs without *transitive* edges. Nowakowski and Parker [27] asked whether the stack number of a planar poset is bounded by a constant. Notice that the question is a special case of Open Problem 1. Several works provide bounds for the stack number of special classes of posets and bounds in terms of various parameters (e.g., height or bump number) [4,20,34]. To our knowledge, there is no indication that the absence of transitive edges simplifies Open Problem 1.

As for the lower bounds on the stack number of DAGs and posets, not many results are known. It is easy to construct a planar poset with stack number 4 [4,20], which for a long time has been the best known lower bound for the stack number of upward planar DAGs. Our Theorem 3 strengthens the result by showing that there exist (maximal) upward planar 3-trees with stack number 5. Merker [26] independently constructed a planar poset with stack number 5. Jungeblut et al. [21] further showed that upward planar graphs of constant width and height have a bounded stack number, and combined the two results to get an $\mathcal{O}(n^{2/3}\log(n)^{2/3})$ upper bound on the stack number of general upward planar graphs. Yet these results do not imply any upper bound for the graph classes considered in Sect. 3 (since oDAGs can be non-upward) or in Sect. 4 (since upward planar 3-trees can have linear width and height).

Some proofs have been omitted. They can be found in Nöllenburg and Pupyrev [28].

2 Preliminaries

Throughout the paper, $G = (V, E)$ is a simple directed graph (digraph) with vertex set V and edge (arc) set E. A *vertex order*, σ, of a digraph G is a linear extension of V. That is, if G contains an edge from a vertex u to a vertex v, denoted $(u, v) \in E$, then $u <_\sigma v$ in any feasible vertex order σ of V. Let F be a set of $k \geq 2$ independent (that is, having no common endpoints) edges

$(s_i, t_i), 1 \leq i \leq k$. If $s_1 <_\sigma \cdots <_\sigma s_k <_\sigma t_1 <_\sigma \cdots <_\sigma t_k$, then F is a *k-twist.* Two independent edges forming a 2-twist are called *crossing*. A *k-stack layout* of G is a pair $(\sigma, \{\mathcal{S}_1, \ldots, \mathcal{S}_k\})$, where σ is a vertex order of G and $\{\mathcal{S}_1, \ldots, \mathcal{S}_k\}$ is a partition of E into *stacks*, that is, sets of pairwise non-crossing edges. The minimum number of stacks in a stack layout of G is its *stack number.*

The size of the largest twist in a vertex order is tied to the number of stacks needed for the edges of the graph under the vertex order. In one direction, a vertex order with a k-twist needs at least k stacks, since each edge of a twist must be in a distinct stack. In the other direction, a vertex order with no $(k+1)$-twist needs at most $\mathcal{O}(k \log k)$ stacks [13], which matches the lower bound of $\Omega(k \log k)$ [24]. An order without a 2-twist (that is, when $k = 1$) corresponds to an outerplanar drawing of a graph, which is a 1-stack layout. For $k = 2$ (an order without a 3-twist), 5 stacks are sufficient and sometimes necessary [1,23].

In the following we use notation $E(V_1 \rightarrow V_2)$ to indicate a subset of E between disjoint subsets $V_1, V_2 \subseteq V$, that is, $(x, y) \in E$ for $x \in V_1, y \in V_2$. Notation $E(V_1 \rightarrow V_2, V_3 \rightarrow V_4, \ldots)$ indicates the union of the edge sets, that is, $E(V_1 \rightarrow V_2) \cup E(V_3 \rightarrow V_4) \cup \ldots$. Similarly, we write $twist(V_1 \rightarrow V_2) \leq k$ to indicate that the maximum twist of the edges $E(V_1 \rightarrow V_2)$ is of size at most k. Slightly abusing the notation, we sometimes write $E(v \rightarrow V_1)$ or $E(V_1 \rightarrow v)$, where $v \in V \setminus V_1$ and $V_1 \subset V$. To specify a relative order between disjoint subsets of vertices, we use $\sigma = [V_1, V_2, \ldots, V_r]$, where $V_i \subseteq V$ for $1 \leq i \leq r$. For the vertex order σ, it holds that $x <_\sigma y$ for all $x \in V_i$, $y \in V_j$ such that $i < j$.

3 Outerplanar DAGs

We study the stack number of oDAGs, that is, directed acyclic outerplanar graphs. We stress that such graphs are planar but not necessarily upward. For example, the graph in Fig. 1b cannot be drawn in an upward fashion. We assume oDAGs are maximal as it is straightforward to augment an oDAG to a maximal one, and the stack number is a monotone parameter under taking subgraphs.

It is well-known that every maximal outerplanar directed acyclic graph can be constructed from an edge, which we call the *base* edge, by repeatedly *stellating* edges [22]; that is, picking an edge, (s, t), on its outerface and adding a vertex x together with two edges connecting x with s and t; see Fig. 2a. In order to keep the graph acyclic, the directions of the new edges must be either *transitive*:

O.1 $(s, x) \in E$ and $(x, t) \in E$, or *monotone*:
O.2 $(s, x) \in E$ and $(t, x) \in E$, **O.3** $(x, s) \in E$ and $(x, t) \in E$.

We emphasize that every edge, including the base edge, in the construction sequence of outerplanar graphs can be stellated at most once; relaxing the condition yields a construction scheme for 2-trees.

We study subclasses of outerplanar DAGs that can be constructed using a subset of the three operations. First observe that so-called *transitive* oDAGs that are constructed from an edge by applying **O.1** have a single source vertex, a single sink vertex, and an edge connecting the source with the sink. Such graphs

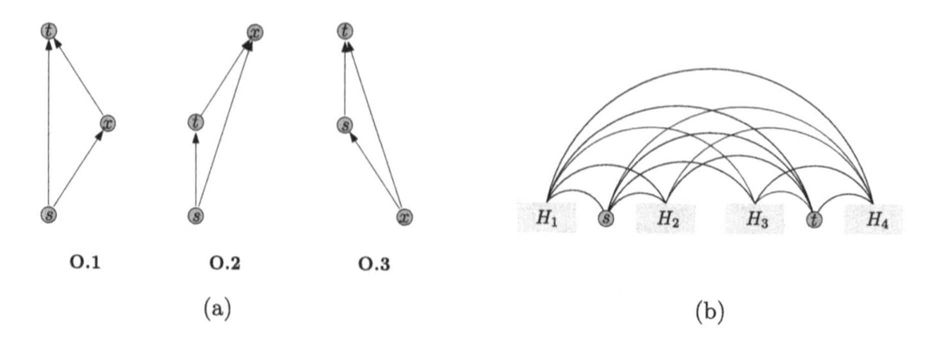

Fig. 2. (a) Possible ways of stellating base edge (s, t) with a vertex x for constructing oDAGs. (b) Vertex order utilized for the inductive schemes in Sect. 3.

are trivially embeddable in one stack. In Sect. 3.1 we observe that single-source oDAGs can be constructed using **O.1** and **O.2**; similarly, single-sink oDAGs can be constructed by **O.1** and **O.3**. We show that single-source (single-sink) oDAGs admit a layout in a constant number of stacks. Furthermore, using monotone operations (**O.2** and **O.3**), one can construct outerplanar graphs with arbitrarily many sources and sinks. Such *monotone* oDAGs admit layouts in a constant number of stacks, as we prove in Sect. 3.2. Finally, we investigate *outerpath* DAGs, that is, oDAGs whose weak dual is a path. In Sect. 3.3 we describe a construction scheme for such graphs and prove that their stack number is constant.

All our proofs are based on an inductive scheme by decomposing a given oDAG into two subgraphs that can be embedded so that a list of carefully chosen invariants is maintained. Then we show how to combine the layouts of the subgraphs and verify the invariants. To this end, we consider a base edge $(s, t) \in E$ and define a vertex order consisting of six vertex-disjoint parts $\sigma = [H_1, s, H_2, H_3, t, H_4]$, where $H_i \subset V, 1 \leq i \leq 4$. For all the considered graph classes, we require that $E(H_2 \to H_3) = \emptyset$; see Fig. 2b. In all figures in the paper all edges are oriented from left to right unless the arrows explicitly indicate edge directions.

3.1 Single-Source oDAGs

Here we consider *single-source* (*single-sink*) outerplanar DAGs that contain only one source (sink) vertex. Single-source oDAGs can be constructed from an edge by applying two of the operations, **O.1** and **O.2**. To this end, choose an edge incident to the source on the outerface of the graph as the base edge, and observe that applying **O.3** would create a predecessor of the source or an additional source. Similarly, single-sink graphs can be constructed by two operations, **O.1** and **O.3**.

Lemma 1. *Every single-source (single-sink) outerplanar DAG admits an order whose twist size is at most* 3.

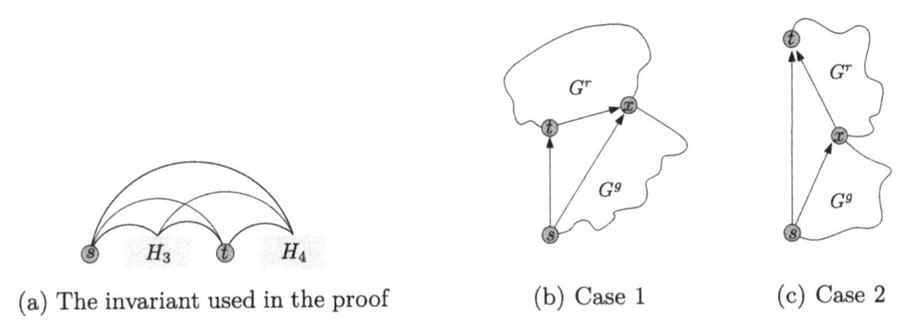

(a) The invariant used in the proof (b) Case 1 (c) Case 2

Fig. 3. An illustration for Lemma 1

Proof (sketch). Let $G = (V, E)$ be a given oDAG with a unique source $s \in V$, and assume that $(s, t) \in E$ is the base edge in the construction sequence of G. We prove the claim by induction on the size of G by using the following invariant (see Fig. 3a): There exists an order of V consisting of four parts, $\sigma = [s, H_3, t, H_4]$ (that is, $H_1 = H_2 = \emptyset$), such that the following holds:

I.1 $twist(s \to H_4, H_3 \to H_4) \leq 1$ **I.2** $twist(E) \leq 3$

Now we prove that these invariants can be maintained. If G is a single edge, then the base of the induction clearly holds. For the inductive case, we consider the base edge (s, t) of G. Let x be the unique neighbor of s and t. Since G is a single-source oDAG, there are two ways the edges between x and s, t are directed, corresponding to operations **O.1** and **O.2**. Consider both cases.

Case 1. First assume $(s, x) \in E$ and $(t, x) \in E$. It is easy to see that G is decomposed into two edge-disjoint subgraphs sharing a single vertex x; denote the graph containing (s, x) by G^g and the graph containing (t, x) by G^r; see Fig. 3b. Since G is a single-source oDAG, G^g is also a single-source oDAG with source s and base edge (s, x). Similarly, G^r is a single-source oDAG with source t and base edge (t, x). By the induction hypothesis, the graphs admit orders σ^g and σ^r satisfying the invariant. Next we combine the orders into a single one for G.

Let $\sigma^g = [s, G_3, x, G_4]$ and $\sigma^r = [t, R_3, x, R_4]$. Then we set

$$\sigma = [s, t, R_3, G_3, x, G_4, R_4]$$

and observe that in the order $H_3 = \emptyset$, $H_4 = [R_3, G_3, x, G_4, R_4]$; see Fig. 4a. It is easy to see that σ is a linear extension of V, that is, $u <_\sigma v$ for all edges $(u, v) \in E$. Next we verify the conditions of the invariant.

I.1. Since $H_3 = \emptyset$, we have $twist(s \to H_4, H_3 \to H_4) = twist(s \to H_4) \leq 1$, where the inequality holds because all edges $E(s \to H_4)$ share a common vertex, s.

I.2. Consider the maximum twist κ in G under vertex order $\sigma = [s, H_3, t, H_4]$, and suppose for contradiction that $|\kappa| \geq 4$. Observe that $H_3 = \emptyset$ in the

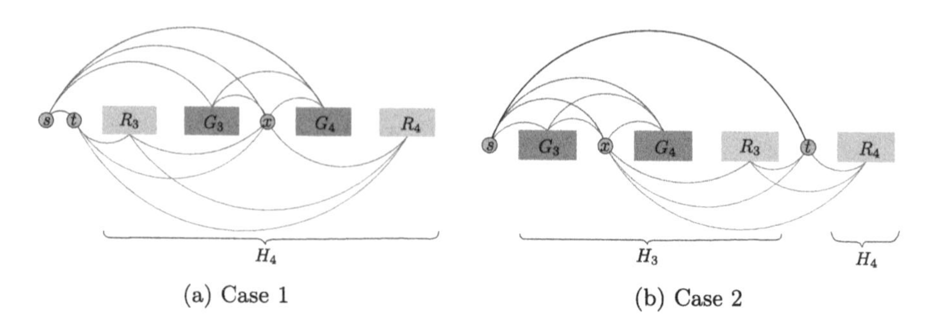

(a) Case 1

(b) Case 2

Fig. 4. An inductive step in the proof of Lemma 1

considered case, and κ may contain at most one edge incident to s and at most one edge incident to t. Thus, at least two of the edges of κ are from $E(H_4 \rightarrow H_4)$; see Fig. 4a. Denote one of the two edges by $e \in \kappa$.
Since $e \in E(H_4 \rightarrow H_4)$, both endpoints of e are in $R_3 \cup G_3 \cup \{x\} \cup G_4 \cup R_4$. Notice that if the two endpoints are in the same part (e.g., R_i or G_i for some i), then all edges of κ have at least one endpoint in that part (since they all cross e), and specifically all edges of κ are either in G^g or G^r, which implies that $|\kappa| \leq 3$ by the induction hypothesis. Hence, we assume that e belongs to $E(R_3 \rightarrow R_4, R_3 \rightarrow x, x \rightarrow R_4, G_3 \rightarrow G_4, G_3 \rightarrow x, x \rightarrow G_4)$ and that none of the edges of κ contains both endpoints in the same part R_i or G_i of V.

- If $e \in E(G_3 \rightarrow x, x \rightarrow G_4)$, then the only edges potentially crossing e are in $E(G_3 \rightarrow G_4, s \rightarrow G_4, s \rightarrow G_3)$, that is, they all belong to G^g. In that case $|\kappa| \leq 3$ by the hypothesis **I.2.** applied to G^g, a contradiction. Therefore, $\kappa \cap E(G_3 \rightarrow x, x \rightarrow G_4) = \emptyset$.
- If $e \in E(G_3 \rightarrow G_4)$, then the edges crossing e are either incident to s, or incident to x, or in $E(G_3 \rightarrow G_4)$. Since $twist(G_3 \rightarrow G_4) \leq 1$, we have that $|\kappa| \leq 3$. Therefore, $\kappa \cap E(G_3 \rightarrow G_4) = \emptyset$.
- If $e \in E(R_3 \rightarrow x)$, then the edges crossing e are in $E(s \rightarrow G_3, t \rightarrow R_3, R_3 \rightarrow R_4)$. Observe that each of the three subsets contributes at most one edge to κ; thus, $|\kappa| \leq 3$. Therefore, $\kappa \cap E(R_3 \rightarrow x) = \emptyset$.
- If $e \in E(x \rightarrow R_4)$, then the edges crossing e are in $E(s \rightarrow G_4, t \rightarrow R_4, R_3 \rightarrow R_4)$. Each of the three subsets contributes at most one edge to κ; thus, $|\kappa| \leq 3$. Therefore, $\kappa \cap E(x \rightarrow R_4) = \emptyset$.
- If $e \in E(R_3 \rightarrow R_4)$, then the edges crossing e are either adjacent to s or t, or in $E(R_3 \rightarrow R_4)$. Since $twist(R_3 \rightarrow R_4) \leq 1$, we have that $|\kappa| \leq 3$, contradicting our assumption.

Case 2. Now assume $(s, x) \in E$ and $(x, t) \in E$; see Fig. 3c. Again, G is decomposed into two edge-disjoint single-source subgraphs sharing a single vertex x; denote the graph containing (s, x) by G^g and the graph containing (x, t) by G^r, where s is the single source of G^g and x is the single source of G^r. By the induction hypothesis, the two graphs admit orders σ^g and σ^r satisfying the invariant. Let $\sigma^g = [s, G_3, x, G_4]$ and $\sigma^r = [x, R_3, t, R_4]$. Then $\sigma = [s, G_3, x, G_4, R_3, t, R_4]$,

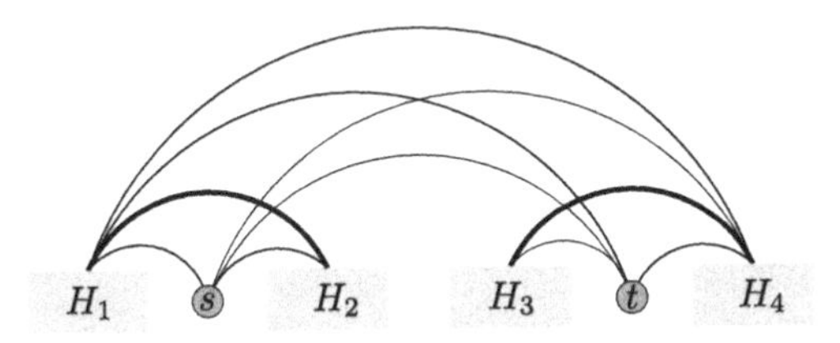

Fig. 5. The invariant used in the proof of Lemma 3

where $H_3 = [G_3, x, G_4, R_3]$, $H_4 = R_4$; see Fig. 4b. In [28] we show that the invariants are maintained. □

The recent result of Davies [13] implies that the stack number of single-source outerplanar DAGs is at most 48. We reduce this upper bound on the stack number to 4 via a similar argument that employs the same recursive decomposition as in Lemma 1. The proof of Lemma 2 is in [28].

Lemma 2. *Every single-source outerplanar DAG admits a* 4*-stack layout.*

It is straightforward to extend Lemma 2 to oDAGs with a constant number of sources (sinks), that is, to construct a layout of an oDAG with $4s$ stacks, where s is the number of sources (sinks) in the graph. Partition the oDAG into s single-source subgraphs and embed each of them in a separate set of 4 stacks.

3.2 Monotone oDAGs

Here we consider *monotone* outerplanar DAGs that are constructed from an edge by applying operations **O.2** and **O.3**. As in the previous section, we assume that the construction sequence along with the base edge is known.

Lemma 3. *Every monotone outerplanar DAG admits an order whose twist size is at most* 4*.*

Proof (sketch). We prove the claim by induction on the size of the given oDAG, $G = (V, E)$, by using the following invariants (see Fig. 5): For a base edge $(s, t) \in E$, there exists a vertex order consisting of six parts, $\sigma = [H_1, s, H_2, H_3, t, H_4]$, such that the following holds:

I.1 $E(H_1 \rightarrow H_3) = E(H_2 \rightarrow H_3) = E(H_2 \rightarrow H_4) = \emptyset$
I.2 $twist(H_1 \cup \{s\} \rightarrow \{t\} \cup H_4) \leq 1$
I.3 $twist(H_1 \rightarrow H_2 \cup \{t\} \cup H_4) \leq 2$
I.4 $twist(H_1 \cup \{s\} \cup H_3 \rightarrow H_4) \leq 2$
I.5 $twist(E) \leq 4$

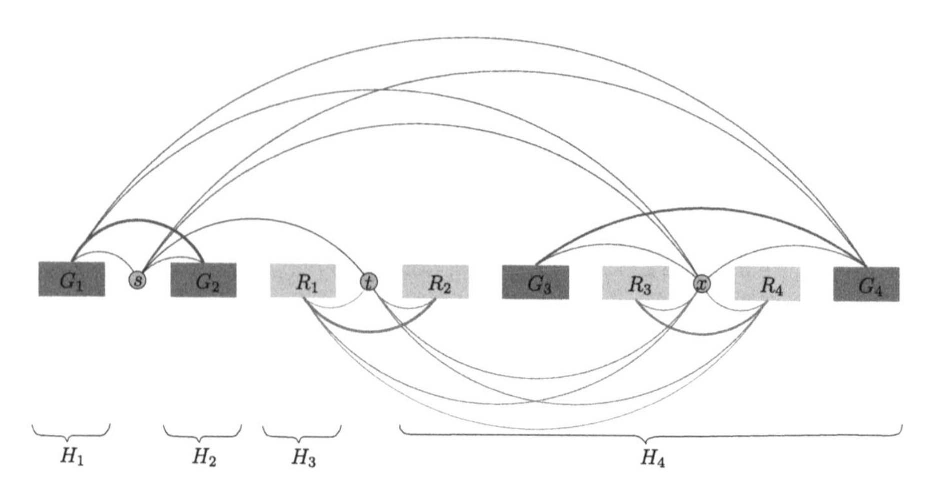

Fig. 6. An inductive step used in the proof of Lemma 3

If G consists of a single edge, then the base of the induction clearly holds. For the inductive case, consider the base edge (s, t) of G and choose the unique common neighbor of s and t, denoted $x \in V$. Since G is monotone and acyclic, there are two ways the edges between x and s, t are directed: either $(s, x) \in E, (t, x) \in E$ (**O.2**) or $(x, s) \in E, (x, t) \in E$ (**O.3**). Observe that, since a (monotone) outerplanar DAG remains (monotone) outerplanar after reversing all edge directions and the described invariants are symmetric with respect to parts H_1, H_2 and parts H_3, H_4, it is sufficient to study only one of the two cases. Therefore we investigate the former case, while the latter case follows from the symmetry.

Assume $(s, x) \in E$ and $(t, x) \in E$. It is easy to see that G is decomposed into two edge-disjoint monotone oDAGs sharing a vertex $x \in V$; denote the graph containing (s, x) by G^g and the graph containing (t, x) by G^r. By the induction hypothesis, the two graphs admit orders σ^g and σ^r satisfying the described invariant. Let $\sigma^g = [G_1, s, G_2, G_3, x, G_4]$ and $\sigma^r = [R_1, t, R_2, R_3, x, R_4]$. Then

$$\sigma = [G_1, s, G_2, R_1, t, R_2, G_3, R_3, x, R_4, G_4]$$

where $H_1 = G_1$, $H_2 = G_2$, $H_3 = R_1$, $H_4 = [R_2, G_3, R_3, x, R_4, G_4]$ in σ; see Fig. 6. In [28] we show that the invariants are maintained under σ. □

3.3 Outerpath DAGs

Let G be an embedded (plane) graph. Recall that a *weak dual* is a graph whose vertices are bounded faces of G and edges connect adjacent faces of G. A graph is an *outerpath* if its weak dual is a path. Consider a face of an outerpath $G = (V, E)$ that corresponds to a terminal of the path, and make an edge on the face adjacent to the outerface of G to be a base edge. It is easy to see that every outerpath can be constructed from such a base edge by repeatedly stellating edges such that the following holds (which keeps the weak dual to be a path): After stellating edge (u, v) with a vertex w, only one of the two newly added edges, $\{u, w\}$ and $\{v, w\}$, can be further stellated. In order to construct an outerpath DAG, the directions of the edges have to follow one of the operations, **O.1**, **O.2**, or **O.3**.

Lemma 4. *Every outerpath DAG admits an order whose twist size is at most* 4.

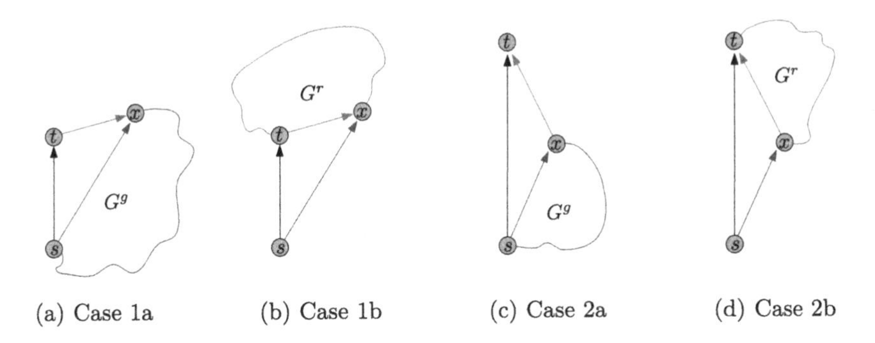

Fig. 7. Cases in Lemma 4: stellating base edge (s, t) with a vertex x

Proof (sketch). We prove the claim by induction on the size of the given outerpath DAG, $G = (V, E)$, by using the following invariants: For a base edge $(s, t) \in E$, there exists a vertex order consisting of six parts, $\sigma = [H_1,\ s,\ H_2,\ H_3,\ t,\ H_4]$, such that the following holds:

I.1 $H_2 = \emptyset$ or $H_3 = \emptyset$, that is, $E(H_2 \rightarrow H_3) = \emptyset$
I.2 $twist(H_2 \rightarrow t, H_2 \rightarrow H_4) \leq 1$
I.3 $twist(H_1 \rightarrow H_3, s \rightarrow H_3) \leq 1$
I.4 $twist(H_1 \cup \{s\} \cup H_2 \rightarrow H_3 \cup \{t\} \cup H_4) \leq 2$
I.5 $twist(H_1 \rightarrow H_2, H_2 \rightarrow \{t\} \cup H_4) \leq 3$
I.6 $twist(H_1 \cup \{s\} \rightarrow H_3, H_3 \rightarrow H_4) \leq 3$
I.7 $twist(H_1 \cup \{s\} \cup H_2 \cup H_3 \rightarrow H_4) \leq 3$
I.8 $twist(H_1 \rightarrow H_2 \cup H_3 \cup \{t\} \cup H_4) \leq 3$
I.9 $twist(E) \leq 4$

If G consists of a single edge, then the base of the induction clearly holds. For the inductive case, consider a base edge $(s, t) \in E$ and choose the unique common neighbor of s and t, denoted $x \in V$. Although all three operations can be applied on (s, t), by symmetry, it is sufficient to study only **O.1** and **O.2**. Depending on which edges of face $\langle s, t, x \rangle$ are utilized for the construction, we distinguish four cases; see Fig. 7. As in earlier proofs we denote the graphs constructed on (s, x) by G^g and the graph on (t, x) by G^r, and assume the graphs admit orders σ^g and σ^r satisfying the invariants. Notice however that, since G is an outerpath, only one of G^g, G^r contains more than two vertices.

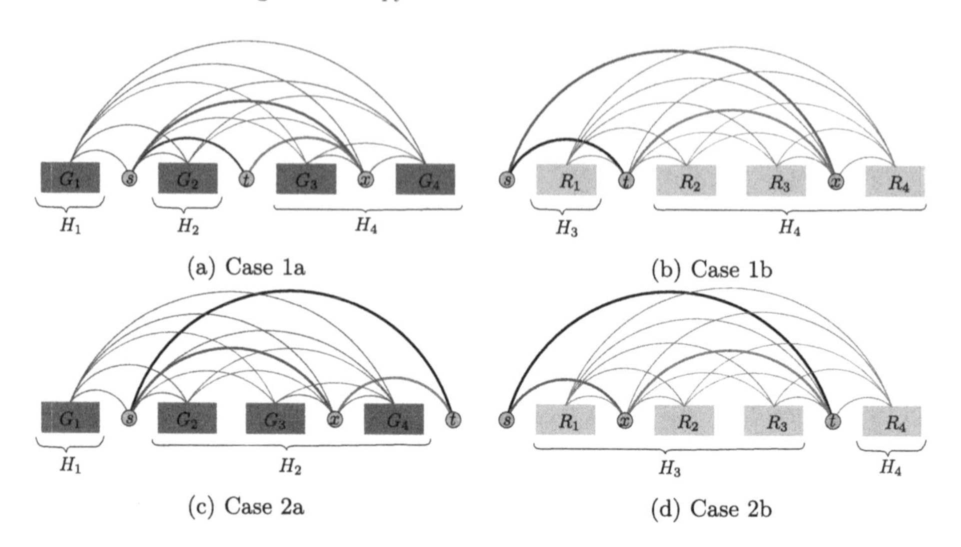

Fig. 8. An illustration for Lemma 4

Case 1a. Assume that $(s, x) \in E$, $(t, x) \in E$, $\sigma^g = [G_1, s, G_2, G_3, x, G_4]$, and $\sigma^r = [t, x]$. We set $\sigma = [G_1, s, G_2, t, G_3, x, G_4]$, where $H_1 = G_1$, $H_2 = G_2$, $H_3 = \emptyset$, $H_4 = [G_3, x, G_4]$ in σ; see Fig. 8a.

Case 1b. Assume $(s, x) \in E$, $(t, x) \in E$, $\sigma^g = [s, x]$, and $\sigma^r = [R_1, t, R_2, R_3, x, R_4]$. We set $\sigma = [s, R_1, t, R_2, R_3, x, R_4]$, where $H_1 = H_2 = \emptyset$, $H_3 = R_1$, $H_4 = [R_2, R_3, x, R_4]$ in σ; see Fig. 8b.

Case 2a. Assume $(s, x) \in E$, $(x, t) \in E$, $\sigma^g = [G_1, s, G_2, G_3, x, G_4]$, and $\sigma^r = [x, t]$. We set $\sigma = [G_1, s, G_2, G_3, x, G_4, t]$, where $H_1 = G_1$, $H_2 = [G_2, G_3, x, G_4]$, $H_3 = H_4 = \emptyset$ in σ; see Fig. 8c.

Case 2b. Assume $(s, x) \in E$, $(x, t) \in E$, $\sigma^g = [s, x]$, and $\sigma^r = [R_1, x, R_2, R_3, t, R_4]$. The case is reduced to *Case 2a* by reversing all edge directions; see Fig. 8d.
In [28] we show that the invariants are maintained in each of the cases. □

4 Upward Planar 3-Trees

Theorem 2. *Every upward planar* 3*-tree admits an order whose twist size is at most* 5.

Proof (sketch). We prove the claim by induction on the size of a given upward planar 3-tree, $G = (V, E)$, by using the following invariants (see Fig. 9a): For the outerface $\langle s, m, t \rangle$ of G, there exists a vertex order consisting of five parts, $\sigma = [s, H_1, m, H_2, t]$, where $H_1, H_2 \subset V$, and the following holds:

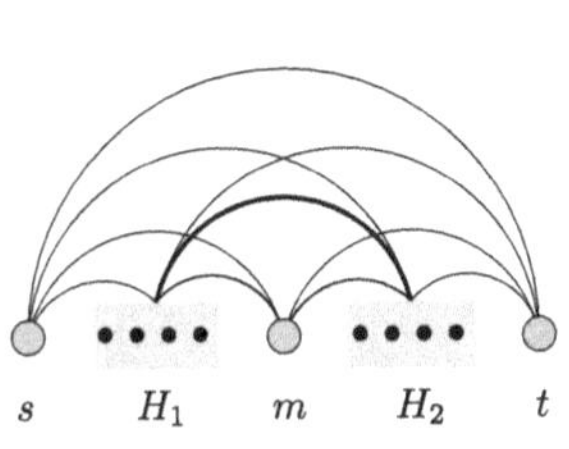

(a) The invariant used in the proof of Theorem 2

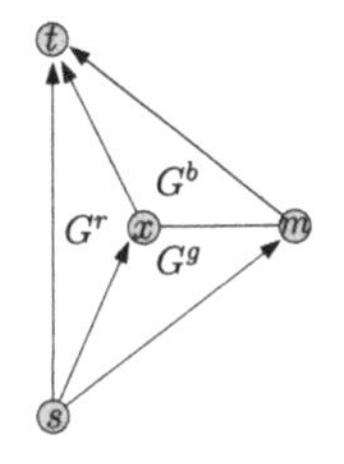

(b) Decomposing an upward planar 3-tree into G^r, G^g, and G^b

Fig. 9. Bounding the twist size of upward planar 3-trees

I.1 $twist(\{s\} \cup H_1 \rightarrow H_2 \cup \{t\}) \leq 2$ **I.2** $twist(E) \leq 5$

The base of the induction clearly holds when G is a triangle. For the inductive case, consider the outerface, $\langle s, m, t \rangle$, of G and identify the unique vertex, $x \in V$, adjacent to s, m, t. Since G is upward planar, we have $(s, x) \in E$, $(x, t) \in E$; for the direction of the edge between x and m, there are two possible cases. We can reduce one case to another one by reversing edge directions, which preserves upward planarity of the graph. Therefore, we study only one of the cases.

Assume $(x, m) \in E$. Then G is decomposed into three upward planar subgraphs bounded by faces $\langle s, x, t \rangle$, $\langle s, x, m \rangle$, and $\langle x, m, t \rangle$; denote the graphs by G^r, G^g, and G^b, respectively; see Fig. 9b. By the induction hypothesis, the three graphs admit orders $\sigma^r, \sigma^g, \sigma^b$ satisfying the described invariants. Let $\sigma^r = [s, R_1, x, R_2, t]$, $\sigma^g = [s, G_1, x, G_2, m]$, and $\sigma^b = [x, B_1, m, B_2, t]$. Then

$$\sigma = [s, R_1, G_1, x, G_2, R_2, B_1, m, B_2, t]$$

where $H_1 = [R_1, G_1, x, G_2, R_2, B_1]$ and $H_2 = B_2$; see [28], where we show that the invariants are maintained under the vertex order. □

5 Lower Bounds

We construct and computationally verify specific graphs that require a minimum number of stacks in every layout utilizing a SAT formulation of the linear layout problem [9,32]. Using a modern SAT solver, one can evaluate small and medium size instances (up to a few hundred of vertices) within a few seconds. An online tool and the source code of the implementation is available at [30].

Using the formulation, we identified a single-source single-sink upward oDAG that requires three stacks; see Fig. 10a. There are only two linear extensions of the graph, $[a, b, c, d, e, f]$ and $[a, b, d, c, e, f]$, and both require three stacks.

Next we found an upward outerplanar DAG with four sources and three sinks whose stack number is 4; see Fig. 10b. This oDAG is upward but not monotone, that is, it requires an addition of transitive edges via operation **O.1**.

Finally, we construct an upward planar 3-tree that requires five stacks; see Fig. 11a. The results are summarized in Theorem 3.

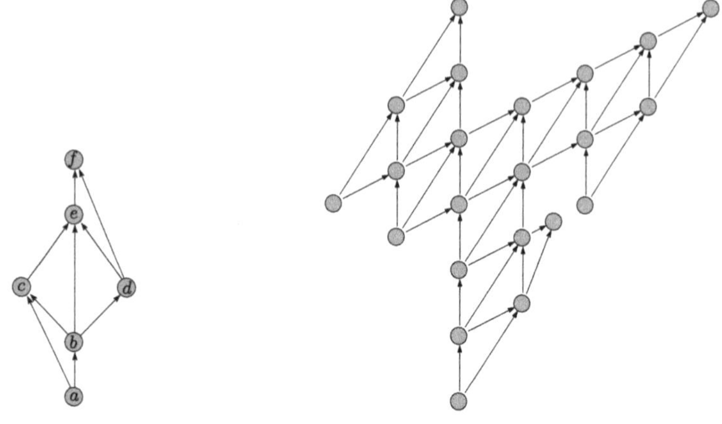

(a) A singles-source single-sink outerplanar DAG that requires three stacks

(b) An upward outerplanar DAG that requires four stacks

Fig. 10. Lower bound examples

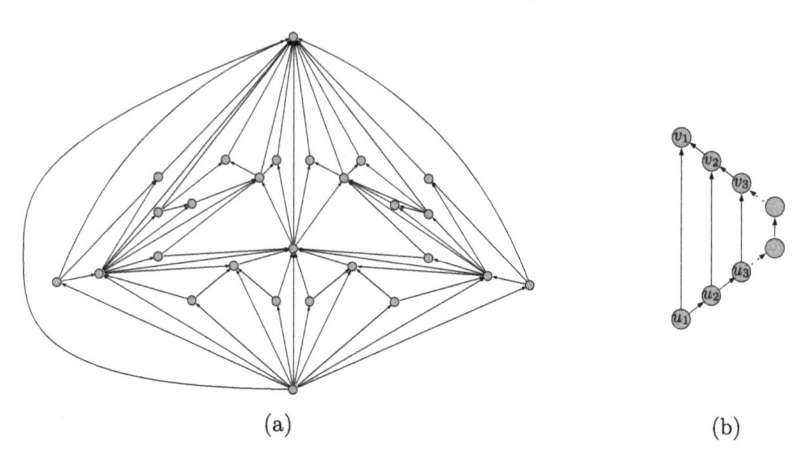

(a)

(b)

Fig. 11. (a) An upward planar DAG that require 5 stacks. (b) An upward outerplanar DAG that requires $n/2$ queues.

6 Conclusions

In this paper we studied the stack number of upward planar and outerplanar DAGs and provided improved upper and lower bounds for some interesting subclasses via their maximum twist sizes. With the recent results of Jungeblut et al. [22] one of the intriguing open questions is to decrease the gap between our lower bound of 4 and their upper bound of 24776 for oDAGs. Moreover, since our upper bounds are mostly based on bounding the twist number, they are likely too large and it would be interesting to decrease them further.

A *queue layout* of DAGs is a related concept, in which a pair of edges cannot nest. While two queues are sufficient for trees and unicyclic DAGs [19], there exist single-source single-sink upward oDAGs that require a linear number of queues; see Fig. 11b. This is in contrast with undirected planar graphs, which have a constant queue number [2,15]. We suggest to investigate mixed stack-queue layouts in which every page is either a stack or a queue [6,12,31]. Another direction is to parameterize the queue number by a graph parameter that is tied to the queue number for undirected graphs, such as the width of a poset [3,33].

Acknowledgments. We thank the organizers and other participants of Dagstuhl seminar 19092 "Beyond-Planar Graphs: Combinatorics, Models and Algorithms", where this work started, in particular F. Frati and T. Mchedlidze. M. Nöllenburg acknowledges funding by the Vienna Science and Technology Fund (WWTF) [10.47379/ICT19035].

References

1. Ageev, A.A.: A triangle-free circle graph with chromatic number 5. Discret. Math. **152**(1–3), 295–298 (1996). https://doi.org/10.1016/0012-365X(95)00349-2
2. Alam, J.M., Bekos, M.A., Gronemann, M., Kaufmann, M., Pupyrev, S.: Queue layouts of planar 3-trees. Algorithmica **82**(9), 2564–2585 (2020). https://doi.org/10.1007/s00453-020-00697-4
3. Alam, J.M., Bekos, M.A., Gronemann, M., Kaufmann, M., Pupyrev, S.: Lazy queue layouts of posets. Algorithmica **85**(5), 1176–1201 (2023). https://doi.org/10.1007/s00453-022-01067-y
4. Alhashem, M., Jourdan, G., Zaguia, N.: On the book embedding of ordered sets. Ars Comb. **119**, 47–64 (2015)
5. Alzohairi, M., Rival, I.: Series-parallel planar ordered sets have pagenumber two. In: North, S. (ed.) GD 1996. LNCS, vol. 1190, pp. 11–24. Springer, Heidelberg (1997). https://doi.org/10.1007/3-540-62495-3_34
6. Angelini, P., Bekos, M.A., Kindermann, P., Mchedlidze, T.: On mixed linear layouts of series-parallel graphs. Theor. Comput. Sci. **936**, 129–138 (2022). https://doi.org/10.1016/j.tcs.2022.09.019
7. Bekos, M.A., et al.: On linear layouts of planar and k-planar graphs. In: Hong, S.H., Kaufmann, M., Pach, J., Tóth, C.D. (eds.) Beyond-Planar Graphs: Combinatorics, Models and Algorithms, vol. 167, p. 144. Dagstuhl Reports (2019)
8. Bekos, M.A., Da Lozzo, G., Frati, F., Gronemann, M., Mchedlidze, T., Raftopoulou, C.N.: Recognizing DAGs with page-number 2 is NP-complete. Theor. Comput. Sci. **946**, 113689 (2023). https://doi.org/10.1016/j.tcs.2023.113689
9. Bekos, M.A., Kaufmann, M., Klute, F., Pupyrev, S., Raftopoulou, C.N., Ueckerdt, T.: Four pages are indeed necessary for planar graphs. J. Comput. Geom. **11**(1), 332–353 (2020). https://doi.org/10.20382/jocg.v11i1a12
10. Bhore, S., Da Lozzo, G., Montecchiani, F., Nöllenburg, M.: On the upward book thickness problem: combinatorial and complexity results. Eur. J. Comb. **110**, 103662 (2023). https://doi.org/10.1016/j.ejc.2022.103662
11. Binucci, C., Da Lozzo, G., Di Giacomo, E., Didimo, W., Mchedlidze, T., Patrignani, M.: Upward book embeddings of st-graphs. In: Barequet, G., Wang, Y. (eds.) International Symposium on Computational Geometry. LIPIcs, vol. 129, pp. 13:1–13:22. Schloss Dagstuhl - Leibniz-Zentrum für Informatik (2019). https://doi.org/10.4230/LIPIcs.SoCG.2019.13

12. de Col, P., Klute, F., Nöllenburg, M.: Mixed linear layouts: complexity, heuristics, and experiments. In: Archambault, D., Tóth, C.D. (eds.) GD 2019. LNCS, vol. 11904, pp. 460–467. Springer, Cham (2019). https://doi.org/10.1007/978-3-030-35802-0_35
13. Davies, J.: Improved bounds for colouring circle graphs. Proc. Am. Math. Soc. **150**(12), 5121–5135 (2022). https://doi.org/10.1090/proc/16044
14. Di Giacomo, E., Didimo, W., Liotta, G., Wismath, S.K.: Book embeddability of series-parallel digraphs. Algorithmica **45**(4), 531–547 (2006). https://doi.org/10.1007/s00453-005-1185-7
15. Dujmović, V., Joret, G., Micek, P., Morin, P., Ueckerdt, T., Wood, D.R.: Planar graphs have bounded queue-number. J. ACM **67**(4), 1–38 (2020). https://doi.org/10.1145/3385731
16. Frati, F., Fulek, R., Ruiz-Vargas, A.J.: On the page number of upward planar directed acyclic graphs. J. Graph Algorithms Appl. **17**(3), 221–244 (2013). https://doi.org/10.7155/jgaa.00292
17. Heath, L.S., Pemmaraju, S.V.: Stack and queue layouts of posets. SIAM J. Discrete Math. **10**(4), 599–625 (1997). https://doi.org/10.1137/S0895480193252380
18. Heath, L.S., Pemmaraju, S.V.: Stack and queue layouts of directed acyclic graphs: Part II. SIAM J. Comput. **28**(5), 1588–1626 (1999). https://doi.org/10.1137/S0097539795291550
19. Heath, L.S., Pemmaraju, S.V., Trenk, A.N.: Stack and queue layouts of directed acyclic graphs: Part I. SIAM J. Comput. **28**(4), 1510–1539 (1999). https://doi.org/10.1137/S0097539795280287
20. Hung, L.: A planar poset which requires 4 pages. Ars Comb. **35**, 291–302 (1993)
21. Jungeblut, P., Merker, L., Ueckerdt, T.: A sublinear bound on the page number of upward planar graphs. In: Symposium on Discrete Algorithms, pp. 963–978. SIAM (2022). https://doi.org/10.1137/1.9781611977073.42
22. Jungeblut, P., Merker, L., Ueckerdt, T.: Directed acyclic outerplanar graphs have constant stack number. In: Foundations of Computer Science (2023). https://arxiv.org/abs/2211.04732, to appear
23. Kostochka, A.: Upper bounds on the chromatic number of graphs. Trudy Inst. Mat. (Novosibirsk) 10(Modeli i Metody Optim.), 204–226 (1988)
24. Kostochka, A., Kratochvíl, J.: Covering and coloring polygon-circle graphs. Discret. Math. **163**(1–3), 299–305 (1997). https://doi.org/10.1016/S0012-365X(96)00344-5
25. Mchedlidze, T., Symvonis, A.: Crossing-free acyclic hamiltonian path completion for planar *st*-digraphs. In: Dong, Y., Du, D.-Z., Ibarra, O. (eds.) ISAAC 2009. LNCS, vol. 5878, pp. 882–891. Springer, Heidelberg (2009). https://doi.org/10.1007/978-3-642-10631-6_89
26. Merker, L.: Ordered covering numbers. Masters thesis, Karlsruhe Institute of Technology (2020)
27. Nowakowski, R., Parker, A.: Ordered sets, pagenumbers and planarity. Order **6**(3), 209–218 (1989). https://doi.org/10.1007/BF00563521
28. Nöllenburg, M., Pupyrev, S.: On families of planar DAGs with constant stack number. CoRR abs/2107.13658 (2023). http://arxiv.org/abs/2107.13658
29. Papakostas, A.: Upward planarity testing of outerplanar DAGs (extended abstract). In: Tamassia, R., Tollis, I.G. (eds.) GD 1994. LNCS, vol. 894, pp. 298–306. Springer, Heidelberg (1995). https://doi.org/10.1007/3-540-58950-3_385
30. Pupyrev, S.: A SAT-based solver for constructing optimal linear layouts of graphs. Source code available at: https://github.com/spupyrev/bob

31. Pupyrev, S.: Mixed linear layouts of planar graphs. In: Frati, F., Ma, K.-L. (eds.) GD 2017. LNCS, vol. 10692, pp. 197–209. Springer, Cham (2018). https://doi.org/10.1007/978-3-319-73915-1_17
32. Pupyrev, S.: Improved bounds for track numbers of planar graphs. J. Graph Algorithms Appl. **24**(3), 323–341 (2020). https://doi.org/10.7155/jgaa.00536
33. Pupyrev, S.: Queue layouts of two-dimensional posets. In: Angelini, P., von Hanxleden, R. (eds.) International Symposium on Graph Drawing and Network Visualization. LNCS, vol. 13764, pp. 353–360. Springer, Cham (2022). https://doi.org/10.1007/978-3-031-22203-0_25
34. Sysło, M.M.: Bounds to the page number of partially ordered sets. In: Nagl, M. (ed.) WG 1989. LNCS, vol. 411, pp. 181–195. Springer, Heidelberg (1990). https://doi.org/10.1007/3-540-52292-1_13
35. Yannakakis, M.: Embedding planar graphs in four pages. J. Comput. Syst. Sci. **38**(1), 36–67 (1989). https://doi.org/10.1016/0022-0000(89)90032-9

On 3-Coloring Circle Graphs

Patricia Bachmann(✉), Ignaz Rutter, and Peter Stumpf

Faculty of Informatics and Mathematics, University of Passau, Passau, Germany
{bachmanp,rutter,stumpf}@fim.uni-passau.de

Abstract. Given a graph G with a fixed vertex order $\prec$, one obtains a circle graph H whose vertices are the edges of G and where two such edges are adjacent if and only if their endpoints are pairwise distinct and alternate in $\prec$. Therefore, the problem of determining whether G has a k-page book embedding with spine order $\prec$ is equivalent to deciding whether H can be colored with k colors. Finding a k-coloring for a circle graph is known to be NP-complete for $k \geq 4$ and trivial for $k \leq 2$. For $k = 3$, Unger (1992) claims an efficient algorithm that finds a 3-coloring in $O(n \log n)$ time, if it exists. Given a circle graph H, Unger's algorithm (1) constructs a 3-SAT formula Φ that is satisfiable if and only if H admits a 3-coloring and (2) solves Φ by a backtracking strategy that relies on the structure imposed by the circle graph. However, the extended abstract misses several details and Unger refers to his PhD thesis (in German) for details.

In this paper we argue that Unger's algorithm for 3-coloring circle graphs is not correct and that 3-coloring circle graphs should be considered as an open problem. We show that step (1) of Unger's algorithm is incorrect by exhibiting a circle graph whose formula Φ is satisfiable but that is not 3-colorable. We further show that Unger's backtracking strategy for solving Φ in step (2) may produce incorrect results and give empirical evidence that it exhibits a runtime behaviour that is not consistent with the claimed running time.

Keywords: book embedding · circle graphs · graph coloring

1 Introduction

Let $G = (V, E)$ be a graph. A *k-page book embedding* of G is a total order $\prec$ of V and a partition of E into k sets $E_1, \dots, E_k$, called *pages* such that no page E_i contains two edges $\{u_1, v_1\}$, $\{u_2, v_2\}$ with $v_1 \prec v_2 \prec u_1 \prec u_2$. The *page number* (also called *stack number*) of a graph is the smallest k such that G admits a k-page booking embedding.

Book embeddings are a central element to graph drawing. They have been studied in the context of VSLI design [7], arc diagrams [14] and circular layouts [3] as well as clustered planarity [10] and simultaneous embedding [1]. There

Funded by the Deutsche Forschungsgemeinschaft (German Research Foundation, DFG) under grant RU-1903/3-1.

M. A. Bekos and M. Chimani (Eds.): GD 2023, LNCS 14465, pp. 152–160, 2023.
https://doi.org/10.1007/978-3-031-49272-3_11

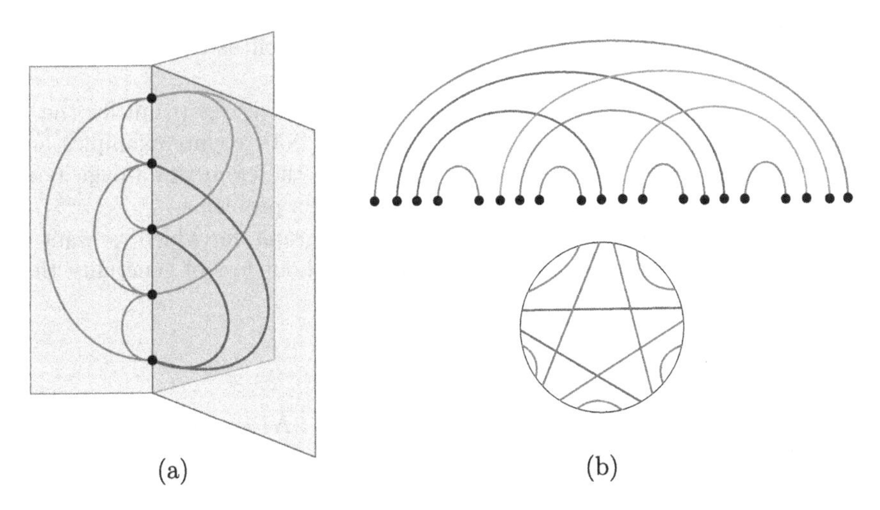

Fig. 1. (a) Book embedding for K_5. (b) Chord diagram of the corresponding circle graph H.

is a plethora of results that aim at bounding the page numbers of various graph classes. For example, the page number of planar graphs is 4 [5,15] and the page number of 1-planar graphs is at most 39 [4].

Since computing the page number is NP-complete [6], often additional restrictions are imposed. One such restriction is to find a k-page book embedding with a fixed order $\prec$. This problem is closely related to the k-coloring problem on circle graphs. A *k-coloring* of a graph G is a function $\text{col} : V(G) \to \{1, \ldots, k\}$ such that $\text{col}(u) \neq \text{col}(v)$ for every $\{u, v\} \in E(G)$. A *circle graph* is an undirected graph H that has an intersection representation with chords of a circle. More precisely, in a *chord diagram* of H we represent each vertex $v \in V(H)$ by a chord C_v such that two chords C_u and C_v intersect if and only if $\{u, v\} \in E(H)$.

Finding a k-page book embedding of a graph $G = (V, E)$ with a fixed vertex order $\prec$ is equivalent to solving the k-coloring problem for circle graphs. To see this, construct a graph H such that $V(H) = E$ and two edges in $V(H)$ are adjacent in H if and only if their endpoints alternate in $\prec$. Then two edges of G can be in the same page of a book embedding with order $\prec$ if and only if they are not adjacent in H. Therefore we obtain a bijection between the k-page book embeddings with order $\prec$ and the k-colorings of H. It is further readily seen that H admits a chord diagram; see Fig. 1 for an illustration.

The k-coloring problem for circle graphs is known to be NP-complete for $k \geq 4$ [11] and efficiently solvable for $k \leq 2$. The case $k = 3$ remained as an open problem until Unger claimed it to be solvable in polynomial time [13]. Unfortunately, in the publication many details and proofs are missing and no journal version followed. Instead, Unger refers to his PhD thesis [12] for a full version, which is written in German and not available online. This bad state of affairs has been pointed out by David Eppstein in a blog post, where he writes that the problem should be considered open [9], as well as by Dujmović and Wood [8].

In this paper we present Unger's ideas for an efficient algorithm for the 3-coloring problem on circle graphs and show, using both counterexamples and empirical results, why 3-coloring circle graphs and therefore the 3-page book embedding problem should indeed be considered open problems.

Throughout this work let $G = (V, E)$ be a circle graph, for which we want to decide the existence of a 3-coloring. We assume without loss of generality that G is connected and contains no induced K_4.

2 Unger's 3-Coloring Algorithm

For a graph H let $\mathcal{P}(H) = \{\{u, v\} \in \binom{V(H)}{2} \setminus E(H) \mid N(u) \cap N(v) \neq \emptyset\}$ be the pairs of non-adjacent vertices that have a common neighbor and let $\mathcal{D} \subseteq \mathcal{P}(H)$. An *auxiliary coloring function for* $\mathcal{D}$ is a function $c_{\text{aux}} \colon \mathcal{D} \to \{\texttt{true}, \texttt{false}\}$. An auxiliary coloring function is *realizable* if there exists a 3-coloring col of H such that for each pair $\{x, y\} \in \mathcal{D}$ we have $\text{col}(x) = \text{col}(y) \iff c_{\text{aux}}(\{x, y\}) = \texttt{true}$.

Let now $G = (V, E)$ be a graph and let $\mathcal{G}$ be a set of induced subgraphs of G and let $\mathcal{P}(\mathcal{G}) = \bigcup_{H \in \mathcal{G}} \mathcal{P}(H)$. An *auxiliary coloring function of* G *with respect to* $\mathcal{G}$ is an auxiliary coloring function for $\mathcal{P}(\mathcal{G})$. Such a function is called *consistent* if its restriction $c_{\text{aux}}|_{\mathcal{P}(H)}$ is realizable for each $H \in \mathcal{G}$.

In his approach, Unger constructs for a given circle graph G a family $\mathcal{G}$ of induced subgraphs, which he calls important subgraphs, such that (i) an auxiliary coloring function c_{aux} with respect to $\mathcal{G}$ is consistent if and only if it is realizable, (ii) the existence of a consistent auxiliary coloring function can be expressed by a 3-SAT formula Φ that (iii) can be solved efficiently with a backtracking algorithm.

Let $\mathcal{G}'$ be the family of induced subgraphs of G that are isomorphic to one of the graphs $G_{\lhd}$, $G_{\square}$, $G_{\pentagon}$, $G_{\boxslash}$ from Fig. 2 or to a cycle C_k with $k \geq 6$. For each $H \in \mathcal{G}'$ there is a formula $\Phi(H)$ whose satisfying truth assignments are the realizable auxiliary coloring functions for H. For the three graphs $G_{\lhd}$, $G_{\square}$ and $G_{\boxslash}$, the corresponding formula $\Phi(H)$ is in fact a 2-SAT formula as shown in Fig. 2, where for C_k with $k \geq 5$, the formula $\Phi(H)$ given by Unger is a 3-SAT formula of size linear in k, which uses some additional variables; we refer to the full version [2] for details. For $G_{\pentagon}$, i.e. cycles with $k = 5$, Unger additionally uses the 2-SAT clauses shown in Fig. 2. Then the existence of a consistent auxiliary coloring function for $\mathcal{G}'$ can be expressed by the formula $\Phi(\mathcal{G}') = \bigwedge_{H \in \mathcal{G}'} \Phi(H)$.

The family $\mathcal{G}'$ is however, still too large, e.g., it may contain an exponential number of cycles. Therefore Unger restricts his important subgraphs $\mathcal{G}$ to a subset of $\mathcal{G}'$ that is defined according to a chord diagram of G. To this end, take a chord diagram of G and consider it to be cut open and rolled out such that the chords form arcs over a straight line; see Fig. 3(c). In what follows we identify each vertex v with its chord C_v.

A chord u *encases* a chord v if the endpoints of v lie between the endpoints of u. Further u *directly encases* v if it encases v and there is no vertex $w \neq u, v$ such that u encases w and w encases v; see Fig. 4 for an example.

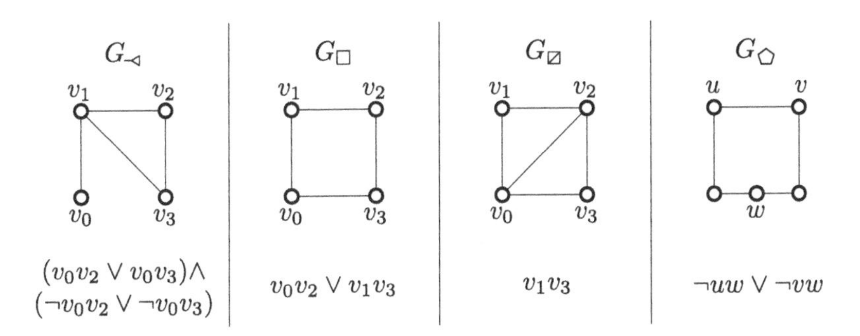

Fig. 2. Important subgraphs and the clauses they contribute to Φ.

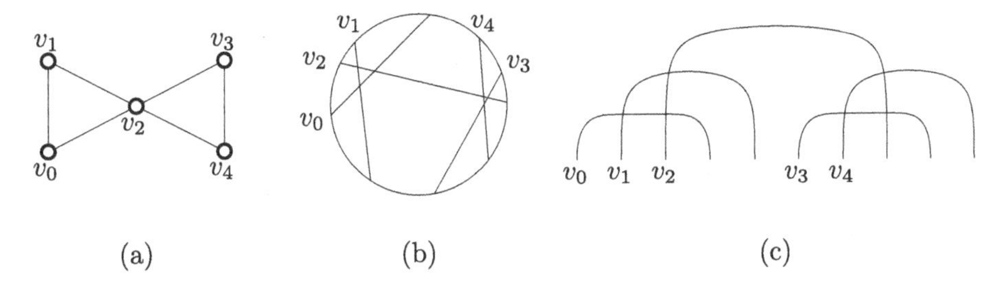

Fig. 3. (a) An undirected graph G. (b) Representation of G of chords on a circle. (c) Alternative circle graph representation for G.

The *levels* of a circle graph G are recursively defined as

$$\text{level}_G(l) = \begin{cases} \{v \in V(G) \colon \nexists u \in V(G) \colon u \text{ encases } v\} & l = 1 \\ \{v \in V(G) \colon \exists u \in \text{level}_G(l-1) \colon u \text{ directly encases } v\} & l > 1 \end{cases}.$$

The set of important subgraphs $\mathcal{G}$ consists of those graphs $H \in \mathcal{G}'$ where if H is isomorphic to $G_{\pentagon}$ or to C_k with $k \geq 6$, then all vertices of H belong to two adjacent levels, and otherwise for each pair $\{u, v\} \in \mathcal{P}(H)$ the vertices u and v are either on the same level or one directly encases the other. Unger's algorithm relies on two claims [13, p.394 ff.]: (1) The graph G is 3-colorable if and only if there exists a consistent auxiliary coloring function with respect to $\mathcal{G}$. (2) The formula $\Phi(\mathcal{G})$ can be solved efficiently by a backtracking algorithm whose search tree has $O(\log n)$ leaves.

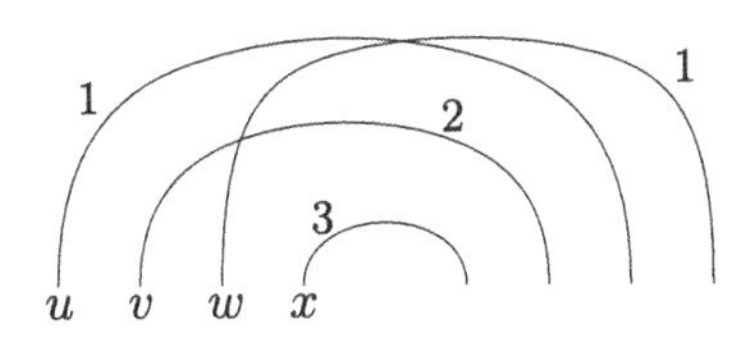

Fig. 4. Chord u encases v and x and directly encases v; v and w directly encase x.

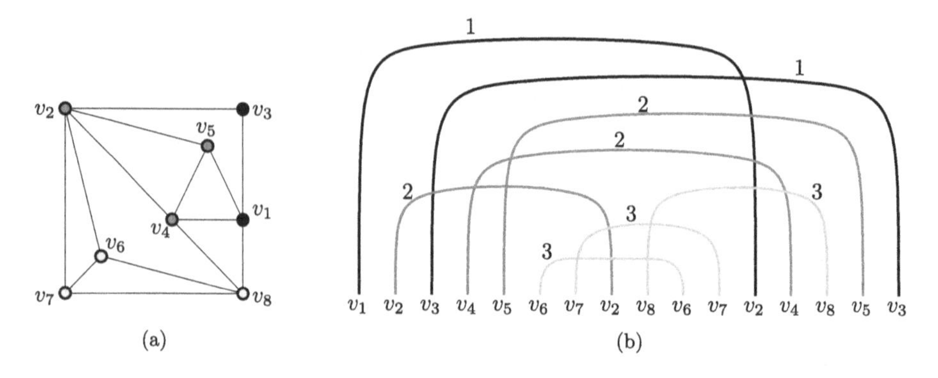

Fig. 5. Counterexample G. Lighter shade indicate higher level.

3 The Counterexample

We show that Unger's claim (1) is false by giving a counterexample. Let G be the graph given in Fig. 5a. Observe that G is a circle graph as witnessed by the chord diagram in Fig. 5b.

We first show that G is not 3-colorable. Namely, the two subgraphs induced by $\{v_1, v_2, v_4, v_5\}$ and $\{v_2, v_6, v_7, v_8\}$ imply $\mathrm{col}(v_1) = \mathrm{col}(v_2)$ and $\mathrm{col}(v_2) = \mathrm{col}(v_8)$, respectively. However v_1 and v_8 are adjacent.

On the other hand, we show that for the family $\mathcal{G}$ of important subgraphs of G with respect to the chord diagram in Fig. 5, the formula $\Phi(\mathcal{G})$ is satisfiable. We first give the important subgraphs of G and then construct the corresponding formula. In Fig. 5a the vertices are colored according to their levels, where lighter colors indicate higher levels. It is not hard to see that G contains no induced cycle isomorphic to $G_{\pentagon}$ or to C_k with $k \geq 6$ that is contained in two adjacent levels. It hence suffices to find the important subgraphs isomorphic to $G_{\lhd}$,$G_{\square}$, and $G_{\boxslash}$.

We start with $G_{\square}$ and $G_{\boxslash}$. Observe that each of the pairs $\{v_2, v_8\}$ and $\{v_3, v_8\}$ neither lies on the same level nor does one the vertices directly encase the other. Hence no important subgraph contains both vertices of these pairs. For $G_{\square}$ this leaves only the subgraphs listed in Table 1a and, similarly, for $G_{\boxslash}$ only the subgraphs listed in Table 1b. Finally, it is straightforward to check that for $G_{\lhd}$ the only subgraphs are listed in Table 1c. A detailed description of all important subgraphs in G is given in the full version [2].

Table 1 shows the clauses of the formula $\Phi := \Phi(\mathcal{G})$. Finally, Table 2 gives a satisfying truth assignment for Φ. The underlined literals in Table 1 are those satisfied by that truth assignment. Since every clause contains a satisfied literal, Φ is satisfiable and the truth assignment defines a consistent auxiliary coloring function for $\mathcal{G}$. However, as G is not 3-colorable, this contradicts Unger's claim (1).

Different Notions of Important Subgraphs. We note that the definition of important subgraphs subtly differs between the extended abstract [13] and Unger's

Table 1. Important subgraphs and the corresponding clauses for each of (a) $G_{\square}$, (b) $G_{\boxslash}$, and (c) $G_{\multimap}$. For the satisfying variable assignment from Table 2 all true literals are underlined.

Important Subgraph $G_{\square}$	Clauses
$\{v_1, v_2, v_3, v_4\}$	$\underline{v_1v_2} \vee \underline{v_3v_4}$
$\{v_1, v_2, v_3, v_5\}$	$\underline{v_1v_2} \vee v_3v_5$

(a)

Important Subgraph $G_{\boxslash}$	Clauses
$\{v_1, v_2, v_4, v_5\}$	$\underline{v_1v_2}$
$\{v_1, v_4, v_5, v_8\}$	$\underline{v_5v_8}$

(b)

Important Subgraph $G_{\multimap}$	Clauses
$\{v_1, v_6, v_7, v_8\}$	$(v_1v_6 \vee \underline{v_1v_7}) \wedge (\underline{\neg v_1v_6} \vee \neg\, v_1v_7)$
$\{v_1, v_3, v_4, v_5\}$	$(\underline{v_3v_4} \vee v_3v_5) \wedge (\neg v_3v_4 \vee \underline{\neg v_3v_5})$
$\{v_2, v_3, v_4, v_5\}$	$(\underline{v_3v_4} \vee v_3v_5) \wedge (\neg v_3v_4 \vee \underline{\neg v_3v_5})$
$\{v_2, v_4, v_6, v_7\}$	$(\underline{v_4v_6} \vee v_4v_7) \wedge (\neg v_4v_6 \vee \underline{\neg v_4v_7})$
$\{v_2, v_5, v_6, v_7\}$	$(v_5v_6 \vee \underline{v_5v_7}) \wedge (\underline{\neg v_5v_6} \vee \neg v_5v_7)$
$\{v_2, v_4, v_5, v_6\}$	$(\underline{v_4v_6} \vee v_5v_6) \wedge (\neg v_4v_6 \vee \underline{\neg v_5v_6})$
$\{v_2, v_4, v_5, v_7\}$	$(v_4v_7 \vee \underline{v_5v_7}) \wedge (\underline{\neg v_4v_7} \vee \neg v_5v_7)$

(c)

Table 2. Satisfying variable assignment for Φ.

True	v_1v_2, v_1v_7, v_2v_5, v_3v_4, v_4v_6, v_5v_7, v_5v_8
False	v_1v_6, v_4v_7, v_5v_6, v_3v_5

PhD thesis [12]. Namely, in the thesis, important subgraphs are not defined only via direct encasing but for some types their vertices also have to belong to (at most) two adjacent levels. The counterexample above refutes the claim from the extended abstract. In the full version [2], we give an example which refutes the analogous claim from Unger's thesis.

4 Unger's Backtracking Algorithm

In addition to the counterexample, we investigated the backtracking algorithm described by Unger. Let Φ_1 be the 2-SAT instance obtained from the important subgraphs $G_{\multimap}$, $G_{\square}$, $G_{\boxslash}$ and $G_{\pentagon}$. Checking if Φ_1 is solvable can be done in polynomial time. For the SAT instance Φ_2 obtained from the remaining important subgraphs $G_{\pentagon}$ and C_k for $k \geq 6$ it is less clear how to solve it efficiently. Unger [13] proposes the following modified backtracking algorithm. When encountering a clause during the variable assignment whose literals are all set to `False`, we may jump in the backtracking tree to a variable which (1) results in a literal of that clause being set to `True` by flipping its assigned value and (2) the child that corresponds to this new assigned value is not included in the backtracking tree yet. We note that we rephrased property (2) from "[t]he new value of [the variable] is not already included in the backtracking tree" [13, p. 396]. If no such variable exists, the algorithm terminates and reports that there is no solution for Φ. If a solution is found, it is used to compute a 3-coloring.

Unger claims that this backtracking tree has at most $O(\log n)$ leaves. To verify this claim we implemented his backtracking algorithm, ran it on randomly generated 3-colorable circle graphs and counted the number of leaves in the backtracking tree; see Fig. 6(a) and further details on the test data are given in the full version [2]. While the algorithm is reasonably fast, the number of leaves scatters quite a bit, e.g. some graphs produce backtracking trees with around 500 leaves even though they have only 250 to 500 vertices. This is likely not consistent with an upper bound of $O(\log n)$. More importantly, while all the graphs are 3-colorable, the backtracking algorithm reports no solution for all but one of the instances. In the full version [2], we show how the algorithm may already fail for a cycle of length 5 if the order of variable assignments is poorly chosen.

Therefore, we also evaluated a regular backtracking algorithm on the same instances. We found that the number of leaves of these backtracking trees grows exponentially. Figure 6(b) shows the percentage of solved instances within a time limit of one hour. Notably, starting at around 50 vertices, the algorithm barely manages to solve any instances. For readability, the Figure shows only the fraction of solved instances with up to 250 vertices as the backtracking algorithm does not terminate on any of the larger instances. This indicates the impracticality of a regular backtracking approach and that crucial insights are still missing to see how a backtracking approach could be modified to be more efficient on circle graphs.

Finally, we also implemented the coloring algorithm that uses the solution for Φ to color the graph. Our result is that less than 20% of the computed colorings were valid. This shows that claim (1) not only fails qualitatively du to the counterexample from Sect. 3 but also quantitatively for the vast majority of instances.

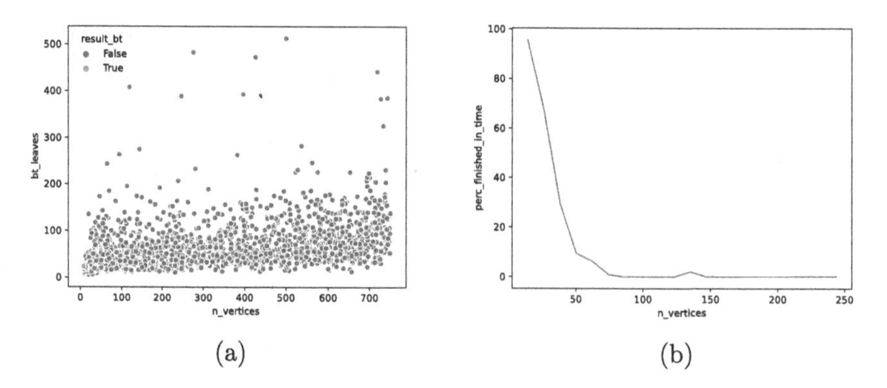

Fig. 6. (a) Number of leaves in Unger's backtracking tree. (b) Percentage of instances solved by regular backtracking within a one-hour time limit.

5 Conclusion

We have shown that the question of whether 3-coloring for circle graphs is possible in polynomial time should be considered open, even though Unger claimed to provide a polynomial time algorithm in a conference paper in 1992 [13]. To this end, we provided two counterexamples: one that contradicts the characterization in terms of the auxiliary coloring function and one that shows that the modified backtracking algorithm may fail to compute a correct solution. We further gave empirical evidence that displays a discrepancy to the claimed running time.

Considering the approach with important subgraphs, it appears that especially large induced cycles increase the difficulty, since for the other important subgraphs only a 2-SAT formula is constructed.

Question 1. What is the complexity of 3-coloring for circle graphs where no C_k with $k > 4$ is an induced subgraph?

We note that when considering all important subgraphs containing four vertices for the 2-SAT instance disregarding levels, we were not able to find a counterexample similar to the one in Sect. 3.

References

1. Angelini, P., Battista, G.D., Frati, F., Patrignani, M., Rutter, I.: Testing the simultaneous embeddability of two graphs whose intersection is a biconnected or a connected graph. J. Discrete Algorithms **14**, 150–172 (2012). https://doi.org/10.1016/j.jda.2011.12.015
2. Bachmann, P., Rutter, I., Stumpf, P.: On 3-coloring circle graphs. CoRR abs/2309.02258 (2023). https://doi.org/10.48550/arXiv.2309.02258
3. Baur, M., Brandes, U.: Crossing reduction in circular layouts. In: Hromkovič, J., Nagl, M., Westfechtel, B. (eds.) WG 2004. LNCS, vol. 3353, pp. 332–343. Springer, Heidelberg (2004). https://doi.org/10.1007/978-3-540-30559-0_28
4. Bekos, M.A., Bruckdorfer, T., Kaufmann, M., Raftopoulou, C.: 1-planar graphs have constant book thickness. In: Bansal, N., Finocchi, I. (eds.) ESA 2015. LNCS, vol. 9294, pp. 130–141. Springer, Heidelberg (2015). https://doi.org/10.1007/978-3-662-48350-3_12
5. Bekos, M.A., Kaufmann, M., Klute, F., Pupyrev, S., Raftopoulou, C.N., Ueckerdt, T.: Four pages are indeed necessary for planar graphs. J. Comput. Geom. **11**(1), 332–353 (2020). https://doi.org/10.20382/jocg.v11i1a12
6. Chung, F., Leighton, F., Rosenberg, A.: A graph layout problem with applications to VLSI design. Manuscript (1985)
7. Chung, F., Leighton, F., Rosenberg, A.: Embedding graphs in books: a layout problem with applications to VLSI design. SIAM J. Algebraic Discrete Methods **8**(1), 33–58 (1987). https://doi.org/10.1137/0608002
8. Dujmović, V., Pór, A., Wood, D.R.: Track layouts of graphs. Discrete Math. Theor. Comput. Sci. **6**(2) (2004). https://doi.org/10.46298/dmtcs.315
9. Eppstein, D.: Three-colorable circle graphs and three-page book embeddings (2014). http://11011110.github.io/blog/2014/08/09/three-colorable-circle-graphs.html. Accessed 6 June 2023

10. Hong, S.H., Nagamochi, H.: Simpler algorithms for testing two-page book embedding of partitioned graphs. Theor. Comput. Sci. **725**, 79–98 (2018). https://doi.org/10.1016/j.tcs.2015.12.039
11. Unger, W.: On the k-colouring of circle-graphs. In: Cori, R., Wirsing, M. (eds.) STACS 1988. LNCS, vol. 294, pp. 61–72. Springer, Heidelberg (1988). https://doi.org/10.1007/bfb0035832
12. Unger, W.: Färbung von Kreissehnengraphen. Ph.D. thesis, University of Paderborn, Germany (1990). http://d-nb.info/920881181
13. Unger, W.: The complexity of colouring circle graphs. In: Finkel, A., Jantzen, M. (eds.) STACS 1992. LNCS, vol. 577, pp. 389–400. Springer, Heidelberg (1992). https://doi.org/10.1007/3-540-55210-3_199
14. Wattenberg, M.: Arc diagrams: visualizing structure in strings. In: Wong, P.C., Andrews, K. (eds.) Proceedings of the IEEE Symposium on Information Visualization (InfoVis 2002), pp. 110–116. IEEE Computer Society (2002). https://doi.org/10.1109/INFVIS.2002.1173155
15. Yannakakis, M.: Embedding planar graphs in four pages. J. Comput. Syst. Sci. **38**(1), 36–67 (1989). https://doi.org/10.1016/0022-0000(89)90032-9

Geometric Aspects

The Complexity of Recognizing Geometric Hypergraphs

Daniel Bertschinger[1], Nicolas El Maalouly[1], Linda Kleist[2], Tillmann Miltzow[3], and Simon Weber[1(✉)]

[1] Department of Computer Science, ETH Zürich, Zürich, Switzerland
{daniel.bertschinger,nicolas.elmaalouly,simon.weber}@inf.ethz.ch
[2] Department of Computer Science, TU Braunschweig, Braunschweig, Germany
kleist@ibr.cs.tu-bs.de
[3] Department of Information and Computing Sciences, Utrecht University, Utrecht, The Netherlands
t.miltzow@uu.nl

Abstract. As set systems, hypergraphs are omnipresent and have various representations ranging from Euler and Venn diagrams to contact representations. In a *geometric representation* of a hypergraph $H = (V, E)$, each vertex $v \in V$ is associated with a point $p_v \in \mathbb{R}^d$ and each hyperedge $e \in E$ is associated with a connected set $s_e \subset \mathbb{R}^d$ such that $\{p_v \mid v \in V\} \cap s_e = \{p_v \mid v \in e\}$ for all $e \in E$. We say that a given hypergraph H is *representable* by some (infinite) family $\mathcal{F}$ of sets in $\mathbb{R}^d$, if there exist $P \subset \mathbb{R}^d$ and $S \subseteq \mathcal{F}$ such that (P, S) is a geometric representation of H. For a family $\mathcal{F}$, we define RECOGNITION($\mathcal{F}$) as the problem to determine if a given hypergraph is representable by $\mathcal{F}$. It is known that the RECOGNITION problem is $\exists\mathbb{R}$-hard for halfspaces in $\mathbb{R}^d$. We study the families of translates of balls and ellipsoids in $\mathbb{R}^d$, as well as of other convex sets, and show that their RECOGNITION problems are also $\exists\mathbb{R}$-complete. This means that these recognition problems are equivalent to deciding whether a multivariate system of polynomial equations with integer coefficients has a real solution.

Keywords: Hypergraph · geometric hypergraph · recognition · computational complexity · convex · ball · ellipsoid · halfplane · halfspace

1 Introduction

As set systems, hypergraphs appear in various contexts, such as databases, clustering, and machine learning. They are also known as *range spaces* (in computational geometry) or *voting games* (in social choice theory). A hypergraph can be represented in various ways, e.g., by a bipartite incidence graph, a simplicial representation (if the set system is closed under taking subsets), Euler or

The full version of this paper can be found on arXiv: 2302.13597v2.

M. A. Bekos and M. Chimani (Eds.): GD 2023, LNCS 14465, pp. 163–179, 2023.
https://doi.org/10.1007/978-3-031-49272-3_12

Venn diagrams etc. Similar to classical graph drawing, one can represent vertices by points and hyperedges by connected sets in $\mathbb{R}^d$ such that each set contains exactly the points of a hyperedge. For the purposes of legibility, uniformity, or also for aesthetic reasons, it is desirable that these sets satisfy additional properties, e.g., being convex or having similar appearance such as being homothetic copies or even translates of each other.

For an introductory example, suppose we are organizing a conference and have a list of accepted talks. Clearly, each participant wants to quickly identify talks of their specific interest. In order to create a good overview, we want to find a good representation. To this end, we label each talk by several tags, e.g., `hypergraphs`, `complexity theory`, `planar graphs`, `beyond planarity`, `straight-line drawing`, `crossing numbers`, etc. Then, we create a representation, where each tag is represented by a unit disk (or another nice geometric object of our choice) containing points representing the talks that have this tag, see Fig. 1 for an example. In other words, we are interested in a geometric representation of the hypergraph where the vertex set is given by the talks and tags define the hyperedges.

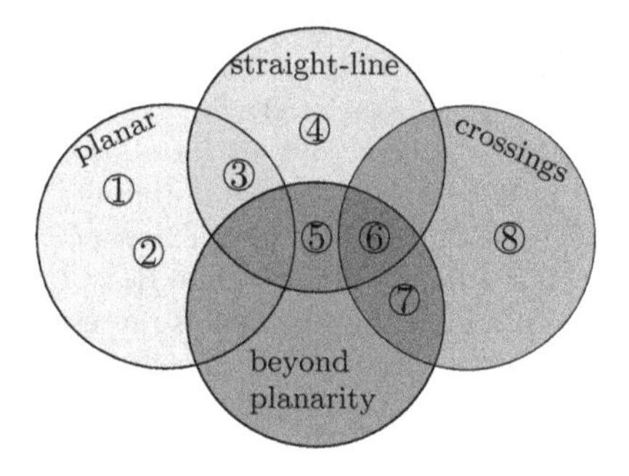

Fig. 1. A geometric representation with unit disks of the abstract hypergraph $H = (V, E)$ with $V = [8]$ and $E = \{\{1,2,3\},\{3,4,5,6\},\{5,6,7\},\{6,7,8\}\}$.

In this work, we investigate the complexity of deciding whether a given hypergraph has such a geometric representation. We start with a formal definition.

Problem Definition. In a *geometric representation* of a hypergraph $H = (V, E)$, each vertex $v \in V$ is associated with a point $p_v \in \mathbb{R}^d$ and each hyperedge $e \in E$ is associated with a connected set $s_e \subset \mathbb{R}^d$ such that $\{p_v \mid v \in V\} \cap s_e = \{p_v \mid v \in e\}$ for all $e \in E$. We say that a given hypergraph H is *representable* by some (possibly infinite) family $\mathcal{F}$ of sets in $\mathbb{R}^d$, if there exist $P \subset \mathbb{R}^d$ and $S \subseteq \mathcal{F}$ such that (P, S) is a geometric representation of H. For a family $\mathcal{F}$ of geometric objects in $\mathbb{R}^d$, we define RECOGNITION($\mathcal{F}$) as the problem to determine whether a given hypergraph is representable by $\mathcal{F}$. Next, we give some definitions describing the geometric families studied in this work.

Bi-curved, Difference-Separable, and Computable Convex Sets. We study convex sets that are bi-curved, difference-separable and computable. While the first two properties are needed for $\exists\mathbb{R}$-hardness, the last one is used to show $\exists\mathbb{R}$-membership.

Let $C \subset \mathbb{R}^d$ be a convex set. We call C *computable* if for any point $p \in \mathbb{R}^d$ we can decide in polynomial time on a real RAM whether p is contained in C. We say that C is *bi-curved* if there exists a unit vector $v \in \mathbb{R}^d$, such that there are two distinct tangent hyperplanes on C with normal vector v; with each of these hyperplanes intersecting C in a single point, and C being *smooth* at both of these intersection points. Informally, a convex set is bi-curved, if its boundary has two smoothly curved parts in which the tangent hyperplanes are parallel. Note that a convex, bi-curved set is necessarily bounded. As a matter of fact, any strictly convex bounded set in any dimension is bi-curved. For such sets, any unit vector v fulfills the conditions. As can be seen in Fig. 2a, being strictly convex is not necessary for being bi-curved.

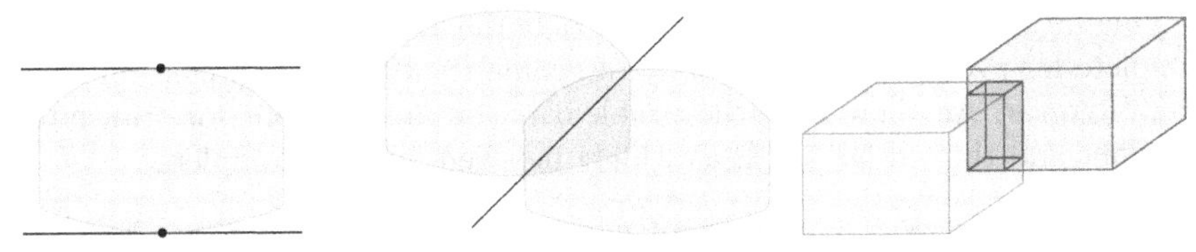

(a) This burger-like set is bi-curved as shown by the two tangent hyperplanes.

(b) A hyperplane separating the symmetric difference of two translates of the burger-like set.

(c) Two cubes in $\mathbb{R}^3$ whose symmetric difference cannot be separated by a plane.

Fig. 2. Illustration for the notions bi-curved and difference-separable.

We call C *difference-separable* if for any two translates C_1, C_2 of C, there exists a hyperplane which strictly separates $C_1 \backslash C_2$ from $C_2 \backslash C_1$. Being difference-separable is fulfilled by any convex set in $\mathbb{R}^2$, see Fig. 2b for an example. For a proof of this fact we refer to [32, Corollary 2.1.2.2]. However, in higher dimensions this is not the case: for a counterexample, consider two 3-cubes as in Fig. 2c. In higher dimensions, the bi-curved and difference-separable families include the balls and ellipsoids. We are not aware of other natural geometric families with those two properties. Note that balls and ellipsoids are naturally computable.

We are now ready to state our results.

1.1 Results

We study the recognition problem of geometric hypergraphs. We first consider the maybe simplest type of geometric hypergraphs, namely those that stem from halfspaces. It is known due to Tanenbaum, Goodrich, and Scheinerman [56] that the RECOGNITION problem for geometric hypergraphs of halfspaces is NP-hard, but their proof actually implies $\exists\mathbb{R}$-hardness as well. We present a slightly different proof of this fact due to two reasons. Firstly, their proof lacks details about extensions to higher dimensions. Secondly, it is a good stepping stone towards our proof of Theorem 2.

Theorem 1 (Tanenbaum, Goodrich, Scheinerman [56]). *For every $d \geq 2$, RECOGNITION($\mathcal{F}$) is $\exists\mathbb{R}$-complete for the family $\mathcal{F}$ of halfspaces in $\mathbb{R}^d$.*

Next we consider families of objects that are translates of a given object.

Theorem 2. *For $d \geq 2$, let $C \subseteq \mathbb{R}^d$ be a convex, bi-curved, difference-separable and computable set, and let $\mathcal{F}$ be the family of all translates of C. Then RECOGNITION($\mathcal{F}$) is $\exists\mathbb{R}$-complete.*

We note that for $d = 1$, the RECOGNITION problems of halfspaces and translates of convex sets can be solved by sorting. Consequently, they can be decided in polynomial time.

One might be under the impression that the RECOGNITION problem is $\exists\mathbb{R}$-complete for every reasonable family of geometric objects of dimension at least two. However, we show that the problem is contained in NP for translates of polygons and thus, if NP $\subsetneq \exists\mathbb{R}$ as widely believed, not $\exists\mathbb{R}$-complete.

Theorem 3. *Let P be a simple polygon with integer coordinates in $\mathbb{R}^2$, and $\mathcal{F}$ the family of all translates of P. Then RECOGNITION($\mathcal{F}$) is contained in NP.*

Organization. We give an overview over our proof techniques in Sect. 1.3. Full proofs of Theorem 3 as well as the membership parts of Theorems 1 and 2 are found in Sect. 2. We introduce the version of pseudohyperplane stretchability used in our hardness reductions in Secton 3. Full proofs of the hardness parts of Theorems 1 and 2 can be found in Sects. 4 and 5, respectively.

1.2 Related Work

In this section we give a concise overview over related work on the complexity class $\exists\mathbb{R}$, geometric intersection graphs, and on other set systems related to hypergraphs.

The Existential Theory of the Reals. The complexity class $\exists\mathbb{R}$ (pronounced as 'ER' or 'exists R') is defined via its canonical complete problem ETR (short for *Existential Theory of the Reals*) and contains all problems that polynomial-time many-one reduce to it. In an ETR instance, we are given a sentence of the form

$$\exists x_1, \ldots, x_n \in \mathbb{R} : \varphi(x_1, \ldots, x_n),$$

where φ is a well-formed and quantifier-free formula consisting of polynomial equations and inequalities in the variables and the logical connectives $\{\wedge, \vee, \neg\}$. The goal is to decide whether this sentence is true.

The complexity class $\exists\mathbb{R}$ gains its importance from its numerous influential complete problems. Important $\exists\mathbb{R}$-completeness results include the realizability of abstract order types [39,51], geometric linkages [44], and the recognition of geometric intersection graphs, as further discussed below.

More results concern graph drawing [20,21,31,45], the Hausdorff distance [27], polytopes [19,42], Nash-equilibria [8,10,11,24,47], training neural networks [3,9], matrix factorization [17,48–50,57], continuous constraint satisfaction problems [38], geometric packing [5], the art gallery problem [2,55], and covering polygons with convex polygons [1].

Geometric Hypergraphs. Many aspects of hypergraphs with geometric representations have been studied. Hypergraphs represented by touching polygons in $\mathbb{R}^3$ have been studied by Evans et al. [23]. Bounds on the number of hyperedges in hypergraphs representable by homothets of a fixed convex set S have been established by Axenovich and Ueckerdt [7]. Smorodinsky studied the chromatic number and the complexity of coloring of hypergraphs represented by various types of sets in the plane [53]. Dey and Pach [18] generalize many extremal properties of geometric graphs to hypergraphs where the hyperedges are induced simplices of some point set in $\mathbb{R}^d$. Haussler and Welzl [25] defined ϵ-nets, subsets of vertices of hypergraphs called range spaces with nice properties. Such ϵ-nets of geometric hypergraphs have been studied quite intensely [6,35,40,41].

While there are many structural results, we are not aware of any research into the complexity of recognizing hypergraphs given by geometric representations, other than the recognition of embeddability of simplicial complexes, as we will discuss in the next paragraph.

Other Representations of Hypergraphs. Hypergraphs are in close relation with abstract simplicial complexes. In particular, an abstract simplicial complex (complex for short) is a set system that is closed under taking subsets. A k-complex is a complex in which the maximum size of a set is k. In a geometric representation of an abstract simplicial complex $H = (V, E)$ each ℓ-set of E is represented by a ℓ-simplex such that two simplices of any two sets intersect exactly in the simplex defined by their intersection (and are disjoint in case of an empty intersection). Note that 1-complexes are graphs and hence deciding the representability in the plane corresponds to graph planarity (which is in P). In stark contrast, Abrahamsen, Kleist and Miltzow recently showed that deciding whether a 2-complex has a geometric embedding in $\mathbb{R}^3$ is $\exists\mathbb{R}$-complete [4]; they also prove hardness for other dimensions. Similarly, piecewise linear embeddings of simplicial complexes have been studied [13–15,33,34,37,52].

Recognizing Geometric Intersection Graphs. Given a set of geometric objects, its intersection graph has a vertex for each object, and an edge between any two intersecting objects. The complexity of recognizing geometric intersection graphs has been studied for various geometric objects. We summarize these results in Fig. 3.

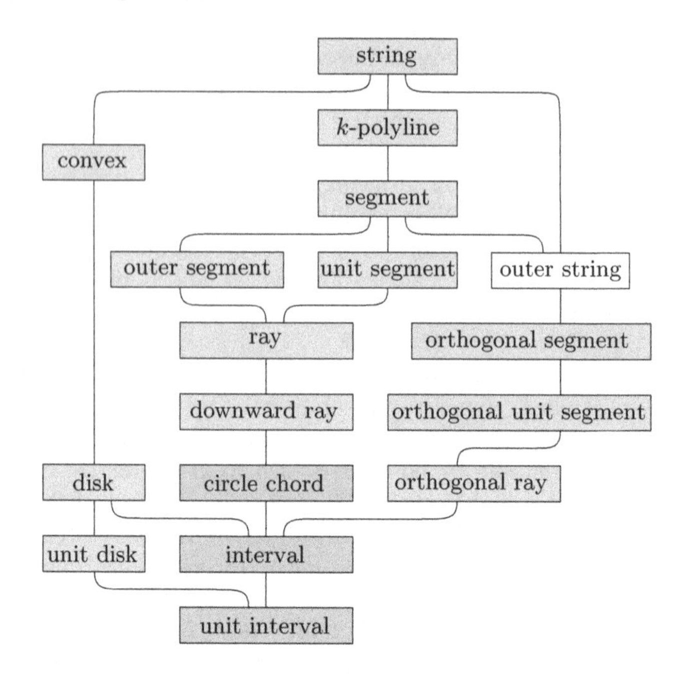

Fig. 3. Containment relations of geometric intersection graphs. Recognition of a green class is in P, of a grey class is NP-complete, of a blue class is $\exists\mathbb{R}$-complete, and of a white class is unknown. (Color figure online)

While intersection graphs of circle chords (Spinnrad [54]), unit intervals (Looges and Olariu [30]) and intervals (Booth and Lueker [12]) can be recognized in polynomial time, recognizing string graphs (Schaefer and Sedgwick [46]) is NP-complete. In contrast, $\exists\mathbb{R}$-completeness of recognizing intersection graphs has been proved for (unit) disks by McDiarmid and Müller [36], convex sets by Schaefer [43], downward rays by Cardinal et al. [16], outer segments by Cardinal et al. [16], unit segments by Hoffmann et al. [26], segments by Kratochvíl and Matoušek [29], k-polylines by Hoffmann et al. [26], and unit balls by Kang and Müller [28].

The existing research landscape indicates that recognition problems of intersection graphs are $\exists\mathbb{R}$-complete in case that the family of objects satisfy two conditions: Firstly, they need to be "geometrically solid", i.e., not strings. Secondly, some non-linearity must be present by either allowing rotations, or by the objects having some curvature. Our results indicate that this general intuition might translate to the recognition of geometric hypergraphs.

1.3 Overview of Proof Techniques

We prove containment in $\exists\mathbb{R}$ and NP using standard arguments, providing witnesses and verification algorithms.

We prove the hardness parts of Theorems 1 and 2 by reduction from stretchability of pseudohyperplane arrangements. The hypergraph we build from the given arrangement differs from the one built in the proof of Theorem 1 given in [56], since we wish to use a single construction which works nicely for both theorems. Given a simple pseudohyperplane arrangement $\mathcal{A}$, we construct a hypergraph H as follows: We double each pseudohyperplane by giving it a parallel *twin*. In this arrangement, we place a point in every d-dimensional cell. These points represent the vertices of H. Every pseudohyperplane ℓ then defines a hyperedge, which contains all of the points on the same side of ℓ as its twin pseudohyperplane. See Fig. 5 for an illustration of this construction.

Because this construction can also be performed on a hyperplane arrangement, it is straightforward to prove that if $\mathcal{A}$ is stretchable, H can be represented by halfspaces. Conversely, we show that the hyperplanes bounding the halfspaces in a representation of H must be a stretching of $\mathcal{A}$.

For Theorem 2, bi-curvedness of a set C implies that locally, C can approximate any halfspace with normal vector close to v as in the definition of bi-curved. This allows us to prove that stretchability of $\mathcal{A}$ implies representability of H by translates of C. The set C being difference-separable is used when reconstructing a hyperplane arrangement from a representation of H.

2 Membership

In this section we show $\exists\mathbb{R}$- and NP-membership.

2.1 Halfspaces

For a given hypergraph H, it is not difficult to formulate an ETR formula describing all needed properties for a geometric representation by halfspaces. Therefore, we get the $\exists\mathbb{R}$-membership part of Theorem 1.

Lemma 1. *Fix $d \geq 1$ and let $\mathcal{F}$ denote the family of halfspaces in $\mathbb{R}^d$. Then* Recognition*($\mathcal{F}$) is contained in $\exists\mathbb{R}$.*

The $\exists\mathbb{R}$-membership part of Theorem 2 is obtained by providing a simple verification algorithm [22] (similar to how NP-membership can be shown), based on the fact that our considered set C is computable.

Lemma 2. *For some $d \geq 1$, let $C \subseteq \mathbb{R}^d$ be a computable set and let $\mathcal{F}$ be the family of all translates of C. Then,* Recognition*($\mathcal{F}$) is contained in $\exists\mathbb{R}$.*

The full proofs of Lemmas 1 and 2 have been omitted due to space constraints and can be found in the full version of this paper.

2.2 Translates of Polygons – Proof of Theorem 3

Here, we show Theorem 3, i.e., NP-membership of Recognition of translates of some simple polygon P.

Theorem 3. *Let P be a simple polygon with integer coordinates in $\mathbb{R}^2$, and $\mathcal{F}$ the family of all translates of P. Then* Recognition*($\mathcal{F}$) is contained in* NP.

Proof. The proof uses a similar argument to the one used to show that the problem of packing translates of polygons inside a polygon is in NP [5]. For an illustration, consider Fig. 4. We first triangulate the convex hull of P, such that each edge of P appears in the triangulation. Then, a representation of a hypergraph H by translates of P gives rise to a certificate as follows: For each pair of a point p and a translate P' of P, we specify whether p lies in the convex hull of P'. If it does, we specify in which triangle p lies. Otherwise, we specify an edge of the convex hull for which p and P' lie on opposite sides of the line through the edge.

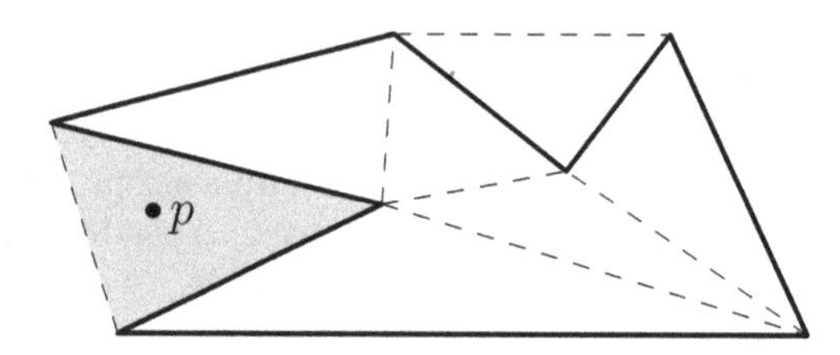

Fig. 4. The polygon P', a triangulation of its convex hull, and the triangle that contains p.

Such a certificate can be tested in polynomial time: we create a linear program whose variables describe the locations of the points p and the translation vectors of each translate of P, and whose constraints enforce the points to lie in the areas described by the certificate. This linear program has a number of constraints and variables polynomial in the size of H, and can be thus solved in polynomial time.

The solution of this linear program gives the location of the points and the translation vectors of the polygons. This implies that these coordinates are all polynomial and could be used as a certificate directly. □

3 Pseudohyperplane Stretchability

A *pseudohyperplane arrangement* in $\mathbb{R}^d$ is an arrangement of *pseudohyperplanes*, where a pseudohyperplane is a set homeomorphic to a hyperplane, and each intersection of pseudohyperplanes is homeomorphic to a plane of some dimension. In the classical definition, every set of d pseudohyperplanes has a non-empty intersection. Here, we consider *partial pseudohyperplane arrangements (*PPHAs*)*, where not necessarily every set of $\leq d$ pseudohyperplanes has a common intersection.

A PPHA is *simple* if no more than k pseudohyperplanes intersect in a space of dimension $d - k$, in particular, no $d + 1$ pseudohyperplanes have a common intersection. We call the 0-dimensional intersection points of d pseudohyperplanes the *vertices* of the arrangement. A simple PPHA $\mathcal{A}$ is called *stretchable* if there exists a hyperplane arrangement $\mathcal{A}'$ such that each vertex in $\mathcal{A}$ also exists in $\mathcal{A}'$ and each (pseudo-)hyperplane splits this set of vertices the same way in $\mathcal{A}$ and $\mathcal{A}'$. In other words, each vertex of $\mathcal{A}$ lies on the correct side of each hyperplane in $\mathcal{A}$'. We then call the hyperplane arrangement $\mathcal{A}'$ a *stretching* of $\mathcal{A}$.

The problem d-STRETCHABILITY is the problem of deciding whether a simple PPHA in $\mathbb{R}^d$ is stretchable. For $d = 2$, d-STRETCHABILITY contains the stretchability of simple pseudoline arrangements which is known to be $\exists\mathbb{R}$-hard [39,51]. It is straightforward to prove $\exists\mathbb{R}$-hardness for all $d \geq 2$; the proof can be found in the full version of this paper.

Theorem 4. *d-STRETCHABILITY is $\exists\mathbb{R}$-hard for all $d \geq 2$.*

Similar extensions of pseudoline stretchability to higher dimensions have been studied in the literature. For example, Mnëv's universality theorem [39] extends to higher dimensions, however we are not aware of any existing proofs that it also implies $\exists\mathbb{R}$-hardness in $d > 2$. Kang and Müller [28] also studied a similar version of stretchability of partial arrangements of pseudohyperplanes.

4 Hardness for Halfspaces – Proof of Theorem 1

We now present the hardness part of Theorem 1.

Theorem 1 (Tanenbaum, Goodrich, Scheinerman [56]). *For every $d \geq 2$, RECOGNITION($\mathcal{F}$) is $\exists\mathbb{R}$-complete for the family $\mathcal{F}$ of halfspaces in $\mathbb{R}^d$.*

Proof (of Theorem 1). We reduce from d-STRETCHABILITY. Let $\mathcal{A}$ be a simple PPHA. For an example consider Fig. 5a. In a first step, we insert a parallel twin ℓ' for each pseudohyperplane ℓ. The twin is close enough to ℓ such that ℓ and

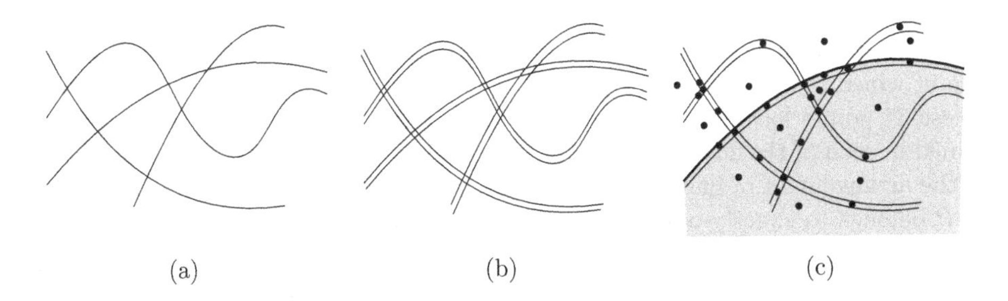

Fig. 5. Construction of the hypergraph H. (a) A simple PPHA $\mathcal{A}$. (b) The arrangement $\mathcal{A}'$ obtained by inserting twins. (c) The vertices of H are the points in the cells, hyperedges of H are defined by the pseudohalfspaces; the gray region shows one of the hyperedges. (Color figure online)

ℓ' have the same intersection pattern. Since ℓ and ℓ' are parallel, they do not intersect each other. This yields an arrangement $\mathcal{A}'$, see Fig. 5b.

In a second step, we introduce a point in each d-dimensional cell of $\mathcal{A}'$; each point represents a vertex in our hypergraph H. Lastly, we define a hyperedge for each pseudohyperplane ℓ of $\mathcal{A}'$: The hyperedge contains all of the points that lie on the side of ℓ that the twin pseudohyperplane ℓ' lies in, see Fig. 5c. Note that we define such a hyperedge for every pseudohyperplane of $\mathcal{A}'$. Thus, for every pseudohyperplane ℓ of the original arrangement $\mathcal{A}$ we define two hyperedges, whose union contains all vertices of H.

It remains to show that H is representable by halfspaces if and only if $\mathcal{A}$ is stretchable. If $\mathcal{A}$ is stretchable, the construction of a representation of H is straightforward: Consider a hyperplane arrangement $\mathcal{B}$ which is a stretching of $\mathcal{A}$. Then, for each hyperplane, we add a parallel hyperplane very close, so that their intersection patterns coincide. This results in a hyperplane arrangement $\mathcal{B}'$. We now prove that every d-dimensional cell of $\mathcal{A}'$ must also exist in $\mathcal{B}'$. First, note that each such cell corresponds to a cell of $\mathcal{A}$, which has at least one vertex on its boundary. All vertices of $\mathcal{A}$ exist in $\mathcal{B}$ by definition of a stretching. Furthermore, the subarrangement of the d hyperplanes in $\mathcal{B}$ intersecting in this vertex must be simple, since their intersection could not be 0-dimensional otherwise. In the twinned hyperplane arrangement $\mathcal{B}'$, all 3^d of the d-dimensional cells incident to this vertex (a cell is given by the following choice for each of the hyperplane pairs: above both hyperplanes, between the hyperplanes, or below both hyperplanes) must exist. This proves that all d-dimensional cells of $\mathcal{A}'$ also exist in $\mathcal{B}'$. Inserting a point in each such d-dimensional cell and considering the (correct) halfspaces bounded by the hyperplanes of $\mathcal{B}'$ yields a representation of H.

We now consider the reverse direction. Let $(P, \mathcal{H})$ be a tuple of points and halfspaces representing H. Let $h_{i,1}$ and $h_{i,2}$ be the two halfspaces associated with a pseudohyperplane ℓ_i of $\mathcal{A}$. Let p_i denote the $(d-1)$-dimensional hyperplane bounding $h_{i,1}$. We show that the family $\{p_i\}_i$ of these hyperplanes is a stretching of $\mathcal{A}$.

For each intersection point q of d pseudohyperplanes $\ell_1, \ldots \ell_d$ in $\mathcal{A}$, we consider the corresponding $2d$ pseudohyperplanes in $\mathcal{A}'$. The PPHA $\mathcal{A}'$ contains 3^d d-dimensional cells incident to their 2^d intersections; each of which contains a point. We first show that the associated halfspaces must induce at least 3^d cells, one of which is bounded and represents the intersection point, see also Fig. 6a: These 3^d points have pairwise distinct patterns of whether or not they are contained in each of the $2d$ halfspaces. Thus, these points need to lie in distinct cells of the arrangement of halfspaces, which proves the claim. Moreover, every point in P belongs to exactly one of these 3^d cells. In particular, the central bounded cell, denoted by $c(q)$, contains exactly one point of P, see Fig. 6b.

Now, we argue that the complete cell $c(q)$ (and thus in particular the intersection point of the hyperplanes representing q) lies on the correct side of each hyperplane p in $\{p_i\}_i$. Note that, by construction of the hypergraph H, the 3^d points of q lie on the same side of p. Suppose for a contradiction that p intersects $c(q)$, see Fig. 6c. Then there exist two unbounded cells incident to $c(q)$ which lie

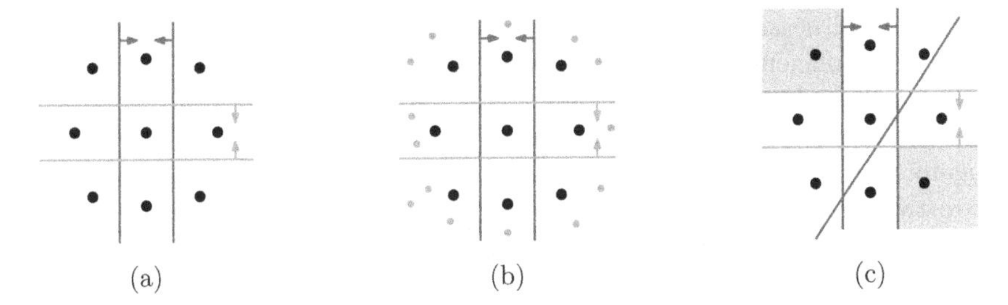

Fig. 6. Illustration for the proof of Theorem 1 for $d = 2$ showing that representability of H implies stretchability of $\mathcal{A}$. (a) Any two pseudolines ℓ_i, ℓ_j in $\mathcal{A}$ have four corresponding lines bounding the respective halfplanes in H; these four lines induce 9 cells, each of which contains a point. (b) Each point in P belongs to exactly one of these 9 cells; the central bounded cell contains a unique point representing the intersection of ℓ_i and ℓ_j. (c) The central bounded cell cannot be intersected by a line p_k with $k \neq i, j$.

on different sides of p; these cells can be identified by translating p until it intersects $c(q)$ only in the boundary. This yields a contradiction to the fact that the 3^d points of q lie on the same side of p.

We conclude that each intersection point of d pseudohyperplanes in $\mathcal{A}$ also exists in the arrangement $\{p_i\}_i$ and lies on the correct side of all hyperplanes. Thus, $\{p_i\}_i$ is a stretching of $\mathcal{A}$ and $\mathcal{A}$ is stretchable. □

5 Hardness for Convex, Bi-curved, and Difference-Separable Sets – Proof of Theorem 2

We are now going to prove the hardness part of Theorem 2.

Theorem 2. *For $d \geq 2$, let $C \subseteq \mathbb{R}^d$ be a convex, bi-curved, difference-separable and computable set, and let $\mathcal{F}$ be the family of all translates of C. Then* RECOGNITION*($\mathcal{F}$) is $\exists\mathbb{R}$-complete.*

To this end, consider any fixed convex, bi-curved, and difference-separable set C in $\mathbb{R}^d$. Note that we can assume C to be fully-dimensional, since otherwise each connected component would live in some lower-dimensional affine subspace, with no interaction between such components. We use the same reduction from the problem d-STRETCHABILITY as in the proof for halfspaces in the previous section and show that the constructed hypergraph H is representable by translates of C if and only if the given PPHA $\mathcal{A}$ is stretchable.

Lemma 3. *If $\mathcal{A}$ is stretchable, H is representable by translates of C.*

The full proof of this lemma can be found in the full version of this paper.

The idea behind the proof is that a stretching $\mathcal{A}$' of $\mathcal{A}$ can be scaled and stretched in such a way that every hyperplane has a normal vector close to the vector v witnessing that C is bi-curved, and such that all the vertices lie within some sufficiently small box. Then, for every halfspace $h_\pm$ bounded by some hyperplane h in $\mathcal{A}$', there exists a translate of C which approximates $h_\pm$ within the small box. This intuition is shown in Fig. 7. Since the hyperplane arrangement is simple, and there is some slack between the hyperplanes bounding the two twin halfspaces (as we argued above in the proof of Theorem 1), such an approximation is sufficient.

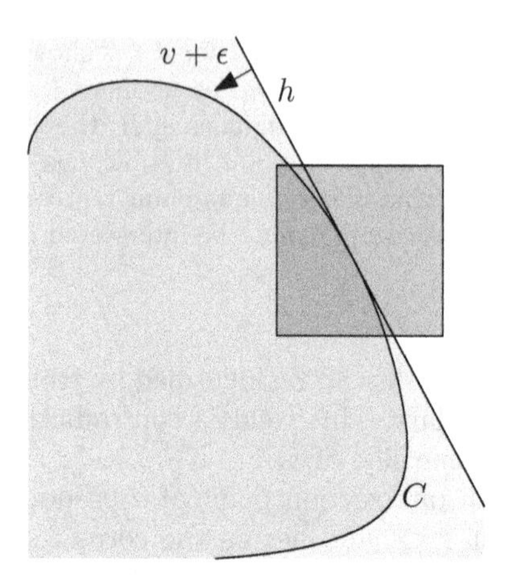

Fig. 7. Illustration for the proof of Lemma 3. Within the small box (dark grey), the translate of C (green) approximates the halfspace (light grey) bounded by h. (Color figure online)

Lemma 4. *If the hypergraph H is representable by translates of C, then $\mathcal{A}$ is stretchable.*

Proof. Assume H is representable. By construction, the two translates $C_{i,r}, C_{i,l}$ of C corresponding to the two hyperedges of each pseudohyperplane ℓ_i must intersect as they contain at least one common point. We call their convex intersection the *lens* of this pseudohyperplane. For each pseudohyperplane ℓ_i of $\mathcal{A}$, we consider some hyperplane p_i which separates $C_{i,r} \setminus C_{i,l}$ from $C_{i,l} \setminus C_{i,r}$. Such a hyperplane exists since C is difference-separable. Let $\mathcal{P} := \{p_i\}_i$ be the hyperplane arrangement consisting of all these separators. We aim to show that $\mathcal{P}$ is a stretching of $\mathcal{A}$.

To this end, consider d pseudohyperplanes $\ell_1, \ldots, \ell_d$ which intersect in $\mathcal{A}$. Figure 8 displays the case $d = 2$. Furthermore, consider one more pseudohyperplane ℓ', and let p', C'_r, C'_l denote the separator hyperplane and translates of C corresponding to ℓ'. We show that the intersection $I_p := p_1 \cap \ldots \cap p_d$ is a single point which lies on the same side of p' as the point $I_\ell := \ell_1 \cap \ldots \cap \ell_d$ lies of ℓ'.

The hyperplane p' divides the space into two halfspaces h_r and h_l such that $C'_r \setminus C'_l \subseteq h_r$ and $C'_l \setminus C'_r \subseteq h_l$. By construction, the two hyperedges defined for ℓ'

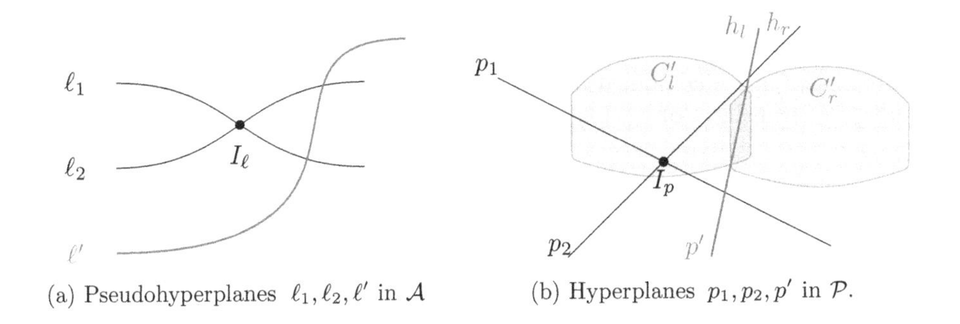

(a) Pseudohyperplanes ℓ_1, ℓ_2, ℓ' in $\mathcal{A}$

(b) Hyperplanes p_1, p_2, p' in $\mathcal{P}$.

Fig. 8. Illustration for the proof of Lemma 4 for $d = 2$. Some pseudohyperplanes in $\mathcal{A}$ and their corresponding hyperplanes in $\mathcal{P}$.

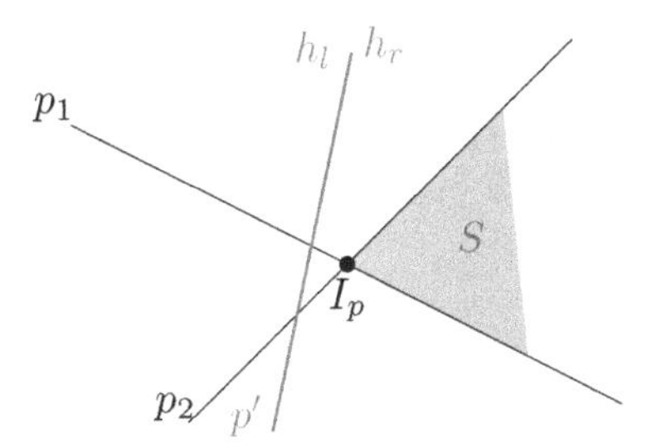

Fig. 9. Illustration for the proof of Lemma 4 for $d = 2$. The cone S must intersect $C'_l \setminus C'_r$, which contradicts I_p lying in h_r.

cover all vertices of H and the vertices in the cells around I_ℓ belong to only one hyperedge. Suppose without loss of generality that these vertices only belong to the hyperedge represented by C'_l. We will show that the intersection I_p must then be a point in h_l.

We first show that the intersection I_p is a point, i.e., 0-dimensional. Consider all 2^d d-dimensional cells of $\mathcal{A}$ around I_ℓ. The construction of H implies that each such cells contains a distinct point, and these points must all lie in distinct cells of the sub-arrangement of the involved hyperplanes $p_1, \ldots, p_d$. Assuming that I_p is not a single point, this sub-arrangement is not simple, and the hyperplanes divide space into strictly fewer than 2^d cells, which results in a contradiction.

Next we prove that I_p is in h_l. Assume towards a contradiction that $I_p \in h_r$, see also Fig. 9. Consider the d lines that are formed by the intersections of subsets of $d-1$ hyperplanes among $p_1, ..., p_d$. Each of these lines is the union of two rays beginning at I_p. Observe that the hyperplane p' can only intersect one of the two rays forming each line. Let S be the convex cone centered at I_p defined by the d non-intersected rays. Observe that S does not intersect p', so S must be fully contained in h_r, i.e., $S \cap h_l = \emptyset$. Note, however, by the construction of the hypergraph, there must be a point that lies in $S \cap (C'_l \setminus C'_r) \subseteq S \cap h_l$, which is a contradiction.

We conclude that $\mathcal{P}$ is a stretching of $\mathcal{A}$, and thus $\mathcal{A}$ is stretchable. □

Lemmas 3 and 4 combined now prove hardness of RECOGNITION($\mathcal{F}$) for the family $\mathcal{F}$ of translates of C. This completes the proof of Theorem 2.

6 Future Directions

We conclude with a list of interesting open problems: As mentioned above, we are not aware of interesting families of bi-curved and difference-separable sets in higher dimensions beyond balls and ellipsoids. The families of translates of a given polygon show the need for some curvature in order to show $\exists\mathbb{R}$-hardness. We wonder if it is sufficient for $\exists\mathbb{R}$-hardness to assume curvature at only one boundary part instead of two opposite ones. Another open question is to consider families that include rotated copies or homothetic copies of a fixed geometric object. Allowing for rotation, it is conceivable that $\exists\mathbb{R}$-hardness even holds for polygons. Allowing for homothetic copies, the next natural question would be to study the family of all balls, not just balls of the same size.

References

1. Abrahamsen, M.: Covering polygons is even harder. In: IEEE 62nd Annual Symposium on Foundations of Computer Science (FOCS 2022), pp. 375–386 (2022). https://doi.org/10.1109/FOCS52979.2021.00045
2. Abrahamsen, M., Adamaszek, A., Miltzow, T.: The art gallery problem is $\exists\mathbb{R}$-complete. In: Proceedings of the 50th Annual ACM SIGACT Symposium on Theory of Computing (STOC 2018), pp. 65–73 (2018). https://doi.org/10.1145/3188745.3188868
3. Abrahamsen, M., Kleist, L., Miltzow, T.: Training neural networks is $\exists\mathbb{R}$-complete. In: Advances in Neural Information Processing Systems (NeurIPS 2021), vol. 34, pp. 18293–18306 (2021). https://proceedings.neurips.cc/paper_files/paper/2021/file/9813b270ed0288e7c0388f0fd4ec68f5-Paper.pdf
4. Abrahamsen, M., Kleist, L., Miltzow, T.: Geometric embeddability of complexes is $\exists\mathbb{R}$-complete. In: Symposium on Computational Geometry (SOCG 2023) (2023, to appear). https://doi.org/10.48550/arXiv.2108.02585
5. Abrahamsen, M., Miltzow, T., Seiferth, N.: Framework for $\exists\mathbb{R}$-completeness of two-dimensional packing problems. In: IEEE 61st Annual Symposium on Foundations of Computer Science (FOCS 2020), pp. 1014–1021 (2020). https://doi.org/10.1109/FOCS46700.2020.00098
6. Aronov, B., Ezra, E., Sharir, M.: Small-size ϵ-nets for axis-parallel rectangles and boxes. SIAM J. Comput. (SiComp) **39**(7), 3248–3282 (2010). https://doi.org/10.1137/090762968
7. Axenovich, M., Ueckerdt, T.: Density of range capturing hypergraphs. J. Comput. Geom. (JoCG) **7**(1) (2016). https://doi.org/10.20382/jocg.v7i1a1
8. Berthelsen, M.L.T., Hansen, K.A.: On the computational complexity of decision problems about multi-player nash equilibria. In: Fotakis, D., Markakis, E. (eds.) SAGT 2019. LNCS, vol. 11801, pp. 153–167. Springer, Cham (2019). https://doi.org/10.1007/978-3-030-30473-7_11

9. Bertschinger, D., Hertrich, C., Jungeblut, P., Miltzow, T., Weber, S.: Training fully connected neural networks is ∃ℝ-complete (2022). https://doi.org/10.48550/arXiv.2204.01368
10. Bilò, V., Mavronicolas, M.: A catalog of ∃ℝ-complete problems about Nash equilibria in multi-player games. In: 33rd Symposium on Theoretical Aspects of Computer Science (STACS 2016), pp. 17:1–17:13. (LIPIcs) (2016). https://doi.org/10.4230/LIPIcs.STACS.2016.17
11. Bilò, V., Mavronicolas, M.: ∃ℝ-complete decision problems about symmetric Nash equilibria in symmetric multi-player games. In: 34th Symposium on Theoretical Aspects of Computer Science (STACS 2017). (LIPIcs), vol. 66, pp. 13:1–13:14 (2017). https://doi.org/10.4230/LIPIcs.STACS.2017.13
12. Booth, K.S., Lueker, G.S.: Testing for the consecutive ones property, interval graphs, and graph planarity using PQ-tree algorithms. J. Comput. Syst. Sci. **13**(3), 335–379 (1976). https://doi.org/10.1016/S0022-0000(76)80045-1
13. Čadek, M., Krčál, M., Matoušek, J., Sergeraert, F., Vokřínek, L., Wagner, U.: Computing all maps into a sphere. J. ACM **61**(3), 1–44 (2014). https://doi.org/10.1145/2597629
14. Čadek, M., Krčál, M., Matoušek, J., Vokřínek, L., Wagner, U.: Time computation of homotopy groups and Postnikov systems in fixed dimension. SIAM J. Comput. **43**(5), 1728–1780 (2014). https://doi.org/10.1137/120899029
15. Čadek, M., Krčál, M., Vokřínek, L.: Algorithmic solvability of the lifting-extension problem. Discrete Comput. Geom. (DCG) **57**(4), 915–965 (2017). https://doi.org/10.1007/s00454-016-9855-6
16. Cardinal, J., Felsner, S., Miltzow, T., Tompkins, C., Vogtenhuber, B.: Intersection graphs of rays and grounded segments. J. Graph Algorithms Appl. **22**(2), 273–294 (2018). https://doi.org/10.7155/jgaa.00470
17. Chistikov, D., Kiefer, S., Marusic, I., Shirmohammadi, M., Worrell, J.: On restricted nonnegative matrix factorization. In: 43rd International Colloquium on Automata, Languages, and Programming (ICALP 2016). LIPIcs, vol. 55, pp. 103:1–103:14 (2016). https://doi.org/10.4230/LIPIcs.ICALP.2016.103
18. Dey, T.K., Pach, J.: Extremal problems for geometric hypergraphs. Discrete Comput. Geom. **19**(4), 473–484 (1998). https://doi.org/10.1007/PL00009365
19. Dobbins, M.G., Holmsen, A., Miltzow, T.: A universality theorem for nested polytopes (2019). https://doi.org/10.48550/arXiv.1908.02213
20. Dobbins, M.G., Kleist, L., Miltzow, T., Rzążewski, P.: Completeness for the complexity class ∀∃ℝ and area-universality. Discrete Comput. Geom. (DCG) (2022). https://doi.org/10.1007/s00454-022-00381-0
21. Erickson, J.: Optimal curve straightening is ∃ℝ-complete (2019). https://doi.org/10.48550/arXiv.1908.09400
22. Erickson, J., van der Hoog, I., Miltzow, T.: Smoothing the gap between NP and ∃ℝ. In: IEEE 61st Annual Symposium on Foundations of Computer Science (FOCS 2020), pp. 1022–1033 (2020). https://doi.org/10.1109/FOCS46700.2020.00099
23. Evans, W., Rzążewski, P., Saeedi, N., Shin, C.-S., Wolff, A.: Representing graphs and hypergraphs by touching polygons in 3D. In: Archambault, D., Tóth, C.D. (eds.) GD 2019. LNCS, vol. 11904, pp. 18–32. Springer, Cham (2019). https://doi.org/10.1007/978-3-030-35802-0_2
24. Garg, J., Mehta, R., Vazirani, V.V., Yazdanbod, S.: ∃ℝ-completeness for decision versions of multi-player (symmetric) Nash equilibria. ACM Trans. Econ. Comput. **6**(1), 1:1–1:23 (2018). https://doi.org/10.1145/3175494
25. Haussler, D., Welzl, E.: ε-nets and simplex range queries. Discrete & Comput. Geom. 127–151 (1987). https://doi.org/10.1007/BF02187876

26. Hoffmann, M., Miltzow, T., Weber, S., Wulf, L.: Recognition of unit segment graphs is $\exists\mathbb{R}$-complete, unpublished (2023, in preparation)
27. Jungeblut, P., Kleist, L., Miltzow, T.: The complexity of the Hausdorff distance. In: 38th International Symposium on Computational Geometry (SoCG 2022). LIPIcs, vol. 224, pp. 48:1–48:17 (2022). https://doi.org/10.4230/LIPIcs.SoCG.2022.48
28. Kang, R.J., Müller, T.: Sphere and dot product representations of graphs. Discrete Comput. Geom. **47**(3), 548–568 (2012). https://doi.org/10.1007/s00454-012-9394-8
29. Kratochvíl, J., Matoušek, J.: Intersection graphs of segments. J. Combin. Theory Ser. B **62**(2), 289–315 (1994). https://doi.org/10.1006/jctb.1994.1071
30. Looges, P.J., Olariu, S.: Optimal greedy algorithms for indifference graphs. Comput. Math. Appl. **25**(7), 15–25 (1993). https://doi.org/10.1016/0898-1221(93)90308-I
31. Lubiw, A., Miltzow, T., Mondal, D.: The complexity of drawing a graph in a polygonal region. In: Biedl, T., Kerren, A. (eds.) GD 2018. LNCS, vol. 11282, pp. 387–401. Springer, Cham (2018). https://doi.org/10.1007/978-3-030-04414-5_28
32. Ma, L.: Bisectors and voronoi diagrams for convex distance functions. Ph.D. thesis, FernUniversität Hagen (2000). https://ub-deposit.fernuni-hagen.de/receive/mir_mods_00000857
33. Matoušek, J., Sedgwick, E., Tancer, M., Wagner, U.: Embeddability in the 3-sphere is decidable. J. ACM **65**(1), 1–49 (2018). https://doi.org/10.1145/2582112.2582137
34. Matoušek, J., Tancer, M., Wagner, U.: Hardness of embedding simplicial complexes in $\mathbb{R}^d$. JEMS **13**(2), 259–295 (2011). https://doi.org/10.4171/JEMS/252
35. Matoušek, J., Seidel, R., Welzl, E.: How to net a lot with little: small ϵ-nets for disks and halfspaces. In: Sixth Annual Symposium on Computational Geometry (SoCG 1990), pp. 16–22 (1990). https://doi.org/10.1145/98524.98530
36. McDiarmid, C., Müller, T.: Integer realizations of disk and segment graphs. J. Combin. Theory Ser. B **103**(1), 114–143 (2013). https://doi.org/10.1016/j.jctb.2012.09.004
37. Mesmay, A.D., Rieck, Y., Sedgwick, E., Tancer, M.: Embeddability in $\mathbb{R}^3$ is NP-hard. J. ACM **67**(4), 20:1–20:29 (2020). https://doi.org/10.1145/3396593
38. Miltzow, T., Schmiermann, R.F.: On classifying continuous constraint satisfaction problems. In: IEEE 62nd Annual Symposium on Foundations of Computer Science (FOCS 2021), pp. 781–791 (2022). https://doi.org/10.1109/FOCS52979.2021.00081
39. Mnev, N.E.: The universality theorems on the classification problem of configuration varieties and convex polytopes varieties. In: Viro, O.Y., Vershik, A.M. (eds.) Topology and Geometry—Rohlin Seminar. LNM, vol. 1346, pp. 527–543. Springer, Heidelberg (1988). https://doi.org/10.1007/BFb0082792
40. Pach, J., Tardos, G.: Tight lower bounds for the size of epsilon-nets. J. Am. Math. Soc. **26**(3), 645–658 (2013). https://doi.org/10.1090/S0894-0347-2012-00759-0
41. Pach, J., Woeginger, G.: Some new bounds for epsilon-nets. In: Sixth Annual Symposium on Computational Geometry (SoCG 1990), pp. 10–15 (1990). https://doi.org/10.1145/98524.98529
42. Richter-Gebert, J., Ziegler, G.M.: Realization spaces of 4-polytopes are universal. Bull. Am. Math. Soc. **32**(4), 403–412 (1995). https://doi.org/10.1090/S0273-0979-1995-00604-X
43. Schaefer, M.: Complexity of some geometric and topological problems. In: Eppstein, D., Gansner, E.R. (eds.) GD 2009. LNCS, vol. 5849, pp. 334–344. Springer, Heidelberg (2010). https://doi.org/10.1007/978-3-642-11805-0_32

44. Schaefer, M.: Realizability of graphs and linkages. In: Pach, J. (ed.) Thirty Essays on Geometric Graph Theory, pp. 461–482. Springer, New York (2013). https://doi.org/10.1007/978-1-4614-0110-0_24
45. Schaefer, M.: Complexity of geometric k-planarity for fixed k. J. Graph Algorithms Appl. **25**(1), 29–41 (2021). https://doi.org/10.7155/jgaa.00548
46. Schaefer, M., Sedgwick, E., Štefankovič, D.: Recognizing string graphs in NP. J. Comput. Syst. Sci. **67**(2), 365–380 (2003). https://doi.org/10.1016/S0022-0000(03)00045-X
47. Schaefer, M., Štefankovič, D.: Fixed points, Nash equilibria, and the existential theory of the reals. Theory Comput. Syst. **60**, 172–193 (2017). https://doi.org/10.1007/s00224-015-9662-0
48. Schaefer, M., Štefankovič, D.: The complexity of tensor rank. Theory Comput. Syst. **62**(5), 1161–1174 (2018). https://doi.org/10.1007/s00224-017-9800-y
49. Shitov, Y.: A universality theorem for nonnegative matrix factorizations (2016). https://doi.org/10.48550/arXiv.1606.09068
50. Shitov, Y.: The complexity of positive semidefinite matrix factorization. SIAM J. Optim. **27**(3), 1898–1909 (2017). https://doi.org/10.1137/16M1080616
51. Shor, P.W.: Stretchability of pseudolines is NP-hard. In: Gritzmann, P., Sturmfels, B. (eds.) Applied Geometry And Discrete Mathematics. DIMACS Series in Discrete Mathematics and Theoretical Computer Science, vol. 4, pp. 531–554 (1991). https://doi.org/10.1090/dimacs/004/41
52. Skopenkov, A.: Extendability of simplicial maps is undecidable (2020). https://doi.org/10.48550/arXiv.2008.00492
53. Smorodinsky, S.: On the chromatic number of geometric hypergraphs. SIAM J. Discrete Math. **21**(3), 676–687 (2007). https://doi.org/10.1137/050642368
54. Spinrad, J.: Recognition of circle graphs. J. Algorithms **16**(2), 264–282 (1994). https://doi.org/10.1006/jagm.1994.1012
55. Stade, J.: Complexity of the boundary-guarding art gallery problem (2022). https://doi.org/10.48550/arXiv.2210.12817
56. Tanenbaum, P.J., Goodrich, M.T., Scheinerman, E.R.: Characterization and recognition of point-halfspace and related orders. In: Tamassia, R., Tollis, I.G. (eds.) GD 1994. LNCS, vol. 894, pp. 234–245. Springer, Heidelberg (1995). https://doi.org/10.1007/3-540-58950-3_375
57. Tuncel, L., Vavasis, S., Xu, J.: Computational complexity of decomposing a symmetric matrix as a sum of positive semidefinite and diagonal matrices (2022). https://doi.org/10.48550/arXiv.2209.05678

On the Complexity of Lombardi Graph Drawing

Paul Jungeblut(✉)

Karlsruhe Institute of Technology, Karlsruhe, Germany
paul.jungeblut@kit.edu

Abstract. In a *Lombardi drawing* of a graph the vertices are drawn as points and the edges are drawn as circular arcs connecting their respective endpoints. Additionally, all vertices have perfect angular resolution, i.e., all angles incident to a vertex v have size $2\pi/\deg(v)$. We prove that it is $\exists\mathbb{R}$-complete to determine whether a given graph admits a Lombardi drawing respecting a fixed cyclic ordering of the incident edges around each vertex. In particular, this implies NP-hardness. While most previous work studied the (non-)existence of Lombardi drawings for different graph classes, our result is the first on the computational complexity of finding Lombardi drawings of general graphs.

Keywords: Graph drawing · Lombardi drawing · Existential theory of the reals

1 Introduction

Inspired by the work of American artist Mark Lombardi [15], a *Lombardi drawing* of a given graph G draws vertices as points and edges as circular arcs or line segments connecting their endpoints. Further, each vertex v has *perfect angular resolution*, meaning that all angles between edges incident to v have size $2\pi/\deg(v)$. Notably, planarity is not required (even for planar graphs) and the crossing angle at intersections may be arbitrary. See Fig. 1 for Lombardi drawings of three well-known graphs.

Introduced by Duncan, Eppstein, Goodrich, Kobourov and Nöllenburg over ten years ago [7], Lombardi drawings have received a lot of attention in the graph drawing community, see the related work in Sect. 1.1 below. While most literature focuses on the construction of Lombardi drawings for different graph classes, the computational complexity to decide whether a Lombardi drawing exists remains largely unknown. To the best of our knowledge, NP-completeness is only known for certain regular graphs under the additional requirement that all vertices must lie on a common circle [7]. No lower or upper bounds on the complexity for general graphs are known (allowing arbitrary vertex placement).

In this paper we consider the case that the graph G comes with a fixed *rotation system* $\mathcal{R}$, i.e., a cyclic ordering of the incident edges around each vertex. Our main result is to determine the exact computational complexity of deciding whether G admits a Lombardi drawing respecting $\mathcal{R}$:

M. A. Bekos and M. Chimani (Eds.): GD 2023, LNCS 14465, pp. 180–194, 2023.
https://doi.org/10.1007/978-3-031-49272-3_13

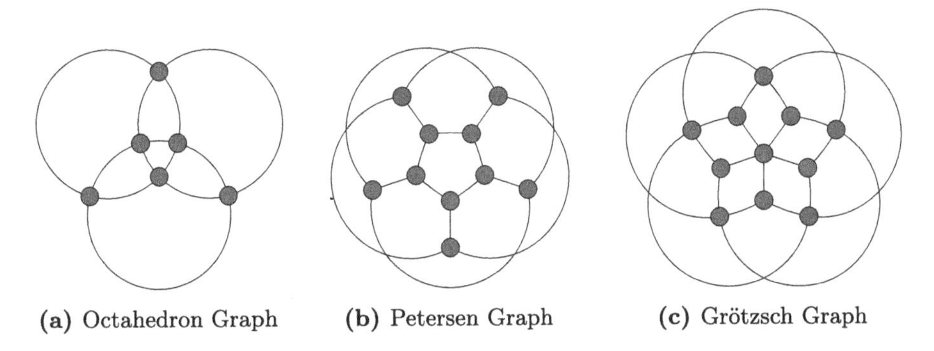

(a) Octahedron Graph (b) Petersen Graph (c) Grötzsch Graph

Fig. 1. Lombardi drawings created with the *Lombardi Spirograph* from [7].

Theorem 1. *Given a graph G with a rotation system $\mathcal{R}$, it is $\exists\mathbb{R}$-complete to decide whether G admits a Lombardi drawing respecting $\mathcal{R}$.*

The complexity class $\exists\mathbb{R}$ contains all problems that can be reduced to solving a system of polynomial equations and inequalities, see Sect. 2.2 for a formal definition. Since $\mathsf{NP} \subseteq \exists\mathbb{R}$, our result also implies NP-hardness.

Previous work frequently utilizes hyperbolic geometry to construct Lombardi drawings [7,9,12], the reason being that straight line segments in the hyperbolic plane $\mathbb{H}^2$ can be visualized by circular arcs in the Euclidean plane $\mathbb{R}^2$ (with the same crossing angles). We take a similar approach: A key ingredient of our $\exists\mathbb{R}$-hardness reduction is a recent observation by Bieker, Bläsius, Dohse and Jungeblut [1] stating that a simple pseudoline arrangement is stretchable in the Euclidean plane $\mathbb{R}^2$ if and only if it is stretchable in the hyperbolic plane $\mathbb{H}^2$ (see Sects. 2.1 and 2.3 for the necessary definitions). Their result allows us on the one hand to construct Lombardi drawings from hyperbolic line arrangements, and on the other hand to prove that sometimes no Lombardi drawing can exist.

1.1 Related Work

Lombardi drawings were introduced by Duncan, Eppstein, Goodrich, Kobourov and Nöllenburg [7], motivated by the network visualizations of Mark Lombardi [15]. While not all graphs admit Lombardi drawings (with or without prescribing the rotation system) [6,7], many graph classes always admit Lombardi drawings. Among them are 2-degenerate (and some 3-degenerate) graphs [7], subclasses of 4-regular graphs [7,18] and many classes of planar graphs that even admit planar Lombardi drawings. These include trees [8], cactus graphs [13], Halin graphs [7,10], subcubic graphs [9] and outerpaths [6]. However, many planar graphs do not admit planar Lombardi drawings in general [6,7,9,11,18].

A user study confirmed that Lombardi drawings are considered more aesthetic than straight line drawings but do not increase the readability [24].

Many variants have been considered: In a *k-circular Lombardi drawing* all vertices lie on one of k concentric circles [7]. Slightly relaxing the perfect angular resolution condition leads to *near Lombardi drawings* [5,18]. Lastly, edges in *k-Lombardi drawings* are drawn as the concatenation of up to k circular arcs [6,18].

Not much is known regarding the computational complexity of deciding whether a given graph admits a Lombardi drawing. Proving that a graph class always admits a Lombardi drawing is usually done constructively (this is the case for all classes mentioned above). In fact, these proofs lead to efficient algorithms (at least in a real RAM model of computation where square roots can be computed exactly). On the other hand, for d-regular graphs with $d \equiv 2 \mod 4$ it is NP-complete to decide whether they have a 1-circular Lombardi drawing [7]. Containment in NP might be surprising as this is in contrast to our main result showing $\exists\mathbb{R}$-hardness for general graphs and "classical" Lombardi drawings. NP-membership follows, because those graphs are yes-instances if and only if they are Hamiltonian.

2 Preliminaries

Let us recall the necessary geometric foundation for our reduction.

2.1 Hyperbolic Geometry

The hyperbolic plane $\mathbb{H}^2$ is an example of a non-Euclidean geometry. In many ways it behaves similar to the Euclidean plane $\mathbb{R}^2$, e.g. two points define a unique line and we can measure distances and angles.

Formally, both $\mathbb{R}^2$ and $\mathbb{H}^2$ can be described by an axiomatic system (like the one from Hilbert for $\mathbb{R}^2$ [14]). In fact, axiomatic systems for $\mathbb{R}^2$ and $\mathbb{H}^2$ are nearly identical, explaining the many similarities between $\mathbb{R}^2$ and $\mathbb{H}^2$. Without going into the technical details, Hilbert's axiomatic system contains the so-called *parallel postulate* stating that for any line ℓ and point p not on ℓ in $\mathbb{R}^2$ there is at most one (indeed exactly one) line through p parallel to ℓ. Negating this axiom turns Hilbert's axiomatic system for $\mathbb{R}^2$ into one that defines $\mathbb{H}^2$.

When working with the hyperbolic plane $\mathbb{H}^2$ we usually avoid working with the axioms directly. Instead we consider so-called *models*, i.e., embeddings of $\mathbb{H}^2$ into (in our case) $\mathbb{R}^2$. Several of these models are used in the literature. Important for us is the *Poincaré disk model*, see Fig. 2, where $\mathbb{H}^2$ is mapped to the interior of a unit disk D called the *Poincaré disk*. We omit how $\mathbb{H}^2$ is mapped into D and instead focus on some useful properties:

- Hyperbolic lines are mapped to either circular arcs orthogonal to D or diameters of D. By a slight perturbation it is actually always possible to obtain a realization in the Poincaré disk in which each hyperbolic line is represented by a circular arc.
- The Poincaré disk model is *conformal*, meaning that the angles in the hyperbolic plane equal the angles in a drawing inside the Poincaré disk D. Conformality is crucial in our reduction to obtain perfect angular resolution.

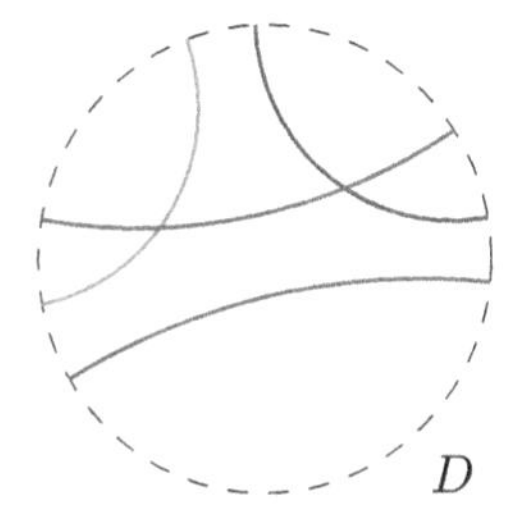

Fig. 2. Hyperbolic lines in the Poincaré disk D.

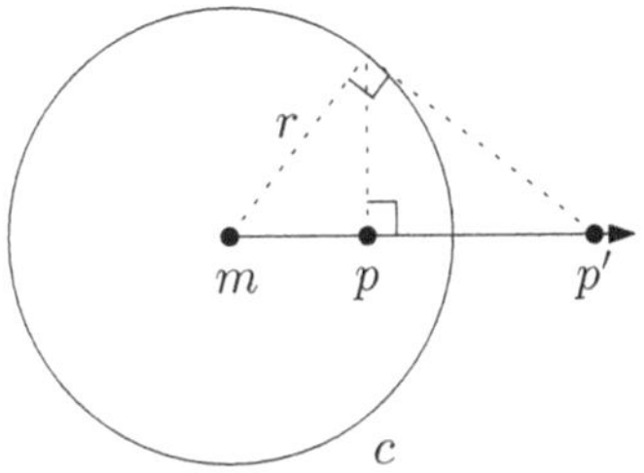

Fig. 3. Circle inversion.

2.2 Complexity Class $\exists\mathbb{R}$

Intuitively, the complexity class $\exists\mathbb{R}$ contains all problems that can be formulated as a system of polynomial equations and inequalities. Formally, it is defined to contain all problems that polynomial-time many-one reduce to the decision problem ETR (short for "existential theory of the reals") which is defined as follows: The input of ETR is a well-formed sentence Φ in the existential fragment of the first-order theory of the reals, i.e., a sentence of the form

$$\Phi \equiv \exists X_1, \ldots, X_n \in \mathbb{R} : \varphi(X_1, \ldots, X_n),$$

where φ is a quantifier-free formula consisting of polynomial equations and inequalities with integer coefficients. The task is to decide whether Φ is true. For example, $\exists X, Y \in \mathbb{R} : XY - 2X = 1 \wedge X + Y = 4$ is a yes-instance of ETR because for $(X, Y) = (1, 3)$ both polynomial equations are satisfied. On the other hand, $\exists X \in \mathbb{R} : X^2 < 0$ is a no-instance: There is no real number X whose square is negative. It is known that $\mathsf{NP} \subseteq \exists\mathbb{R} \subseteq \mathsf{PSPACE}$ and both inclusions are conjectured to be strict [3,31,33].

Many problems from computational geometry and especially graph drawing have been shown to be $\exists\mathbb{R}$-complete. Examples include RAC-drawability [30], geometric k-planarity [28], the recognition of many types of geometric intersection graphs [1,19,26], simultaneous graph embedding [4,20,29], variants of the segment number [23], as well as extending a partial planar straight line drawing of a planar graph inside a polygonal region to a drawing of the full graph [21].

2.3 Stretchability of Pseudolines

A *pseudoline arrangement* $\mathcal{A} = \{\ell_1, \ldots, \ell_n\}$ is a set of x-monotone curves in $\mathbb{R}^2$ such that each pair of curves intersects at most once. We say that $\mathcal{A}$ is *simple* if each pair intersects exactly once and no three pseudolines in $\mathcal{A}$ intersect in a point. See Fig. 4a for a simple pseudoline arrangement.

We always assume that the pseudolines are labeled such that a vertical line to the left of all intersections crosses ℓ_i below ℓ_j (for $i, j \in \{1, \ldots, n\}$) if and only

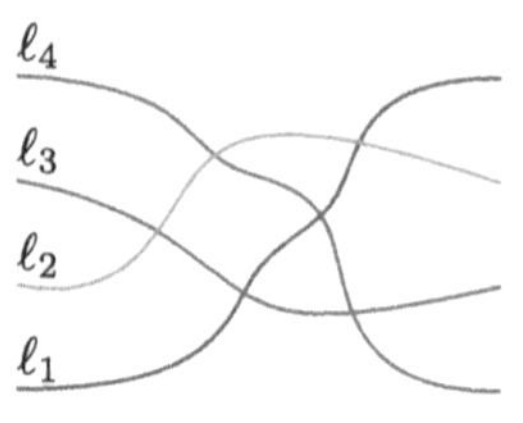

(a) A simple arrangement of four pseudolines.

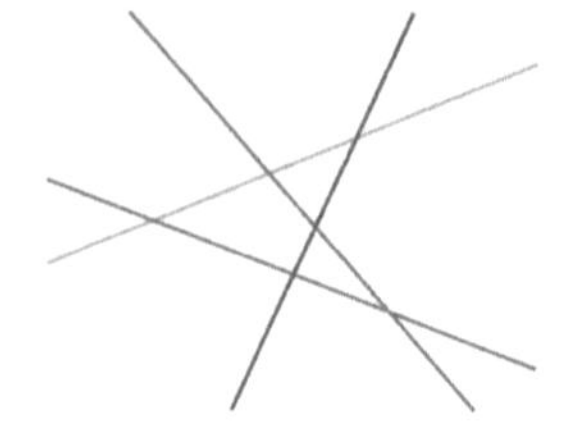

(b) A corresponding line arrangement in $\mathbb{R}^2$.

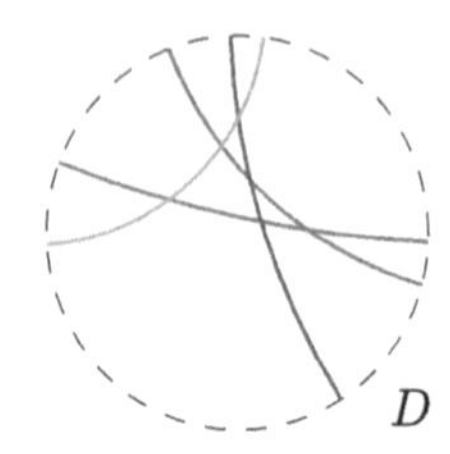

(c) A corresponding line arrangement in $\mathbb{H}^2$ in the Poincaré disk D.

Fig. 4. Stretchability of a simple pseudoline arrangement in $\mathbb{R}^2$ and $\mathbb{H}^2$.

if $i < j$. There are several equivalent ways to describe the intersection pattern of a given pseudoline arrangement. For us, a *combinatorial description* $\mathcal{D}$ of $\mathcal{A}$ is a list of n lists, one for each pseudoline, listing the order of intersections along it from left to right. For example, the list of intersections for ℓ_1 in Fig. 4a contains (in this order) ℓ_3, ℓ_4 and ℓ_2.

We say that a pseudoline arrangement $\mathcal{A}$ is *stretchable (in $\mathbb{R}^2$)* if there is a line arrangement that is homeomorphic to $\mathcal{A}$, i.e., having exactly the same intersection pattern. Figure 4b shows one way to stretch the pseudolines from Fig. 4a. Given a combinatorial description $\mathcal{D}$ of a (simple) pseudoline arrangement $\mathcal{A}$, we denote by (SIMPLE)STRETCHABILITY the decision problem whether $\mathcal{A}$ is stretchable. It is well-known that both problems are $\exists\mathbb{R}$-complete [22,25,33] and many $\exists\mathbb{R}$-hardness results are by a reduction from one of them [2,16,19,26,27].

Instead of asking for stretchability in the Euclidean plane $\mathbb{R}^2$, one may also consider stretchability in the hyperbolic plane $\mathbb{H}^2$. For example, Fig. 4c shows a hyperbolic line arrangement in the Poincaré disk model with the same intersection pattern as the pseudolines from Fig. 4a. Recently, Bieker, Bläsius, Dohse and Jungeblut observed that being a yes- or no-instance of SIMPLESTRETCHABILITY is independent of the underlying plane being $\mathbb{R}^2$ or $\mathbb{H}^2$ [1], thereby allowing us to use the term "stretchable" without specifying whether we consider $\mathbb{R}^2$ or $\mathbb{H}^2$:

Theorem 2 ([1]). *Let $\mathcal{D}$ be the combinatorial description of a simple pseudoline arrangement. Then $\mathcal{D}$ is stretchable in $\mathbb{R}^2$ if and only if $\mathcal{D}$ is stretchable in $\mathbb{H}^2$.*

2.4 Circle Geometry

A *circle inversion* with respect to a circle c with midpoint m and radius r swaps the interior and exterior of c. Each point $p \in \mathbb{R}^2 \setminus \{m\}$ is mapped to another point $p' \in \mathbb{R}^2 \setminus \{m\}$ such that both lie on the same ray originating from m and such that $\mathrm{d}(p, m) \cdot \mathrm{d}(p', m) = r^2$ (here $\mathrm{d}(\cdot, \cdot)$ is the Euclidean distance). By adding a single *point at infinity* (denoted by ∞) to $\mathbb{R}^2$ we obtain the so-called *extended plane*, allowing us to extend the definition of a circle inversion to $\mathbb{R} \cup \{\infty\}$. Now m is mapped to ∞ and vice versa. See Fig. 3 for an example.

Circle inversions map circles and straight lines to other circles and straight lines. Further, they are conformal, i.e., they preserve the angles between crossing lines and circles. In particular, they map Lombardi drawings to other Lombardi drawings. See [32] for a more thorough introduction.

3 Complexity of Lombardi Drawability

We prove that deciding whether a graph admits a Lombardi drawing respecting a fixed rotation system is $\exists\mathbb{R}$-complete by providing a polynomial-time many-one reduction from SIMPLESTRETCHABILITY.

Our reduction is split into two parts: We start by transforming a combinatorial description $\mathcal{D}$ of a simple pseudoline arrangement into a graph G with rotation system $\mathcal{R}$. The reduction is such that $\mathcal{D}$ is stretchable if and only if G admits a Lombardi drawing Γ respecting $\mathcal{R}$ under the additional restriction that certain cycles in G must be drawn as circles in Γ. Only then we extend our construction so to enforce the additional restrictions "automatically" in each Lombardi drawing.

3.1 Restricted Lombardi Drawings

The following construction is illustrated in Fig. 5. Let $\mathcal{D}$ be a combinatorial description of a simple arrangement $\mathcal{A} = \{\ell_1, \ldots, \ell_n\}$ of $n \geq 2$ pseudolines, i.e., an instance of the $\exists\mathbb{R}$-complete SIMPLESTRETCHABILITY problem. To recall, this means that for each pseudoline $\ell_i \in \mathcal{A}$ we have an ordered list containing the intersections with other pseudolines. As $\mathcal{A}$ is simple, for each pseudoline this list contains exactly $n-1$ intersections, one for each other pseudoline.

The first step of the construction is to extend the pseudoline arrangement $\mathcal{A}$ by a simple closed curve γ intersecting every pseudoline in $\mathcal{D}$ exactly twice, such that γ contains all intersections of $\mathcal{A}$ in its interior, see Fig. 5a. Let us denote the resulting arrangement by $\mathcal{A}_\gamma$. The combinatorial description $\mathcal{D}_\gamma$ of $\mathcal{A}_\gamma$ can be obtained by adding for each pseudoline $\ell_i \in \mathcal{A}$ (for $i \in \{1, \ldots, n\}$) one intersection with γ to the beginning and to the end of its list of intersections. Further, $\mathcal{D}_\gamma$ contains a list of intersections for γ containing every pseudoline in $\mathcal{A}$ exactly twice and whose cyclic ordering is $\ell_1, \ldots, \ell_n, \ell_1, \ldots, \ell_n$.

Now let G_γ be the following graph: We start by adding two vertices v_i^l and v_i^r per pseudoline ℓ_i corresponding to the left- and rightmost intersections of ℓ_i (these are the ones with γ). These $2n$ vertices are then connected to a cycle in the order they appear on γ. Next, we connect each pair v_i^l and v_i^r by an edge e_i. The rotation system $\mathcal{R}_\gamma$ shall be such that all edges e_i are on the same side of C_γ. Now for each pseudoline ℓ_i we add a path P_i from v_i^l to v_i^r by iterating through the list of its intersections from left to right. For each intersection with another pseudoline ℓ_j we add (in this order) two new vertices $v_{i,j}^l$ and $v_{i,j}^r$ to the path. In $\mathcal{R}_\gamma$, path P_i and edge e_i should be on opposite sides of C_γ. Let us denote by C_i the cycle formed by concatenating P_i with e_i. Lastly, for each intersection of two pseudolines ℓ_i and ℓ_j with $i < j$ we connect (in this order) $v_{i,j}^l$, $v_{j,i}^l$, $v_{i,j}^r$ and $v_{j,i}^r$

into a 4-cycle $C_{i,j}$. In $\mathcal{R}_\gamma$ the circular ordering around each of the four vertices should contain alternately an edge of $C_{i,j}$ and an edge of C_i respectively C_j. See Fig. 5b for the complete construction.

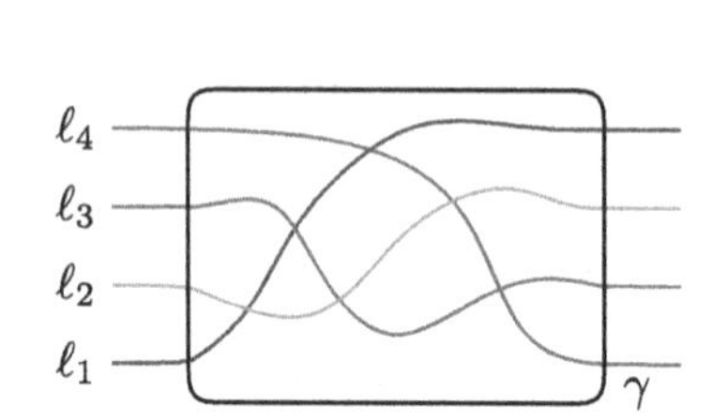

(a) An arrangement of four pseudolines enclosed by a curve γ.

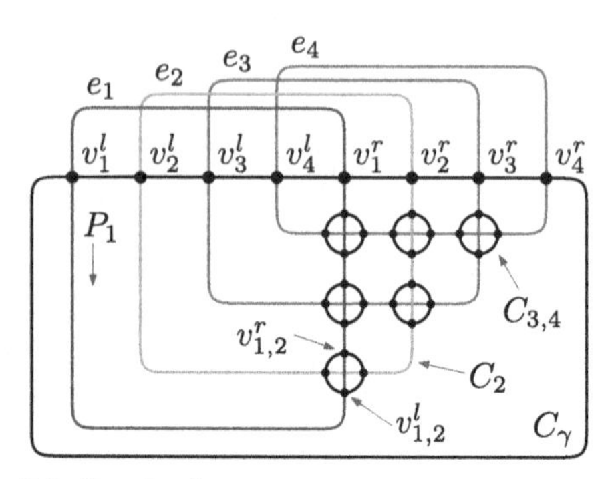

(b) Graph G_γ drawn such that it respects rotation system $\mathcal{R}_\gamma$.

Fig. 5. Example construction of G_γ and $\mathcal{R}_\gamma$ from a pseudoline arrangement $\mathcal{A}$.

In the two lemmas below we restrict ourselves to drawings of G_γ in which some cycles must be drawn as circles. A cycle C is said to be *drawn as a circle* c if all vertices and edges of C lie on c and the drawing is non-degenerate[1]. In particular this fixes the ordering of the vertices and edges of C along c (the only degree of freedom is whether this ordering is clockwise or counterclockwise).

Lemma 3. *If $\mathcal{D}$ is stretchable, then G_γ has a Lombardi drawing Γ respecting $\mathcal{R}_\gamma$ such that the cycles C_γ, all C_i (for $i \in \{1, \ldots, n\}$) and all $C_{i,j}$ (for $i, j \in \{1, \ldots, n\}$ with $i < j$) are drawn as circles in Γ.*

Proof. By Theorem 2 we can obtain a hyperbolic line arrangement realizing $\mathcal{D}$ in the Poincaré disk model. Further, each pseudoline ℓ_i is drawn as a circular arc a_i (with underlying circle c_i) inside and orthogonal to the Poincaré disk. From this, we construct a Lombardi drawing Γ of G_γ respecting $\mathcal{R}_\gamma$.

We denote by c_γ the circle representing the Poincaré disk and draw all vertices of cycle C_γ on it, such that v_i^l and v_i^r are placed at the left and right intersection of a_i with c_γ. Next, we draw the edges e_i outside of c_γ on $c_i \setminus a_i$ and the paths P_i inside c_γ on a_i. As v_i^l and v_i^r have degree four and c_i is orthogonal to c_γ, all vertices on C_γ have perfect angular resolution.

Next, for each pair of intersecting pseudolines ℓ_i and ℓ_j (with $i < j$) we place the vertices $v_{i,j}^l$ and $v_{i,j}^r$ to the left, respectively to the right, of the intersections on a_i (and similar $v_{j,i}^l$ and $v_{j,i}^r$ on a_j) such that they lie on a common circle $c_{i,j}$ which is orthogonal to a_i and a_j. Here, the orthogonality of $c_{i,j}$ with a_i and a_j

[1] A drawing is *degenerate* if two vertices are drawn at the same point or a vertex is drawn in the interior of an edge.

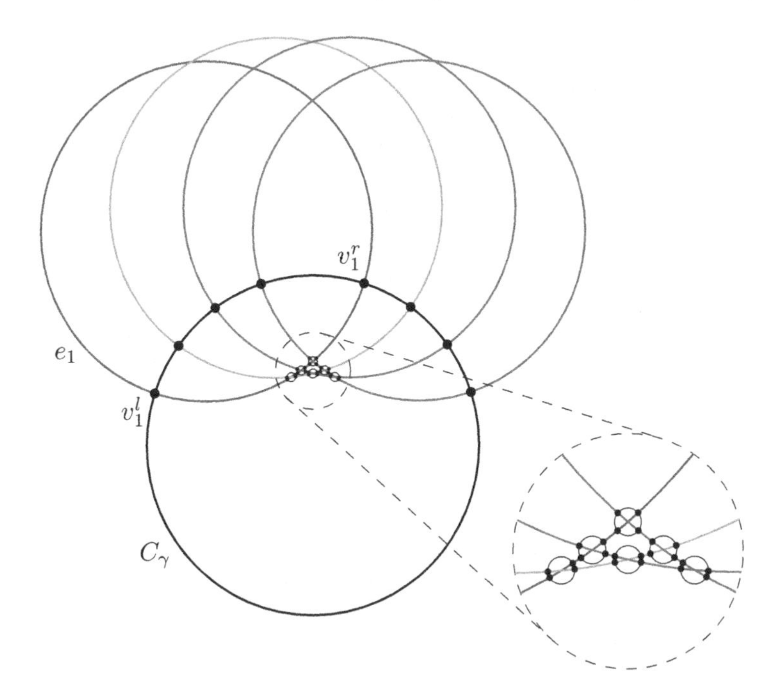

Fig. 6. A Lombardi drawing of the graph constructed in Fig. 5.

guarantees perfect angular resolution at the four involved vertices. (We prove in Lemma 11 in the full version [17] that such a circle indeed exists). Further, we can choose $c_{i,j}$ small enough so that no two such circles intersect, touch or contain each other.

The drawing is non-degenerate, respects $\mathcal{R}_\gamma$, has C_γ, all C_i and all $C_{i,j}$ drawn as circles and perfect angular resolution, i.e., it is a Lombardi drawing. □

See Fig. 6 for a Lombardi drawing of the graph shown in Fig. 5b that is constructed as described in the proof of Lemma 3.

Lemma 4. *If G_γ has a Lombardi drawing Γ respecting $\mathcal{R}_\gamma$ such that C_γ, all C_i (for $i \in \{1, \ldots, n\}$) and all $C_{i,j}$ (for $i, j \in \{1, \ldots, n\}$ with $i < j$) are drawn as circles in Γ, then $\mathcal{D}$ is stretchable.*

Proof. Let c_γ be the circle that C_γ is drawn as. We can assume without loss of generality that all edges e_i are drawn outside of c_γ and all paths P_i are drawn inside of c_γ, as we can otherwise consider the drawing obtained from Γ by a circle inversion with respect to c_γ. Recall that all vertices on C_γ have degree four. Let c_i be the circles that the cycles C_i are drawn as (for $i \in \{1, \ldots, n\}$). All c_i are orthogonal to c_γ because Γ has perfect angular resolution.

This allows us to interpret c_γ as a Poincaré disk and each circular arc a_i of c_i containing the drawing of P_i in Γ as a hyperbolic line. Therefore, the interior of c_γ induces a hyperbolic line arrangement. We prove that this hyperbolic line arrangement has combinatorial description $\mathcal{D}$.

To this end consider an arbitrary but fixed path P_i. Along P_i from left to right we encounter pairs of vertices $v^l_{i,j}$ and $v^r_{i,j}$ that, together with $v^l_{j,i}$ and $v^r_{j,i}$ form a 4-cycle corresponding to the intersection between pseudolines ℓ_i and ℓ_j (assume $i < j$). As c_i and c_j are circles orthogonal to c_γ, their circular arcs a_i and a_j intersect at most once. Further, as the vertices on $C_{i,j}$ are alternately on P_i and P_j, there must be an odd number of intersections between a_i and a_j inside the drawing of $C_{i,j}$ in Γ. It follows, that a_i and a_j intersect exactly once and they do so between $v^l_{i,j}$ and $v^r_{i,j}$. Thus, each pseudoline intersects each other pseudoline exactly once and in the order described by $\mathcal{D}$, i.e., our Lombardi drawing Γ induces a hyperbolic line arrangement with combinatorial description $\mathcal{D}$. Because $\mathcal{D}$ is simple and by Theorem 2 it follows that $\mathcal{D}$ is then also stretchable in $\mathbb{R}^2$. □

Summarizing the results so far, we see that Lemmas 3 and 4 give a reduction from SIMPLESTRETCHABILITY to a restricted form of Lombardi drawing. In what follows, we see how to incorporate these restrictions into the reduction itself.

3.2 Enforcing the Circles

In Lemmas 3 and 4 above we assumed that certain cycles in G_γ are drawn as circles. Below we describe how we can omit this explicit restriction by enforcing all possible Lombardi drawings to "automatically" satisfy it.

By an *arc-polygon* we denote a set of points $v_0, \ldots, v_k$ such that v_i and v_{i+1} (with $v_{k+1} = v_0$) are connected by a circular arc or line segment. An arc-polygon is simple if it does not self-touch or self-intersect. In case of two or three vertices we speak of a *bigon* and an *arc-triangle*, respectively. We utilize a lemma by Eppstein, Frishberg and Osegueda [13] who characterized simple arc-triangles. We follow their notation: The vertices v_0, v_1 and v_2 are numbered in clockwise order such that the interior of the arc-triangle is to the right when going from v_i to $v_{(i+1) \bmod 3}$. The vertices enclose internal angles θ_0, θ_1 and θ_2. If the vertices do not lie on a common line, then they define a unique circle c. In this case we denote by ϕ_i the internal angle of the bigon enclosed by c and the circular arc a_i between $v_{(i-1) \bmod 3}$ and $v_{(i+1) \bmod 3}$. Negative (positive) values of ϕ_i mean that a_i is outside (inside) of c and $\phi_i = 0$ means that a_i is on c. See Fig. 7 for an illustration.

Lemma 5 (**[13, Lemma** 4 **and Corollary** 5**]).** *Let v_0, v_1 and v_2 be a simple arc-triangle as above, not on a common line. Then for $\psi = \left(\pi - \sum_{i=0}^{2} \theta_i\right)/2$ it holds that $\phi_i = \psi + \theta_i$.*

We use Lemma 5 to prove that prescribing the interior angles of an arc-triangle to certain values is enough to guarantee that one of its vertices lies on the underlying circle of the circular arc connecting the other two vertices:

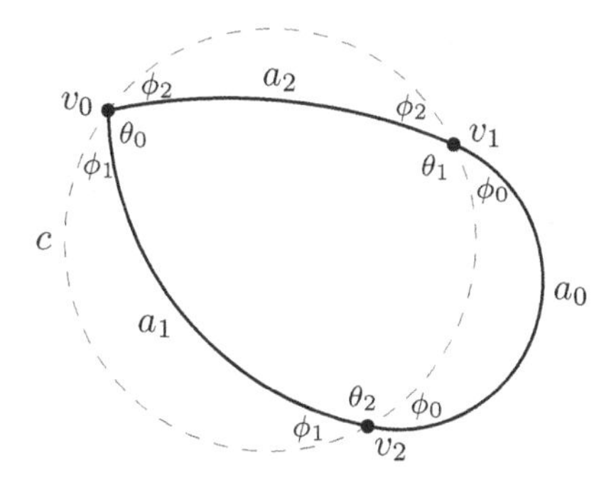

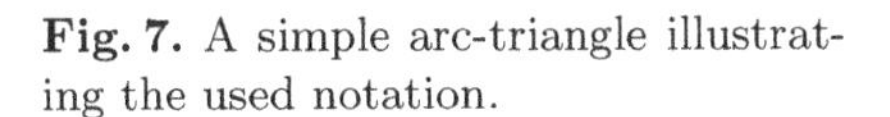
Fig. 7. A simple arc-triangle illustrating the used notation.

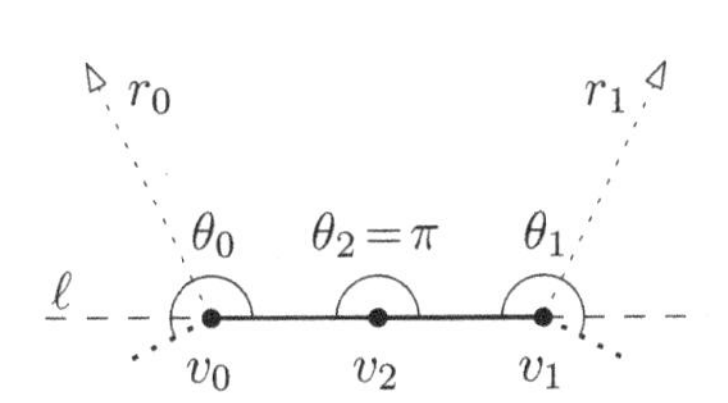

Fig. 8. Ruling out the collinear case in Lemma 6.

Lemma 6. *Let v_0, v_1 and v_2 be a simple arc-triangle as above such that $\theta_0 = \theta_1 \in [\pi, 3\pi/2)$ and $\theta_2 = \pi$. Further, edge v_0v_2 is drawn as a circular arc a_1 (and not as a line segment) with underlying circle c_1. Then v_0, v_1 and v_2 do not lie on a common line and c_1 is the unique circle through them, with v_1 on $c_1 \setminus a_1$.*

Proof. We first rule out the case that all three vertices lie on a common line ℓ, see Fig. 8: As the internal angle at v_2 has size π, vertices v_0 and v_1 must be on opposite sides of ℓ. The internal angle θ_0 at v_0 defines a ray r_0 that must contain the center of the underlying circle of a_2 (the circular arc connecting v_0 and v_1). Similarly, θ_1 defines another ray r_1 that must contain the center of a_2. However, $r_0 \cap r_1 = \emptyset$, because $\theta_0 = \theta_1 \in [\pi, 3\pi/2)$. We conclude that the three vertices cannot lie on a common line and therefore lie on a unique circle c.

We use Lemma 5 to compute the internal angle of the bigon enclosed by a_1 and c: We get that $\psi = (\pi - \sum_{i=0}^{2} \theta_i)/2 = -\theta_1$ and with that $\phi_1 = -\theta_1 + \theta_1 = 0$, i.e., a_1 must lie on c and in particular $c_1 = c$. As simple arc-triangles do not self-intersect or self-touch, it follows that v_1 lies on $c_1 \setminus a$. □

For our reduction we construct, for a given pseudoline arrangement $\mathcal{D}$, a graph $G_{\mathcal{D}}$ with rotation system $\mathcal{R}_{\mathcal{D}}$, which extend G_γ and $\mathcal{R}_\gamma$ by new vertices and edges. As we will see, this forms several arc-triangles in $G_{\mathcal{D}}$ that fulfill the conditions of Lemma 6. Iteratively applying this lemma will allow us to prove that C_γ as well as all C_i (for $i \in \{1, \dots, n\}$) and all $C_{i,j}$ (for $i, j \in \{1, \dots, n\}$ with $i < j$) must be drawn as circles in any Lombardi drawing of $G_{\mathcal{D}}$ respecting $\mathcal{R}_{\mathcal{D}}$.

Let us introduce some notation for cycles $C = (v_1, \dots, v_k)$. We denote by e_i the edge of the cycle connecting v_i and v_{i+1} (where $v_{k+1} = v_1$). If each v_i (for $i \in \{1, \dots, k\}$) has degree four, then the incident edges form four angles between them, which in case of perfect angular resolution must all have size $\pi/2$. We call these the *quadrants* of v_i and label them by $q_i^1, \dots, q_i^4$ in counterclockwise order such that q_i^1 is left of e_i when traversing it from v_i to v_{i+1}. By a *half-edge* we denote an incident edge to a vertex whose other endpoint is not yet specified.

A *circle gadget* for a cycle $C = (v_1, \dots, v_k)$ as above together with $k-3$ additional half-edges in each quadrant of all $v \in V(C)$ is the following set of edges: For $j \in \{1, \dots, k-3\}$, the j-th half-edge of q_1^1 in clockwise order is joined with the j-th half-edge of q_{k-j}^2 in counterclockwise order, see Fig. 9. The following lemma shows that these edges enforce that C is drawn as a circle.

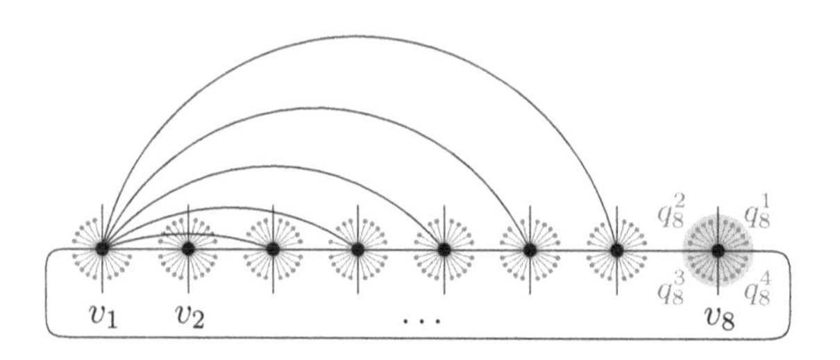

Fig. 9. Circle gadget for a cycle $C = (v_1, \dots, v_8)$. The quadrants at v_8 are labeled.

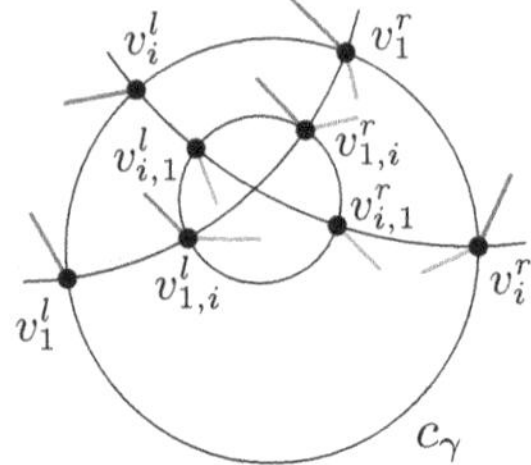

Fig. 10. How to combine different circle gadgets.

Lemma 7. *Let G be a graph containing a cycle $C = (v_1, \dots, v_k)$ with the edges of a circle gadget as described above. Then in every Lombardi drawing Γ that maps edge e_k to a circular arc a (i.e., not to a line segment), all vertices and edges of C are drawn onto the underlying circle c of a.*

Proof. First note that each vertex has equally many incident (half-)edges in each of its four quadrants, so by the perfect angular resolution of Γ, each quadrant spans an angle of $\pi/2$. In particular, between any two consecutive cycle edges there is an angle of π in Γ. Now consider the three cycle vertices v_1, v_{k-1} and v_k which form a simple arc-triangle whose internal angles satisfy the conditions of Lemma 6. It follows that v_{k-1} and e_{k-1} are drawn onto $c \setminus a$.

As we now know that the path from v_1 counterclockwise via v_k to v_{k-1} follows a single circular arc a' in Γ, the same argument can be repeated for the simple arc-triangle formed by the points v_1, v_{k-2} and v_{k-1}. It follows that v_{k-2} and e_{k-2} lie on c. Iterating the argument until we reach the simple arc-triangle formed by v_1, v_2 and v_3 proves the statement. □

With the circle gadget at hand, we can finally construct $G_\mathcal{D}$ and $\mathcal{R}_\mathcal{D}$. Recall that G_γ is 4-regular. We add $2n-3$ half-edges into each quadrant of every vertex $v \in V(G_\gamma)$ and then the following circle gadgets:

- For the cycle $C_\gamma = (v_1^l, \dots, v_n^l, v_1^r, \dots, v_n^r)$ on $2n$ vertices. Here v_1^l takes the role of v_1 and v_n^r takes the role of v_k in the circle gadget.
- For each cycle C_i on $2n$ vertices (for $i \in \{1, \dots, n\}$). Here v_i^r takes the role of v_1 and v_i^l takes the role of v_k in the circle gadget.
- For each cycle $C_{i,j}$ (for $i, j \in \{1, \dots, n\}$ with $i < j$). Here $v_1 = v_{i,j}^l$, $v_2 = v_{j,i}^l$, $v_3 = v_{i,j}^r$ and $v_4 = v_{j,i}^r$.

Note that several vertices are involved in multiple circle gadgets in our construction. This is not a problem because we carefully placed the circle gadgets such that no two circle gadgets operate in the same quadrant on each vertex. See Fig. 10 for a visualization (for simplicity, just C_1 and one C_i are drawn): Green half-edges show which quadrants are used by the circle gadget for C_γ. Orange half-edges belong to the circle gadgets of C_1 and C_i. Lastly, blue half-edges belong to the circle gadget of $C_{1,i}$.

As the last step of the reduction, all remaining half-edges are terminated with a new vertex of degree one. The resulting graph and rotation system are $G_\mathcal{D}$ and $\mathcal{R}_\mathcal{D}$.

Lemma 8. *If $\mathcal{D}$ is stretchable, then $G_\mathcal{D}$ has a Lombardi drawing respecting $\mathcal{R}_\mathcal{D}$.*

Proof. We start by applying Lemma 3 to obtain a Lombardi drawing Γ_γ of the subgraph G_γ of $G_\mathcal{D}$ respecting $\mathcal{R}_\gamma$ and in which cycle C_γ, all C_i (for $i \in \{1, \dots, n\}$) and all $C_{i,j}$ (for $i, j \in \{1, \dots, n\}$ with $i < j$) are drawn as circles.

It remains to draw the vertices and edges added by the circle gadgets. Recall that equally many edges were added into each quadrant of all vertices $v \in V(G_\gamma)$. Thus, the angles between edges in $E(G_\gamma)$ remain unchanged and the new edges must be drawn with equal angles between them into their quadrants to obtain perfect angular resolution.

Let e be an edge of a circle gadget with endpoints u and v and let c be the circle that u and v lie on in Γ_γ. By construction, e was obtained by joining the j-th half-edge in clockwise order in the first quadrant of u with the j-th half-edge in counterclockwise order in the second quadrant of v for some j. Thus the two angles between c and the arc representing e at u and v are equal and in $[0, \pi/2)$. There is exactly one circular arc that e can be drawn onto [7, Property 1].

Lastly, we need to make sure that the vertices of degree 1 that resulted from unjoined half-edges are drawn such that they do not lie on any other edge. This can be achieved by drawing the half-edges sufficiently short. □

Lemma 9. *If $G_\mathcal{D}$ has a Lombardi drawing Γ respecting $\mathcal{R}_\mathcal{D}$, then $\mathcal{D}$ is stretchable.*

Proof. Recall that edges of $G_\mathcal{D}$ are mapped to circular arcs or line segments in Γ and that each vertex has perfect angular resolution. We begin by analyzing how the cycle $C_\gamma = (v_1^l, \dots, v_n^l, v_1^r, \dots, v_n^r)$ in $G_\mathcal{D}$ must be drawn in Γ. We can assume that v_1^l, v_n^r and v_{n-1}^r do not lie on a common line and that the edge between v_1^l and v_n^r is mapped to a circular arc a (by a suitable circle inversion). Then by Lemma 7 all vertices and edges of C_γ lie on the underlying circle c_γ of a.

Next, we consider the circles C_i for $i \in \{1, \dots, n\}$. By applying a suitable circle inversion with respect to c_γ if necessary, we can assume that the paths P_i are drawn inside c_γ. In converse, the edges e_i are drawn outside of c_γ and therefore must be drawn as circular arcs (because a line segment would be inside c_γ). Applying Lemma 7 to each C_i yields that it is drawn as a circle c_i in Γ.

It remains to consider the circles $C_{i,j}$ for $i, j \in \{1, \dots, n\}$ with $i < j$. If they are not already drawn as a circle $c_{i,j}$ we can replace their drawing by a circle with sufficiently small radius by Lemma 11 (in the full version [17]).

Now it follows from Lemma 4 that $\mathcal{D}$ is stretchable. □

At this point we can finally prove our main result, Theorem 1:

Proof (of Theorem 1). Let $\mathcal{D}$ be a combinatorial description of a simple pseudoline arrangement. Construct $G_{\mathcal{D}}$ and $\mathcal{R}_{\mathcal{D}}$ as described above. By Lemmas 8 and 9, $\mathcal{D}$ is stretchable if and only if $G_{\mathcal{D}}$ admits a Lombardi drawing respecting $\mathcal{R}_{\mathcal{D}}$, proving $\exists\mathbb{R}$-hardness. We prove $\exists\mathbb{R}$-membership in Lemma 12 (in the full version [17]). □

4 Conclusion and Open Problems

In this paper we proved that it is $\exists\mathbb{R}$-complete to decide whether a given graph G with a fixed rotation system $\mathcal{R}$ admits a Lombardi drawing respecting $\mathcal{R}$. To the best of our knowledge, this is the first result on the complexity of Lombardi drawing for general graphs.

In fact, Lombardi drawing is just a special case of the more general problem where instead of enforcing perfect angular resolution we want to draw a graph with circular arc edges and angles of prescribed size. Our reduction immediately proves $\exists\mathbb{R}$-hardness for this problem as well:

Corollary 10. *Let G be a graph with a rotation system $\mathcal{R}$ and let Θ be an angle assignment prescribing the size of all angles in $\mathcal{R}$. Then it is $\exists\mathbb{R}$-complete to decide whether G admits a drawing with edges as circular arcs or line segments respecting $\mathcal{R}$ and Θ.*

On the other hand, several interesting questions remain open: Our reduction heavily relies on fixing the rotation system $\mathcal{R}$. By the perfect angular resolution requirement this fixes all angles in every Lombardi drawing. We wonder whether the problem remains $\exists\mathbb{R}$-complete without fixing $\mathcal{R}$:

Open Problem 1. *What is the computational complexity of deciding whether a graph admits any Lombardi drawing (without fixing a rotation system $\mathcal{R}$)?*

Given a planar graph, one usually asks for a planar Lombardi drawing. The graphs constructed in our reduction are in general not planar. In fact, they contain arbitrarily large clique minors. This motivates our second open problem:

Open Problem 2. *What is the complexity of deciding whether a planar graph admits a planar Lombardi drawing (with or without fixing a rotation system $\mathcal{R}$)?*

Acknowledgements. We thank Torsten Ueckerdt, Laura Merker and three anonymous reviewers for carefully reading this manuscript and providing valuable feedback.

References

1. Bieker, N., Bläsius, T., Dohse, E., Jungeblut, P.: Recognizing unit disk graphs in hyperbolic geometry is ∃ℝ-complete. In: Proceedings of the 39th European Workshop on Computational Geometry (EuroCG 2023), pp. 35:1–35:8 (2023). https://doi.org/10.48550/arXiv.2301.05550
2. Bienstock, D.: Some provably hard crossing number problems. Discrete Comput. Geom. **6**(3), 443–459 (1991). https://doi.org/10.1007/BF02574701
3. Canny, J.: Some algebraic and geometric computations in PSPACE. In: Proceedings of the Twentieth Annual ACM Symposium on Theory of Computing, STOC 1988, pp. 460–467. Association for Computing Machinery, New York (1988). https://doi.org/10.1145/62212.62257
4. Cardinal, J., Kusters, V.: The complexity of simultaneous geometric graph embedding. J. Graph Algorithms Appl. **19**(1), 259–272 (2015). https://doi.org/10.7155/jgaa.00356
5. Chernobelskiy, R., Cunningham, K.I., Goodrich, M.T., Kobourov, S.G., Trott, L.: Force-directed Lombardi-style graph drawing. In: van Kreveld, M., Speckmann, B. (eds.) GD 2011. LNCS, vol. 7034, pp. 320–331. Springer, Heidelberg (2012). https://doi.org/10.1007/978-3-642-25878-7_31
6. Duncan, C.A., Eppstein, D., Goodrich, M.T., Kobourov, S.G., Löffler, M., Nöllenburg, M.: Planar and poly-arc Lombardi drawings. J. Comput. Geom. **9**(1), 328–355 (2018). https://doi.org/10.20382/jocg.v9i1a11
7. Duncan, C.A., Eppstein, D., Goodrich, M.T., Kobourov, S.G., Nöllenburg, M.: Lombardi drawings of graphs. J. Graph Algorithms Appl. **16**(1), 85–108 (2012). https://doi.org/10.7155/jgaa.00251
8. Duncan, C.A., Eppstein, D., Goodrich, M.T., Kobourov, S.G., Nöllenburg, M.: Drawing trees with perfect angular resolution and polynomial area. Discrete Comput. Geom. **49**(2), 157–182 (2013). https://doi.org/10.1007/s00454-012-9472-y
9. Eppstein, D.: A Möbius-invariant power diagram and its applications to soap bubbles and planar Lombardi drawing. Discrete Comput. Geom. **52**(3), 515–550 (2014). https://doi.org/10.1007/s00454-014-9627-0
10. Eppstein, D.: Simple recognition of Halin graphs and their generalizations. J. Graph Algorithms Appl. **20**(2), 323–346 (2016). https://doi.org/10.7155/jgaa.00395
11. Eppstein, D.: Bipartite and series-parallel graphs without planar Lombardi drawings. J. Graph Algorithms Appl. **25**(1), 549–562 (2021). https://doi.org/10.7155/jgaa.00571
12. Eppstein, D.: Limitations on realistic hyperbolic graph drawing. In: Purchase, H.C., Rutter, I. (eds.) GD 2021. LNCS, vol. 12868, pp. 343–357. Springer, Cham (2021). https://doi.org/10.1007/978-3-030-92931-2_25
13. Eppstein, D., Frishberg, D., Osegueda, M.C.: Angles of arc-polygons and Lombardi drawings of cacti. Comput. Geom. **112** (2023). https://doi.org/10.1016/j.comgeo.2023.101982
14. Hilbert, D.: Grundlagen der Geometrie. Teubner, 13 edn. (1968). https://doi.org/10.1007/978-3-322-92726-2
15. Hobbs, R.: Mark Lombardi: global networks. Independent Curators International (2003)
16. Hoffmann, U.: On the complexity of the planar slope number problem. J. Graph Algorithms Appl. **21**(2), 183–193 (2017). https://doi.org/10.7155/jgaa.00411
17. Jungeblut, P.: On the complexity of Lombardi graph drawing (2023). https://doi.org/10.48550/arXiv.2306.02649. arXiv preprint

18. Kindermann, P., Kobourov, S.G., Löffler, M., Nöllenburg, M., Schulz, A., Vogtenhuber, B.: Lombardi drawings of knots and links. J. Comput. Geom. **10**(1), 444–476 (2018). https://doi.org/10.20382/jocg.v10i1a15
19. Kratochvíl, J., Matoušek, J.: Intersection graphs of segments. J. Combin. Theory Ser. B **62**(2), 289–315 (1994). https://doi.org/10.1006/jctb.1994.1071
20. Kynčl, J.: Simple realizability of complete abstract topological graphs in P. Discrete Comput. Geom. **45**(3), 383–399 (2011). https://doi.org/10.1007/s00454-010-9320-x
21. Lubiw, A., Miltzow, T., Mondal, D.: The complexity of drawing a graph in a polygonal region. J. Graph Algorithms Appl. **26**(4), 421–446 (2022). https://doi.org/10.7155/jgaa.00602
22. Mnev, N.E.: The universality theorems on the classification problem of configuration varieties and convex polytopes varieties. In: Viro, O.Y., Vershik, A.M. (eds.) Topology and Geometry — Rohlin Seminar. LNM, vol. 1346, pp. 527–543. Springer, Heidelberg (1988). https://doi.org/10.1007/BFb0082792
23. Okamoto, Y., Ravsky, A., Wolff, A.: Variants of the segment number of a graph. In: Archambault, D., Tóth, C.D. (eds.) GD 2019. LNCS, vol. 11904, pp. 430–443. Springer, Cham (2019). https://doi.org/10.1007/978-3-030-35802-0_33
24. Purchase, H.C., Hamer, J., Nöllenburg, M., Kobourov, S.G.: On the usability of Lombardi graph drawings. In: Didimo, W., Patrignani, M. (eds.) GD 2012. LNCS, vol. 7704, pp. 451–462. Springer, Heidelberg (2013). https://doi.org/10.1007/978-3-642-36763-2_40
25. Richter-Gebert, J.: Mnëv's universality theorem revisited (2002). https://geo.ma.tum.de/_Resources/Persistent/3/e/a/2/3ea2ad59228a1a24a67d1e994fa77266a599e73a/15_MnevsUniversalityhTheorem.pdf
26. Schaefer, M.: Complexity of some geometric and topological problems. In: Eppstein, D., Gansner, E.R. (eds.) GD 2009. LNCS, vol. 5849, pp. 334–344. Springer, Heidelberg (2010). https://doi.org/10.1007/978-3-642-11805-0_32
27. Schaefer, M.: Realizability of graphs and linkages. In: Pach, J. (ed.) Thirty Essays on Geometric Graph Theory, pp. 461–482. Springer, New York (2013). https://doi.org/10.1007/978-1-4614-0110-0_24
28. Schaefer, M.: Complexity of geometric k-planarity for fixed k. J. Graph Algorithms Appl. **25**(1), 29–41 (2021). https://doi.org/10.7155/jgaa.00548
29. Schaefer, M.: On the complexity of some geometric problems with fixed parameters. J. Graph Algorithms Appl. **25**(1), 195–218 (2021). https://doi.org/10.7155/jgaa.00557
30. Schaefer, M.: RAC-drawability is $\exists\mathbb{R}$-complete. In: Purchase, H.C., Rutter, I. (eds.) GD 2021. LNCS, vol. 12868, pp. 72–86. Springer, Cham (2021). https://doi.org/10.1007/978-3-030-92931-2_5
31. Schaefer, M., Štefankovič, D.: Fixed points, Nash equilibria, and the existential theory of the reals. Theory Comput. Syst. **60**, 172–193 (2017). https://doi.org/10.1007/s00224-015-9662-0
32. Schwerdtfeger, H.: Geometry of Complex Numbers. Dover Publications, Inc. (1979)
33. Shor, P.W.: Stretchability of pseudolines is NP-Hard. In: Gritzmann, P., Sturmfels, B. (eds.) Applied Geometry And Discrete Mathematics, Proceedings of a DIMACS Workshop, Providence, Rhode Island, USA, September 18, 1990. DIMACS Series in Discrete Mathematics and Theoretical Computer Science, vol. 4, pp. 531–554 (1991). https://doi.org/10.1090/dimacs/004/41

Tree Drawings with Columns

Jonathan Klawitter[1(✉)] and Johannes Zink[2]

[1] School of Computer Science, University of Auckland, Auckland, New Zealand
jo.klawitter@gmail.com
[2] Institut für Informatik, Universität Würzburg, Würzburg, Germany
zink@informatik.uni-wuerzburg.de

Abstract. Our goal is to visualize an additional data dimension of a tree with multifaceted data through superimposition on vertical strips, which we call *columns*. Specifically, we extend upward drawings of unordered rooted trees where vertices have assigned heights by mapping each vertex to a column. Under an orthogonal drawing style and with every subtree within a column drawn planar, we consider different natural variants concerning the arrangement of subtrees within a column. We show that minimizing the number of crossings in such a drawing can be achieved in fixed-parameter tractable (FPT) time in the maximum vertex degree Δ for the most restrictive variant, while becoming NP-hard (even to approximate) already for a slightly relaxed variant. However, we provide an FPT algorithm in the number of crossings plus Δ, and an FPT-approximation algorithm in Δ via a reduction to feedback arc set.

Keywords: tree drawing · multifaceted graph · feedback arc set · NP-hardness · fixed-parameter tractability · approximation algorithm

1 Introduction

Visualizations of trees have been used for centuries, as they provide valuable insights into the structural properties and visual representation of hierarchical relationships [15]. Over time, numerous approaches have been developed to create tree layouts that are both aesthetically pleasing and rich in information. These developments have extended beyond displaying the tree structure alone to also encompass different *facets* (dimensions) of the underlying data. As a result, researchers have explored various layout styles beyond layered node-link diagrams and even higher-dimensional representations [17,18]. Hadlak, Schumann, and Schulz [10] introduced the term of a *multifaceted graph* for a graph with associated data that combines various facets (also called aspects or dimensions) such as spatial, temporal, and other data. An example of a multifaceted tree is a *phylogenetic tree*, that is, a rooted tree with labeled leaves where edge lengths represent genetic differences or time estimates by assigning heights to

J.K. was supported by MBIE grant UOAX1932, J.Z. by DFG project Wo758/11-1.

M. A. Bekos and M. Chimani (Eds.): GD 2023, LNCS 14465, pp. 195–210, 2023.
https://doi.org/10.1007/978-3-031-49272-3_14

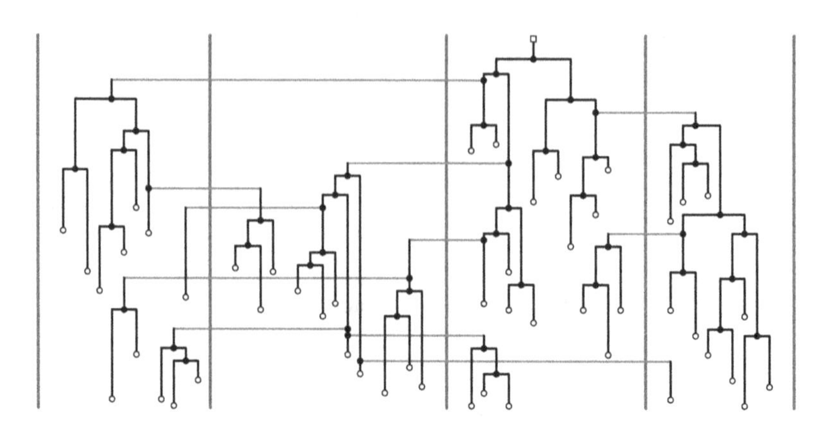

Fig. 1. A binary column tree with four columns.

the vertices. In this paper, we introduce an extension of classical node-link drawings for rooted phylogenetic trees, where vertices are mapped to distinct vertical strips, referred to as *columns*. This accommodates not only an additional facet of the data but also introduces the possibility of crossings between edges, thus presenting us with new algorithmic challenges.

Motivation. Our new drawing style is motivated by the visualization of transmission trees (i.e., the tree of who-infected-who in an infectious disease outbreak), which are phylogenetic trees that often have rich multifaceted data associated. In these visualizations, geographic regions associated with each case are commonly represented by coloring the vertex and its incoming edge [9]. However, there are scenarios where colors are not available or suitable. In such a case and in the context of a transmission tree, our alternative approach maps each, say, region or age group to a separate column (see Fig. 1). The interactive visualization platform Nextstrain [9] for tracking of pathogen evolution partially implements this with a feature that allows the mapping of the leaves to columns based on one facet. Yet, since they then leave out inner vertices and edges, the topology of the tree is lost. We hope that our approach enables users to quickly grasp group sizes and the number of edges (transmissions) between different groups.

Related Work. In visualization methods for so-called reconciliation trees [2,4] and multi-species coalescent trees [6,12], a guest tree P is drawn inside a space-filling drawing of a host tree H. The edges of the host tree can thus closely resemble columns, though the nature of and relationship between P and H prohibit a direct mapping to column trees. Spatial data associated with a tree has also been visualized by juxtaposition for so-called phylogeographic trees [13,16]. Betz et al. [1] investigated orthogonal drawings with column assignments, where each column is a single vertical line (i.e. analogous to a layer assignment).

Setting. The input for our drawings consists of an unordered rooted tree T, where each vertex $v \in V(T)$ has an assigned *height* $h(v)$, and a surjective *column mapping* $c\colon V(T) \rightarrow [1, \ell]$ for some $\ell \geq 2$. Together, we call $\langle T, h, c\rangle$ a *column tree*. We mostly assume that the order of the columns is *fixed* (from 1 to ℓ left-to-right), but in a few places we also consider the case that it is *variable*. For an edge uv in T, u is the *parent* of v, and v is the *child* of u. We call u and v the *source* and *target* (vertex) of uv, respectively. We call uv an *intra-edge* if $c(u) = c(v)$ and an *inter-edge* otherwise. The degree of u is the number of children of u.

Visualization of Column Trees. We draw a column tree $\langle T, h, c\rangle$ with a *rectangular cladogram* style, that is, each edge is drawn orthogonally and (here) downward with respect to the root; hence we assume that h corresponds to y-coordinates where the root has the maximum value and every parent vertex has a strictly greater value than its children. Each edge uv has at most one bend such that the horizontal segment (if existent) has the y-coordinate of the parent vertex u. Each column $\gamma \in [1, \ell]$ is represented by a vertical strip C_γ of variable width and a vertex v with $c(v) = \gamma$ must be placed within C_γ. We need a few definitions to state further drawing conventions.

A *column subtree* is a maximal subtree within a column. Note that each column subtree (except the one containing the root of T) has an incoming inter-edge to its root and may have various outgoing inter-edges to column subtrees in other columns. The *width* of a column subtree at height η is the number of edges of the column subtree intersected by the horizontal line at $y = \eta$. We say that two column subtrees A and B *interleave* in a drawing of a column tree if there is a horizontal line that intersects first an edge of A, then one of B, and then again one of A (or with A and B in reversed roles).

In graph drawing, we usually forbid overlaps. Here, however, if a vertex has more than one child to, say, its right, then the horizontal segments of the edges to these children overlap. As we permit more than three children, we allow these overlaps and we do not count them as crossings. On the other hand, we do not want overlaps to occur between non-neighboring elements. Thus, we assume that each vertex v being the source of an inter-edge has a unique height $h(v)$ as otherwise we cannot always avoid overlaps between an edge and a vertex and between two edges with distinct endpoints. Furthermore, we require that *no two intra-edges cross.* Hence, each column subtree is drawn planar and no two

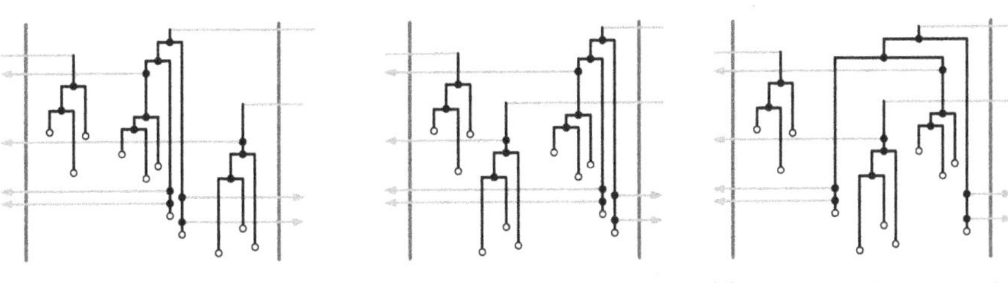

(a) V1 – subtree "stick" to column borders (13 crossings) (b) V2 – non-interleaving subtrees (11 cr.) (c) V3 – interleaving subtrees allowed (6 cr.)

Fig. 2. The three variants for subtree arrangements.

column subtrees intersect. However, since for inter-edges it is not always possible to avoid all edge crossings, we distinguish three drawing conventions concerning inter-edges and column subtree relations. These increasingly trade clarity of placement with the possibility to minimize the number of crossings (see Fig. 2):

(V1) No inter-edge uv intersects an intra-edge in column $c(v)$.
(V2) No two column subtrees interleave.
(V3) Column subtrees may interleave.

We remark that V1 is based on the idea that going from top to bottom, the column subtree rooted at v is greedily placed as soon as possible in $c(v)$. Thus, the subtrees "stick" to the column border closer to their root's parent. Note that V1 implies V2 as interleaving a column subtree B inside a column subtree A requires that the incoming inter-edge of B crosses an intra-edge of A.

Combinatorial Description. We are less interested in the computation of x-coordinates but in the underlying algorithmic problems regarding the number of crossings. Hence, it suffices to describe a column tree drawing combinatorially in terms of what we call the *subtree embeddings* and the *subtree arrangement* of the drawing. The subtree embedding is determined by the order of the children of each vertex in the subtree. The subtree arrangement of a column represents the relative order between column subtrees that overlap vertically. From a computational perspective, one possible approach to specify the subtree arrangement is by storing, for each root vertex v of a column subtree, which edges of which column subtrees lie horizontally to the left and to the right of v. This requires only linear space and all necessary information, e.g., for an algorithm that computes actual x-coordinates, can be recovered in linear time. A fixed column order, a subtree embedding of each column subtree, and a subtree arrangements of each column collectively form what we refer to as an *embedding* of a column tree $\langle T, h, c\rangle$.

Problem Definition. An instance of the problem TreeColumns|Vi, $i \in \{1, 2, 3\}$, is a column tree $\langle T, h, c\rangle$. The task is to find an embedding of $\langle T, h, c\rangle$ with the minimum total number of crossings among all possible embeddings of T under the drawing convention Vi. In the decision variant, an additional integer k is given, and the task is to find a column embedding with at most k crossings. If the column order is not fixed, we get another three problem versions.

We define three types of crossings. Consider an inter-edge uv with $c(u) < c(v)$. In any drawing, uv spans over the columns $c(u)+1, \dots, c(v)-1$ and crosses always the same set of edges in these columns. We call the resulting crossings *inter-column* crossings, denoted by k_{inter}. Within the column $c(u)$ $[c(v)]$, uv may cross edges of the column subtree of u $[v]$ and of other column subtrees. We call these crossings *intra-subtree* and *intra-column* crossings, denoted by k_{subtree} and k_{column}, respectively. The total number of crossings is then $k = k_{\mathsf{subtree}} + k_{\mathsf{column}} + k_{\mathsf{inter}}$.

It would also be natural to seek an embedding or drawing with minimum width, yet this is known to be NP-hard even for a single phylogenetic tree [19].

Observe that for TreeColumns|V1 and V2, modifying the embedding of a single column subtree can only change the intra-subtree crossings. On the other hand, permuting the order of the column subtrees within a column can only change the intra-column crosssings. Finding a minimum-crossing embedding can thus be split into two tasks, namely, *embedding the subtrees* and *finding a subtree arrangement*. Furthermore, note that V1 mostly enforces a subtree arrangement since in general each column subtree is the left-/rightmost in its column at the height of its root. Only for column subtrees that are in the same column and whose roots share a parent, is the relative order not known. The main task for V1 is thus to find subtree embeddings. For V3, in contrast, the two tasks are intertwined since a minimum-crossing embedding may use locally suboptimal subtree embeddings (see Fig. 2c). We briefly discuss heuristics in the full version of this paper on arXiv [14], where we also provide the proofs statements marked with a clickable ($\star$).

Contribution. We first give a fixed-parameter tractable (FPT) algorithm in the maximum vertex degree Δ for the subtree embedding task, the main task for V1 and a binary column tree (Sect. 2). We remark that the previously described applications usually use binary phylogenetic trees. This makes our algorithm, which uses a sweep line to determine the crossing-minimum child order at each vertex, a practically relevant polynomial-time algorithm. On the other hand, if we have a large vertex degree, minimizing the number of crossings is NP-hard, even to approximate, for all variants (Sect. 3). This holds true for the less restrictive variants V2 and V3 even for binary trees. Leveraging a close relation between the task of finding a column subtree arrangement and the feedback arc set (FAS) problem, we devise for V2 an FPT algorithm in the number of crossings k plus Δ and an FPT-approximation algorithm in Δ (Sect. 4). However, for the most general variant V3, the tasks of embedding the column subtrees and arranging them cannot be considered separately, and hence we only suggest heuristics to address this challenge (see the full version [14]).

2 Algorithm for Subtree Embedding

In this section, we describe how to find an optimal subtree embedding for a single column subtree. The algorithm has an FPT running time in the maximum vertex degree Δ making it a polynomial-time algorithm for bounded-degree trees like binary trees, ternary trees, etc.

Our algorithm is based on the following observation, described here for a binary column subtree S. Fix an embedding for S, i.e., the child order for each vertex. Consider an inter-edge $e = vw$ with $c(w) < c(v)$. Let u be a vertex of the path from the parent of v to the root of B. Observe that if v is in the left subtree of u, then e does not intersect the right subtree of u. Yet, if v is in the right subtree of u, then e intersects the left subtree of u (if it extends vertically beyond $h(v)$). On the other hand, if u is not on the path to the root, then the child order of u has no effect on whether e intersects its subtrees. Thus, to find an optimal child order of a vertex, it suffices to consider the directions (left/right) towards which the inter-edges of its descendants extend. Our algorithm is illustrated in Fig. 3.

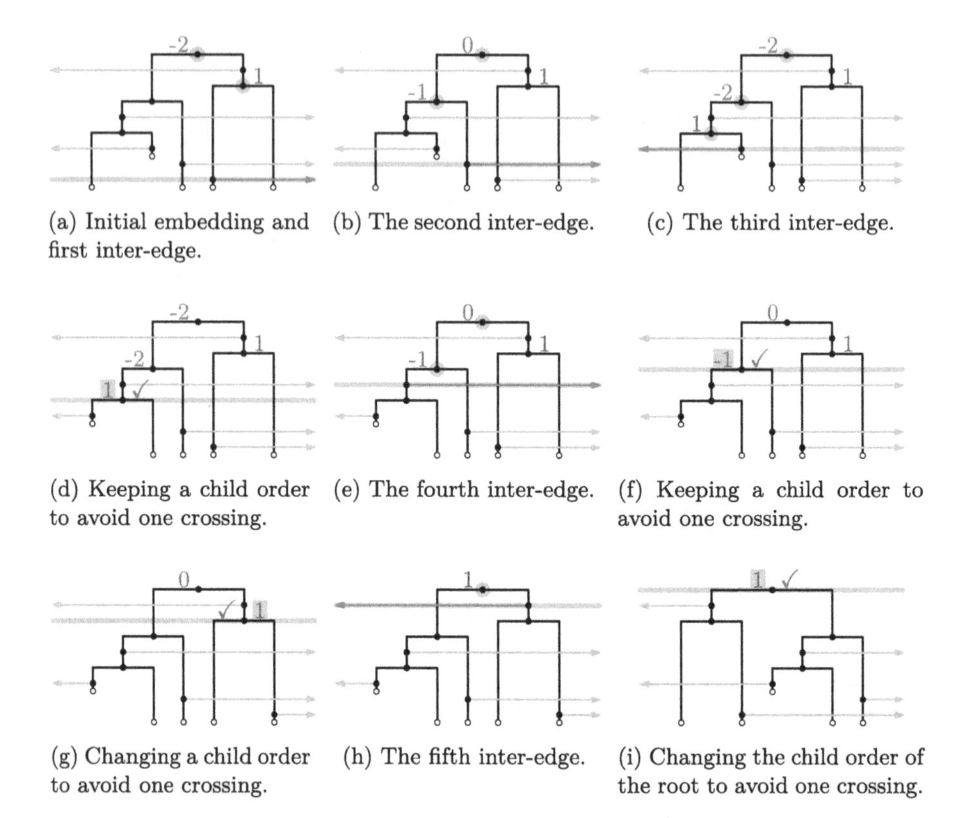

(a) Initial embedding and first inter-edge.
(b) The second inter-edge.
(c) The third inter-edge.
(d) Keeping a child order to avoid one crossing.
(e) The fourth inter-edge.
(f) Keeping a child order to avoid one crossing.
(g) Changing a child order to avoid one crossing.
(h) The fifth inter-edge.
(i) Changing the child order of the root to avoid one crossing.

Fig. 3. Steps of the sweep-line algorithm to find, for a given (here, binary) column subtree, a subtree embedding with the minimum number of intra-subtree crossings. A number at a vertex v reflects the count of crossings for the two possible permutations of v (positive number $\Leftrightarrow$ more crossings if v's child order is kept). In steps considering inter-edges, affected ancestor vertices are marked orange. (Color figure online)

Lemma 1. *For an n-vertex column subtree S with maximum degree Δ and t inter-edges, a minimum-crossing subtree embedding of S can be computed in $\mathcal{O}(\Delta!\Delta tn + n \log n)$ time.*

Proof. In a bottom-to-top sweep-line approach over S, we apply the following greedy-permuting strategy that finds, for each vertex, the child order that causes the fewest crossings. To this end, first sort the vertices of S by ascending height in $\mathcal{O}(n \log n)$ time. Also, each inner vertex v with d_v ($d_v \leq \Delta$) children in S has $d_v!$ counters to store for each of the $d_v!$ possible child orders (of its children in S) the computed number of crossings induced by that order (as described below). Initially, all counter are set to zero.

Having started at the bottom, let v be the next encountered vertex. First, process each inter-edge e whose source is v. Let u be a vertex on the path P from the parent of v to the root. Let T_v be the subtree rooted at a child of u that contains v. Compute for each subtree T rooted at a child of u (except for T_v) the width at $h(v)$ as follows.

Initialize a counter for the width of T at $h(v)$ with zero. Traverse T and whenever we encounter an edge of T that has one endpoint above and the other endpoint on or below $h(v)$, we increment that counter. Hence, we can determine the width of T at $h(v)$ in $\mathcal{O}(n_T)$ time, where n_T is the number of vertices in T, and we can determine the width of all subtrees rooted at children of vertices from P in $\mathcal{O}(n)$ time.

Recall that u has $d_u!$ (with $d_u \leq \Delta$) counters for its $d_u!$ possible child orders. For each such child order, add, if e goes to the left [right], the width at $h(v)$ of all subtrees rooted at children of u that are left [right] of T_v to the corresponding counter. Note that the resulting number is exactly the number of intra-subtree crossings induced by e, as our observation still holds for the general case because the number of intra-subtree crossings induced by e only depends on the child orders of the vertices of P. Of course, as the numbers are independent of other inter-edges, we can add up these numbers in our child order counters. Updating these counters for a vertex u on P takes $\mathcal{O}(d_u!d_u) \subseteq \mathcal{O}(\Delta!\Delta)$ time. Since the length of P can be linear, we can handle e in $\mathcal{O}(\Delta!\Delta n)$ time.

Second, since the counters of v only depend on the already processed inter-edges below v, pick the child order of v with the lowest counter for v.

In total, this sweeping phase of the algorithm runs in $\mathcal{O}(\Delta!\Delta tn)$ time, where t is the number of inter-edges. This is because, although the sweep-line algorithm has $\Theta(n)$ event points (the vertices of S), we apply the previously described $\mathcal{O}(\Delta!\Delta n)$-time subroutine only if we encounter an inter-edge. □

Note that for Lemma 1, if $t = 0$, any embedding is crossing free, and if Δ is constant, the running time is at most quadratic in n.

To solve an instance of TreeColumns|V1, we first apply Lemma 1 to each column subtree separately. Then for each vertex with multiple outgoing inter-edges to the same column, we try all possible orders for the respective column subtrees and keep the best. Hence, we get the following result.

Theorem 2. TreeColumns|V1 *is fixed-parameter tractable in Δ. More precisely, given an instance $\langle T, h, c\rangle$ with n vertices and maximum vertex degree Δ, there is an algorithm computing an embedding of $\langle T, h, c\rangle$ with the minimum number of crossings in $\mathcal{O}(\Delta!\Delta n^2)$ time.*

3 NP-Hardness and APX-Hardness

In this section, we show that TreeColumns becomes NP-hard (even to approximate) when finding a subtree arrangement is non-trivial or if we have large vertex degrees. We use a reduction from the (unweighted) Feedback Arc Set (FAS) problem, where we are given a digraph G and the task is to find a minimum-size set of edges that if removed make G acyclic. We assume, without loss of generality, that G is biconnected. FAS is one of Karp's original 21 NP-complete problems [11]. It is also NP-hard to approximate within a factor less than 1.3606 [5] and, presuming the unique games conjecture, even within any constant factor [8].

Theorem 3. *The* TreeColumns *problem is NP-complete for both fixed and variable column orders already for two columns under V1 if the degree is unbounded, and under V2 and V3 even for a binary column tree. Moreover, it is NP-hard to approximate within a factor of* $1.3606 - \varepsilon$ *for any* $\varepsilon > 0$, *and NP-hard to approximate within any constant factor presuming the unique games conjecture.*

Proof. The problem is in NP, since given an embedding for a column tree, it is straightforward to check whether it has at most k crossings in polynomial time.

To prove NP-hardness (of approximation), we use a reduction from FAS. Let G be an instance of FAS; let $n = |V(G)|$ and $m = |E(G)|$, and fix an arbitrary vertex order via the indices, so $V(G) = \{v_1, \ldots, v_n\}$, and an arbitrary edge order. We construct a column tree $\langle T, h, c\rangle$ such that there is a bijection between vertex orders of G (where each vertex order implies a solution set for the FAS problem) and solutions of $\langle T, h, c\rangle$, whose number of crossings depends on the size of the FAS solution and vice versa. We show this first for V1 with unbounded degree; see Fig. 4.

Give T two columns and, assuming for now that their order is fixed, call them left and right column. The total height of T is $a = 3(m+1)$. The root column subtree of T is in the left column, and has a single inner vertex v at $a-1$. For each $v_i \in V(G)$, there is an inter-edge from v to a column subtree T_i, the *vertex gadget* of v_i, in the right column. The idea is that a subtree arrangement of $\{T_1, \ldots, T_n\}$ corresponds to a topological order of $G \setminus S$ where each edge whose *edge gadget* induces a large number of crossings is in the FAS solution set S.

Vertex Gadget. The subtree T_i has a *backbone* path on $\deg(v_i) + 1$ vertices with its leaf at height 0. Attached to the backbone of T_i, there are inter-edges and subtrees depending on the edges incident to v_i. We describe them next.

Edge Gadget. The edge gadget for an edge $v_i v_j \in E(G)$ consists of an inter-edge $e_{i,j}$ to the left column attached to the backbone of T_i and of a star $S_{i,j}$ with n^3 leaves attached to the backbone of T_j. Both are contained in a horizontal strip of height three. So if $v_i v_j$ is, say, the t-th edge, then we use the heights between $b = a - 3t$ to $b - 2$. The root of $S_{i,j}$ is at height b and the leaves of $S_{i,j}$ are at height $b-2$. The source and the target of $e_{i,j}$ are at heights $b-1$ and $b-2$, respectively. Note that every T_i (on its own) can always be drawn planar.

Analysis of Crossings. Note that, for each $v_i v_j \in E(G)$, the inter-edge $e_{i,j}$ induces at most $n-1$ crossings with the backbones of $\{T_1, \ldots, T_n\}$ independent of the subtree arrangement. With $m \le n(n-1)/2$, there can thus be at most $n(n-1)/2 \cdot (n-1) = n^3/2 - n^2 + n/2$ crossings that do not involve a star subtree. Now, if T_i is to the left of T_j, then $e_{i,j}$ does not intersect $S_{i,j}$. On the other hand, if T_i is to the right of T_j, then $e_{i,j}$ intersects $S_{i,j}$, which causes n^3 crossings. So, for a subtree arrangement with s edge gadgets causing crossings, the embedding of T contains sn^3 plus at most rn^3 crossings, where $r = 1/2 - 1/n + 1/(2n^2) \le 1$.

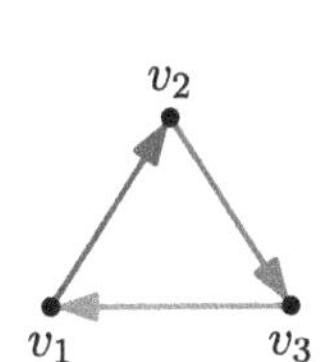

(a) A FAS instance on three edges.

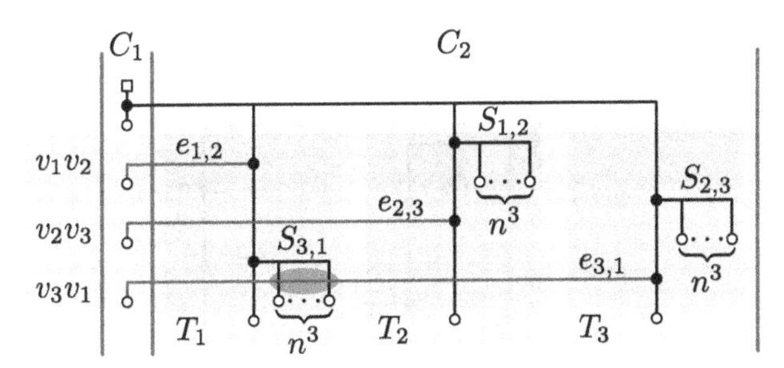

(b) In the corresponding column subtree for V1, each edge gadget occupies a horizontal strip and we need to find a subtree arrangement of the vertex gadgets (grey background).

Fig. 4. Reduction of a FAS instance to a TreeColumns instance.

Bijection of Vertex Orders and Subtree Arrangements. As each columns subtree T_i ($i \in \{1, \ldots, n\}$) corresponds to vertex v_i of G, each subtree arrangement of $\langle T, h, c\rangle$ implies exactly one vertex order of G and vice versa. A vertex order Π of G, in turn, implies the FAS solution set containing all edges whose target precedes its source in Π. In the other direction, for a FAS solution set S, we find a corresponding vertex order by computing a topological order of $G \setminus S$.

Consequently, a subtree arrangement of $\langle T, h, c\rangle$ whose number k of crossings lies in $[sn^3, (s+r)n^3]$ implies a FAS solution of size s and vice versa. In particular, a minimum-size FAS solution of size s^* corresponds to an optimal subtree arrangement with $k_{\min} = (s^* + r^*)n^3$ crossings, where $0 \le r^* \le r \le 1$.

Hardness of Approximation. If it is NP-hard to approximate FAS by a factor less than α, then it is NP-hard to approximate TreeColumns|V1 by a factor less than β due to the previous bijection, where β can be bounded as follows.

$$\beta \cdot k_{\min} = \beta \cdot (s^* + r^*)n^3 \ge \alpha s^* n^3 \quad \Leftrightarrow \quad \beta \ge \frac{\alpha s^*}{s^* + r^*} = \alpha - \frac{\alpha r^*}{s^* + r^*}$$

Since s^* is unbounded, $\alpha r^*/(s^* + r^*)$ can be arbitrarily close to zero. Hence, it is NP-hard to approximate TreeColumns|V1 with unbounded maximum degree within any factor $\alpha - \varepsilon$ for any $\varepsilon > 0$.

Other Variants. For a variable column order, note that, for our bijection, the vertex order is the same but mirrored if we swap the left and right column.

For a binary column tree under V2 and V3, we let the sources of the inter-edges in the root column subtree, which have as targets the roots of the T_i, form a path in the root column subtree; see Fig. 5. These edges can form at most n^2 pairwise crossings, which changes r and r^* only slightly and has thus no effect on our bijection. Similarly, the star subtrees can simply be substituted with binary subtrees where all inner vertices have a height between b and $b-1$. Also note that the T_i cannot interleave since they span from above the first edge gadgets all the way to the bottom. □

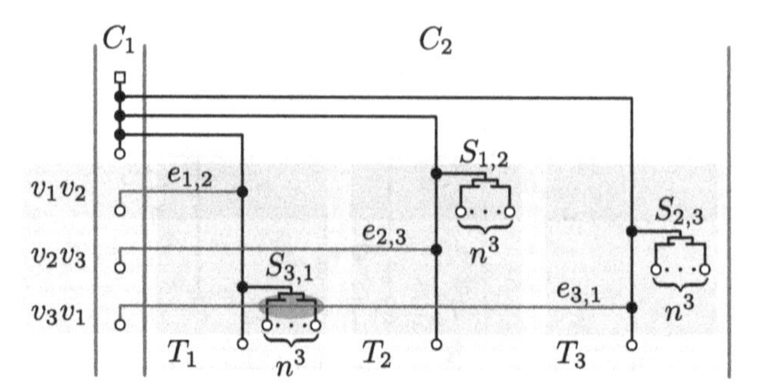

Fig. 5. Reduction for V2 & V3 with binary trees.

Note that in the proof of Theorem 3, the variant with binary columns trees does not work under V1 (see Fig. 5), since then it would no longer be possible to permute the T_i; there would only be one possible subtree arrangement.

4 Algorithms for Subtree Arrangement

In the previous section, we have seen that TreeColumns|V2 is NP-hard, even to approximate, by reduction from FAS. Next, we show that we can also go the other way around and reduce TreeColumns|V2 to FAS to obtain an FPT algorithm in the number of crossings and the maximum vertex degree Δ, and to obtain an FPT-approximation algorithm in Δ.

Integer-Weighted Feedback Arc Set. As an intermediate step in our reduction, we use the Integer-Weighted Feedback Arc Set (IFAS) problem which is defined for a digraph G as FAS but with the additional property that each edge e has a positive integer weight $w(e)$, $w\colon E(G) \to \mathbb{N}$, and the objective is to find a minimum-weight set of edges whose removal results in an acyclic graph. Clearly, IFAS is a generalization of FAS because FAS is IFAS with unit weights. However, an instance $\langle G, w\rangle$ of IFAS can also straightforwardly be expressed as an instance G' of FAS: Obtain G' from G by substituting each edge uv with $w(uv)$ length-two directed paths from u to v. For an illustration, see Fig. 6.

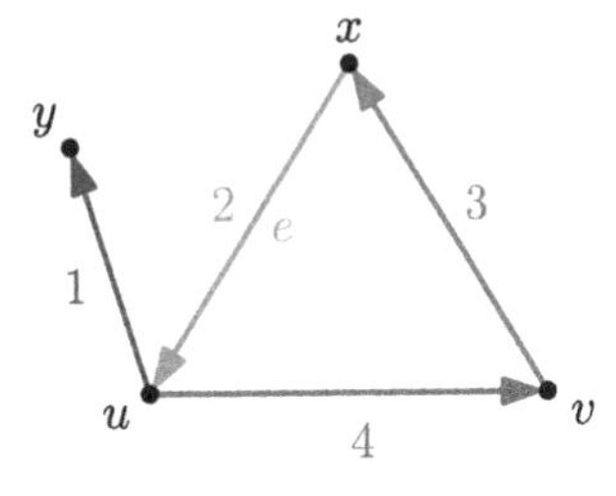

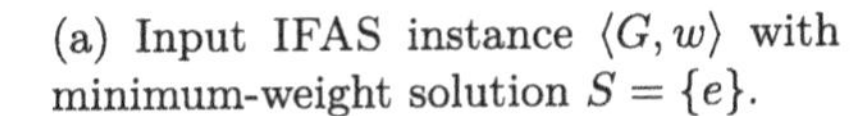
(a) Input IFAS instance $\langle G, w\rangle$ with minimum-weight solution $S = \{e\}$.

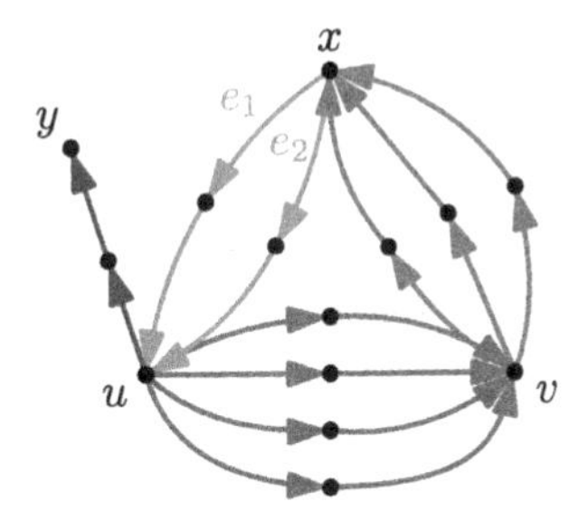

(b) Output FAS instance G' with minimum-size solution $S' = \{e_1, e_2\}$.

Fig. 6. Reduction of an IFAS instance to a FAS instance.

Lemma 4 ($\star$). *We can reduce an instance $\langle G, w\rangle$ of* IFAS *with m edges and maximum weight $w_{\max}$ to an instance G' of* FAS *in time $\mathcal{O}(mw_{\max})$ such that*

- *G' has size in $\mathcal{O}(mw_{\max})$,*
- *the size of a minimum-size solution of G' equals the weight of a minimum-weight solution of $\langle G, w\rangle$, and*
- *we can transform a solution of G' with size s in $\mathcal{O}(s)$ time to a solution of G with weight at most s.*

Reduction to Feedback Arc Set. Next, we show that we can express every instance of TreeColumns|V2 as an instance of IFAS. Note, however, that we have split the problem into the tasks of embedding the subtrees and finding a subtree arrangement. Here, we are only concerned with finding a subtree arrangement for every column, and we assume that we separately solve the problem of embedding every column subtree, e.g., by employing the algorithm from Lemma 1.

Lemma 5. *We can reduce an instance $\langle T, h, c\rangle$ of* TreeColumns|V2 *on n vertices to an instance $\langle G, w\rangle$ of* IFAS *in time $\mathcal{O}(n^2)$ such that*

- *G has size in $\mathcal{O}(n^2)$ and maximum weight $w_{\max} \in \mathcal{O}(n^2)$,*
- *the number of intra-column crossings in a minimum-crossing solution of $\langle T, h, c\rangle$ equals the weight of a minimum-weight solution of $\langle G, w\rangle$ plus t where t is some integer in $\mathcal{O}(n^2)$ depending only on $\langle T, h, c\rangle$, and*
- *we can transform a solution of $\langle G, w\rangle$ with weight s in $\mathcal{O}(n^2)$ time to a solution of $\langle T, h, c\rangle$ with at most $s + t$ intra-column crossings.*

Proof. To each column in c, we apply the following reduction; see Fig. 7. Let $T_1, \ldots, T_r$ be the set of column subtrees of c. For each pair T_i, T_j of subtrees ($i, j \in \{1, \ldots, r\}, i \neq j$), we consider the number of intra-column crossings where only edges with an endpoint in T_i and T_j are involved – once if T_i is placed to the left of T_j and once if T_j is placed to the left of T_i; see Figs. 7a and 7b. We let these numbers be k_{ij} and k_{ji}, respectively; see Fig. 7c. We then construct

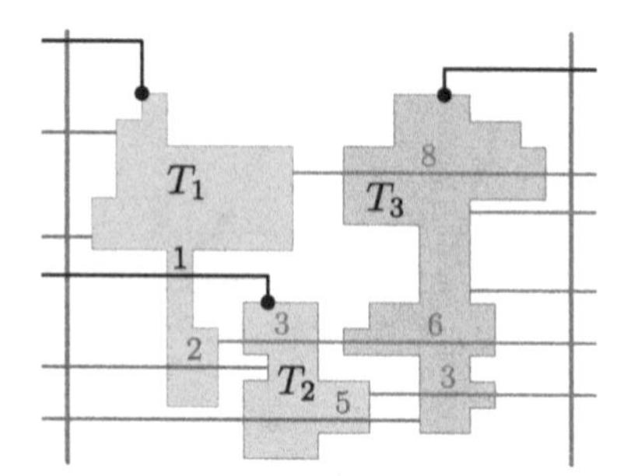

(a) Possible permutation Π of the subtrees.

(b) Reverse permutation of Π.

k_{ij}	1	2	3
1	–	6	14
2	2	–	8
3	17	5	–

(c) Table of pairwise intra-column crossings k_{ij} occurring if subtree T_i is placed to the left of subtree T_j.

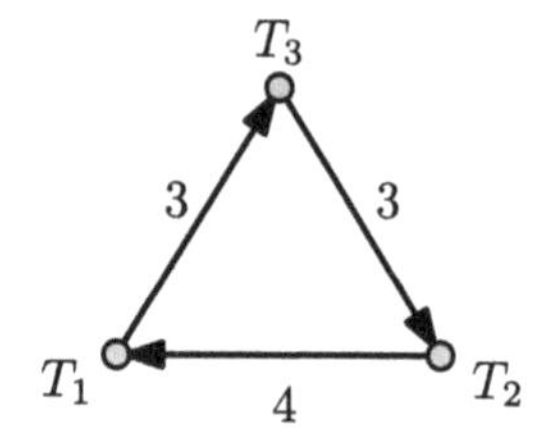

(d) Resulting IFAS instance where the difference $k_{ij} - k_{ji}$ determines the direction and the weight of an edge.

Fig. 7. Reduction of an TreeColumns|V2 instance to an IFAS instance.

an IFAS instance $\langle G, w \rangle$ where G has a vertex for every column subtree. For each pair T_i, T_j, G has the edge T_iT_j if $k_{ij} - k_{ji} < 0$, or G has the edge T_jT_i if $k_{ij} - k_{ji} > 0$, or there is no edge between T_i and T_j otherwise. The weight of each edge is $|k_{ij} - k_{ji}|$. So we have an edge in the direction where the left-to-right order yields fewer intra-column crossings, and no edge, if the number of intra-column crossings is the same, no matter their order.

Note that instead of considering each column separately, we can also think of G as a single graph having at least one connected component for each column.

Implementation and Running Time. For counting the number of crossings between pairs of subtrees, we initialize $\mathcal{O}(r^2)$ variables with zero. Then, we use a horizontal sweep line traversing all trees in parallel top-down while maintaining the current widths of $T_1, \dots, T_r$ in variables $b_1, \dots b_r$. Our event points are the heights of the vertices (given by h) and the heights of the parents of the subtree roots (where we have the horizontal segment of an edge entering the column).

Whenever we encounter an inter-edge uv, we update the crossing variables. More precisely, if $u \in T_i$ for some $i \in \{1, \dots, r\}$, we consider each $j \in \{1, \dots, r\} \setminus \{i\}$ and we set k_{ji}+= b_j if v is in a column on the left, and, symmetrically, we set k_{ij}+= b_j if v is in a column on the right.

The running time of this approach is $\mathcal{O}(r(n_{\text{column}} + m_{\text{column}}))$ per column where n_{column} and m_{column} are the numbers of vertices and edges in the column, respectively. Over all columns, this can be accomplished in $\mathcal{O}(n^2)$ time.

Size of the Instance. The resulting graph G has $\mathcal{O}(n)$ vertices and $\mathcal{O}(n^2)$ edges as we may have an edge for every pair of subtrees in a column and we may have a linear number of column subtrees. The maximum weight $w_{\max}$ in w is in $\mathcal{O}(n^2)$ since we have $\mathcal{O}(n)$ inter-edges and the maximum width of a tree is in $\mathcal{O}(n)$.

Comparison of Optima. For each column $\gamma \in \{1, \dots, \ell\}$, we have the following lower bound L_γ on the number k^γ_{column} of intra-column crossings:

$$L_\gamma = \sum_{i=1}^{r-1} \sum_{j=i+1}^{r} \min\{k_{ij}, k_{ji}\} \leq k^\gamma_{\text{column}}.$$

Now consider a minimum-crossing solution of $\langle T, h, c\rangle$. In column γ, the column subtrees have a specific order, which we can associate with a permutation Π^* of the column subtrees. For simplicity, we rename the column subtrees as $T_1, \dots, T_r$ according to Π^*. Then, the number of intra-column crossings is

$$k^\gamma_{\text{column}} = \sum_{i=1}^{r-1} \sum_{j=i+1}^{r} k_{ij}.$$

Because it is a minimum-crossing solution, the number δ_γ of additional crossings (i.e., the deviation of k^γ_{column} from L_γ) due to "unfavorably" ordered pairs of subtrees is minimized. We can express δ_γ in terms of all k_{ij} by

$$\delta_\gamma = k^\gamma_{\text{column}} - L_\gamma = \sum_{i \neq j, k_{ij} > k_{ji}} k_{ij} - k_{ji}.$$

Now consider a minimum-weight solution S of $\langle G, w\rangle$ with weight s. After removing S from G, we have an acyclic graph for which we can find a topological order Π of its vertices, which in turn correspond to the column subtrees in $\langle T, h, c\rangle$. Again, we rename these subtrees as $T_1, \dots, T_r$ according to Π. If we arrange G linearly according to Π, only the edges of S point backward and, by definition of the edge directions, for exactly these edges $k_{ij} > k_{ji}$ holds. The sum

$$s_\gamma = \sum_{i \neq j, k_{ij} > k_{ji}} k_{ij} - k_{ji}$$

of weights in S whose vertices represent pairs of column subtrees in γ is minimized because G has independent components for all columns in c. Therefore, $s_\gamma = \delta_\gamma$. Over all columns $\{1, \dots, \ell\}$, we set $t = \sum_{i=1}^{\ell} L_i$, we have $s = \sum_{i=1}^{\ell} s_i$, and we conclude

$$k_{\text{column}} = \sum_{i=1}^{\ell} k^i_{\text{column}} = \sum_{i=1}^{\ell} (L_i + \delta_i) = \sum_{i=1}^{\ell} (L_i + s_i) = s + t.$$

Note that $t \in \mathcal{O}(n^2)$ as we have $\mathcal{O}(n^2)$ pairs of edges, which cross at most once.

Transforming a Solution Back. Similarly, we can find for any solution S of $\langle G, w\rangle$ with size s a topological order, which corresponds to a subtree arrangement in $\langle T, h, c\rangle$. There, together with the t unavoidable crossings, we have at most s additional crossings due to "unfavorably" ordered pairs. We can find the topological order in linear time in the size of G, which is in $\mathcal{O}(n^2)$. □

When we combine Lemmas 4 and 5, we obtain Corollary 6.

Corollary 6. *We can reduce an instance $\langle T, h, c\rangle$ of* TREECOLUMNS|V2 *on n vertices to an instance G of* FAS *in time $\mathcal{O}(n^4)$ such that*

- *G has size in $\mathcal{O}(n^4)$,*
- *the number of intra-column crossings in a minimum-crossing solution of $\langle T, h, c\rangle$ equals the size of a minimum-size solution of G plus t where t is some integer in $\mathcal{O}(n^2)$ depending only on $\langle T, h, c\rangle$, and*
- *we can transform a solution of G with size s in $\mathcal{O}(n^2)$ time to a solution of $\langle T, h, c\rangle$ with $s + t$ crossings.*

Fixed-Parameter Tractable Algorithm. For TREECOLUMNS|V2, one of the most natural parameters is the number of crossings, which is also the objective value. With our reduction to FAS at hand, it is easy to show that TREECOLUMNS|V2 is fixed-parameter tractable (FPT) in this parameter for bounded-degree column trees. This follows from the fact that FAS is FPT in its natural parameter (the solution size) as first shown by Chen, Liu, Lu, O'Sullivan, and Razgon [3].

Theorem 7 ($\star$). TREECOLUMNS|V2 *is fixed-parameter tractable in the number k of crossings plus Δ. More precisely, given an instance $\langle T, h, c\rangle$ with n vertices and maximum vertex degree Δ, there is an algorithm computing an embedding with the minimum number k of crossings in $\mathcal{O}(\Delta!\Delta n^2 + n^{16}4^k k^3 k!)$ time.*

Approximation Algorithm. Similar to our FPT result, we can use any approximation algorithm for FAS to approximate TREECOLUMNS|V2. It is an unresolved problem whether FAS admits a constant-factor approximation. In case such an approximation is found, this immediately propagates to TREECOLUMNS|V2. Currently, we can employ the best known approximation algorithm of FAS due to Even, Naor, Schieber, and Sudan [7], which has an approximation factor of $\mathcal{O}(\log n' \log\log n')$, where n' is the number of vertices in the FAS instance. We remark that this algorithm involves solving a linear program and the authors do not write much about precise running time bounds, which we also avoid here.

Theorem 8 ($\star$). *There is an approximation algorithm that, for a given instance $\langle T, h, c\rangle$ of* TREECOLUMNS|V2 *with n vertices and maximum degree Δ, computes in $\mathcal{O}(\mathsf{poly}(n) \cdot \Delta!\Delta)$ time an embedding where the number of crossings is at most $\mathcal{O}(\log n \log\log n)$ times the minimum number of crossings.*

References

1. Betz, G., Gemsa, A., Mathies, C., Rutter, I., Wagner, D.: Column-based graph layouts. J. Graph Algorithms Appl. **18**(5), 677–708 (2014). https://doi.org/10.7155/jgaa.00341
2. Calamoneri, T., Di Donato, V., Mariottini, D., Patrignani, M.: Visualizing co-phylogenetic reconciliations. Theoret. Comput. Sci. **815**, 228–245 (2020). https://doi.org/10.1016/j.tcs.2019.12.024
3. Chen, J., Liu, Y., Lu, S., O'Sullivan, B., Razgon, I.: A fixed-parameter algorithm for the directed feedback vertex set problem. J. ACM **55**(5), 21:1–21:19 (2008). https://doi.org/10.1145/1374376.1374404
4. Chevenet, F., Doyon, J., Scornavacca, C., Jacox, E., Jousselin, E., Berry, V.: SylvX: a viewer for phylogenetic tree reconciliations. Bioinformatics **32**(4), 608–610 (2016). https://doi.org/10.1093/bioinformatics/btv625
5. Dinur, I., Safra, S.: On the hardness of approximating vertex cover. Ann. Math. **162**(1), 439–485 (2005). https://doi.org/10.4007/annals.2005.162.439
6. Douglas, J.: UglyTrees: a browser-based multispecies coalescent tree visualizer. Bioinformatics **37**, 268–269 (2020). https://doi.org/10.1093/bioinformatics/btaa679
7. Even, G., Naor, J., Schieber, B., Sudan, M.: Approximating minimum feedback sets and multicuts in directed graphs. Algorithmica **20**(2), 151–174 (1998). https://doi.org/10.1007/PL00009191
8. Guruswami, V., Håstad, J., Manokaran, R., Raghavendra, P., Charikar, M.: Beating the random ordering is hard: every ordering CSP is approximation resistant. SIAM J. Comput. **40**(3), 878–914 (2011). https://doi.org/10.1137/090756144
9. Hadfield, J., et al.: Nextstrain: real-time tracking of pathogen evolution. Bioinformatics **34**(23), 4121–4123 (2018). https://doi.org/10.1093/bioinformatics/bty407
10. Hadlak, S., Schumann, H., Schulz, H.: A survey of multi-faceted graph visualization. In: Borgo, R., Ganovelli, F., Viola, I. (eds.) Eurographics Conference on Visualization(EuroVis 2015), pp. 1–20. Eurographics Association (2015). https://doi.org/10.2312/eurovisstar.20151109
11. Karp, R.M.: Reducibility among combinatorial problems. In: Miller, R.E., Thatcher, J.W. (eds.) Symposium on the Complexity of Computer Computations. The IBM Research Symposia Series, pp. 85–103. Plenum Press, New York (1972). https://doi.org/10.1007/978-1-4684-2001-2_9
12. Klawitter, J., Klesen, F., Niederer, M., Wolff, A.: Visualizing multispecies coalescent trees: drawing gene trees inside species trees. In: Gasieniec, L. (ed.) SOFSEM 2023. LNCS, vol. 13878, pp. 96–110. Springer, Cham (2023). https://doi.org/10.1007/978-3-031-23101-8_7
13. Klawitter, J., Klesen, F., Scholl, J.Y., van Dijk, T.C., Zaft, A.: Visualizing geophylogenies - internal and external labeling with phylogenetic tree constraints. In: International Conference Geographic Information Science (GIScience). LIPIcs, Schloss Dagstuhl - LZI (2023, to appear)
14. Klawitter, J., Zink, J.: Tree drawings with columns. Arxiv report (2023)
15. Lima, M.: The Book of Trees: Visualizing Branches of Knowledge. Princeton Architectural Press (2014)
16. Parks, D.H., et al.: GenGIS 2: geospatial analysis of traditional and genetic biodiversity, with new gradient algorithms and an extensible plugin framework. PLoS ONE **8**(7), 1–10 (2013). https://doi.org/10.1371/journal.pone.0069885

17. Rusu, A.: Tree drawing algorithms. In: Tamassia, R. (ed.) Handbook on Graph Drawing and Visualization, chap. 3, pp. 155–192. Chapman and Hall/CRC (2013)
18. Schulz, H.: Treevis.net: a tree visualization reference. IEEE Comput. Graphics Appl. **31**(6), 11–15 (2011). https://doi.org/10.1109/MCG.2011.103
19. Besa, J.J., Goodrich, M.T., Johnson, T., Osegueda, M.C.: Minimum-width drawings of phylogenetic trees. In: Li, Y., Cardei, M., Huang, Y. (eds.) COCOA 2019. LNCS, vol. 11949, pp. 39–55. Springer, Cham (2019). https://doi.org/10.1007/978-3-030-36412-0_4

Visualization Challenges

On the Perception of Small Sub-graphs

Jacob Miller[1,2,3,4(✉)], Mohammad Ghoniem[1,2,3,4], Hsiang-Yun Wu[1,2,3,4], and Helen C. Purchase[1,2,3,4]

[1] Department of Computer Science, University of Arizona, Tucson, USA
jacobmiller1@arizona.edu
[2] Luxembourg Institute of Science and Technology, Esch-sur-Alzette, Luxembourg
[3] St. Pölten University of Applied Sciences, Sankt Pölten, Austria
[4] Department of Human-Centred Computing, University of Monash, Melbourne, Australia
helen.purchase@monash.edu

Abstract. Interpreting a node-link graph is enhanced if similar sub-graphs (or 'motifs') are depicted in a similar manner – that is, they have the same visual form. Small motifs within graphs may be perceived to be identical when they are structurally dissimilar, or may be perceived to be dissimilar when they are identical. This issue primarily relates to the Gestalt principle of similarity, but may also include an element of quick, low-level pattern-matching. We believe that if motifs are identical, they should be depicted identically; if they are nearly-identical, they should be depicted nearly-identically. This principle is particularly important in domains where motifs hold meaning and where their identification is important. We identified five small motifs: bi-cliques, cliques, cycles, double-cycles, and stars. For each, we defined visual variations on two dimensions – same or different structure, same or different shape. We conducted a crowd-sourced empirical study to test the perception of similarity of these varied motifs, and found that determining whether motifs are identical or similar is affected by both shape and structure.

Keywords: Perception · Graph Motifs · Gestalt Principles · User Study

1 Introduction

As they attempt to understand complex natural, technical or social phenomena, application domain experts use graph models to analyze intricate relationships. They often resort to graph visualization tools to navigate in and make sense of such data [22]. Much work has also looked at the automatic extraction of graph motifs [29], and at counting them [18], to characterize important graph-level properties. Graph motifs are simply local connectivity patterns, or small sub-graphs, lying within graphs; see Fig. 1(a). They can be elementary motifs, like triangles, or assembled into higher-order motifs, like chains and cycles [16]. Given their role in anticipating the behavior of the graph at hand, there is so

M. A. Bekos and M. Chimani (Eds.): GD 2023, LNCS 14465, pp. 213–230, 2023.
https://doi.org/10.1007/978-3-031-49272-3_15

far little work dedicated to the visual perception of graph motifs. In particular, existing graph visualization tools are usually not designed to ensure that similar graph motifs take a similar graphical form.

Figure 1 uses a five-cycle motif to exemplify the problem of visual matching of graph motifs. Motif nodes are highlighted in black, while other nodes are grey. In Fig. 1(a), the base shape places the motif nodes on the vertices of a regular pentagon. In Fig. 1(b), the same structure is also drawn as a regular pentagon, subject to a rotation. Compared to the drawing of the base shape, one might hypothesize that users will recognize that the two motifs are identical. In Fig. 1(c), the motif is drawn as a quadrilateral, with the fifth node lying within, making it harder to match it to the base shape. In Fig. 1(d), the pentagon shape is preserved despite the addition of an edge to the motif. The rationale is that a small structural change should be commensurate to the induced change of shape. In Fig. 1(e), both the structure and the shape are different to the base motif and its shape, which users should see easily. Section 3.1 discusses the full list of motifs considered in this study (see Figs. 3, 4, 5, 6 and 7).

In this paper, we explore the perception of (identical or similar) sub-graphs by conducting an empirical study which asks participants to compare (identical or similar) depictions of sub-graphs. Our results include evidence that depicting identical sub-graphs differently makes them harder to recognise as the same, and depicting different sub-graphs similarly results in false identically judgements.

2 Background and Related Work

Similarity is essential in knowledge development as it allows us to organize principles to classify, form, and generalize concepts [40]. It also serves as a common measure for comparison purposes or tasks in data visualization [23]. Specific shapes in visualization, such as *Star Glyphs* [12], *Scatterplots* [30], and *Directed Acyclic Graphs (DAGs)* [2], have been considered as factors that influence human similarity perception. One particular usage of similarity in graph analytics is *motif analysis*, because structural motifs in graphs often act as influential building blocks in many domains [37], such as biology [33], social science [45], internet communication, and others. Therefore, visually identifying these sub-graph structures (motifs) facilitates effective comparisons among various data.

2.1 Perception of Similarity and Shape

Our research questions draw on well-established research on human perception. Ware's three levels of perceptual processing [44] start with a 'bottom-up' stage that processes low-level visual properties in parallel, identifying, e.g., colour, texture, movement etc.; it is quick, automatic and data-driven. The 'pattern recognition' second level comprises sequential processing of the scene, recognising patterns, contours, and regions. The slower third level 'top-down' phase sequentially explores the scene to identify objects, typically engaging cognition

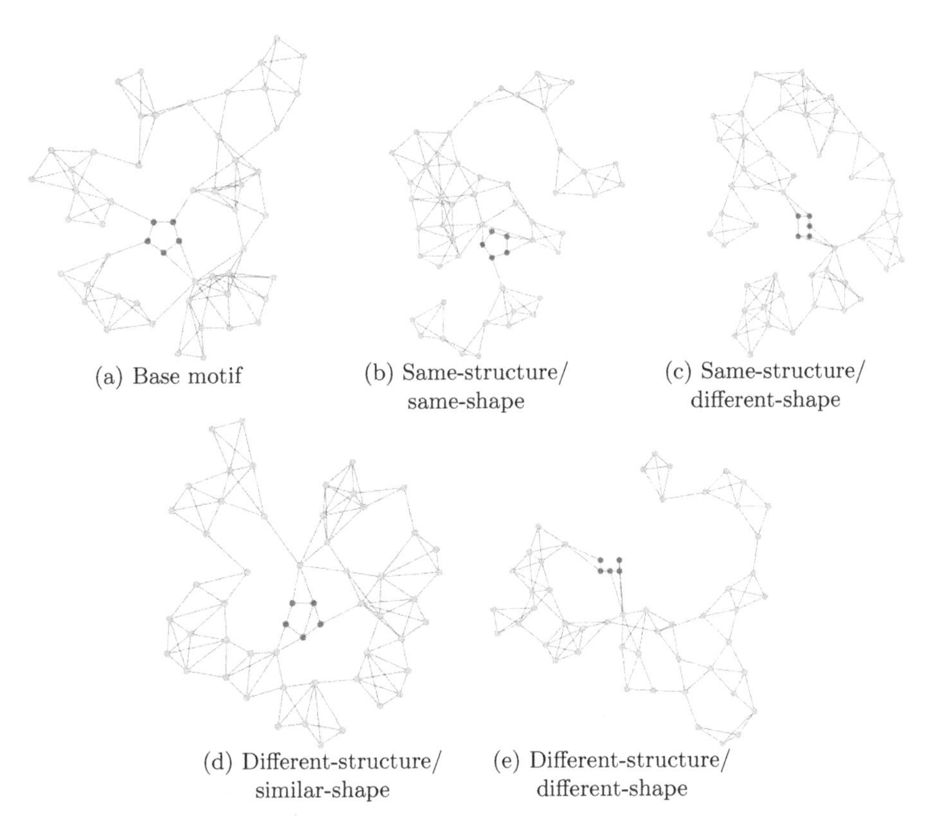

Fig. 1. Example 'cycle' motif drawings. (a) shows the base shape and structure are regular and well-formed; nodes are placed on a pentagon. (b)–(e) show variations of shape (visual form) and/or structure (connectivity).

for the completion of a specific task. The first level includes the immediate identification of prominent objects which 'pop-out', being of obvious different visual form to those surrounding them. The visual features that result in pop-out are of varying effectiveness: colour is the most obvious one; others include texture, orientation, size, shape, curvature [39]. The Gestalt laws [8,20] describe how we see patterns and groups, and relate to the second of Ware's levels. Distinct objects may be seen to form a group by being close together (proximity), looking similar (similarity), or moving together (common fate), etc. This paper considers the quick recognition of shapes and the Gestalt law of similarity in graph drawing: if two sub-graphs are of the same structure, then depicting them in the same visual form will ensure that they can be quickly recognised as the same; if they are of similar structure, then depicting them in the same (or similar) visual form will highlight their similarity. We investigate the immediate recognition of same (or similar) sub-graphs rather than serial processing requiring the use of cognition because making similarities immediately prominent can help in gaining a better (and quicker) overall understanding of the structure of a graph at a glance. Comparing sub-graphs with enough time to engage cognitive processes is simple; making such comparisons quickly may not be. Our research questions focus on determining whether depicting identical sub-graphs using the same (or similar) shape facilitates the recognition of their identicality. We also explore what happens when the sub-graph structure is slightly different, and when its presentation is distorted by additional forces within a force-directed algorithm.

2.2 Related Work

Prior work has investigated the perception of graph properties, such as graph density, clustering coefficient [36] and layout types [21]. Although the perception of shapes in sub-graphs *per se* has not yet been fully investigated with respect to sub-graph structure, some studies of shapes in visualization have been conducted. Gogolou et al. investigated if time series visualizations generated from automatic similarity measures are aligned with readers' similarity constraints [14]. They concluded that the selection of visualizations influences the patterns that readers consider as similar. Ballweg et al. researched the influencing factors of directed acyclic graphs (DAGs) when they are drawn as layered drawings [2]. Their study suggests that the similarity perception of DAGs is mainly affected by the number of levels in the drawings, the number of nodes on a level, and the corresponding overall shape (i.e., convex hull of the DAG), while Wallner et al. [41,42] did not find significance in the perception of overall shapes. Tarr and Pinker studied the effect of mental rotation in shape recognition, focusing on letter-like asymmetrical characters [38]. They found that trained readers can recognize the characters almost as fast at all familiar angles, while the performance varies when the character appears at novel angles.

Besides perceptual studies, several graph drawing algorithms incorporate shapes (i.e., motifs from different graph classes [1]) to improve layout readability. Yuan et al. developed an algorithm based on Laplacian constrained distance embedding to control the shapes of sub-graphs input by the users [48]. Wang et al. generalized the classical stress majorization approach [43], to ease sub-graph shape manipulation. Meidiana et al. extended the classical force-directed and stress minimization algorithms to optimize shape-based metrics measure [11] to achieve faithful drawings [28]. Interactions through aggregation [24] and simplification [10] also consider motif properties to reduce layout visual complexity.

In practice, many applications consider the visual form of motifs, including transcriptional regulation networks analysis [17], phylogenetic trees comparison [6], biological pathways diagrams [48], metro maps creation [3,46], dynamic graph analysis [7], and visual graph matching [15], for example, across distinct layers of a multilayer graph [27]. In this paper, we demonstrate the benefit of depicting motifs within graphs in a consistent, well-formed manner.

3 Methodology

3.1 Stimuli

Structural motifs often bind with semantics in applications for analysis purposes (Sect. 2.2) [16]. For example, a clique (or a bi-clique) in social networks represents strong mutual connections [45], and a cycle in biological networks indicates a circular biochemical reaction [26,34]. We hence identified five 'motifs': small sub-graphs with well-defined graph structure, distinct visual form, easy description and clear definition. We denote these motifs (the graph structure along with node positions) as our 'base' motifs; see Fig. 2. For each base motif, we created 12 variations according to two dimensions: change of structure (Yes/No) and change of shape (Yes/No), where shape is determined by the convex hull of nodes (smallest convex set containing all nodes). There are hence four quadrants.

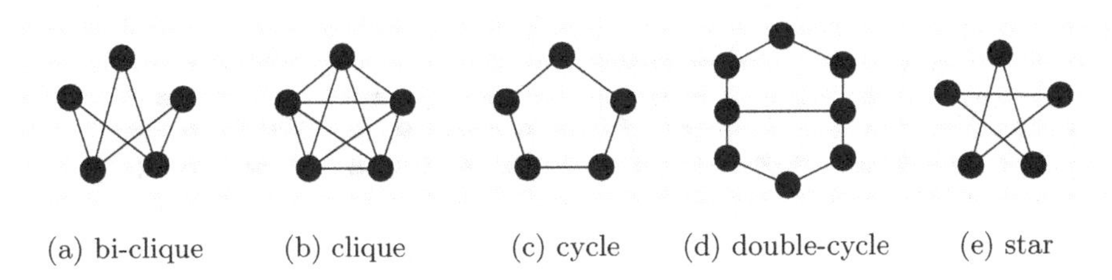

(a) bi-clique (b) clique (c) cycle (d) double-cycle (e) star

Fig. 2. The five base motifs.

The **same-structure/same-shape (SS)** variations depict the same graph structure and same visual form as the base, rotated by 90, 180, 270° (45, 90 and 135° for double-cycles which have two axes of symmetry (Fig. 5 (top left)).

The **same-structure/different-shape (SD)** variations use the same graph structure as the base motif but a different shape; we generated three different drawings by varying the relative positions of nodes (e.g. Fig. 3 (bottom left)).

The **different-structure/similar-shape (DS)** variants adopt a different graph structure to the motif; we change the number of edges (deleting one, deleting two, or adding one) while keeping node positions as in the base – giving them a similar (but not identical) shape (e.g. Fig. 3 (top right)). As edges can not be added to the clique motif, we remove three edges (Fig. 6 (top right)).

The final **different-structure/different-shape (DD)** collection adapts the original motif sub-graph as for the DS variants, but depicts them using different node positions from the base motif. The three DD variants of the base motif differed both by structure and visual form (e.g. Fig. 3 (bottom right)).

The motif variations are: *bi-clique* (Fig. 2(a)), *clique* (Fig. 2(b)), *cycle* (Fig. 2(c)), *double-cycle* (Fig. 2(d)), and *star* (Fig. 2(e)).

Each motif set therefore comprises 13 different visual motif drawings which were integrated into graphs, a total of 65 graph drawings. The graphs and graph drawings were subject to both structural and layout constraints (Sect. 3.4). Motif nodes were coloured black, all others grey.

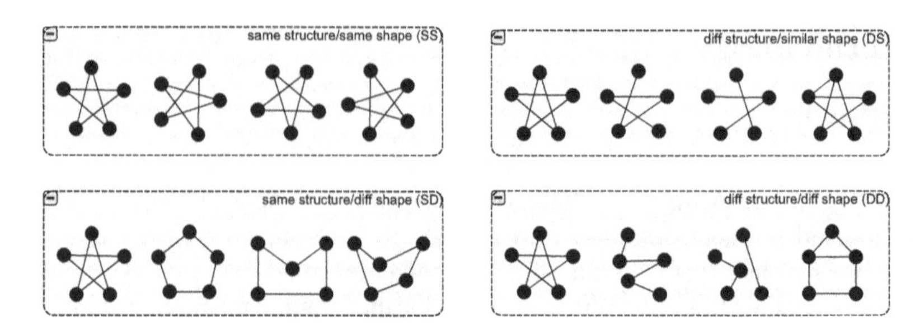

Fig. 3. 12 variations in shape and/or structure of the *star* motif.

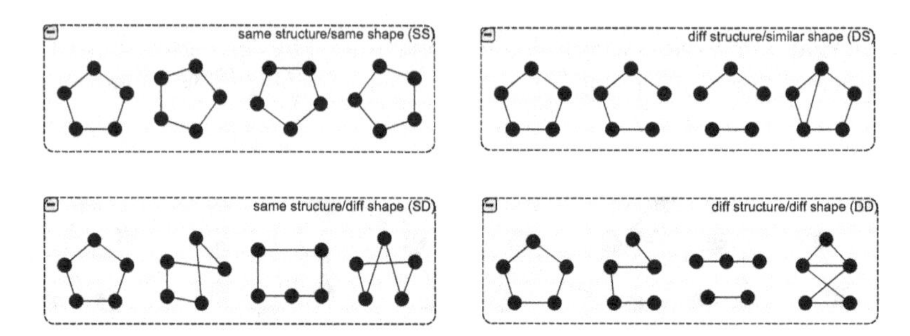

Fig. 4. 12 variations in shape and/or structure of the *cycle* motif.

3.2 Experimental Design

For each motif, we explored participants' ability to assess the identicality or similarity of motifs in two graph drawings, with each participant working with only one motif. In Experiment 1, we explored the effect of motif shape; in Experiment 2, we explored the effect of motif shape distortion. Although the 48 trials for Experiment 1 were interspersed with the 40 trials for Experiment 2, since their aims and stimuli were different, this first section focuses on Experiment 1.

Each trial consisted of a pair of graph drawings displayed side-by-side, with each drawing containing a highlighted motif. Participants were asked to indicate the degree of similarity between the structure of the two motifs. For each motif, the base motif graph drawing (or one of its rotations) was paired with all other 12 motif variant drawings, giving a total of 48 trials. The position of the two drawings in each trial (left or right) was randomly determined. Figure 8 shows the case of a rotated star motif on the right, with the base star motif on the left.

Since we are interested in bottom-up immediate processing, and how known shapes can be identified quickly, we limited the length of time allocated for each trial to 4 s – a duration determined by pilot testing of the complete experiment with 15 participants which considered 3, 4, and 5 s as possibilities: 4 s produced sufficient variability in accuracy to avoid floor (too hard) and ceiling (too easy) effects. Participants judged similarity on a scale by choosing among: '*identical*', '*similar*', '*not so similar*' and '*completely different*'.

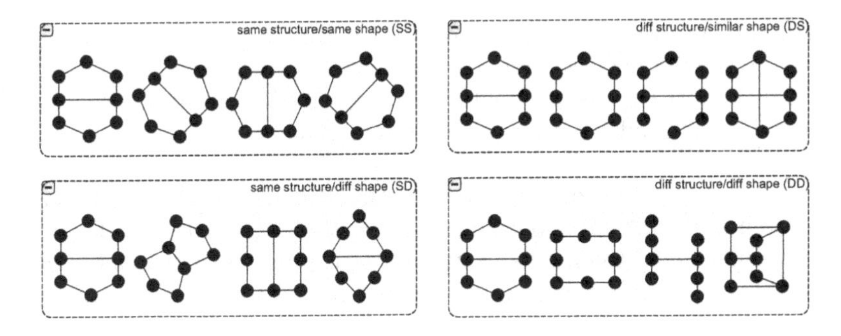

Fig. 5. 12 variations in shape and/or structure of the *double cycle* motif.

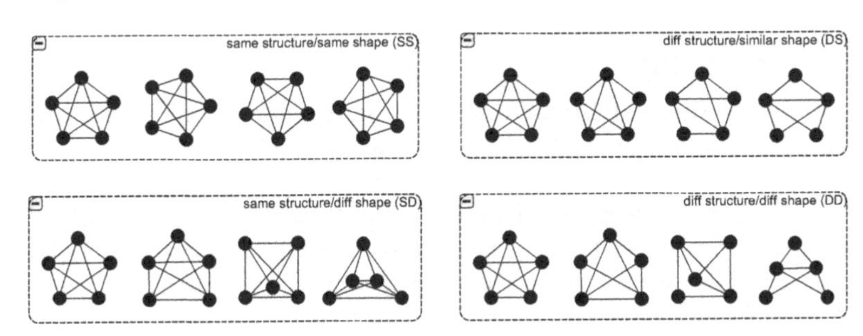

Fig. 6. 12 variations in shape and/or structure of the *clique* motif.

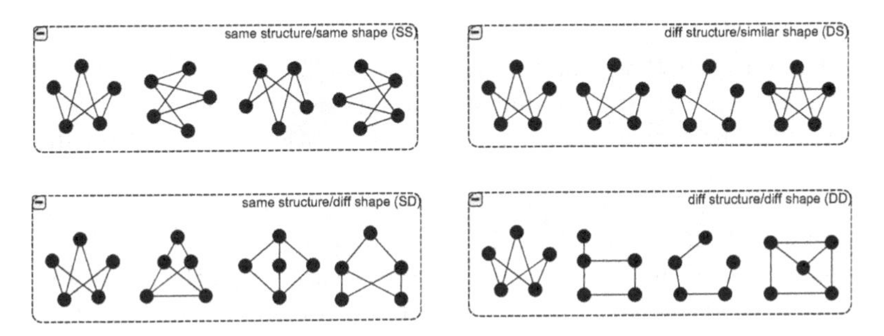

Fig. 7. 12 variations in shape and/or structure of the *bi-clique* motif.

3.3 Experimental Procedure

We used Prolific [31] to recruit participants for our study and Qualtrics [32] to implement the survey. The simplicity of our experiment makes it suited for crowd-sourcing [5], enabling collection from a large number of participants per motif. Participants were paid £9 per hour, with a median completion time of 10.4 min. They were recruited from the UK and required to use a desktop machine (no mobile devices). We ran five within-subject studies, one for each motif, and 30 participants per motif. 46% identified as women, 52% as men, 1% as non-binary/gender diverse, 1% declining to answer. 38% were 18–35, 28% were 36–45, 15% were 46–55, and 19% over the age of 55.

Participants were shown six 'practice' stimuli pairs for which data was not collected. These helped mitigate against the learning effect that can affect within-subject studies, as did the randomization of trials for each participant. The data from participants with response rate less than 85% were discarded, and missing responses from the remaining participants were not included in the calculation of mean accuracy data. Each participant saw each trial exactly once.

3.4 Implementation

Graph Generation and Motif Integration. The motifs were integrated into larger 50-node graphs, generated using a k-nearest neighbor model with k = 3 [9,25]. Each graph motif was 'stitched' into the larger graph by connecting each motif node to up to three non-motif nodes, chosen uniformly at random (though we modify the graph post-layout). Since the drawings focus on the highlighted motifs, the specific properties of the overall graphs (beyond their size) is not relevant. A different graph was created for each of the 13 stimuli, for each motif. The constraints on the graph generation were: **(GC1)** All motif nodes must have at least one edge to a non-motif node. **(GC2)** The graph must be connected.

Graph Drawing. We use multidimensional scaling (MDS) for graph layout [13,49] for its ease of use in adding constraints and reliability in representing geometric structures. The constraints on the creation of the graph drawings were: **(GD1)** Motif nodes are fixed on the plane, so they keep the desired visual form, with the rest of the graph laid out around them. This is the basis of our research. **(GD2)** No motif node may be on the periphery. **(GD3)** Edges connecting non-motif nodes must not intersect edges within the motif, since this would change the visual form of the motif and will hamper its recognition in an uncontrolled manner.

We modify the MDS algorithm with an extra parameter F, standing for the set of motif nodes whose position is fixed: $F = \{v | v \text{ is a node in a motif}\}$. Each member of F is 'pinned' in the plane so that their positions will not be updated, but the surrounding graph drawing will still be based on graph-theoretic distances. More formally, given a graph-theoretic distance matrix d, and a set of nodes with fixed position F, we want to find

$$\min_{X_1,\ldots X_n} \sum_{i \notin F} \sum_{j \notin F} (||X_i - X_j|| - d_{i,j})^2 \quad (1)$$

If a drawing produced by Eq. 1 does not meet these requirements, we attempt to modify the drawing to fit. Since the constraints are not always simultaneously

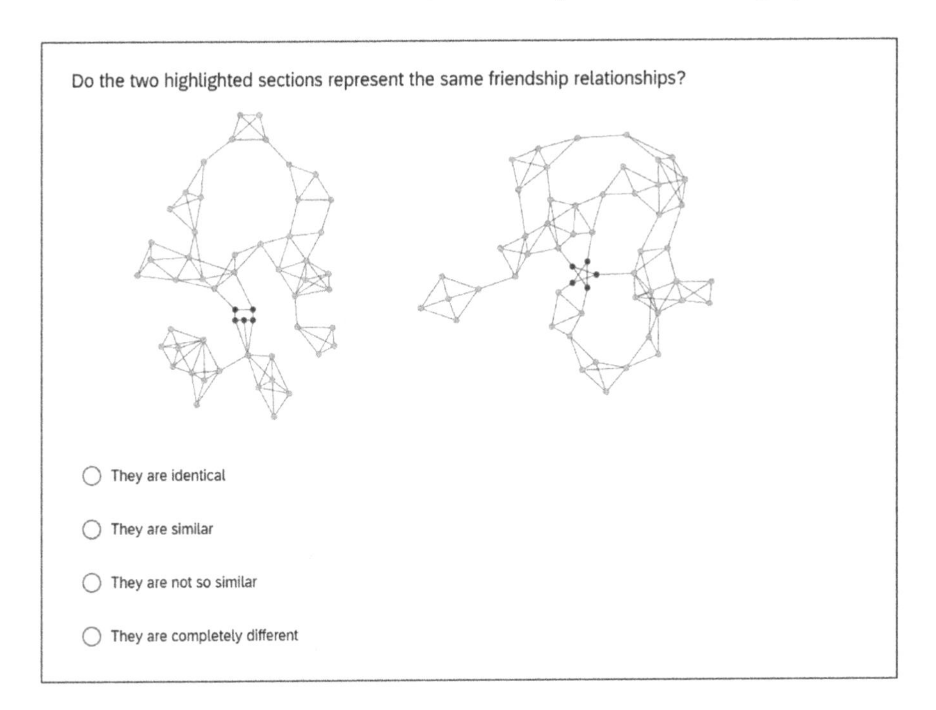

Fig. 8. Example trial – star: same-structure/different-shape (left); base (right).

satisfiable, when they conflict we discard that drawing and try again with a new random graph. To satisfy **GD2**, we compute the convex hull of the layout, and if any motif node lies on the hull, we try again.

GC1 and **GD3** often conflict. To resolve, we first look at all possible $O(|V|^2)$ line segments and categorize them as ***invalid*** if they violate **GD3**, i.e., if they intersect edges within the motif, or as ***valid*** otherwise. Then, we count and remove any ***invalid*** non-motif edge (an edge with an endpoint not in a motif). If some motif nodes are left without non-motif edges, we satisfy **GC1** by adding to each such motif node the edge that is ***valid*** and whose endpoint is nearest. Clearly, **GC1** is satisfied as we have added edges to each node which did not satisfy it, and we can verify that we did not violate **GD3** in the process.

The removal of ***invalid*** edges might yield a graph that is too sparse to be realistic. We add as many ***valid*** edges as the ***invalid*** edges that were removed, shortest length first. Finally, we check if **GC2** is met; if not, we iterate until all constraints are met. All code is available at https://github.com/Mickey253/graph-rep-sym.

Table 1. Participant responses (rows) mapped to graph edit distance (columns). If participant response was *similar* for GED = 2, accuracy is recorded as 0.6.

	GED = 0	GED = 1	GED = 2	GED = 3
identical	1	0.6	0.3	0
similar	0.6	1	0.6	0.3
not so similar	0.3	0.6	1	0.6
completely different	0	0.3	0.6	1

4 Results

4.1 Dependent Variable

Each trial asked the participant whether the motifs were *identical*, *similar*, *not so similar* or *completely different.* In the presented social media context, we asked: "*Do the two highlighted sections represent the same friendship relationships?*".

Measuring accuracy as a binary result (the structure is identical or not) removes subjectivity from the response, but ignores the fact that some structures are more different than others. We therefore use an accuracy measure based on graph edit distance (GED) [35]. We determine the GED between a pair of sub-graphs: identical (GED = 0); one edge difference (GED = 1); two edges difference (GED = 2); three edges difference (GED = 3). For each of these GEDs, we score the participants' accuracy responses as shown in Table 1.

Experiment 1: Fixed Shape Motifs. The pairs of graph drawing stimuli presented to the participants included the base motif in one drawing, paired with each of the 12 other stimuli. For each of the five motifs, we address the questions:

Q1: Does rotation affect the perception of the same sub-graph represented using the same visual form?

We consider the *same-structure/same-shape* variations where the base motif is compared to three of its own rotations. We expect that there will be no difference in accuracy. A repeated measure ANOVA test (Table 2(a)) shows no significant differences, suggesting that the participants were able to correctly recognise the same motif, presented in the same-shape, regardless of its orientation. This result accords with that of Tarr and Pinker who studied the effect of mental rotation in shape recognition, focusing on letter-like asymmetrical characters [38].

Q2: If the motif is adapted to create a sub-graph of different structure, does depicting it in a similar manner affect the ability to distinguish the difference (or would it be better to depict it using a clearly different visual form?)

We compare the results of *different-structure/similar-shape* with *different-structure/different-shape*, expecting that using a similar shape will help in identifying

Table 2. Summary of results for Experiment 1.

(a) Q1: Effect of rotation on determining whether two motifs are identical					
	bi-clique	clique	cycle	double-cycle	star
F (df=2)	0.476	0.783	0.492	0.129	0.146
p	0.623	0.459	0.613	0.879	0.865
(b) Q2: Effect of depicting different structure with similar or different shapes					
	bi-clique	clique	cycle	double-cycle	star
t (df=29)	3.289	2.168	4.230	0.923	3.780
p (1-tailed)	0.001	0.019	<0.001	0.181	<0.001
(c) Q3: Effect of using different shapes for the same sub-graphs					
	bi-clique	clique	cycle	double-cycle	star
t (df=29)	11.071	11.215	13.360	7.807	19.059
p (1-tailed)	<0.001	<0.001	<0.001	<0.001	<0.001
(d) Q4: Effect of different edit differences in different sub-graph structures. The trend line shows accuracy from no edit distance (left) to maximum edit distance (right); the accuracy scales are different for each motif					
	bi-clique	clique	cycle	double-cycle	star
F (df=3)	11.817	19.719	3.07	28.850	36.84
p	<0.001	<0.001	0.031	<0.001	<0.001
Trend line					

structural differences in the sub-graphs. A repeated measures t-test (Table 2(b)) showed that the accuracy for *different-structure/similar-shape* was significantly better than *different-structure/different-shape* for all motifs, except double-cycle.

> **Q3:** Does using a different layout affect perception of motifs of the same structure? If the motifs are identical, does it matter if they are depicted using different visual form – or should they be depicted in the same form?

We compare *same-structure/different-shape* and *same-structure/same-shape*, expecting that using the same shape for the same structure will produce better accuracy than using a different shape for identical structures. A repeated measures t-test (Table 2(c)) showed that accuracy was significantly better for *same-structure/same-shape* than for *same-structure/different-shape*, for all motifs.

> **Q4:** If different sub-graphs are depicted in similar shape, are near-similar sub-graphs incorrectly assessed as being similar?

That is, does the edit-distance between pairs of sub-graphs depicted in similar visual form affect accuracy when determining whether they are identical or not?

Here, we focus on the three different variations in the *different-structure/similar-shape stimuli*, each of which has a different GED with respect to the base

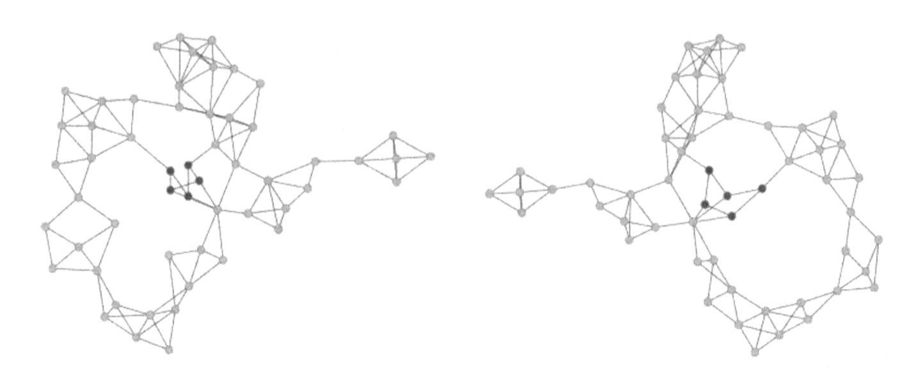

Fig. 9. (left) Cycle motif *same-structure/different-shape* variation introducing unnecessary edge crossings; (right) the same graph drawn using MDS.

motif. We also include one of the *same-structure/same-shape comparisons*, where GED = 0. We expect that where the edit distances are small, the participants were more likely to misjudge two different sub-graphs depicted similarly as identical. A repeated measures ANOVA (Table 2(d)) confirmed that there was an effect of edit distance, for all motifs, where accuracy generally increased with edit distance, with the greatest edit distance accuracy being at least that of the comparisons where GED = 0 (those where the base motif was rotated).

This result confirms our suggestion that the comparisons were made spontaneously – that is, in parallel, rather than serially. If they were serial, participants would have compared the motifs edge-by-edge, and accuracy for the smallest edit distances would have been higher. The spontaneous, parallel pattern matching means that stimuli of the most similar structure are incorrectly seen as identical.

There are a few noticeable data points in this trend analysis: for double-cycle and star, adding two edges leads to lower accuracy – both these extended motifs include at least one edge crossing. Yet, this conclusion does not extend to the bi-clique motif, although we note that the starting point of the bi-clique trend (when GED = 0) is already a low accuracy score.

4.2 Experiment 2: Unconstrained Layout

Here, we used stimuli where the motif nodes positions were 'fixed', but were distorted by MDS forces applied by surrounding nodes. For all Experiment 1 stimuli, we created a paired 'flexible' alternative, removing the three GD constraints of Sect. 3.4, using the unmodified algorithm from [49] (Fig. 9).

Our trials comprised the 'fixed' version of a motif paired with all variants of the 'flexible' version, and participants were asked to judge similarity (as in Experiment 1). In each of the four (shape $\times$ structure) quadrants (Sect. 3.1), we pair the three variants with the drawings of the same graphs rendered by an unconstrained layout, resulting in 36 pairs (4×3 variants (fixed) $\times$ 3 variants (flexible)). We also include the four base motif orientations: a total of 40 pairs for each motif. For each motif, our research question is:

Table 3. The effect of using well-formed shapes to match sub-graphs.

same-structure/same-shape					
	bi-clique	clique	cycle	double-cycle	star
t (df = 29)	8.692	11.611	7.380	10.485	22.254
p (one-tailed)	<0.001	<0.001	<0.001	<0.001	<0.001
same-structure/different-shape					
t (df = 29)	6.649	2.582	1.658	5.477	8.847
p (one-tailed)	<0.001	0.007	**0.054**	<0.001	<0.001

Q5: Is it easier to match identical small motifs if they have both been given a well-formed, regular shape (as opposed to one of them having their shape distorted by forces applied to nodes in the rest of the graph)?

We focus on same-structure stimuli, since the question relates to identical structures, considering the shape dimension. The *same-structure/same-shape* stimuli are rotations of the base; the *same-structure/different-shape* stimuli, while using a different shape, are still regular and well-formed. We expect that the accuracy in matching two identical motifs with well-formed shapes will be higher than for the flexible condition (when the motifs are visually distorted). A repeated measures t-test (see Table 3) showed that the accuracy when comparing two motifs with well-formed, regular shape was always significantly better than when one of the motifs had a distorted form, except in the case of the cycle. We note that the alternative well-formed, regular form of the cycle was one where two versions included unnecessary edge crossings; see Fig. 9.

5 Discussion

Our research questions addressed how visual form (shape) affects the interpretation of the abstract structure of a graph. We showed that by depicting two similar sub-graphs similarly, participants achieved high accuracy for all motifs but the double-cycle (**Q2**). The double-cycle is the largest motif of the five, and our shape variations for it are all roughly two polygons sharing a segment. One possible cause for this is that participants were able to mentally transform the base motif (a mirrored pentagon) to the other shapes (triangle, rectangle). As a result, to gain the benefit of representing similar structures as similar shapes, it may be enough that there is a simple transformation between them, e.g., rotation or translation.

The clear recommendation for visualization designers is that when supporting a task to identify sub-graphs one should ensure that identical sub-graphs are drawn similarly. We have shown that rotation does not have any effect (see **Q1**), so this is a degree of freedom in visualization design, e.g., rotation of fixed motifs can be done to reduce edge crossings while maintaining shape. However, the shape of a sub-graph *does* have an effect; depicting identical sub-graphs using different shapes hampers the recognition of similarity (**Q3**). This means that even if two sections of a node-link diagram represent the same relationships, if they are drawn completely differently, one might never notice. When supporting this type of task, shape cannot be overlooked. Designers can also remember that evaluation of similarity is done quickly (see **Q4**), so should be careful not to depict dissimilar structures using the same shape or one might wrongly interpret the data. The results of both **Q3** and **Q4** seem to indicate that a viewer will interpret similar shapes as having similar structures, regardless of ground truth.

The results of **Q3** and **Q4** can apply beyond the simple five-node motifs we have demonstrated. For instance, graphs with large defined clusters are often depicted with each cluster being drawn as large roughly circular dense regions. If the shape of these regions is roughly the same, a viewer is likely to interpret these large structures as being equivalent. If within-cluster structure is very different between clusters, a designer would want to make this apparent by ensuring these clusters make dissimilar shapes.

The MDS layout algorithm tends to depict cycle motifs as (roughly) circles similar to our defined motif, and this is reflected in the data; see Table 3. Though the star motif is isomorphic to the cycle, comparing the star shape introduced noise that caused participants to not recognize the same structure, another point to show that it is important to represent similar structures with similar shapes. We note as well the star motif is not a planar drawing, where the cycle is. Further motif comparison can be found in the appendix.

6 Conclusion

To better understand the role of spontaneous processing on the representation of sub-graphs in node-link diagrams, we conducted a user experiment to address research questions relating to the perception of both structure (relations in the sub-graph) and shape (visual form). Amongst other findings, we conclude that presenting identical motifs using identical (or rotated) visual motif drawings makes a difference to the ease with which they can be matched.

We considered small motifs (5–8 nodes) in sparse, relatively small graphs. Further work includes determining whether the same results hold for larger motifs or graphs, or for other graph representation idioms (e.g., adjacency matrices or arc diagrams), since the same pattern-matching and bottom-up spontaneous processing issues also apply. Larger, different motifs could be studied, integration into denser graphs, or considering more (and more radical) variations of shape and structure.

Since we have shown that the shape of a motif affects how an individual recognizes it, layout algorithms that extract similar graph motifs and draw them using a similar form would highlight their similarity. Techniques exist to accommodate user constraints such as node positions [4,19,47], but combining this idea with motif extraction may present algorithmic problems and perceptual outcomes.

Acknowledgement. This paper is the result of a collaboration initiated at Dagstuhl Seminar 23051, "Perception in Network Visualization", February 2023.

References

1. de Ridder et al., H.: Information system on graph classes and their inclusions (ISGCI) (2023). https://www.graphclasses.org. Accessed 19 May 2023
2. Ballweg, K., Pohl, M., Wallner, G., von Landesberger, T.: Visual similarity perception of directed acyclic graphs: a study on influencing factors. In: Frati, F., Ma, K.-L. (eds.) GD 2017. LNCS, vol. 10692, pp. 241–255. Springer, Cham (2018). https://doi.org/10.1007/978-3-319-73915-1_20
3. Batik, T., Terziadis, S., Wang, Y.S., Nöllenburg, M., Wu, H.Y.: Shape-guided mixed metro map layout. CGF **41**(7), 495–506 (2022). https://doi.org/10.1111/cgf.14695
4. Böhringer, K.F., Paulisch, F.N.: Using constraints to achieve stability in automatic graph layout algorithms. In: Proceedings of the SIGCHI Conference on Human Factors in Computing Systems, pp. 43–51 (1990). https://doi.org/10.1145/97243.97250
5. Borgo, R., et al.: Crowdsourcing for information visualization: promises and pitfalls. In: Archambault, D., Purchase, H., Hoßfeld, T. (eds.) Evaluation in the Crowd. Crowdsourcing and Human-Centered Experiments. LNCS, vol. 10264, pp. 96–138. Springer, Cham (2017). https://doi.org/10.1007/978-3-319-66435-4_5
6. Bremm, S., von Landesberger, T., Heß, M., Schreck, T., Weil, P., Hamacherk, K.: Interactive visual comparison of multiple trees. In: 2011 IEEE Conference on Visual Analytics Science and Technology (VAST), pp. 31–40 (2011). https://doi.org/10.1109/VAST.2011.6102439
7. Cakmak, E., Fuchs, J., Jackle, D., Schreck, T., Brandes, U., Keim, D.: Motif-based visual analysis of dynamic networks. In: 2022 IEEE Visualization in Data Science (VDS), pp. 17–26 (2022). https://doi.org/10.1109/VDS57266.2022.00007
8. Cherry, K.: What are the gestalt principles? An overview of the gestalt laws of perceptual organization (2023). https://www.verywellmind.com/gestalt-laws-of-perceptual-organization-2795835. Accessed 19 May 2023
9. Dong, W., Moses, C., Li, K.: Efficient k-nearest neighbor graph construction for generic similarity measures. In: Proceedings of the 20th International Conference on World Wide Web, pp. 577–586 (2011). https://doi.org/10.1145/1963405.1963487
10. Dunne, C., Shneiderman, B.: Motif simplification: improving network visualization readability with fan, connector, and clique glyphs. In: Proceedings of the SIGCHI Conference on Human Factors in Computing Systems, CHI 2013, pp. 3247–3256. Association for Computing Machinery, New York (2013). https://doi.org/10.1145/2470654.2466444
11. Eades, P., Hong, S.-H., Klein, K., Nguyen, A.: Shape-based quality metrics for large graph visualization. In: Di Giacomo, E., Lubiw, A. (eds.) GD 2015. LNCS, vol. 9411, pp. 502–514. Springer, Cham (2015). https://doi.org/10.1007/978-3-319-27261-0_41
12. Fuchs, J., Isenberg, P., Bezerianos, A., Fischer, F., Bertini, E.: The influence of contour on similarity perception of star glyphs. IEEE TVCG **20**(12), 2251–2260 (2014). https://doi.org/10.1109/TVCG.2014.2346426

13. Gansner, E.R., Koren, Y., North, S.: Graph drawing by stress majorization. In: Pach, J. (ed.) GD 2004. LNCS, vol. 3383, pp. 239–250. Springer, Heidelberg (2005). https://doi.org/10.1007/978-3-540-31843-9_25
14. Gogolou, A., Tsandilas, T., Palpanas, T., Bezerianos, A.: Comparing similarity perception in time series visualizations. IEEE TVCG **25**(1), 523–533 (2019). https://doi.org/10.1109/TVCG.2018.2865077
15. Hascoët, M., Dragicevic, P.: Interactive graph matching and visual comparison of graphs and clustered graphs. In: Proceedings of the International Working Conference on Advanced Visual Interfaces, AVI 2012, pp. 522–529. Association for Computing Machinery, New York (2012). https://doi.org/10.1145/2254556.2254654
16. Hu, Y., Brunton, S.L., Cain, N., Mihalas, S., Kutz, J.N., Shea-Brown, E.: Feedback through graph motifs relates structure and function in complex networks. Phys. Rev. E **98**, 062312 (2018). https://doi.org/10.1103/PhysRevE.98.062312
17. Huang, W., Murray, C., Shen, X., Song, L., Wu, Y.X., Zheng, L.: Visualisation and analysis of network motifs. In: Ninth International Conference on Information Visualisation (IV 2005), pp. 697–702 (2005)
18. Jerrum, M., Meeks, K.: The parameterised complexity of counting connected subgraphs and graph motifs. J. Comput. Syst. Sci. **81**(4), 702–716 (2015). https://doi.org/10.1016/j.jcss.2014.11.015
19. Kamps, T., Kleinz, J., Read, J.: Constraint-based spring-model algorithm for graph layout. In: Brandenburg, F.J. (ed.) GD 1995. LNCS, vol. 1027, pp. 349–360. Springer, Heidelberg (1996). https://doi.org/10.1007/BFb0021818
20. Koffka, K.: Principles of Gestalt Psychology. Routeledge (1935)
21. Kypridemou, E., Zito, M., Bertamini, M.: Perception of node-link diagrams: the effect of layout on the perception of graph properties. In: Giardino, V., Linker, S., Burns, R., Bellucci, F., Boucheix, J.M., Viana, P. (eds.) Diagrams 2022. LNCS, vol. 13462, pp. 364–367. Springer, Cham (2022). https://doi.org/10.1007/978-3-031-15146-0_32
22. von Landesberger, T.: Visual analysis of large graphs: state-of-the-art and future research challenges. CGF **30**(6), 1719–1749 (2011). https://doi.org/10.1111/j.1467-8659.2011.01898.x
23. von Landesberger, T.: Insights by visual comparison: the state and challenges. IEEE Comput. Graphics Appl. **38**(3), 140–148 (2018). https://doi.org/10.1109/MCG.2018.032421661
24. von Landesberger, T., Görner, M., Rehner, R., Schreck, T.: A system for interactive visual analysis of large graphs using motifs in graph editing and aggregation. In: 14th International Workshop on Vision, Modeling, and Visualization, pp. 331–340 (2009)
25. Lenhof, H.P., Smid, M.: Sequential and parallel algorithms for the k closest pairs problem. Int. J. Comput. Geom. Appl. **5**(03), 273–288 (1995). https://doi.org/10.1142/S0218195995000167
26. Leontis, N., Lescoute, A., Westhof, E.: The building blocks and motifs of RNA architecture. Curr. Opin. Struct. Biol. **16**, 279–87 (2006). https://doi.org/10.1016/j.sbi.2006.05.009
27. McGee, F., Ghoniem, M., Melançon, G., Otjacques, B., Pinaud, B.: The state of the art in multilayer network visualization. CGF **38**(6), 125–149 (2019). https://doi.org/10.1111/cgf.13610
28. Meidiana, A., Hong, S.H., Eades, P.: Shape-faithful graph drawings. In: Angelini, P., von Hanxleden, R. (eds.) GD 2022. LNCS, vol. 13764, pp. 93–108. Springer, Cham (2023). https://doi.org/10.1007/978-3-031-22203-0_8

29. Micale, G., Giugno, R., Ferro, A., Mongiovì, M., Shasha, D., Pulvirenti, A.: Fast analytical methods for finding significant labeled graph motifs. Data Min. Knowl. Disc. **32**, 504–531 (2018). https://doi.org/10.1007/s10618-017-0544-8
30. Pandey, A.V., Krause, J., Felix, C., Boy, J., Bertini, E.: Towards understanding human similarity perception in the analysis of large sets of scatter plots. In: Proceedings of the 2016 CHI Conference on Human Factors in Computing Systems, pp. 3659–3669 (2016). https://doi.org/10.1145/2858036.2858155
31. Prolific (2014). https://www.prolific.co
32. Qualtrics (2005). https://www.qualtrics.com
33. Redhu, N., Thakur, Z.: Chapter 23 - network biology and applications. In: Singh, D.B., Pathak, R.K. (eds.) Bioinformatics, pp. 381–407. Academic Press (2022). https://doi.org/10.1016/B978-0-323-89775-4.00024-9
34. Royer, L., Reimann, M., Andreopoulos, B., Schroeder, M.: Unraveling protein networks with power graph analysis. PLoS Comput. Biol. **4**(7), 1–17 (2008). https://doi.org/10.1371/journal.pcbi.1000108
35. Sanfeliu, A., Fu, K.S.: A distance measure between attributed relational graphs for pattern recognition. IEEE Trans. Syst. Man Cybern. **SMC-13**(3), 353–362 (1983). https://doi.org/10.1109/TSMC.1983.6313167
36. Soni, U., Lu, Y., Hansen, B., Purchase, H.C., Kobourov, S., Maciejewski, R.: The perception of graph properties in graph layouts. CGF **37**(3), 169–181 (2018). https://doi.org/10.1111/cgf.13410
37. Stone, L., Simberloff, D., Artzy-Randrup, Y.: Network motifs and their origins. PLoS Comput. Biol. **15**(4), e1006749 (2019). https://doi.org/10.1371/journal.pcbi.1006749
38. Tarr, M.J., Pinker, S.: Mental rotation and orientation-dependence in shape recognition. Cogn. Psychol. **21**(2), 233–282 (1989). https://doi.org/10.1016/0010-0285(89)90009-1
39. Treisman, A.: Preattentive processing in vision. Comput. Vision, Graph. Image Process. **31**(2), 156–177 (1985). https://doi.org/10.1016/S0734-189X(85)80004-9
40. Tversky, A.: Features of similarity. Psychol. Rev. **84**(4), 327–352 (1977). https://doi.org/10.1037/0033-295X.84.4.327
41. Wallner, G., Pohl, M., von Landesberger, T., Ballweg, K.: Perception of differences in directed acyclic graphs: influence factors & cognitive strategies. In: Proceedings of the 31st European Conference on Cognitive Ergonomics, pp. 57–64 (2019). https://doi.org/10.1145/3335082.3335083
42. Wallner, G., Pohl, M., Graniczkowska, C., Ballweg, K., von Landesberger, T.: Influence of shape, density, and edge crossings on the perception of graph differences. In: Pietarinen, A.-V., Chapman, P., Bosveld-de Smet, L., Giardino, V., Corter, J., Linker, S. (eds.) Diagrams 2020. LNCS (LNAI), vol. 12169, pp. 348–356. Springer, Cham (2020). https://doi.org/10.1007/978-3-030-54249-8_27
43. Wang, Y., et al.: Revisiting stress majorization as a unified framework for interactive constrained graph visualization. IEEE TVCG **24**(1), 489–499 (2018). https://doi.org/10.1109/TVCG.2017.2745919
44. Ware, C.: Visual Thinking for Information Design. Morgan Kaufmann (2021)
45. Wasserman, S., Faust, K.: Social Network Analysis: Methods and Applications. Structural Analysis in the Social Sciences. Cambridge University Press (1994). https://doi.org/10.1017/CBO9780511815478
46. Wu, H.Y., Niedermann, B., Takahashi, S., Roberts, M.J., Nöllenburg, M.: A survey on transit map layout - from design, machine, and human perspectives. CGF **39**(3), 619–646 (2020). https://doi.org/10.1111/cgf.14030

47. Yu, J., Hu, Y., Yuan, X.: UNICON: a UNIform CONstraint based graph layout framework. In: 2022 IEEE 15th Pacific Visualization Symposium (PacificVis), pp. 61–70. IEEE (2022). https://doi.org/10.1109/PacificVis53943.2022.00015
48. Yuan, X., Che, L., Hu, Y., Zhang, X.: Intelligent graph layout using many users' input. IEEE TVCG **18**(12), 2699–2708 (2012). https://doi.org/10.1109/TVCG.2012.236
49. Zheng, J.X., Pawar, S., Goodman, D.F.: Graph drawing by stochastic gradient descent. IEEE TVCG **25**(9), 2738–2748 (2018). https://doi.org/10.1109/TVCG.2018.2859997

TimeLighting: Guidance-Enhanced Exploration of 2D Projections of Temporal Graphs

Velitchko Filipov[1(✉)], Davide Ceneda[1], Daniel Archambault[2], and Alessio Arleo[1]

[1] TU Wien, Vienna, Austria
{velitchko.filipov,davide.ceneda,alessio.arleo}@tuwien.ac.at
[2] Newcastle University, Newcastle, UK
daniel.archambault@newcastle.ac.uk

Abstract. In temporal (or *event-based*) networks, time is a continuous axis, with real-valued time coordinates for each node and edge. Computing a layout for such graphs means embedding the node trajectories and edge surfaces over time in a $2D + t$ space, known as the space-time cube. Currently, these space-time cube layouts are visualized through animation or by slicing the cube at regular intervals. However, both techniques present problems ranging from sub-par performance on some tasks to loss of precision. In this paper, we present `TimeLighting`, a novel visual analytics approach to visualize and explore temporal graphs embedded in the space-time cube. Our interactive approach highlights the node trajectories and their mobility over time, visualizes node "aging", and provides guidance to support users during exploration. We evaluate our approach through two case studies, showing the system's efficacy in identifying temporal patterns and the role of the guidance features in the exploration process.

Keywords: Temporal Graphs · Space-time cube · Dynamic Network Visualization · Visual Analytics

1 Introduction

Temporal (or *event-based*) networks [24] are dynamic graphs where the temporal dynamics, such as node/edge additions and removals, have real-time coordinates. These have been characterized and studied extensively [21], as they are used in many applications to model phenomena of commercial and academic interest, such as interactions in social media [24], communication networks [16], and contact tracing [30], to name a few. In "traditional" dynamic graph drawing [11,21], where time is discretized (or *timesliced*), creating a visualization for such networks poses different challenges. Juxtaposition (or small multiples) would require first identifying suitable timeslices, which inevitably leads to quantization errors that, in turn, obscure the fine temporal details that might be crucial in some

M. A. Bekos and M. Chimani (Eds.): GD 2023, LNCS 14465, pp. 231–245, 2023.
https://doi.org/10.1007/978-3-031-49272-3_16

domains (e.g., the exact order of personal contacts in contact tracing networks). On the other hand, animation would not suffer from such artifacts, and it has been used in previous work on event-based graph drawing as a visualization metaphor to display the computed layouts [7,29]. Animation is a more natural way to encode time; however, it is not perceptually effective for many tasks involving dynamic networks [3,19]. Moreover, the vast majority of research on animation has been done with timesliced graphs (see, e.g., [3,6,11,18]), and its application to temporal networks is still in its infancy.

Temporal networks can also be drawn in a 3D space, namely, the "space-time cube" ($2D + t$). In this case, a drawing algorithm computes the node trajectories over time and space. Existing research [7,28,29] provides evidence that this drawing approach yields better quality drawings of temporal graphs, compared to their timesliced counterparts (e.g., Visone [10]), and for discrete time graphs when many changes occur between timeslices. Despite this, research on visually depicting these trajectories and, in turn, obtaining insights from their behavior (i.e., exploring the network in time) is still a largely under-investigated topic - a gap that we intend to address in this paper.

On these premises, we present `TimeLighting`, a guidance-enhanced Visual Analytics (VA) solution for exploring node trajectories in the space-time cube keeping the *full* temporal resolution of the network. `TimeLighting` supports understanding temporal patterns and behaviors and, in general, extracting insights from datasets with complex temporal dynamics.

The design of `TimeLighting` is inspired by the "time-coloring" operation [8], whereby time is mapped to color to visualize the evolution of nodes and edges through the space-time cube; and is loosely motivated by *transfer functions* used in direct volume rendering [26] to emphasize features of interest in the data. Conceptually, in our approach, we "shine" light through the space-time cube down along its time axis (hence the system's name – `TimeLighting`) in a manner that resembles the behavior of transfer functions for volume rendering. As the light interacts with the node trajectories, they are visualized and colored differently according to the age and persistence of the nodes (i.e., applying the time-coloring operation), generating a 2D visualization of the 3D embedding (see also Fig. 1). The resulting visualization is an explorable 2D map of the nodes' densities, with visible individual node movement and "aging" over time. We complement this visualization with several interactive controls to explore the data and introduce a simple mobility metric, based on the length of each node's trajectory, to rank and identify the more and less stable parts of the graph. We designed `TimeLighting` introducing multiple elements of visual guidance [14] to enhance (and possibly ease) the network exploration process. Finally, we describe two case studies demonstrating how our guidance-enhanced approach supports users in achieving the system design tasks.

2 Related Work

We now illustrate related literature on which we ground our research.

Visualization of Dynamic Networks. Visualizing temporal networks differs significantly from how typically we draw and visualize dynamic graphs in the graph drawing and visualization community where the time axis is a discrete series of timeslices. Each individual time point is called a timeslice, a snapshot that represents the state of the graph over a time interval. This simple yet powerful simplification is used as the basis of visualization [9,11], for layout algorithm design [12,17], and in user studies [4,6,18,20]. The problem with time slicing is that many networks of scientific interest do not have natural timeslices. Therefore, choosing the right time sampling and duration for each can be a complex task, which eventually results in some loss of temporal information. In this regard, visualization techniques have been presented to suggest interesting timeslice selection. Wang et al. [31] present a technique for non-uniform time slicing (that is, selecting slices of different duration) based on histogram equalization to produce timeslices with the same number of events and, in turn, similar visual complexity. Lee et al. [25] experimented with a visualization tool, called "Dynamic Network Plaid" for interactive time slicing on large displays. Users can select interesting time intervals based on the event distribution over time and visualize the corresponding status of the network in those intervals. Recently, A number of algorithms have been proposed to draw temporal networks directly in the space-time cube [7,29]. These approaches do not divide the data into a series of timeslices to draw the graph but directly embed the network in the space-time cube. However, the only way to visualize such $2D + t$ drawings on a 2D plane was to select timeslices or present an animation of the data over time.

Guidance for Graph Exploration. Due to the challenge of analyzing complex events such as those modeled by temporal networks, researchers also investigated approaches to provide support and ease the analysis for users. The resulting approaches fall under the definition of "guidance" [14]. Guidance is characterized as *active* support in response to a *knowledge gap* which hinders the completion of an interactive visual data analysis session. Over the years, several approaches have been devised, providing different types of guidance in different phases of the analysis [13,15]. For instance, May et al. [27] describe a method to enhance the exploration of large graphs using glyphs. While the user explores a given area of interest (the focus), the system automatically highlights the path to other possibly off-screen interesting nodes (the context). Gladisch et al. [23] provide support during the navigation through large hierarchical graphs by suggesting what to explore next. Thanks to a user-customizable degree-of-interest function, the system can suggest how to navigate the graphs, both horizontally and vertically, adjusting the level of abstraction of the hierarchy. Despite the work in this area, applying guidance to temporal networks is uncharted territory. Given a temporal network modeled in a space-time cube, our goal is to provide guidance to support the identification of interesting time intervals and nodes requiring further attention and analysis from the user.

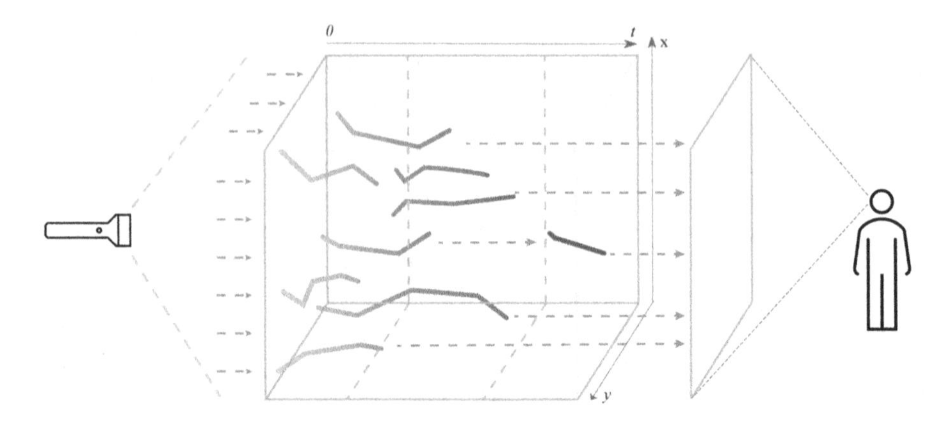

Fig. 1. Exemplification of the `TimeLighting` metaphor. Rays of light (depicted as dashed lines), coming from $t = 0$ travel through the space-time cube and reach the observer at t_{max}. The rays interact with the node trajectories and will carry this information to the projection plane.

3 Design Considerations

In this section, we discuss the most relevant aspects that influenced the design of `TimeLighting`, namely, the data characteristics, the user's tasks, and the time-coloring paradigm used to provide guidance.

Data, Tasks. The data we aim to visualize and explore with `TimeLighting` represents a temporal network [7,29]. In a temporal network $D = (V, E, T)$, with V the set of nodes, E the set of edges, and $T \in \Re$ represents the time-dependant *attributes*. These take the form of functions in the $V \times T$ and $E \times T$ domains for nodes and edges, respectively. For simplicity, and in accordance with existing literature, we consider all attribute functions as piece-wise linear functions. The *appearance* A_x attribute, for example, models the intervals in time in which nodes and edges exist:

$$A_v : V \times T \rightarrow [true, false]$$
$$A_e : E \times T \rightarrow [true, false]$$

A_v and A_e map to the node and edge insertion and deletion *events*, respectively. For this reason, these graphs are also called *event-based* networks, and the terms, temporal and event-based, will be used interchangeably in the remainder of this paper. The *position* $P_v : V \times T \rightarrow \Re^2$ attribute describes the nodes' position over time. The following is an example of how each node's $(v \in V)$ position is computed by an event-based layout algorithm describing the movement over time and space:

$$P_v(t) = \begin{cases} (5,6) \rightarrow (12,11) & \text{for } t \in [0,1] \\ (8,3) \rightarrow (1,7) & \text{for } t \in [5,7] \\ \dots & \\ (0,0) & \text{otherwise} \end{cases}$$

In this paper, we use the `MultiDynNoS` [7] event-based layout algorithm to generate the drawings of the graphs. `TimeLighting` is designed for a number of user tasks, which we characterize using the task taxonomy for network evolution analysis by Ahn et al. [2] and the taxonomy of operations on the space-time cube by Bach et al. [8], and are described in the following.

T1: Overview. `TimeLighting` should provide an overview of the temporal information at a glance. Providing an overview is typically the first necessary step in any VA process.

T2: Tracking Events. Understanding the temporal dynamics and the events' frequency helps the user isolate interesting occurrences in time. Events also cause the nodes' trajectories to bend, i.e., make the node change direction. Understanding the *shape* [2] of changes in node movement over time (e.g., speed, repetition, etc.) would provide further insights during the exploration of the data.

T3: Investigate Relationships. Each edge event occurrence perturbs the trajectories. Identifying which relationships have the most impact or how often they occur might help the user explain the formation of clusters, or, in general, the phenomenon at hand.

Time-Coloring. To highlight interesting nodes and trajectories and more generally to visualize the network's dynamics, we took inspiration from the time-coloring operation described by Bach et al. [8] in their survey. Time-coloring is a content transformation operation applied to the space-time cube. Given a timesliced graph, the procedure consists in coloring each timeslice based on a uniform linear color scale so that it would be possible to identify the "age" of the data points easily. To apply this technique to temporal graphs, we had to find a way to visually represent two features of each node: their *persistence* (i.e., its behavior during its appearance intervals) and their *aging*, that is how their movements are distributed in time from the point of view of the observer, in a continuous time axis scenario. Such network features can be visualized as if we projected light through the cube along the time axis: the interaction with light will be visible from the observer, watching from the other side of the cube (at t_{max}, see Fig. 1).

Guidance. In addition to visualizing the temporal network, we designed guidance to support its exploration and analysis. In general, the degree of support provided to the user may vary significantly, and at least three guidance degrees can be identified [14]: *Orienting* helps users keep an overview of the problem and the alternative analytical paths they can choose to move forward. *Directing* guidance provides users with a set of options and orders them according to their importance for solving current tasks, and *prescribing* guidance (as the name suggests) prescribes a series of actions to take to conclude the task. Given the similarities between our problem and the transfer functions used in volume rendering (although we apply them to a 2D visualization), the general idea at the base of our guidance-enhanced approach is to support and ease the identification of time intervals and nodes with specific desirable characteristics, as well as, investigate their relationships and how they interact (i.e., analyzing the moment of the trajectories), making them stand out from the rest for the user's convenience. Considering our set of design tasks, we highlight the nodes that have the longest trajectories. Long trajectories, in fact, represent nodes that are associated with many events and with high persistence in the currently selected temporal interval (to support **T2**). This type of guidance can be classified as directing, as trajectories are ranked based on their length, and, is necessary to ease or solve the system design tasks. In addition, we highlight temporal intervals in which nodes defined by the user interact with each other, thus providing guidance to **T3**.

4 `TimeLighting`

In this section, we describe our system in detail and how we implemented it considering the design requirements described in the previous section. `TimeLighting` is a guidance-enhanced VA system comprised of two linked views and a focus+context approach. An overview of the prototype is shown in Fig. 2. For more details and resources we refer to the online supplementary material [22].

4.1 Main View

In the main view of `TimeLighting` (see Fig. 2-B), we show a 2D projection of the complete temporal graph (i.e., an overview—**T1**). We discuss the details of our approach in the following. We employ a number of different encodings to highlight features of nodes and edges.

Node Positions are represented as trajectories encoded as a trail of circles (see Fig. 3). First, we place each interval's start and end position within every node's P_v attribute (see Sect. 3). These come directly from the computed drawing and have an orange stroke in order to make these distinguishable from the sampled nodes. To ease the comprehension of the movement flow (**T2**), between the start and end positions of each interval, we place a number of sampled nodes, where the coordinates are interpolated. The user can fine-tune the number of interpolated positions by choosing an appropriate "sampling frequency". The resulting sampled nodes are positioned along the node trajectory but are encoded as smaller circles with no stroke to differentiate them from non-sampled nodes. We calculate and visualize the node *aging* as follows: for each node visualized on screen (nodes both in P_v and interpolated are considered), its age is computed as the difference between its time coordinate and the time of the node's first appearance. We use a linear opacity scale to visually represent the node's aging

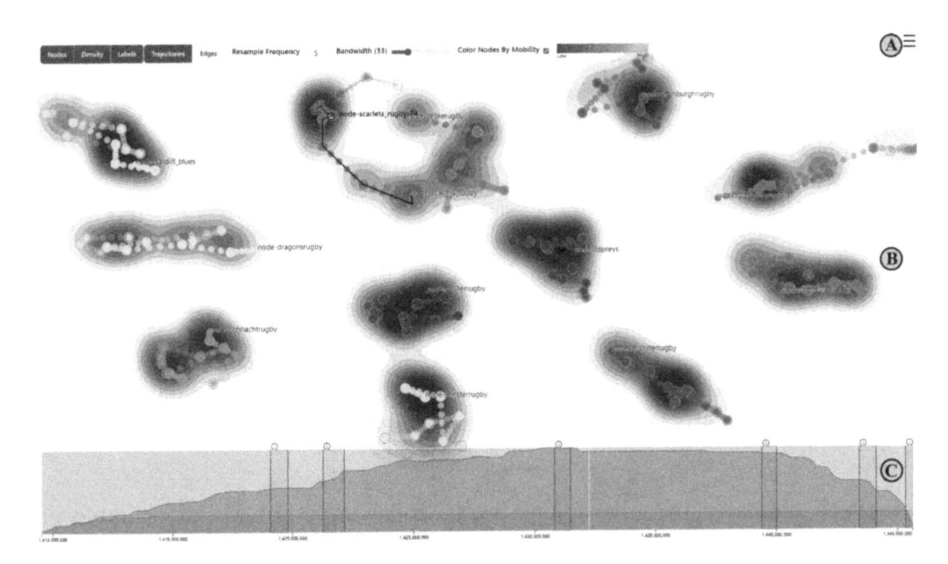

Fig. 2. `TimeLighting` overview. The view is comprised of the (A) toolbar and sidebar, (B) main view, and (C) event timeline. The yellow bar in the timeline shows the absolute age of the hovered node (visible in the top-center area). (Color figure online)

process. This encoding makes it easier to understand the information about the progression and movement of each node over time, providing an overview of the evolution of the network. We use relative aging in this context as it is focused on the individual node's trajectory. Hovering over the node makes it possible to see its position in the timeline (in the context of the full temporal extent of the network) as a yellow bar, corresponding to its time coordinate. Typically, nodes are visualized in gray. However, users can change the visual appearance of the nodes to reflect the cumulative amount of movement (see Sect. 3, *Guidance* paragraph). Activating this type of guidance changes the coloring from gray-scale to a continuous color scale making nodes with higher mobility visually distinct from the more stationary ones (**T2/T3**).

Edges are represented as solid straight lines that connect pairs of nodes belonging to two distinct trajectories. Edges have a pivotal role because these interactions eventually cause node movements and the creation of temporal clusters (**T3**). Edges might also appear or disappear within P_v movement intervals. This justifies the choice of introducing sampled nodes: edges can appear between all the points belonging to a trajectory, including interpolated ones. This allows the user to keep an overview of the finer temporal details, as we can display edges closest to their exact time coordinates. In turn, this could generate visual clutter as each edge is shown once for every node pair between each trajectory, and, depending on the sampling frequency, a trajectory can be comprised of several nodes. We mitigate this by showing edges on-demand and related to one trajectory at a time (the selected one). The user, by hovering on any node belonging to a trajectory, will make all edges incident to those nodes appear. Edge aging

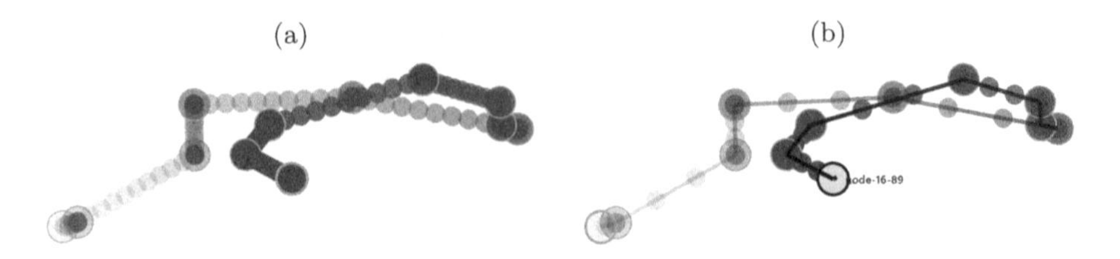

Fig. 3. Examples of trajectory visualizations. In (a) a higher sampling frequency is selected, and aging is clearly visible thanks to the change in opacity. In (b) sampling is lowered to a third (3 points per movement segment), and the trajectory is shown superimposed as the mouse is hovering over one of the nodes, indicated as a yellow node. The orange stroke and size difference indicates coordinates coming from the data compared to interpolated nodes. (Color figure online)

is encoded similarly to node aging. In the following Fig. 6-B, an example of how edges connect the sampled nodes within the current temporal selection is depicted for the currently hovered node (Munster Rugby).

Movement is visualized using a polyline connecting each position in the nodes' P_v attribute. It represents how the node's movement changes over time due to the bends in the trajectories computed by the layout process. We add this encoding as the nodes' opacity alone might not be sufficient in showing how trajectories evolve over time (**T2/T3**). We calculate the age of each trajectory segment and apply a similar linear opacity scale as with the nodes and edges. The trajectory's age is calculated as the mean of the ages of each pair of nodes in that given polyline segment. This information is also shown on-demand by hovering over a trajectory, and, either the edges or the movement of a node can be shown (depending on the selection in the top bar—see Fig. 2-A).

Density is represented as a contour map (see the dark-blue areas in the center of Fig. 2), providing a quick visual indicator that emphasizes locations where a larger number of nodes have existed (**T2**). This kind of encoding also provides a first glance at the trajectories' "shape", which eases keeping an overview of the events (**T1**). To calculate the density map, we translate the original set of nodes for each point in time into a series of objects with x, y coordinates and relative age. The x, y coordinates determine the contours of the density map. The age acts as a weighing function such that older nodes would contribute less to the density map compared to more recent nodes. The "bandwidth" sets the standard deviation of the Gaussian kernel, with lower values showing a sharper picture and higher values more distributed, but also more blurred, representation.

Interactions. are also supported in the main view of `TimeLighting`. Common interactions such as using the mouse scroll wheel to *zoom* and rescale the main view, and *panning* or *dragging* to reposition it.

4.2 Other Views and Guidance

Side Panel shows a list of nodes ordered according to their mobility (as a form of guidance, see Sect. 3). Each node in this list is also accompanied by a small bar chart visualizing the differences in the mobility scores of the nodes. From here, the user can select and "lock" trajectories in the main view. A locked trajectory is always shown regardless of the current temporal interval selection in the timeline (see next paragraph). Locked nodes are colored in bright red if they are in the current temporal selection, while nodes that are out of the current temporal selection are colored in a less saturated hue (see Fig. 4). The encoding and ordering of node trajectories in the side panel serve as visual guidance to support the tracking of specific events (i.e., guidance to ease **T2**). Additionally, when loading a new graph, the three nodes with the highest mobility are locked by default (this can later be refined or changed).

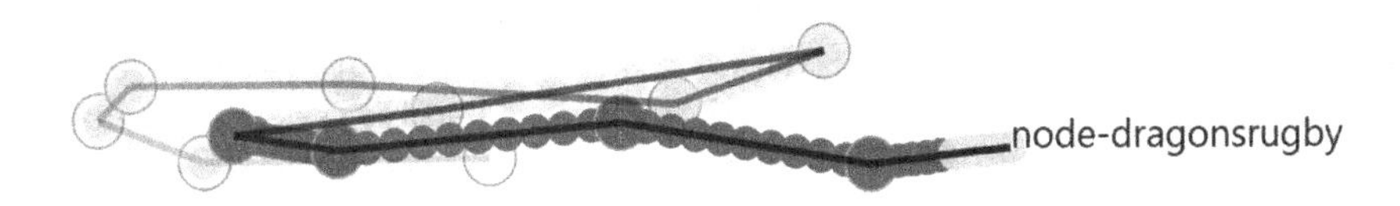

Fig. 4. Example of a locked trajectory. The circles in red are fully within the temporal selection of the users, whereas the less saturated ones are outside of the selected interval. (Color figure online)

Timeline, shown in (Fig. 2-C), allows to select and explore specific temporal intervals as well as keep an overview (**T1**) of the number of nodes and edges that are visible over time. This is obtained by considering the net number of node/edge additions and removals and representing this information as two overlapping area charts (in red—the nodes and in blue—the edges). The timeline serves two purposes: first, the user can brush to select a specific interval within the available data. As a result, only the subgraph existing during the newly selected time interval will be shown in the main view (temporal filtering). This also affects movement coloring, relative age, and density calculation, as well as, limits the edges shown on-screen to those existing in the current selection (these do not apply to locked trajectories). Second, `TimeLighting` uses the timeline to provide guidance and suggest specific time intervals for further inspection. Specifically, the system highlights intervals in time when all the currently locked trajectories interact with each other. These intervals are represented as orange rectangles drawn on top of the timeline (see Fig. 2-C). Clicking one of these intervals will snap onto that temporal selection, helping the user to keep track of and investigate relationships (i.e., guidance to ease **T3**).

5 Case Studies

In this section, we discuss two case studies on a real temporal network. We show how insights can be extracted from the data using `TimeLighting` and how the design tasks are achieved and supported with guidance.

We build our case studies on the *Rugby* dataset, which is a collection of 3151 tweets posted during the Pro12 rugby competition of the 2014–2015 season [1], specifically from September 2014 to October 2015. The network has a node for each team participating in the competition (12 teams in total), and an edge exists between two teams when a tweet from one mentions another. While nodes will stay visible from the moment they appear until the end, edges appear at the exact moment the tweet was posted. To improve the visibility of the edges during the layout process (as tweets do not have a "duration"), edges are given a 24-hour duration. For example, if an edge $\bar{e}$ has a timestamp t, then $A_{\bar{e}} = [t-12h, t+12h]$. Multiple edges between the same teams are merged together if their appearances overlap by a duration of less than one day. This simplification has already been applied in previous work using this dataset [29], and discretization of this dataset with similar resolution would require 417 timeslices. This dataset is particularly interesting as we have ground truth data to validate our findings.

Case Study 1: The First and Second Half of the Season: Examining Trajectories. We begin our use case with an *Overview* task (**T1**), examining the trajectories in the first (see Fig. 5-A) and the second half of the 2014–2015 season (see Fig. 5-B). We can immediately observe from the timeline the trend of the events. There is a steady increase in the number of tweets from the beginning of the season that peaks around the beginning of the second round of the league. This peak remains until the season's final and significantly impacts the nodes' mobility. We can also see that nodes move less in the second half of the season compared to the first half. Specifically, in Fig. 5-B the majority of the nodes are within the purple to orange range of the scale—lower mobility—whereas in Fig. 5-A they are in the yellow to green range—encoding higher mobility. In the first half, instead, tweets are sparser, meaning that the influence of an edge on the movement of nodes is kept (as there is no inertia) until another one changes its trajectory. Continuing the analysis of the network, tweets happen at a much higher rate in the second half of the season, and, since this network is a clique (all teams eventually play against one another), they tend to be "locked" in place by the attractive forces exerted by the other nodes. It must be considered that the layout algorithm attempts to optimize (and reduce) node movement, placing the nodes in an area of the plane where they will likely remain. This behavior can be seen in the density map too, where hot spots are larger and more numerous (i.e., nodes tend to linger more in the same areas) in the second half compared to the first half. Nonetheless, the amount of attractive force will depend on the public interest (i.e., the number of tweets) about individual matches.

Case Study 2: Tracking the Two Least Winning Teams. In this second case study, we *track* (**T2**) the trajectories and *investigate* (**T3**) the relationships

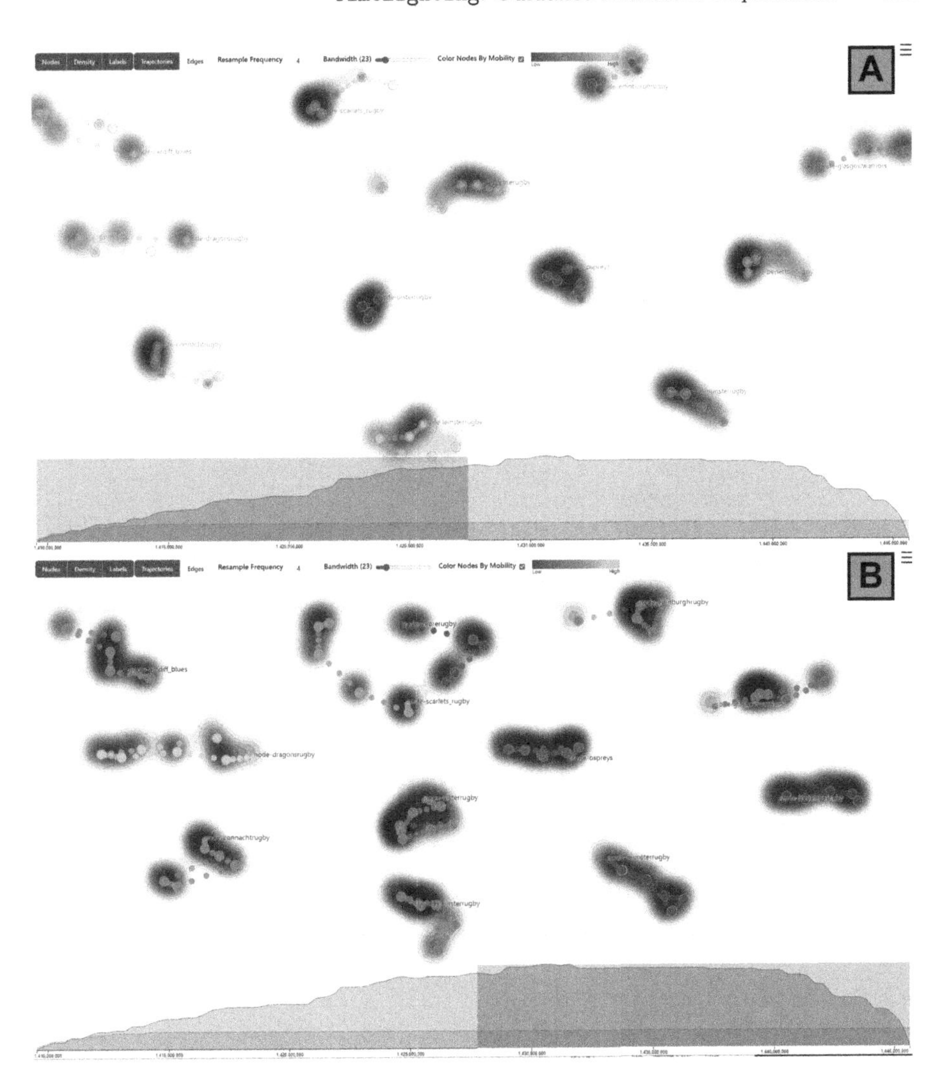

Fig. 5. Illustrations from Case Study 1. It is possible to see how the mobility of nodes changes in the two halves of the season. **(A)** The first half of the season; **(B)** The second half of the season. Timeline brushing is used to filter out events. Trajectory sampling is set at 4 points.

between the two least winning teams of the season (according to the historical information available), namely the "Zebre" (Z) and "Benetton" (B) teams. We begin by selecting them in the sidebar so that the whole trajectory is locked permanently on screen. Guidance shows us the different moments in time when the two teams interact, and we focus on the period of time when the two teams play against each other around the midpoint of the season. The teams played two matches (during the first and second leg of the tournament) in adjacent rounds

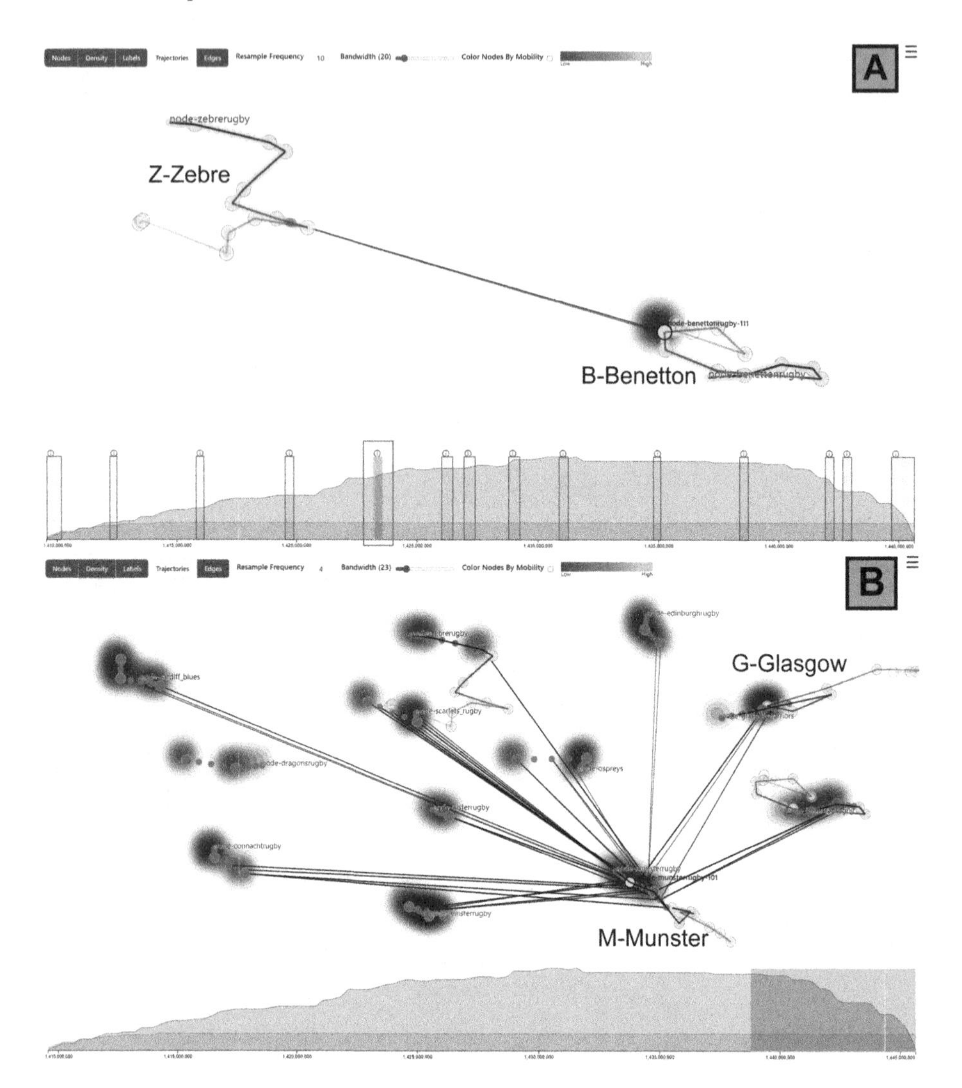

Fig. 6. Illustrations from Case Study 2. **(A)** The two trajectories of the teams in guidance intervals are visible in the timeline, and the one selected (see blue rectangle) relates to the matchup between the two teams in the first round of the competition. **(B)** it is possible to identify the final match and the connections between the 2nd best team and the other teams of interest in the case study. (Color figure online)

(12 and 13). The status of the interface is reported in Fig. 6-A. It is possible to see how the Z trajectory bends significantly towards B at this point in time, and a similar effect is visible the other way around. This attraction strength can be interpreted as the "hype" of the matches building up, as Z and B are the only two teams coming from Italy in the competition. Finally, we compare the relationships between the last two teams in the ranking and the first two,

"Glasgow Warriors" (G), the winners, and "Munster Rugby" (M), referring to the time around the tournament finals (see Fig. 6-B). If we focus on M, it is easy to identify the time the final was played, with mostly all teams connected to it. The B trajectory is strongly influenced by M, as it was one of the final matches before the final; Z, instead, is not largely influenced and drifts away.

6 Conclusion and Future Work

In this paper, we presented and described TimeLighting, a guidance-enhanced VA approach to support the analysis and exploration of temporal networks embedded in a space-time cube. We augmented the visualization approach with guidance to better support users in the visual analysis of the data. We demonstrated the effectiveness of our approach with two use cases, depicting a scenario where the aim is to explore and extract insights from a temporal network describing the events of a rugby season and the relationships between teams. The potential of visualization in exploring temporal graphs in the space-time cube is an opportunity for the entire graph drawing and visualization communities. Future work should primarily be oriented on conducting a formal evaluation of the method, both from a visual quality perspective, i.e., through a selection of metrics that capture the readability, scalability, and expressiveness of the visualization, and from a user perspective, also assessing the impact of the guidance features included in the system, extending experimentation on other real datasets (e.g., [5,16]). Further work includes investigating how TimeLighting supports users when sampling and identifying temporal network features and patterns.

Acknowledgements. For the purpose of open access, the author has applied a Creative Commons Attribution (CC-BY) license to any Author Accepted Manuscript version arising from this submission. This work was conducted within the projects WWTF grant [10.47379/ICT19047], FFG grant DoRIAH [#880883], and FWF grant ArtVis [P35767].

References

1. PRO-12 Rugby Competition 2014–2015 Season Standings. http://rd.pro12rugby.com/matchcentre/table.php?includeref=11189&season=2014-2015. Accessed 03 June 2023
2. Ahn, J.W., Plaisant, C., Shneiderman, B.: A task taxonomy for network evolution analysis. IEEE Trans. Vis. Comput. Graphics **20**(3), 365–376 (2014). https://doi.org/10.1109/TVCG.2013.238
3. Archambault, D., Purchase, H., Pinaud, B.: Animation, small multiples, and the effect of mental map preservation in dynamic graphs. IEEE Trans. Visual Comput. Graphics (2011). https://doi.org/10.1109/TVCG.2010.78
4. Archambault, D., Purchase, H.C.: Mental map preservation helps user orientation in dynamic graphs. In: Didimo, W., Patrignani, M. (eds.) GD 2012. LNCS, vol. 7704, pp. 475–486. Springer, Heidelberg (2013). https://doi.org/10.1007/978-3-642-36763-2_42

5. Archambault, D., Purchase, H.C.: The "map" in the mental map: experimental results in dynamic graph drawing. Int. J. Hum.-Comput. Stud. **71**(11), 1044–1055 (2013)
6. Archambault, D., Purchase, H.C.: Can animation support the visualisation of dynamic graphs? Inf. Sci. **330**, 495–509 (2016). https://doi.org/10.1016/j.ins.2015.04.017
7. Arleo, A., Miksch, S., Archambault, D.: Event-based dynamic graph drawing without the agonizing pain. Comput. Graphics Forum **41**(6), 226–244 (2022). https://doi.org/10.1111/cgf.14615
8. Bach, B., Dragicevic, P., Archambault, D., Hurter, C., Carpendale, S.: A descriptive framework for temporal data visualizations based on generalized space-time cubes. Comput. Graphics Forum **36**(6), 36–61 (2017). https://doi.org/10.1111/cgf.12804
9. Bach, B., Pietriga, E., Fekete, J.D.: GraphDiaries: animated transitions and temporal navigation for dynamic networks. IEEE Trans. Visual Comput. Graphics **20**(5), 740–754 (2014). https://doi.org/10.1109/TVCG.2013.254
10. Baur, M., et al.: Visone software for visual social network analysis. In: Mutzel, P., Jünger, M., Leipert, S. (eds.) GD 2001. LNCS, vol. 2265, pp. 463–464. Springer, Heidelberg (2002). https://doi.org/10.1007/3-540-45848-4_47
11. Beck, F., Burch, M., Diehl, S., Weiskopf, D.: A taxonomy and survey of dynamic graph visualization. Comput. Graphics Forum **36**(1), 133–159 (2017). https://doi.org/10.1111/cgf.12791
12. Brandes, U., Mader, M.: A quantitative comparison of stress-minimization approaches for offline dynamic graph drawing. In: van Kreveld, M., Speckmann, B. (eds.) GD 2011. LNCS, vol. 7034, pp. 99–110. Springer, Heidelberg (2012). https://doi.org/10.1007/978-3-642-25878-7_11
13. Ceneda, D., Arleo, A., Gschwandtner, T., Miksch, S.: Show me your face: towards an automated method to provide timely guidance in visual analytics. IEEE Trans. Visual Comput. Graphics **28**(12), 4570–4581 (2022). https://doi.org/10.1109/TVCG.2021.3094870
14. Ceneda, D., et al.: Characterizing guidance in visual analytics. IEEE Trans. Visual Comput. Graphics **23**(1), 111–120 (2016). https://doi.org/10.1109/TVCG.2016.2598468
15. Ceneda, D., Gschwandtner, T., Miksch, S.: A review of guidance approaches in visual data analysis: a multifocal perspective. Comput. Graphics Forum **38**(3), 861–879 (2019). https://doi.org/10.1111/cgf.13730
16. Eagle, N., Pentland, A.S.: Reality mining: sensing complex social systems. Pers. Ubiquit. Comput. **10**(4), 255–268 (2006)
17. Erten, C., Harding, P.J., Kobourov, S.G., Wampler, K., Yee, G.: GraphAEL: graph animations with evolving layouts. In: Liotta, G. (ed.) GD 2003. LNCS, vol. 2912, pp. 98–110. Springer, Heidelberg (2004). https://doi.org/10.1007/978-3-540-24595-7_9
18. Farrugia, M., Quigley, A.: Effective temporal graph layout: a comparative study of animation versus static display methods. J. Inf. Vis. **10**(1), 47–64 (2011). https://doi.org/10.1057/ivs.2010.10
19. Farrugia, M., Hurley, N., Quigley, A.: Exploring temporal ego networks using small multiples and tree-ring layouts. In: Proceedings of the International Conference on Advances in Computer-Human Interactions (2011)
20. Filipov, V., Arleo, A., Bögl, M., Miksch, S.: On network structural and temporal encodings: a space and time odyssey. IEEE Trans. Visual Comput. Graphics (2023). https://doi.org/10.1109/TVCG.2023.3310019

21. Filipov, V., Arleo, A., Miksch, S.: Are we there yet? A roadmap of network visualization from surveys to task taxonomies. In: Computer Graphics Forum. Wiley Online Library (2023). https://doi.org/10.1111/cgf.14794
22. Filipov, V., Ceneda, D., Archambault, D., Arleo, A.: TimeLighting: guidance-enhanced exploration of 2D projections of temporal graphs (2023). https://arxiv.org/abs/2308.12628
23. Gladisch, S., Schumann, H., Tominski, C.: Navigation recommendations for exploring hierarchical graphs. In: Bebis, G., et al. (eds.) ISVC 2013, Part II. LNCS, vol. 8034, pp. 36–47. Springer, Heidelberg (2013). https://doi.org/10.1007/978-3-642-41939-3_4
24. Holme, P., Saramäki, J.: Temporal networks. Phys. Rep. **519**(3), 97–125 (2012). https://doi.org/10.1007/978-3-642-36461-7
25. Lee, A., Archambault, D., Nacenta, M.: Dynamic network plaid: a tool for the analysis of dynamic networks. In: Proceedings of the 2019 CHI Conference on Human Factors in Computing Systems, pp. 1–14 (2019). https://doi.org/10.1145/3290605.3300360
26. Ljung, P., Krüger, J., Groller, E., Hadwiger, M., Hansen, C.D., Ynnerman, A.: State of the art in transfer functions for direct volume rendering. Comput. Graphics Forum **35**(3), 669–691 (2016). https://doi.org/10.1111/cgf.12934
27. May, T., Steiger, M., Davey, J., Kohlhammer, J.: Using signposts for navigation in large graphs. Comput. Graphics Forum **31**(3pt2), 985–994 (2012). https://doi.org/10.1111/j.1467-8659.2012.03091.x
28. Simonetto, P., Archambault, D., Kobourov, S.: Drawing dynamic graphs without timeslices. In: Frati, F., Ma, K.-L. (eds.) GD 2017. LNCS, vol. 10692, pp. 394–409. Springer, Cham (2018). https://doi.org/10.1007/978-3-319-73915-1_31
29. Simonetto, P., Archambault, D., Kobourov, S.: Event-based dynamic graph visualisation. IEEE Trans. Visual Comput. Graphics **26**(7), 2373–2386 (2018). https://doi.org/10.1109/TVCG.2018.2886901
30. Sondag, M., Turkay, C., Xu, K., Matthews, L., Mohr, S., Archambault, D.: Visual analytics of contact tracing policy simulations during an emergency response. Comput. Graphics Forum **41**(3), 29–41 (2022). https://doi.org/10.1111/cgf.14520
31. Wang, Y., Archambault, D., Haleem, H., Moeller, T., Wu, Y., Qu, H.: Nonuniform timeslicing of dynamic graphs based on visual complexity. In: 2019 IEEE Visualization Conference (VIS), pp. 1–5. IEEE (2019). https://doi.org/10.1109/VISUAL.2019.8933748

Evaluating Animation Parameters for Morphing Edge Drawings

Carla Binucci[1], Henry Förster[2], Julia Katheder[2], and Alessandra Tappini[1](✉)

[1] University of Perugia, Perugia, Italy
carla.binucci@unipg.it
[2] University of Tübingen, Tübingen, Germany
{henry.foerster,julia.katheder}@uni-tuebingen.de,
alessandra.tappini@unipg.it

Abstract. Partial edge drawings (PED) of graphs avoid edge crossings by subdividing each edge into three parts and representing only its stubs, i.e., the parts incident to the end-nodes. The morphing edge drawing model (MED) extends the PED drawing style by animations that smoothly morph each edge between its representation as stubs and the one as a fully drawn segment while avoiding new crossings. Participants of a previous study on MED (Misue and Akasaka, GD19) reported eye straining caused by the animation. We conducted a user study to evaluate how this effect is influenced by varying animation speed and animation dynamic by considering an easing technique that is commonly used in web design. Our results provide indications that the easing technique may help users in executing topology-based tasks accurately. The participants also expressed appreciation for the easing and a preference for a slow animation speed.

Keywords: morphing edge drawings · readability · user study · easing function

1 Introduction

Edge crossings are well-known to reduce the readability and the perceived aesthetics of graph drawings; see, e.g., [25,27]. Bruckdorfer and Kaufmann [6] suggested a quite rigorous solution that avoids crossings in straight-line drawings by drawing edges only partially. More precisely, in a *partial edge drawing* (PED) of a

Research partially supported by: (*i*) University of Perugia, Ricerca di Base 2021, Proj. "AIDMIX—Artificial Intelligence for Decision Making: Methods for Interpretability and eXplainability"; (*ii*) MUR PRIN Proj. 2022TS4Y3N - "EXPAND: scalable algorithms for EXPloratory Analyses of heterogeneous and dynamic Networked Data"; (*iii*) MUR PRIN Proj. 2022ME9Z78 - "NextGRAAL: Next-generation algorithms for constrained GRAph visuALization"; (*iv*) University of Perugia, Ricerca di Base, grant RICBA22CB; (*v*) DFG grant Ka 812-18/2.

M. A. Bekos and M. Chimani (Eds.): GD 2023, LNCS 14465, pp. 246–262, 2023.
https://doi.org/10.1007/978-3-031-49272-3_17

graph, each edge is subdivided into three parts and the middle part is not drawn; the drawn parts are called *stubs*. While the model has received some attention in follow-up studies (see, e.g., [3,5,7–9,11,16,29]), its practical applicability seems to be hindered by the fact that PEDs cannot be read as quickly as traditional node-link diagrams [7]. On the other hand, there are indications that they can be interpreted more accurately than drawings where edges are fully drawn [7].

Recently, Misue and Akasaka [21] suggested to enhance the PED drawing style by a sequence of animations that smoothly morph each edge between its partial representation and its representation as a fully drawn segment. With the resulting drawing style, known as *morphing edge drawing* (MED), the goal is to help the users read drawings more quickly while maintaining the readability offered by PED. A user study by Misue and Akasaka [21] indicates that the model can achieve these goals.

An unfortunate side effect of the addition of the animation appears to be eye fatigue experienced by the users: Namely, while the users of Misue's and Akasaka's study [21] reported that in fact "[morphing] made it easy to confirm the exact adjacency", they also stated that "[their] eyes [were] strained." The users also indicated potential reasons for this negative effect stating that "[it was] messy and difficult to [focus on]", that "the stubs [changed] too fast", and that "the time for stubs to connect [was] too short." The latter two aspects have been considered in Misue's follow-up study [20], where the speed of the morphing animation was substantially reduced while there was a small delay added to the animation at the time that the two stubs meet. It is worth noting that – aside from these two changes – both previous studies on MED [20,21] have mainly focused on minimizing the total animation duration, which is proposed to be a main factor in user response time. However, the existing user study's focus [21] was not to investigate different animation speeds, but to establish the validity of MED in comparison to PED and traditional straight-line drawings.

We also remark that MEDs may be regarded as *staggered* animations between PEDs and traditional straight-line drawings. These types of animations have been previously investigated for morphs between two 2D scatterplots of the same data points [12]. Curiously, in contrast to Misue and Akasaka's findings for MED [21], Chevalier et al. [12] did not observe indications for benefits of staggered morphing animations between two scatterplots.

Our Contribution. We study MEDs from a user-centric point of view. To this end, we investigate in a user study to which extent animation speed influences user response time and accuracy in the execution of tasks. Moreover, to counteract the eye straining reported by the participants of the previous user study [21], we smoothen the animation by an approach called *easing* that is widely used in webdesign [15,18] and experimentally evaluate its impact[1]. We remark that to the best of our knowledge, easing has not been evaluated in the context of graph animations before. Our results indicate that fast animation and easing can help users in performing topology-based tasks. Also, while participants appreciated the easing technique, they preferred a slower speed.

[1] Research data and all stimuli available at https://github.com/tuna-pizza/MEDSupplementaryMaterial.

The paper is structured as follows. Section 2 introduces MEDs and the animation parameters we consider. Section 3 describes the design of our user study. Section 4 discusses the quantitative and the subjective results of our experiment, as well as its limitations. Section 5 lists some future research directions.

2 Morphing Edge Drawings

Model Description. A MED of a graph $G = (V, E)$ is a *dynamic* drawing, i.e., a function $\Gamma : G \times T \rightarrow \mathbb{R}^2$ where T is the time interval of the drawing animation. In this paper, we study straight-line MEDs, where each vertex $v \in V$ is time-independently mapped to a point in the plane and each edge $(u, v) \in E$ is mapped to two fragments of the segment $\overline{uv}$, one incident to u and one incident to v. We call these two subsegments *stubs*, which change length smoothly over time.

Depending on the length changing process, MEDs are *symmetric* if at any time the two stubs of any edge have the same length. Previous studies show that symmetry allows the user to efficiently match stubs belonging to the same edge [3]. We can describe the stub length of an edge (u, v) in a symmetric MED by a *stub length ratio* function $\delta_{(u,v)}(t)$ that describes the ratio between the stub length and the length of the segment $\overline{uv}$. If the minimum of the stub length ratio function of all edges is the same value δ_0, we call the drawing *homogeneous*. We consider only *symmetric* and *homogeneous* MEDs, also known as δ_0-SHMEDs, as it is known that symmetric and homogeneous PEDs are the easiest to read [3].

Similar to previous studies [20,21], we consider only stub length ratio functions that morph stub lengths between two distinct states (namely δ_0 and $1/2$) such that the transition between both states is *strictly monotone*. Namely, we allow such a smooth transition to happen several times for each edge, starting from specific time frames $T_{start}(u, v) \subset T$ during which $\delta_{(u,v)}(t) = \delta_0$. After some predefined elapsed time $\tau_{(u,v)}$ the edge is represented as a straight-line segment, i.e., $\delta_{(u,v)}(t) = 1/2$. Afterwards, for some duration $\tau_{1/2}$, edge (u, v) is completely visible ensuring that the user can see the full edge for some frames in a rendered animation as suggested by Misue [20]. Following this small delay, the animation reverts and $\delta_{(u,v)} = \delta_0$ again after time $\tau_{(u,v)}$. Due to the relation of both parts of the animation, in the following we only describe the first part of the animation. As in previous studies [20], we use the same type of animation for all morphs of an edge, i.e., $\tau_{(u,v)}$ is assumed to be constant. To facilitate distinguishing between several runs of an animation, we allow a new animation to start only after $\tau_{distinct}$ time has elapsed.

If we consider the animations of two edges e_1 and e_2, we may observe that an *avoidable* crossing (between two stubs of length ratio strictly greater than δ_0) can occur if their animations are not scheduled accordingly. We want to avoid that the stubs of e_1 and e_2 ever meet at their avoidable crossing as otherwise they can be perceived as the same object according to Gestalt principles [28]. To counteract this from happening, we also allow e_1 to pass through the same point e_2 has passed through (and vice-versa) only after $\tau_{distinct}$ time has elapsed.

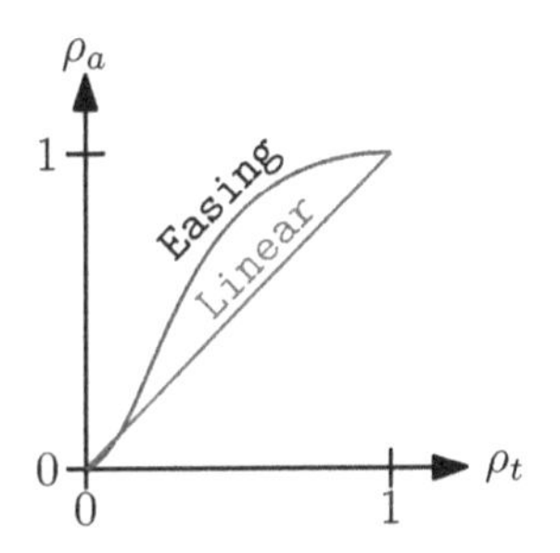

Fig. 1. Easing functions.

Table 1. Parameters used in our experiments.

Model:	SlowLin	SlowEas	FastLin	FastEas
σ_a	100 px/s	100 px/s	200 px/s	200 px/s
η	Linear	Easing	Linear	Easing
$\tau_{1/2}$	100 ms			
$\tau_{distinct}$	50 ms			
δ_0	1/4			

Animation Parameters. We want to describe the morph from $\delta_{(u,v)}(t) = \delta_0$ to $\delta_{(u,v)}(t + \tau_{(u,v)}) = 1/2$ with two easier to understand parameters that we can fix for the *entire* MED so that the aesthetic effect of all edge morphs is similar.

The first parameter is the *speed* of the stub length morphing animation. In the previous studies on MED [20,21], the speed was a constant σ holding for all edges. We can extend the speed parameter to morphings with *non-constant speed* by bounding the *average morphing speed* σ_a or the *maximum morphing speed* σ_m. It is straight-forward to keep σ_a fixed as for fixed σ_a, $\tau_{(u,v)}$ simply depends on the length of segment $\overline{uv}$. Bounding σ_m is more tricky as it requires $\delta_{(u,v)}$ to be differentiable to be efficient to compute. Thus, in this paper, we only consider the average morphing speed σ_a as a parameter.

In addition, we may want to consider animations with a non-constant speed. This may be desirable as humans are used to movements in nature that usually are at first gradually accelerating before reaching a maximum speed before slowing again before coming to a halt. This observation has been considered in multimedia design for a long time [24] and gave rise to so-called *easing* of animations [15,24]. Namely, an *easing function* $\eta : [0,1] \rightarrow [0,1]$ expresses for a given time elapsed ratio ρ_t the progress of the animation ρ_a. In this regard, $\rho_t = 0$ means that at the current time-frame the animation starts (in our case, $t = t_s$ for a $t_s \in T_{start}(u,v)$) whereas $\rho_t = 1$ means that at the current time-frame the animation finishes (in our case, $t = t_s + \tau_{(u,v)}$). Meanwhile $\rho_a = 0$ means that the animation has just started (in our case, the stub length ratio is δ_0) whereas $\rho_a = 1$ means that it has been fully displayed (in our case, the stub length ratio has become 1/2). For our purposes, we want that η is *strictly monotone* and *invertible on the interval* $[0,1]$ to be able to apply the algorithm of Misue and Akasaka [21](refer to [2] for a description of the algorithm).

3 User Study

3.1 Experimental Conditions

We investigate how morphing speed and easing function influence the readability of MEDs. In the first study on MED, σ_a was chosen at $10°$/s to be as fast as possible[2] while remaining human-tractable [21] whereas in the follow-up study [20] it was significantly reduced to 100 px/s. We intend to evaluate MED as a generally applicable tool for graph visualizations in practical applications, where the concrete usage depends on user preferences. As a result, we believe that px/s is a more useful metric than °/s as the latter metric requires to control the relative positioning of the user towards the screen[3]. We investigate two speeds:

`Slow`. Here, we set $\sigma_a = 100$ px/s as in [20].

`Fast`. Here, we set $\sigma_a = 200$ px/s so to have a significantly faster speed as comparison which allows to check how a shorter animation duration at the cost of higher speed influences readability.

In previous studies on MED [20,21], as an easing function, the function $\rho_a = \eta(\rho_t) = \rho_t$, which is also known as the `Linear` easing function, has been used. We believe that the aesthetic appeal of MEDs can be improved by having a non-constant speed allowing for a more natural appearance of the movement of the stubs. As a side effect, edges can be displayed *almost fully drawn* for a longer duration which may allow users to more efficiently observe them; see Fig. 1. There is a plethora of different easing functions in the literature [24]. Currently, functions described by *cubic Bézier curves* appear as the web-design standard [15,18] with a particular focus on the function described by points $P_0 = (0,0)$, $P_1 = (0.25, 0.1)$, $P_2 = (0.25, 1)$ and $P_3 = (1,1)$ which is commonly known also as `Easing`; see Fig. 1. It is worth noting that cubic Bézier curves are defined as a function that maps a parameter p to the tuple (ρ_t, ρ_a). As a result, it is not straight-forward to compute ρ_a in terms of ρ_t [18] and implementations typically solve this problem numerically [17]. For our purposes, we oriented ourselves at the implementation used in the FIREFOX web browser [13].

The combination of speed (`Slow`/`Fast`) and easing function (`Linear`/`Easing`) gives rise to the four models, i.e., the experimental conditions, that we compare in our study; see Table 1. Note that in all models we set $\tau_{1/2} = 100$ ms (as in [20]) and $\tau_{distinct} = 50$ ms. Further, we used $\delta_0 = 1/4$ as in [21].

3.2 Tasks and Hypotheses

We defined five tasks, reported in Table 2, for which we also provide the classification according to the taxonomy by Lee et al. [19]. Figure 2 shows some examples of trials. We designed the tasks so that they require to explore the drawing locally and globally, are easy to explain, can be executed in a reasonably short time, and can be easily measured. Also, most of them have already

[2] The metric °/s refers to the temporal change of the visual angle, which is the size of the image of the object on the observer's retina. Naturally, this metric requires a controlled environment which we believe to conflict with a variance in user preferences.

[3] The concrete effect of a speed measured in px/s vastly depends on the screen resolutions. Hence, we fixed the speed assuming a resolution of 1920 px × 1080 px and scaled the drawings so to cover the same fraction of the screen size on any resolution.

been used in previous graph visualization user studies (e.g., [1,14,23,26]). The main purpose of our study is to evaluate the differences between our experimental conditions in terms of readability, understandability, and effectiveness. For this reason we defined *interpretation tasks*, according to the top-level classification by Burch et al. [10].

Table 2. Tasks used in our experiment.

Task	Classification	Description
T1 (Adjacency)	topology-based (adjacency)	Is there an edge that connects the two blue nodes?
T2 (NeighborhoodSize)	topology-based (adjacency)	How many blue nodes are connected with the orange node?
T3 (PathLength)	topology-based (connectivity)	Is there a path of length at most k that connects the blue node with the orange node?
T4 (CommonNeighbors)	topology-based (common connection)	How many nodes are connected with both the blue nodes?
T5 (InterRegionEdges)	overview	How many edges directly connect the two highlighted parts?

We formulated three experimental hypotheses about the effectiveness of the different conditions to support users in the execution of analysis tasks:

H1. `Fast` speed and cubic Bézier curve `Easing` give lower response time than `Slow` speed and `Linear` easing.
H2. `Slow` speed and cubic Bézier curve `Easing` give lower error rate than `Fast` speed and `Linear` easing.
H3. `Slow` speed and cubic Bézier curve `Easing` are preferred by the users rather than `Fast` speed and `Linear` easing.

Our rationale behind H1 is that the time used for morphing all edges with `Fast` speed is shorter than with `Slow` speed, and cubic Bézier curve `Easing` makes stubs touch for a longer time than with `Linear` easing. About H2, we think that the user can benefit from cubic Bézier curve `Easing` and `Slow` speed in terms of accuracy, since the movement of the stubs is more gradual and connections are visualized for a longer time than with `Linear` easing and `Fast` speed. Finally, about H3 we think that users may find a `Fast` animation more irritating and distracting than a `Slow` one, and the movement of stubs with `Linear` easing less smooth than with cubic Bézier curve `Easing`.

3.3 Stimuli

We investigated our hypotheses on visualizations of real-world networks, more specifically drawn from the annual GD Contest[4]:

[4] See https://mozart.diei.unipg.it/gdcontest/ for further information.

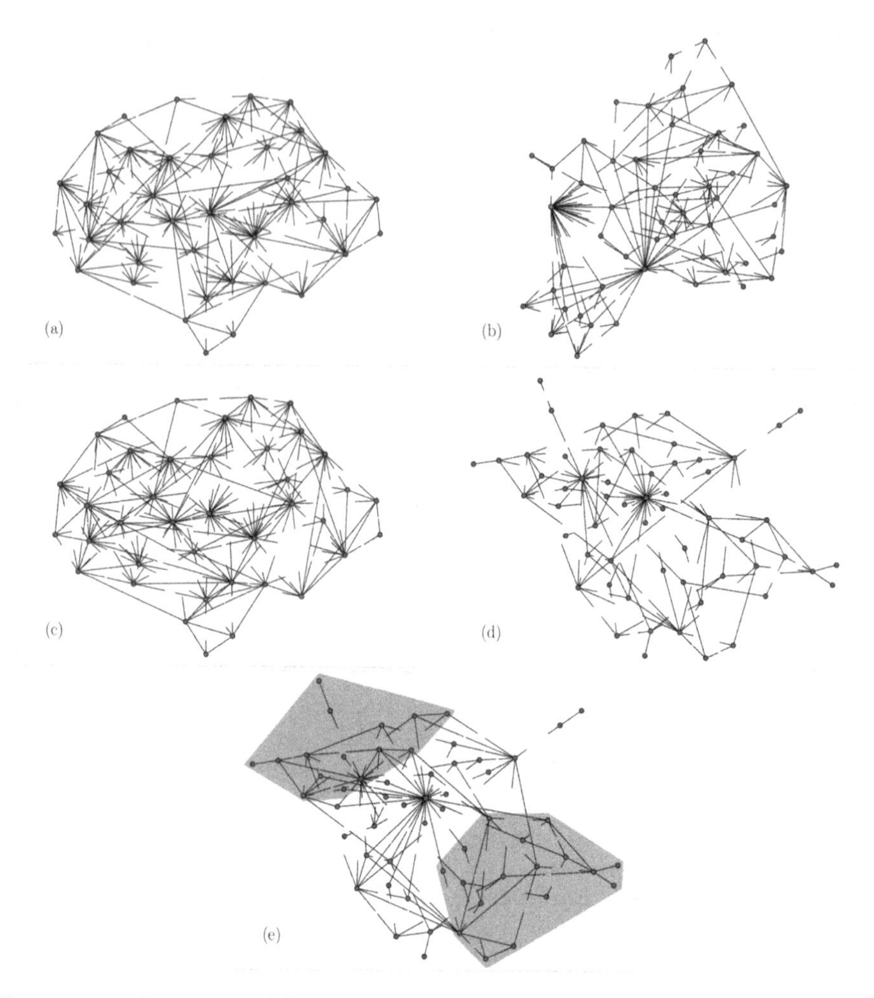

Fig. 2. Examples of trials: (a) T1 (Adjacency), (b) T2 (NeighborhoodSize), (c) T3 (PathLength) with $k = 2$, (d) T4 (CommonNeighbors), (e) T5 (InterRegionEdges).

`boardgames`. This graph from the GD Contest 2023 contains the 40 most popular boardgames as nodes which are connected if fans of one game also liked the other one (data according to https://boardgamegeek.com/).

`marvel`. This bipartite graph from the GD Contest 2019 represents which characters from the Marvel Cinematic Universe are featured in which film.

`kpop`. This graph consists of a subset of the data from the GD Contest 2020 and shows the nodes with betweenness centrality at least 0.3 from the largest connected component of the subgraph induced by K-Pop bands and labels. Connections in this network showcase cooperations between the entities.

We chose the specific networks as they all have a moderate size reasonably displayable on a standard screen (40 to 58 nodes) and offer various densities (ranging from 2.12 to 5.35). We embedded the stimuli graphs using a force directed algorithm available in the `D3.js` library [4]. The resulting drawings roughly required the same area; see Table 3.

Table 3. Data about the stimuli used in our experiment.

Stimulus	Nodes	Edges	Density	Resolution	`SlowLin`	`SlowEas`	`FastLin`	`FastEas`
`boardgames`	40	214	5.35	1030×820 px	7.79s	8.74s	4.25s	4.87s
`marvel`	52	152	2.92	846×974 px	4.92s	5.61s	2.66s	3.08s
`kpop`	58	123	2.12	1142×858 px	5.31s	7.09s	3.43s	3.55s

In our drawings, nodes are represented as disks of radius 7 px filled gray, blue or orange (where the colored variants are used to highlight nodes of interest for the tasks), while the boundaries of the disks and edges are drawn black and 2 px wide. The animations were scheduled using the algorithm of Misue and Akasaka [21] with an improvement of [20] (see [2] for details) and rendered at 30 fps. The resulting animations for `SlowEas` (`FastEas`) were on average 20% (11%, resp.) slower than the corresponding animations for `SlowLin` (`FastLin`, resp.). Comparing the different speeds, the animations for `SlowLin` (`SlowEas`) were on average 74% (87%, resp.) slower than the ones for `FastLin` (`FastEas`, resp.); see Table 3. We produced a trial for each combination of network and task, so to obtain a total of 15 trials for each experimental condition.

3.4 Experimental Process

For our experiment, we opted for a between-subject design where each participant is exposed to one of the four conditions and hence to 15 trials. Indeed, a within-subject design would imply that each user sees the same experimental object 20 times, which makes it difficult to avoid the learning and the fatigue effect. In addition, the between-subject study design allowed us to query subjective feedback from the users about a specific condition after a series of trials. The users performed an online test prepared with the LimeSurvey tool (https://www.limesurvey.org/).

The phases of the survey are the following:

1. we collect some information about the user;
2. we assign the condition to the user by adopting a round robin approach;
3. we present a video tutorial and some training questions;
4. we present the 15 trials in random order;
5. we ask for some subjective feedback: (*i*) Two 5-point Likert scale questions about the beauty of the drawings and the easiness of the questions, (*ii*) one question about the perceived speed of the movements, (*iii*) a question asking users if they found the tasks tiring, and (*iv*) an optional free-form feedback.

Table 4. Data about the participants.

Gender				**Age**		**Screen size**	
Woman	*Man*	*Transgender*	*Prefer not to say*	*18-24*	21%	*<13"*	2%
31%	63%	1%	5%	*25-29*	26%	*13"*	13%
Educational level				*30-34*	21%	*14"*	21%
High school	*Bachelor*	*Master*	*Doctoral degree*	*35-39*	8%	*15"*	20%
14%	27%	30%	29%	*40-44*	7%	*>15"*	38%
Expertise				*45-49*	5%	*No answer*	5%
None	*Low*	*Medium*	*High*	*50-59*	5%		
12%	31%	26%	31%	*60+*	6%		

We used the gdnet, ieee_vis, and infovis mailing lists to recruit participants. Also, we involved our engineering and CS students of the universities of Perugia and Tübingen.

4 Results and Discussion

We collected questionnaires from 84 participants (21 per condition); see Table 4 for some details.

4.1 Quantitative Results

For each question, we recorded answers and response times of all participants. For T1 (Adjacency) and T3 (PathLength), the error rate of a user is computed as the ratio between the number of wrong answers and the total number of questions, that is 3 for each task. For T2 (NeighborhoodSize), T4 (CommonNeighbors), and T5 (InterRegionEdges), the error on a question is computed as $1 - \frac{1}{1+|v_u - v_c|}$, where v_u is the value given by the user and v_c is the correct value.

We performed a Shapiro-Wilk test [30,31] and we found that the populations were normally distributed for tasks T2 (NeighborhoodSize), T4 (CommonNeighbors), and T5 (InterRegionEdges). We thus performed a one-way ANOVA test and post-hoc pairwise comparisons of the considered placement strategies by using Tukey's HSD test [22]. Regarding T1 (Adjacency) and T3 (PathLength), we performed the non-parametric Kruskal-Wallis test. For all the above tests, the significance level was set to $\alpha = 0.05$. From the tests, we did not observe any statistically significant difference between the four experimental conditions. However, by looking at the box-plots of the response time and error rate, which are reported in Figs. 3 and 4, one can observe interesting phenomena[5].

Regarding response time, the plots in Fig. 3 show that for all tasks but T5 (InterRegionEdges), `Fast` speed and cubic Bézier curve `Easing` perform better than `Slow` speed and `Linear` easing, respectively. This behavior

[5] Observe that the scales on the y-axes in Figs. 3 and 4 differ between subfigures.

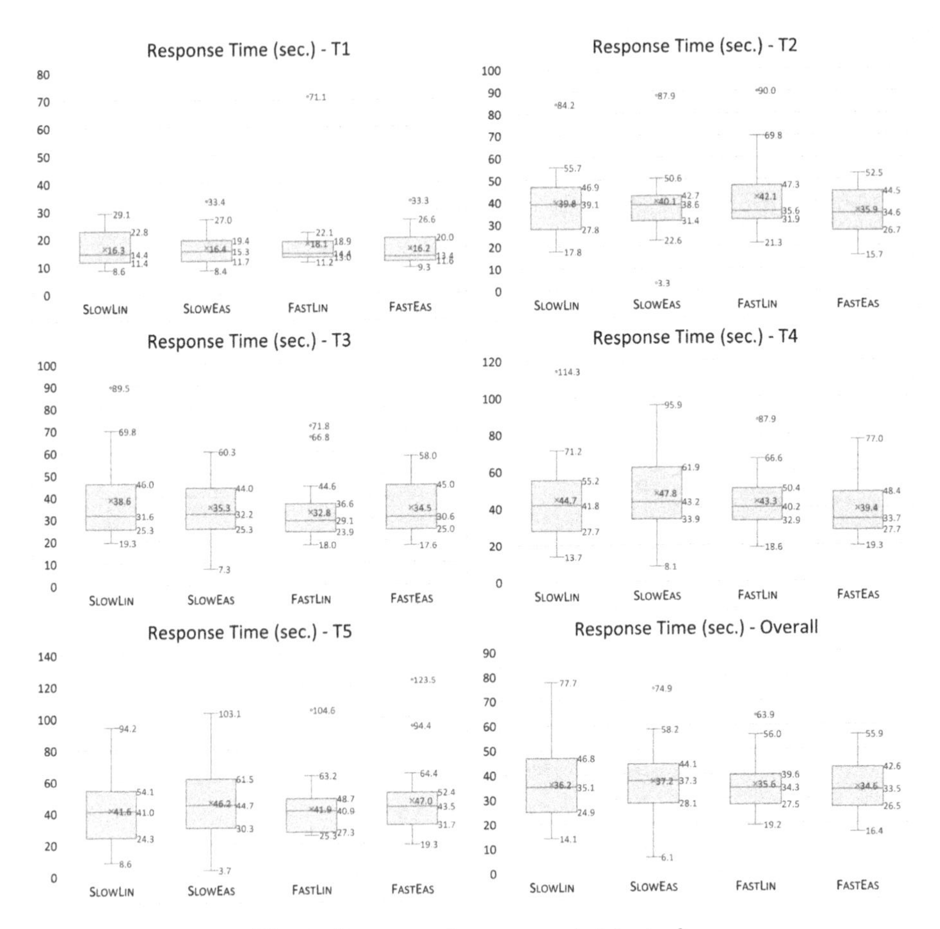

Fig. 3. Response time aggregated by task.

can also be observed over all tasks, but cubic Bézier curve `Easing` seems to perform better than `Linear` easing only if combined with `Fast` speed (`FastEas` model). A similar trend is exhibited by the error rate reported in Fig. 4. In particular, for tasks T2 (NeighborhoodSize), T3 (PathLength), and T4 (CommonNeighbors) cubic Bézier curve `Easing` performs better than `Linear` easing in the case of equal speed. Although average values are very low for T1 (Adjacency), they show slightly better performance with `Fast` speed and cubic Bézier curve `Easing` than with `Slow` speed and `Linear` easing, respectively. This behavior can also be observed over all tasks.

4.2 Subjective Results

The percentage of answers to the Likert scale questions are reported in Figs. 5(a) and 5(b) (the answer distributions as box-plots can be found in [2]); we assigned a score from 1 (lowest) to 5 (highest) to each answer. While there is no statistically significant difference among the conditions, we can observe the following.

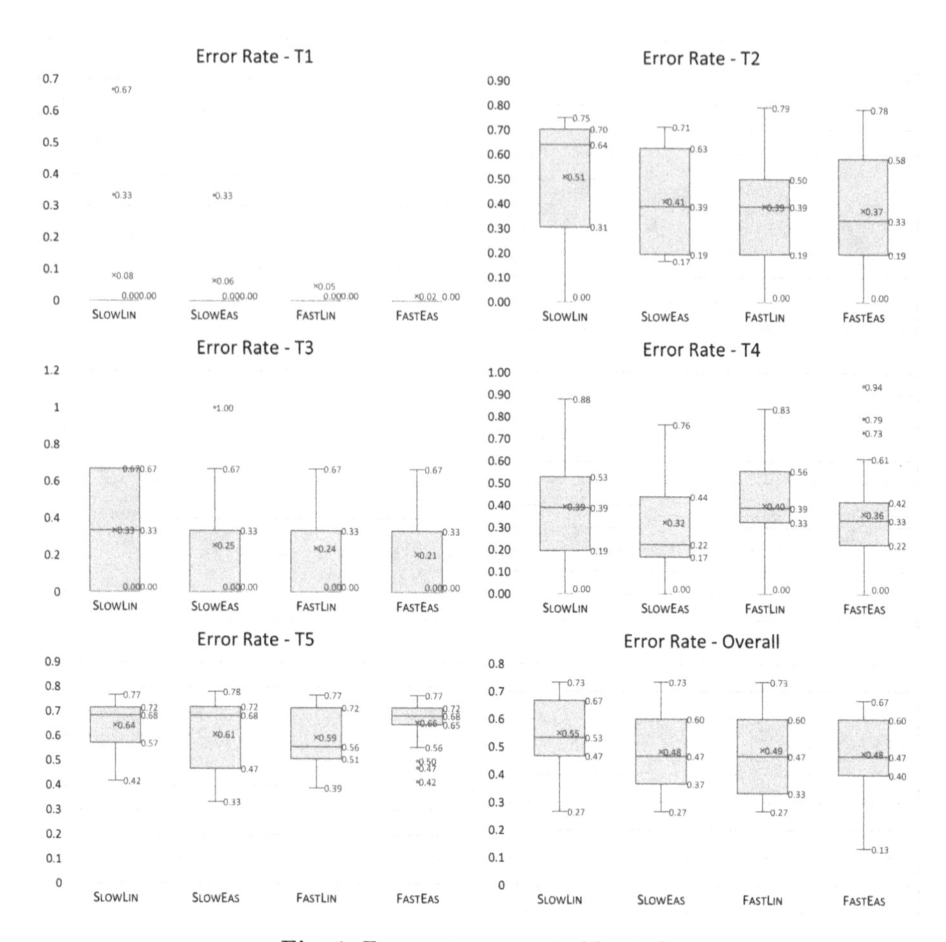

Fig. 4. Error rate aggregated by task.

Regarding beauty, users seem to prefer `Slow` speed and cubic Bézier curve `Easing` rather than `Fast` speed and `Linear` easing. Indeed, the models with `Slow` speed received a percentage of positive appreciations (4 and 5) that is higher than the one obtained by the models with `Fast` speed, and the models with cubic Bézier curve `Easing` received a higher percentage of positive appreciations than the models with `Linear` easing. Also, the average ratings suggest that the preferred model is `SlowEas`. About easiness, on average `FastEas` and `SlowLin` are the models that were found to be most difficult and easiest, respectively.

The percentage of users who considered the speed adequate is on average higher for the models with `Slow` speed than for the ones with `Fast` speed; see Fig. 5(c). The percentage of users who found it tiring to work with `Fast` speed is higher than the one of users who worked with `Slow` speed; see Fig. 5(d). Among the slow models, `Linear` easing seems to produce visualizations that are rated as less tiring than cubic Bézier curve `Easing`.

Additional user feedback. Overall 42% of users submitted additional feedback amounting to 35 submissions, while the proportion of submissions in each group

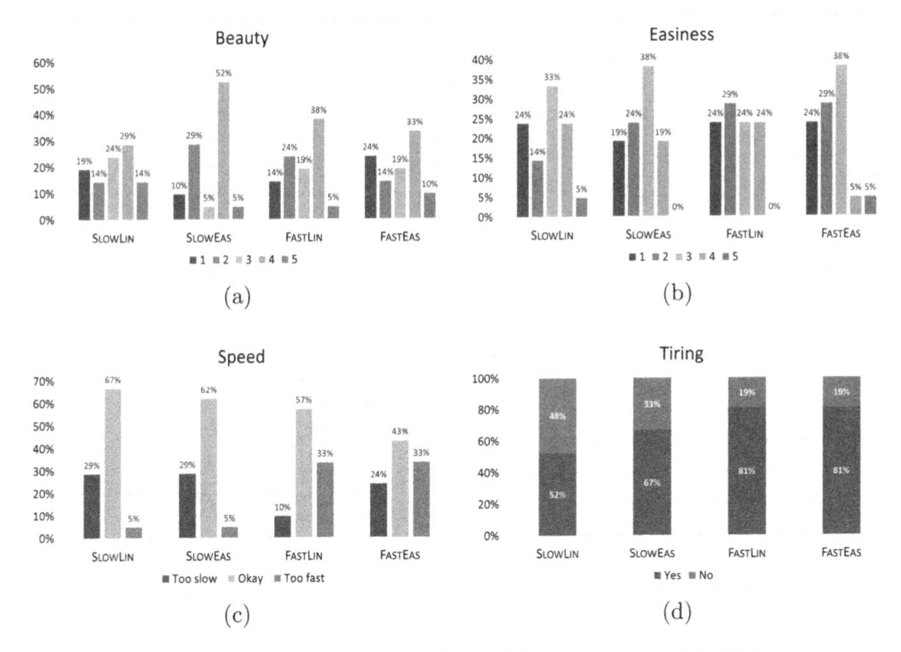

Fig. 5. Subjective results. In (a) and (b), 1=lowest and 5=highest.

was rather balanced, with `SlowEas` and `FastLin` giving the most comments (10 each); see Fig. 6(a). We performed a thematic analysis on the users' feedback and we identified four recurring themes, namely *subjective impressions and emotional impact* (30), closely followed by *challenges and obstructions* when performing the study (29), *suggestions for improvements* (17), and *positive feedback to the proposed model* (11); refer to Fig. 6(b). The different groups have largely made similar points, mostly with no clear trend to be noted. For a breakdown of each theme by model, see Figs. 6(c)–6(d).

Subjective Impressions and Emotional Impact. Negative emotions (5 for `Easing`, 19 for `Linear`) were reported more often than positive ones (6). The dominant factor was a sense of difficulty/frustration (10), followed by exhaustion/nausea (5), and doubting practicality (4, `Slow` only). Other negative sentiments are irritation/confusion (3, `Slow` only) and stress/overwhelm (2, `Fast` only). On the positive side, users reported a liking (3), a sense of learning (2), and fun (1).

Challenges and Obstructions. Users mostly described that waiting for the animation of an edge hindered them when performing the tasks (7, mostly `Easing`). Second most reported were technical issues (5, mostly delay or glitches in the animation), followed by a distraction due to the simultaneous movement (4). Other challenges were the general difficulty to find graph structures requested by the tasks (3), miscounting or losing progress (3, `Fast` only), discerning adjacency

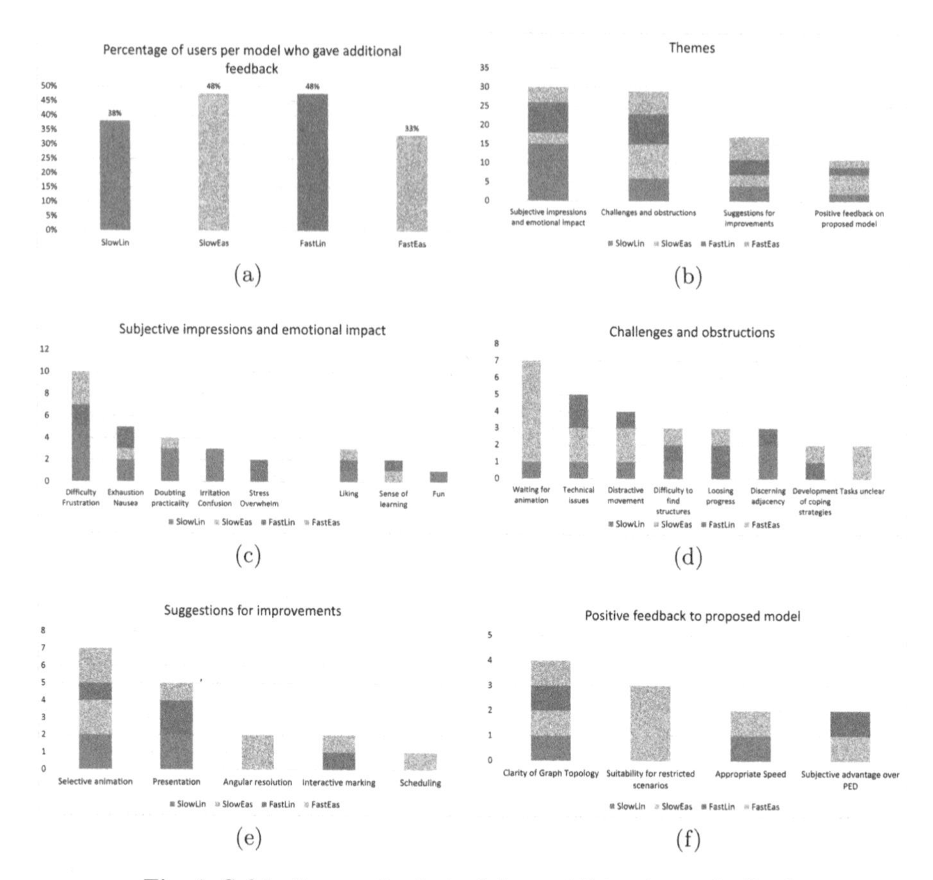

Fig. 6. Subjective results derived from additional user feedback.

due to nodes placement (3), finding coping mechanisms to alleviate difficulty (2, `Fast` only) and not understanding task descriptions (2).

Suggestions for Improvements. Users most frequently requested control over the animations (7, e.g., through hovering or clicking). Further, it was suggested to use more colors and different line-widths to highlight relevant structures (5). Concerning network layout, two users in the `FastEas` group suggested that a better angular resolution might improve task performance. The second requested interactive feature was the possibility to mark nodes for tracking progress (2, `Fast` only). Lastly, there was one call for better animation scheduling.

Positive Feedback to Proposed Model. Most often it was mentioned that the animation made the recognition of the graph topology clearer (4), followed by the potential of the model in restricted scenarios, i.e., for certain graphs or tasks (3, `Easing` only). For the `Slow` speed, two users found the speed to be appropriate. Finally, two users experienced technical issues with animation: They continued solving the tasks as for PEDs, which they described as much harder than MEDs.

4.3 Discussion of Our Results

We did not observe statistically significant differences between the four models in terms of response time and error rate, which suggests that none of the models clearly outperforms the others in terms of readability. However, an analysis of the aggregate data provides some initial indications that cubic Bézier curve `Easing` may help users in executing topology-based tasks in an accurate manner. We do not have any indications for the overview task (i.e., T5 (InterRegionEdges)), whose behavior seems not to be in line with the other tasks and the error rate is very high for all models (the percentages of wrong answers for the `marvel`, `boardgames`, and `kpop` networks are 100%, 99%, and 75%, respectively). Our interpretation is that this task is not suitable for MEDs, since the connections between the boundaries of the highlighted parts of the network appear and disappear repeatedly, thus distracting the user and making it hard to keep track of the previously considered edges. Moreover, while we observed slightly lower response times for the `Fast` animation speed and the cubic Bézier curve `Easing` the collected data does not suggest a model clearly outperforming the others in terms of response time.

Referring to our hypotheses, the data thus provide indications that only partially support H1 and H2. In particular, `Slow` speed does not give a lower error rate than `Fast` speed, while cubic Bézier curve `Easing` behaves better than `Linear` easing. Our interpretation for this phenomenon is that, when analyzing a connection between two nodes, the waiting time for the edge to appear might be long, and this is more evident with a slow animation. Further, the differences between `Linear` and cubic Bézier curve `Easing` may be caused by the circumstance that the longer time when stubs are drawn almost fully in cubic Bézier curve `Easing` helped the users when performing the tasks. Our subjective results support H3. The analysis of the users' free-form feedback suggests that the cubic Bézier curve `Easing` is preferred over `Linear` easing. Concerning the speed, the subjective feedback favors the `Slow` speed for beauty, speed appropriateness and tiredness, while the easiness of the tasks shows no clear trend. We believe that there are two different key effects at play here: First, the higher waiting times for the `Slow` speed are perceived by the users as a distracting factor as also mentioned in the free-form answers. Second, the `Fast` speed shows animations in different parts of the drawing in rapid succession, which may be overwhelming and lead to miscounting or losing progress. Thus, we think that an intermediate speed between 100 and 200 px/s would be most appropriate. This seems to be also partially supported by the answers about speed; see Fig. 5(c). Also, note that `FastEas` has the highest maximum animation speed and the lowest ratings as an *Okay* speed.

4.4 Limitations of Our Experiment

First, we did not observe statistically significant differences between the four models in terms of response time and error rate, which indicates that none of the models outperforms the others in terms of readability. Possible differences between the models could emerge with a higher number of participants or by considering a different study design, i.e., a within-subject experiment.

Second, we noticed that the only overview task that we considered (T5 (InterRegionEdges)) turned out to be very challenging for the users, which made us conclude that MEDs are not a good paradigm to perform such a task. Perhaps, the choice of different overview tasks would have allowed us a better evaluation.

Moreover, the choice of not allowing interaction implied to use networks that fit into a standard screen, thus facilitating the execution of an online test. We think that enabling interaction may allow the evaluation of our models on larger networks and could facilitate the users in the execution of tasks, e.g. by employing selective animation and marking. On the other hand, we believe that this would preferably require a controlled experiment study design.

Finally, the readability of MEDs may be sensitive to the scheduling algorithm used for the animation of edges and to the one used to produce the layout. This justifies further investigation with different scheduling and layout algorithms.

5 Conclusions and Future Research Directions

We presented a user study that evaluates the readability of MEDs by comparing different speeds and easing functions for the animation of edges. Since our results do not exhibit statistically significant differences from a task performance point of view, we suggest to follow user preferences. Namely, users tend to favor the cubic Bézier curve `Easing` and rate slower animation speed as more beautiful and appropriate. Thus, choosing an intermediate animation speed may be the preferred option in practical applications.

Our study has some limitations and cannot be generalized to settings significantly different from ours. This motivates further experiments with larger networks, additional tasks, different scheduling algorithms for the animation of edges, and interaction features. Regarding the last aspect, we believe that the incorporation of user interaction can significantly improve the model by only showing animations of edges or nodes that are selected by the users. We emphasize that this has also been suggested by some participants of our experiment.

References

1. Binucci, C., Didimo, W., Kaufmann, M., Liotta, G., Montecchiani, F.: Placing arrows in directed graph layouts: algorithms and experiments. Comput. Graph. Forum **41**(1), 364–376 (2022). https://doi.org/10.1111/cgf.14440
2. Binucci, C., Förster, H., Katheder, J., Tappini, A.: Evaluating animation parameters for morphing edge drawings. CoRR 2309.00456 (2023). http://arxiv.org/abs/2309.00456
3. Binucci, C., Liotta, G., Montecchiani, F., Tappini, A.: Partial edge drawing: Homogeneity is more important than crossings and ink. In: Bourbakis, N.G., Tsihrintzis, G.A., Virvou, M., Kavraki, D. (eds.) 7th International Conference on Information, Intelligence, Systems & Applications, IISA 2016, pp. 1–6. IEEE (2016). https://doi.org/10.1109/IISA.2016.7785427

4. Bostock, M., Ogievetsky, V., Heer, J.: D^3 data-driven documents. IEEE Trans. Vis. Comput. Graph. **17**(12), 2301–2309 (2011). https://doi.org/10.1109/TVCG.2011.185
5. Bruckdorfer, T., et al.: Progress on partial edge drawings. J. Graph Algorithms Appl. **21**(4), 757–786 (2017). https://doi.org/10.7155/jgaa.00438
6. Bruckdorfer, T., Kaufmann, M.: Mad at edge crossings? Break the edges! In: Kranakis, E., Krizanc, D., Luccio, F. (eds.) FUN 2012. LNCS, vol. 7288, pp. 40–50. Springer, Heidelberg (2012). https://doi.org/10.1007/978-3-642-30347-0_7
7. Bruckdorfer, T., Kaufmann, M., Leibßle, S.: PED user study. In: Di Giacomo, E., Lubiw, A. (eds.) GD 2015. LNCS, vol. 9411, pp. 551–553. Springer, Cham (2015). https://doi.org/10.1007/978-3-319-27261-0_47
8. Bruckdorfer, T., Kaufmann, M., Montecchiani, F.: 1-bend orthogonal partial edge drawing. J. Graph Algorithms Appl. **18**(1), 111–131 (2014). https://doi.org/10.7155/jgaa.00316
9. Burch, M.: A user study on judging the target node in partial link drawings. In: 21st International Conference Information Visualisation, IV 2017, pp. 199–204. IEEE Computer Society (2017). https://doi.org/10.1109/iV.2017.43
10. Burch, M., Huang, W., Wakefield, M., Purchase, H.C., Weiskopf, D., Hua, J.: The state of the art in empirical user evaluation of graph visualizations. IEEE Access **9**, 4173–4198 (2021). https://doi.org/10.1109/ACCESS.2020.3047616
11. Burch, M., Vehlow, C., Konevtsova, N., Weiskopf, D.: Evaluating partially drawn links for directed graph edges. In: van Kreveld, M., Speckmann, B. (eds.) GD 2011. LNCS, vol. 7034, pp. 226–237. Springer, Heidelberg (2012). https://doi.org/10.1007/978-3-642-25878-7_22
12. Chevalier, F., Dragicevic, P., Franconeri, S.: The not-so-staggering effect of staggered animated transitions on visual tracking. IEEE Trans. Vis. Comput. Graph. **20**(12), 2241–2250 (2014). https://doi.org/10.1109/TVCG.2014.2346424
13. gecko dev: Parametric bézier curves. https://github.com/mozilla/gecko-dev/blob/master/servo/components/style/bezier.rs. Accessed 23 May 2023
14. Di Giacomo, E., Didimo, W., Liotta, G., Montecchiani, F., Tappini, A.: Comparative study and evaluation of hybrid visualizations of graphs. IEEE Trans. Visual. Comput. Graph., 1–13 (2022). https://doi.org/10.1109/TVCG.2022.3233389
15. MDN web docs: <easing-function>, https://developer.mozilla.org/en-US/docs/Web/CSS/easing-function. Accessed 28 Apr 2023
16. Hummel, M., Klute, F., Nickel, S., Nöllenburg, M.: Maximizing ink in partial edge drawings of k-plane graphs. In: Archambault, D., Tóth, C.D. (eds.) GD 2019. LNCS, vol. 11904, pp. 323–336. Springer, Cham (2019). https://doi.org/10.1007/978-3-030-35802-0_25
17. Łukasz Izdebski: Bezier curve as easing function (2020). https://asawicki.info/articles/Bezier_Curve_as_Easing_Function.htm
18. Izdebski, Ł, Kopiecki, R., Sawicki, D.: Bézier curve as a generalization of the easing function in computer animation. In: Magnenat-Thalmann, N., et al. (eds.) CGI 2020. LNCS, vol. 12221, pp. 382–393. Springer, Cham (2020). https://doi.org/10.1007/978-3-030-61864-3_32
19. Lee, B., Plaisant, C., Parr, C.S., Fekete, J., Henry, N.: Task taxonomy for graph visualization. In: Proceedings of the 2006 AVI Workshop on BEyond Time and Errors: Novel Evaluation Methods for Information Visualization, BELIV 2006, pp. 1–5 (2006). https://doi.org/10.1145/1168149.1168168
20. Misue, K.: Improved scheduling of morphing edge drawing. In: Angelini, P., von Hanxleden, R. (eds.) GD 2022. LNCS, vol. 13764, pp. 336–349. Springer, Cham (2022). https://doi.org/10.1007/978-3-031-22203-0_24

21. Misue, K., Akasaka, K.: Graph drawing with morphing partial edges. In: Archambault, D., Tóth, C.D. (eds.) GD 2019. LNCS, vol. 11904, pp. 337–349. Springer, Cham (2019). https://doi.org/10.1007/978-3-030-35802-0_26
22. Montgomery, D.C. (ed.): Design and Analysis of Experiments. Wiley, Hoboken (2012)
23. Okoe, M., Jianu, R., Kobourov, S.G.: Node-link or adjacency matrices: old question, new insights. IEEE Trans. Vis. Comput. Graph. **25**(10), 2940–2952 (2019). https://doi.org/10.1109/TVCG.2018.2865940
24. Penner, R.: Motion, tweening, and easing. In: Robert Penner's Programming Macromedia Flash MX. Osborne/McGraw-Hill (2002)
25. Purchase, H.C.: Which aesthetic has the greatest effect on human understanding? In: Graph Drawing, 5th International Symposium, GD 1997, Rome, Italy, 18–20 September 1997, Proceedings, pp. 248–261 (1997). https://doi.org/10.1007/3-540-63938-1_67
26. Purchase, H.C.: Performance of layout algorithms: comprehension, not computation. J. Vis. Lang. Comput. **9**(6), 647–657 (1998). https://doi.org/10.1006/jvlc.1998.0093
27. Purchase, H.C., Carrington, D.A., Allder, J.: Empirical evaluation of aesthetics-based graph layout. Empir. Softw. Eng. **7**(3), 233–255 (2002)
28. Rusu, A., Fabian, A.J., Jianu, R., Rusu, A.: Using the gestalt principle of closure to alleviate the edge crossing problem in graph drawings. In: 2011 15th International Conference on Information Visualisation (IV), pp. 488–493. IEEE (2011)
29. Schmauder, H., Burch, M., Weiskopf, D.: Visualizing dynamic weighted digraphs with partial links. In: Braz, J., Kerren, A., Linsen, L. (eds.) IVAPP 2015 - Proceedings of the 6th International Conference on Information Visualization Theory and Applications, pp. 123–130. SciTePress (2015). https://doi.org/10.5220/0005303801230130
30. Shapiro, S.S., Wilk, M.B.: An analysis of variance test for normality (complete samples). Biometrika **52**(3–4), 591–611 (1965)
31. Thode, H.C.: Testing for normality. Marcel Dekker (2002)

Balancing Between the Local and Global Structures (LGS) in Graph Embedding

Jacob Miller(✉), Vahan Huroyan, and Stephen Kobourov

University of Arizona, Tucson, USA
jacobmiller1@arizona.edu, vahanhuroyan@math.arizona.edu,
kobourov@cs.arizona.edu

Abstract. We present a method for balancing between the Local and Global Structures (LGS) in graph embedding, via a tunable parameter. Some embedding methods aim to capture global structures, while others attempt to preserve local neighborhoods. Few methods attempt to do both, and it is not always possible to capture well both local and global information in two dimensions, which is where most graph drawing live. The choice of using a local or a global embedding for visualization depends not only on the task but also on the structure of the underlying data, which may not be known in advance. For a given graph, LGS aims to find a good balance between the local and global structure to preserve. We evaluate the performance of LGS with synthetic and real-world datasets and our results indicate that it is competitive with the state-of-the-art methods, using established quality metrics such as stress and neighborhood preservation. We introduce a novel quality metric, cluster distance preservation, to assess intermediate structure capture. All source-code, datasets, experiments and analysis are available online.

Keywords: Graph embedding · Graph Visualization · Local and global structures · Dimensionality Reduction · Multi-dimensional Scaling

1 Introduction

Graphs and networks are a powerful tool to encode relationships between objects. Graph embeddings, which map the vertices of a graph to a set of low dimensional vectors (real valued coordinates), are often used in the context of data visualization to produce node-link diagrams. While many layout methods exist [27], dimension reduction (DR) techniques have had success in providing desirable layouts, by capturing graph structure in reasonable computation times. DR methods are used to project high-dimensional data into low-dimensional space and some of these methods only rely on the relationships between the datapoints, rather than *datapoint coordinates* in higher dimension. These techniques are applicable for both graph embeddings and visualization. Further, local DR algorithms attempt to preserve the local neighborhoods, while global DR algorithms attempt to retain all pairwise distances.

M. A. Bekos and M. Chimani (Eds.): GD 2023, LNCS 14465, pp. 263–279, 2023.
https://doi.org/10.1007/978-3-031-49272-3_18

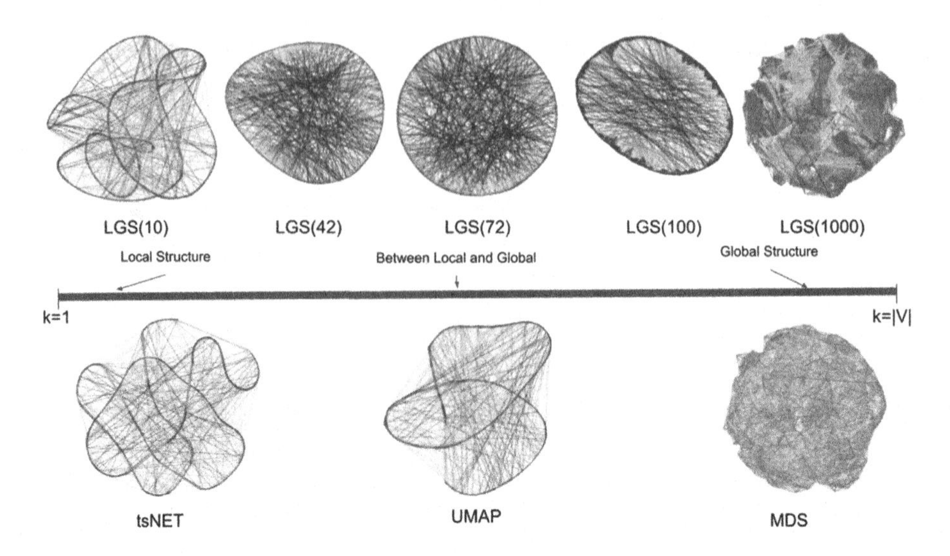

Fig. 1. Embeddings of the connected_watts_1000 graph; see Sect. 4. The top row shows LGS embeddings – from local to global – with varying neighborhood sizes (k). The LGS(72) layout captures the correct underlying model. The bottom row shows tsNET [15], UMAP [18], and MDS [30] embedding of the same graph.

Two popular techniques that are adapted in graph visualization are (metric) Multi-Dimensional Scaling (MDS) [5,16] and t-distributed stochastic neighbor embedding (t-SNE) [17]. The goals of these two algorithms are somewhat orthogonal: MDS focuses on preserving all pairwise distances, while t-SNE aims to preserve the likelihood of points being close in the embedding if they were close in the original space. MDS is said to preserve *global* structure, while t-SNE is said to preserve *local* neighborhoods [7]. These ideas are directly applicable to graph visualization, where we can define the distances as the graph theoretic distances, e.g., via all-pairs shortest paths (APSP) computation. In the graph layout literature, MDS is often referred to as stress minimization [11,30], and t-SNE has been adapted to graph layout in an algorithm known as tsNET [15] and later DRGraph [31]. Choosing the "best" graph embedding algorithm depends on the graph structure and the task. MDS is effective for structured/mesh-like graphs, while t-SNE works better for clustered/dense graphs. This phenomenon also applies to local and global force-directed layouts as well [15].

Automating the selection of the "best" embedding algorithm is challenging due to its dependency on graph structure. We introduce the Local-to-Global Structures (LGS) algorithm which provides a parameter-tuneable framework that can produces embeddings that span the spectrum from local optimization to global optimization.

Smaller values of the LGS parameter prioritize local structure, while larger values emphasize global structure. LGS enables exploration of the trade-off, revealing meaningful middle ground solutions. We introduce a new metric called *cluster distance* to measure how well this intermediate structure is preserved. Everything described in this paper is available on Github: https://github.com/Mickey253/L2G. We provide a video and additional layouts and analysis in [19].

2 Background

Dimensionality Reduction (DR) refers to a large family of algorithms that map a set of high-dimensional datapoints in lower-dimensionsal space. Different DR algorithms aim to preserve various properties of the dataset, such as total variance, global distances, local distances, etc. In visualization contexts, the dataset is typically projected onto 2D or 3D Euclidean space. DR algorithms generally accept input of two types: sample or distance. Sample-based algorithms, such as Principal Component Analysis (PCA) [8,13] project the high dimensional data down to the embedding space. For distance-based inputs, the algorithms directly work with distance metrics. In the case of graph embeddings, the graph-theoretic distance is used, often all-pairs shortest path (APSP).

Popular techniques in the local category include t-SNE [17], UMAP [18], LLE [25], IsoMap [28], etc. For global structure, methods such as PCA [8] and MDS [5,16] are used. MDS has variants, but here we mean metric MDS which minimizes stress [26]. Few techniques attempt to capture both global and local structure. Chen and Buja [3] adapt MDS to capture local structure by selectively preserving distances between a subset of pairs using kNN. The underlying idea is similar to ours , but it does not provide a framework to cover the spectrum from local to global as our method does. While t-SNE's perplexity parameter aims to imitate the size of neighborhood to be preserved, in general increasing its value does not lead to a global structure preservation [29]. Anchor-t-SNE improves the global structure preservation by anchoring a set of points to use as a skeleton for the rest of the embedding [9], however, it does not provide a framework to cover the spectrum from local to global. UMAP [12,18] also aims to preserve the local structures of a dataset. While UMAP claims to preserve the global structures better than t-SNE, we show that this is not universally true for graph data in Sect. 5.

Graph Embedding is a problem to assign vectors to graph vertices, capturing the graph structure. More formally, given a graph G = (V, E), find a d-dimensional vector representation of V that optimally preserves properties [2] (e.g., pairwise distances in MDS [5,16]). We restrict ourselves to 2D node-link visualization with edges represented by straight-line segments, so the problem is reduced to finding a 2D embedding for the vertices. Aesthetic criteria are often used to evaluate the quality of a graph embedding: the number of edge crossings, average edge length, overall symmetry, etc. [24]. Aesthetic criteria enhance readability and task facilitation, but information *faithfulness* is equally important. It ensures that the embedding accurately represents all underlying data, regardless of the task [20,21] and graph embeddings provide a nice benefit by directly optimizing graph structure preservation. *Graph structure* is a nebulous term; referring to inherent properties of the underlying graph such as local/global distances. *Global distance* preservation methods capture the graph's topological structure by closely aligning embedded distances with graph-theoretic distances. This approach is ideal for connectivity-based tasks and offers insights into the global scale and shape of the data. *Local structure* preservation methods preserve the immediate neighborhood of each vertex, effectively capturing clusters or densely connected subgraphs. While nearby vertices in the embedding can be considered similar, distant vertices may have irrelevant distances. This can be observed in the presence of long edges in the local embedding column in Fig. 2.

Graph Embedding by Dimensionality Reduction: In a good embedding, the drawn distance should closely match the graph-theoretic distance between vertices [14]. This observation led to the use of stress function, which MDS aims to optimize, to obtain a graph embedding [11]. Stress can be minimized by majorization [11], stochastic gradient descent (SGD) [30], etc. The MDS approach suffers from an APSP computation, which usually relies on Floyd-Warshall's $O(|V|^3)$, or on Johnson's $O(|V|^2 \log |V| + |E||V|)$ algorithms. The maximum entropy model (MaxEnt) [10] adds a negative entropy between vertices in the graph. The motivation for MaxEnt is to improve the asymptotic complexity. The MaxEnt model places neighbor nodes closer while maximizing the distance between all vertices. This is conceptually similar to the LMDS of Chen and Buja [3]. Our approach differs from the MaxEnt model in motivation: Our LGS captures local structure, global structure, or balances between the two, whereas MaxEnt is primarily concerned with speed. We cannot avoid an APSP computation, and make use of SGD to optimize our objective function in lieu of majorization.

Optimizing stress creates effective layouts, but may neglect local structures; see Fig. 2. tsNET [15] captures local structure by also adding a repulsive force between vertices to achieve cluster separation. tsNET has been sped up by making use of negative sampling and sparse approximation to avoid the APSP computation [31]. Nocaj et al. [22] achieve effects similar to tsNET by weighting edges based on "edge embeddedness" and perform MDS on the weighted graph.

3 The Local-to-Global Structures (LGS) Algorithm

Local methods (e.g., t-SNE) preserve local neighborhoods, while global methods (e.g., MDS) capture all pair-wise distances. We propose the Local-to-Global Structures (LGS) algorithm that achieves the following 3 goals:

G1 A single parameter controlling local-global embedding balance
G2 When this parameter is small, the embedding preserves local neighborhoods
G3 When this parameter is large, the embedding preserves the global structure

By "local neighborhood" of a vertex we refer to the immediate neighbors of the vertex being considered. If the nearest neighbors of each vertex in an embedding match well with the nearest neighbors in the actual graph, then the embedding accurately preserves the local structures. By "global structure" we refer to the

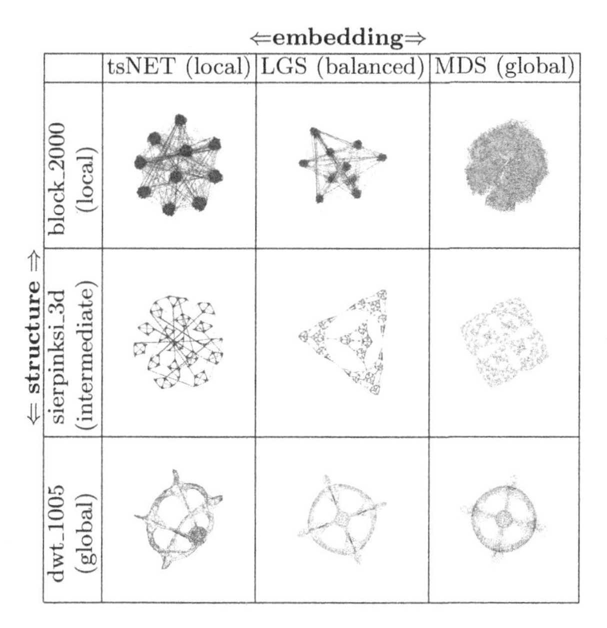

Fig. 2. Local embedding methods perform well on graphs with distinct local structure (block_2000), but they can distort the global shape of the graph (dwt_1005). Global methods capture the overall shape (e.g., dwt_1005), but may miss important local structures (block_2000). LGS(100) performs well for graphs with both local and global structure, such as sierpinski_3d, allowing us to see its fractal nature.

preservation of all pairwise graph distances (including long ones) in the embedding. Finally, "intermediate structure" refers to capturing both local neighbors and global structure. Figure 2 shows graphs exemplifying local, intermediate, and global structures and Sect. 3.2 defines formal embedding measures: neighborhood error, cluster distance, and stress. In Sect. 3.1 we explain the selection process for the balance parameter k and the objective function to ensure that the solution aligns with the stated goals. For **G1**, we modify MDS to preserve distances in a neighborhood defined by a parameter k). Thus, preserving distances for large neighborhoods satisfies **G3**. This leaves a question for **G2**: Does applying distance preservation to a subset of pairs result in locally faithful embeddings?

3.1 Adapting Stress Minimization for Local Preservation

We define a parameter, k, that represents the size of a neighborhood surrounding each vertex. A straightforward approach would involve simply selecting the k-nearest vertices for every given vertex (as in [3]). However, the graph-theoretic distance in an undirected graph is a discrete measure, which can create complications. For example, consider the local structure graph (top row) in Fig. 2. Although the within-cluster density is high, there are many edges between different clusters. Unfortunately, there is no simple way to test if an edge is within clus-

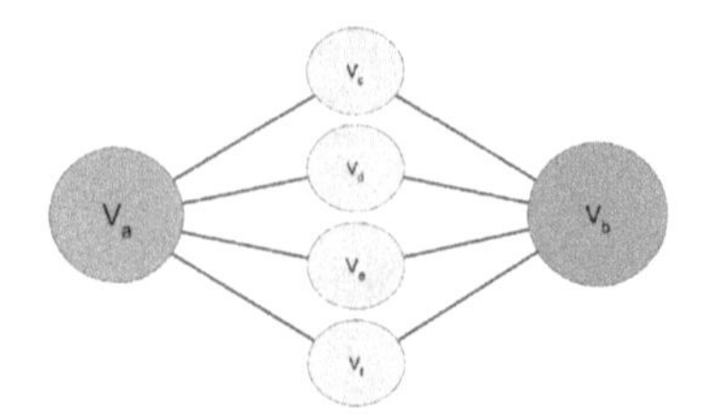

Fig. 3. An example of how we may skip over immediate neighbors when selecting neighborhoods to preserve. In this case, $c = 2$. There is only one unique walk of length ≤ 2 from v_a to v_c, v_d, v_e, v_f, but there are 4 such walks from v_a to v_b. In this case, v_b would be the first vertex added to v_a's most connected neighborhood.

ter or out-of-cluster. In order to produce tsNET-like embeddings, which should pay more attention to local structures, we must avoid preserving out-of-cluster edges.

Instead of considering distances directly, we find the top k most connected vertices for each vertex based on the hypothesis that more possible walks between vertices indicate greater similarity; see Fig. 3. Despite v_A and v_B not sharing an edge, they have the same set of neighbors. When v_A and v_B are both neighbors to a set of vertices, we can confidently state their similarity, confirmed by their shared proximity to vc, v_d, etc. [6]. The c-th power of an adjacency matrix $(A_G^c)_{i,j}$ encodes the number of c-length walks from vertex i to vertex j. To find the top k "most connected" vertices for each vertex, follow this procedure: Given an adjacency matrix of an undirected graph, A_G, raise it to the c-th power, take the sum of all powers $\mathbf{A}^* = \sum_{\mathbf{1} \leq \mathbf{i} \leq \mathbf{c}} \mathbf{A}_\mathbf{G}^\mathbf{i}$, to obtain a matrix whose (i, j)-th element shows the number of walks from i to j of length less or equal than c. Since each row in $\mathbf{A}^*$ corresponds to a vertex, we find the k largest values in row i (by sorting). We define these top k vertices to be the "most connected" neighborhood $N_k(v_a)$ of vertex v_a; see Fig. 3. We further weight the power of the matrix with a decaying weight factor, $s, 0 < s < 1$, such that $A^* = \sum_{1 \leq i \leq c} s^i A_G^i$. We investigate a range of values for s, and set $s = 0.1$; see [19]. We propose a procedure to reduce the number of matrix multiplications which we used in our experiments; see [19].

Objective Function. We remark, that only preserving distances of a subset of pairs will result in poor embeddings: e.g., two vertices that cannot "see" each other can be placed arbitrarily close with no penalty. A second term is needed in the objective function to prevent this, and we add an entropy repulsion term as in [3,10], to force pairs of vertices away from each other. For a given pairwise distance matrix $[d_{ij}]_{i,j=1}^n$ we define the following generalized stress function as an objective function:

$$\sigma(X) = \sum_{(i,j) \in N_k} (\|X_i - X_j\| - d_{ij})^2 - \alpha \sum_{(i,j) \notin N_k} \log \|X_i - X_j\|, \tag{1}$$

where X_i is the embedded point in $\mathbb{R}^d$, α is a fixed constant parameter that controls the weight of the logarithmic term, and N_k corresponds to the neighborhood that we aim to preserve in the embedded space. This objective function ensures that distances are preserved between the most-connected neighborhoods, while maximizing entropy. We use the negative logarithm of the distance between points, so that the repulsive force is relatively strong at small distances, but quickly decays (so that distant points are not forced to be too distant from each other). While similar to LMDS [3] and MaxEnt [10], the proposed objective function in Eq. 1 differs in (1) how the set N_k is selected (LMDS uses a kNN search

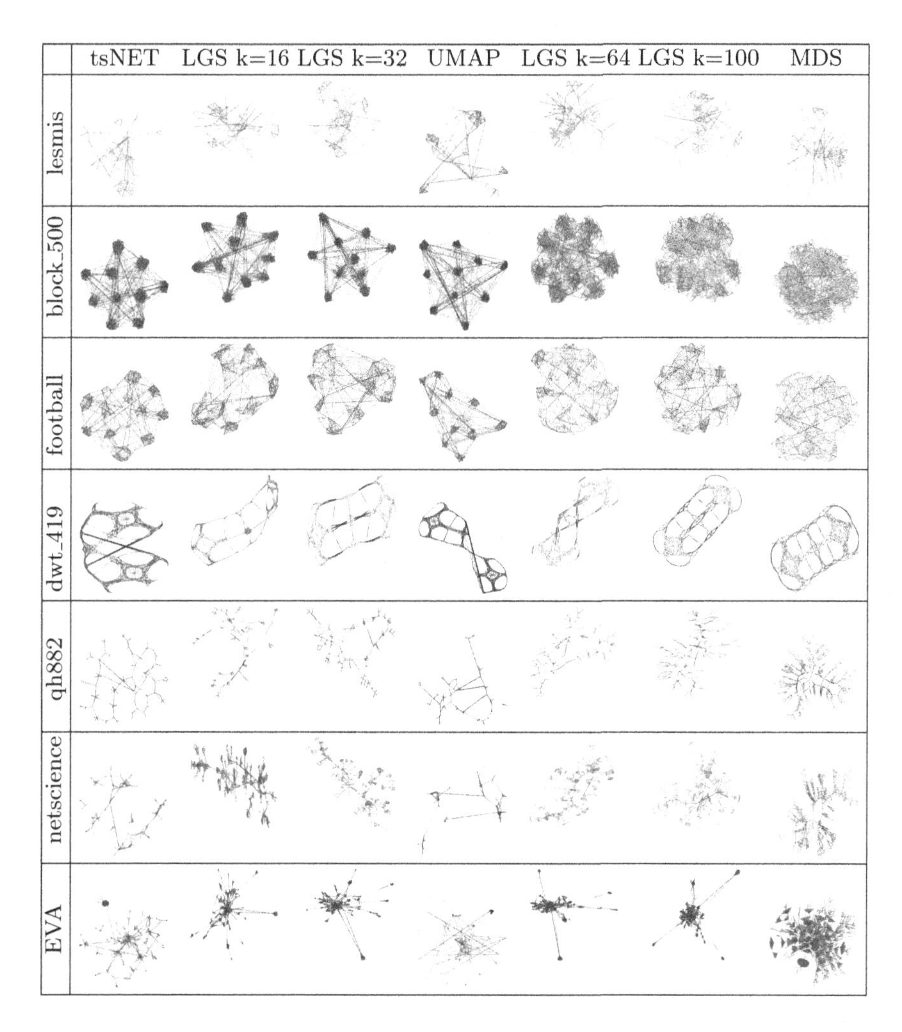

Fig. 4. Example embeddings. The first and last columns show the two extremes tsNET (local) and MDS (global); the middle column shows UMAP. The remaining columns show a gradual increase of LGS's k parameter, moving from local to global distance preservation (left to right). Note LGS outputs are vertically higher in each row.

and MaxEnt preserves distances between two vertices if and only if they share an edge) and (2) LMDS and MaxEnt cannot be easily parameterized to balance local and global structure preservation. We minimize the objective function by SGD which works well for stress minimization [1,30]. The parameter space of the algorithm is discussed in the supplemental material [19].

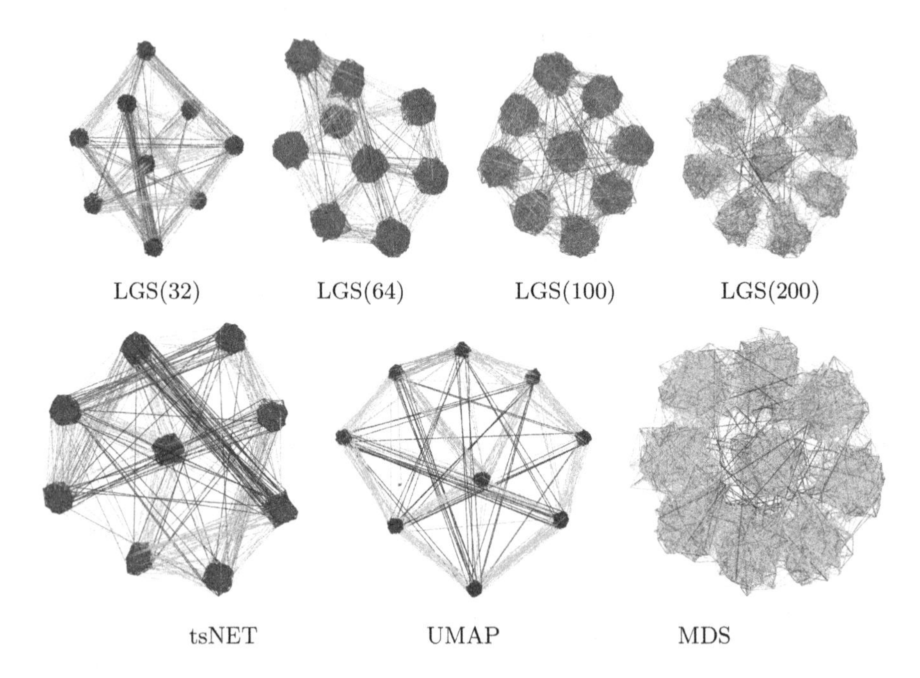

Fig. 5. The grid_cluster graph is generated so that each cluster has many out-of-cluster edges to its neighbors in a 3×3 lattice, providing a recognizable intermediate structure. tsNET and UMAP do not place clusters on a grid, MDS mixes the clusters; LGS(100) captures the 3×3 grid and shows distinct clusters.

3.2 Evaluation Metrics

We discuss the evaluation metrics for embedding algorithms: local neighborhood error (NE) score, intermediate structure (CD), and global distances (Stress).

NE Metric: Neighborhood hits (NH) measures how well an embedding preserves local structures [3,7]. NH is the average Jaccard similarity of the neighbors in the high-dimensional and low-dimensional embedding. Let Y be an $n \times d$ dimensional dataset, X be its $n \times 2$-dimensional embedding, and a radius r defines the size of the neighborhood one intends to measure. NH is defined as:

$$NH(Y, X, r) = \frac{1}{n} \sum_{i=1}^{n} \frac{|N_Y(p_i, r) \cap N_X(p_i, r)|}{|N_Y(p_i, r) \cup N_X(p_i, r)|} \tag{2}$$

where $N_Y(p_i, r)$ denotes the r nearest points to point p_i in Y and $N_X(p_i, r)$ the r nearest points to point p_i in X. For graph embeddings, this notion is called neighborhood preservation (NP) [10,15,31], with the main difference being that the radius r now refers to graph-theoretic distance: all vertices with shortest path distance $\leq r$ from vertex v_i. Specifically, NP measures the average Jaccard similarity of a vertex's graph-theoretic neighborhood of radius r and an equally sized neighborhood of that vertex's closest embedded neighbors. Since NH and NP measure accuracy, it is desirable to maximize these values. To facilitate comparison with the other two metrics (where lower scores mean better embeddings), we use Jaccard dissimilarity instead and refer to it as Neighborhood Error (NE).

Cluster Distance Metric: We introduce a new metric to measure how well intermediate structures are captured in an embedding. Since the distances between clusters in t-SNE cannot be interpreted as actual distances [29], while clusters in MDS embeddings are often poorly separated, we measure how faithful the relative distances between cluster centers are represented in the embedding. When cluster labels are given as part of the input (e.g., labels, classes), we can use them to define distances between the clusters. When cluster information is not given, we use k-means clustering in the high-dimensional data case, and modularity clustering in the graph case. The distances between clusters in the high-dimensional case is given by the Euclidean distance between the cluster centers. For graphs, we measure the distance between clusters by first taking the normalized count of edges between them, then subtracting the normalized count to convert similarity into dissimilarity. This produces a cluster-distance matrix, δ. Let $C_1, \ldots, C_n$ be the set of vertices belonging to cluster $1, \ldots, n$, then

$$\delta_{i,j} = 1 - \frac{1}{|E|} \sum_{u \in C_i, v \in C_j} \mathbb{1}(u, v \in E)$$

where $\mathbb{1}$ is the indicator function (1 if (u, v) is an edge and 0 otherwise). Once δ is computed, we compute the geometric center of each embedded cluster and compute the cluster-level stress between the graph-level-cluster and realized-cluster distances. This measure is small when similar clusters are placed closer and dissimilar clusters are placed far apart. The cluster distance (CD) is:

$$CD(\delta, \chi) = \sum_{i,j} \left(\frac{\delta_{i,j} - ||\chi_i - \chi_j||}{\delta_{i,j}} \right)^2 \quad (3)$$

where $\delta_{i,j}$ is the dissimilarity measure between cluster i and cluster j and χ_i is the geometric center of cluster i in the embedding. Although there are several existing metrics to measure cluster accuracy, such as silhouette distance and between/within-cluster sum of squares, they are not well suited to measure the quality of intermediate embeddings. Ideally, we would need a measure that checks how well the clusters are preserved and also verifies that the relative placement of the clusters is meaningful. The CD metric verifies meaningful cluster placements by measuring all pairwise distances between cluster centers. We remark that the

CD metric works best when the clusters have convex shapes (or shapes similar to spheres). For arbitrary non-convex shapes, such as half-moons or donuts, the CD metric might not provide meaningful insights.

Stress Metric: Stress has been used in many graph embedding evaluations.

$$\text{stress}(d, X) = \sum_{i,j} \left(\frac{d_{i,j} - ||X_i - X_j||}{d_{i,j}} \right)^2 \tag{4}$$

where d is the given distance matrix and X is the embedding. Embeddings are scaled to ensure fair comparisons in computing stress [10,15,31].

4 LGS Embedding of Graphs

We start with a visual analysis and discussion of layouts produced by LGS. Following the convention, several embeddings of the same graph are displayed side-by-side with increasing the value of k from left to right, going from local to global. Underneath the LGS embeddings, we place t-SNE, UMAP, and MDS embeddings of the same graph. For all graph embeddings provided in this paper, we use the jet color scheme to encode edge length. An edge length of 1 (ideal for unweighted graphs) is drawn in green, while red indicates that edge has been compressed (length < 1) and blue indicates the edge is stretched (length > 1). This makes clusters easy to spot as bundles of red edges, and global structure preservation apparent when most edges are green. Similar to tsNET, low values of k capture local neighborhoods well, by allowing some longer edges. As a result, clusters tend to be well separated. Note that tsNET allows even longer edges in an embedding, occasionally breaking the topology; see Fig. 1. Higher k values make LGS similar to MDS, with more uniform edge lengths. This reveals global structures (e.g., mesh, grid, lattice) but may overlook clusters.

Grid_cluster is a synthetic example with 900 vertices (9 clusters of size 100 each) and 10108 edges, created by stochastic block model (SBM) to illustrate the notion of cluster distance preservation. Within cluster edges are created with probability 0.8. We distinguish between two types of out-of-cluster edges. Clusters are first placed on a lattice. Out-of-cluster edges are created with probability 0.01 if they are adjacent in the lattice (no diagonals) and 0.001 otherwise. The layouts of this graph are in Fig. 5. Note the visual similarities between LGS(32) and tsNET; both seperate each cluster into dense sub-regions and place them seemingly randomly in the plane. Also note the similarities between LGS(200) and MDS, where both methods tend to miss the clusters. UMAP also fails to capture the intermediate structure built into this network: while there is a single cluster placed in the middle of the other eight, the surrounding shape is not a square. LGS(100) accurately places each cluster in the appropriate position, making it the only one that clearly shows the 3×3 underlying lattice. Although less dense and separable the clusters are more faithful in terms of placement.

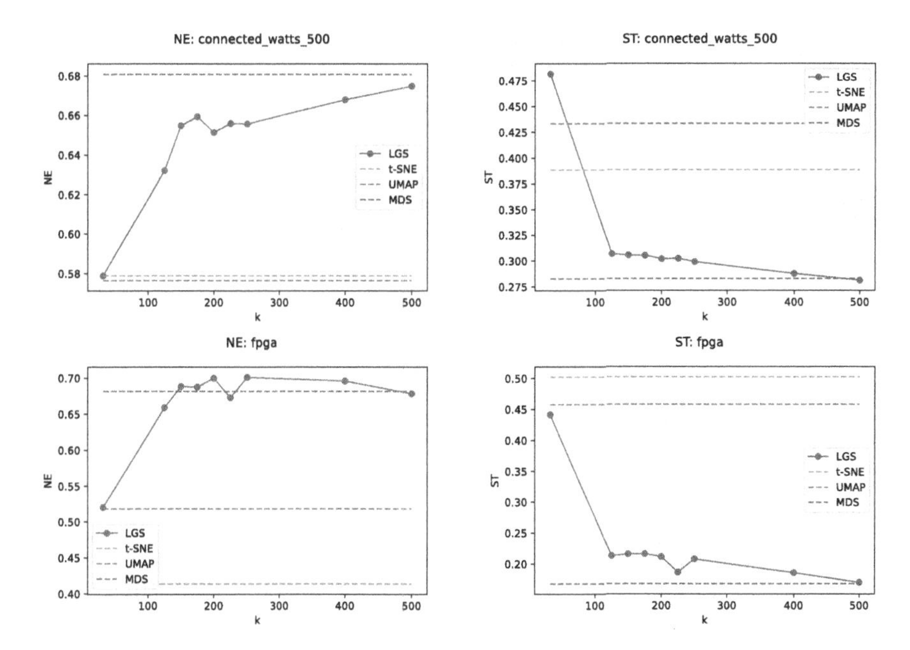

Fig. 6. Behavior of NE and stress: as k increases NE gets worse and stress gets better (LGS transitions from preserving local to global structure); tsNET, UMAP, and MDS values are shown as dotted lines for comparison. Note that in general, we expect to see an upward trend in NE, a downward trend for stress, and a parabola shape for CD.

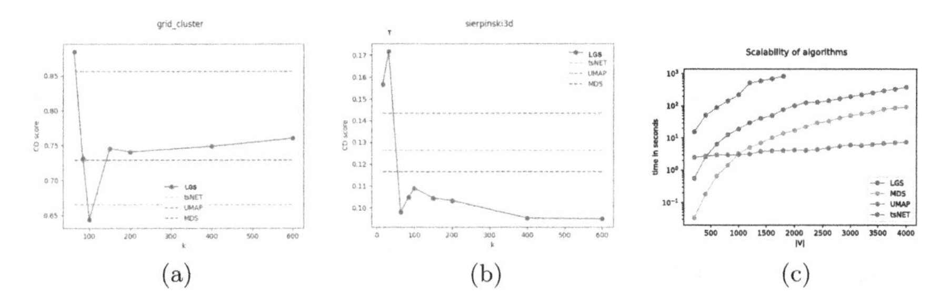

Fig. 7. (a–b) CD metric on the grid_cluster and sierpinkski3d graphs. Note that in these examples there are values of k which outperform competing algorithms. (d) Running time of each tested algorithm.

Connected_watts_1000 is a Watts-Strogatz random graph on a 1000 vertices and 11000 edges. It first assigns the vertices evenly spaced around a cycle with the nearest (7) vertices connected by an edge. Then, with low probability, some random 'chords' of the cycle are added by rewiring some of the local edges to other random vertices. This type of graph models the small-world phenomenon seen in real-world examples, such as social networks. The embeddings of connected_watts_1000, obtained by LGS, tsNET, UMAP, and MDS are in Fig. 1. We observe that tsNET and UMAP embeddings accurately capture the existence of a one-dimensional structure, but twist and break the circle to varying degrees. Meanwhile, MDS overcrowds the space, forming a classic 'hairball' where there is no discernible structure. For intermediate values of k in LGS, the circular structure in the data and the numerous chord connections become clearly visible.

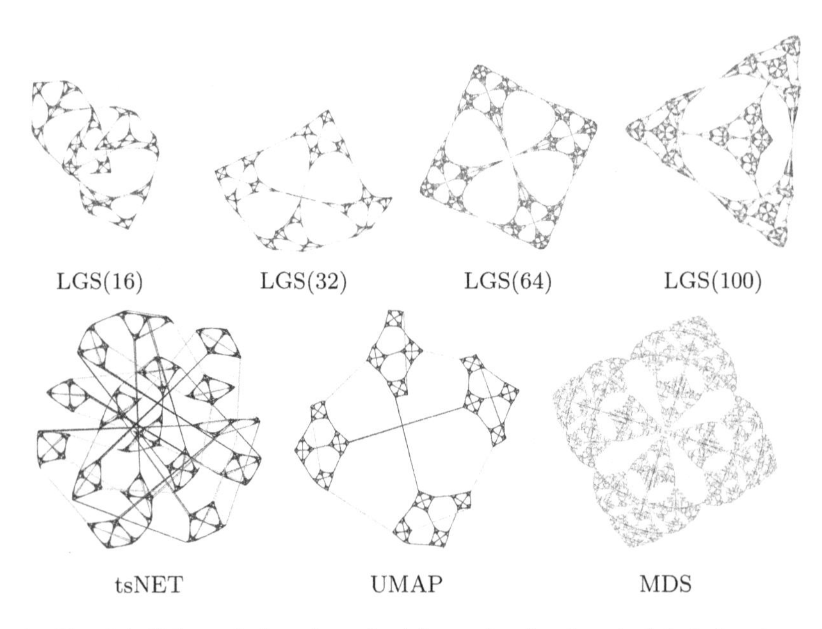

Fig. 8. Sierpinksi3d graph is a fractal with regular local and global structure. LGS manages to capture the recursive nature of the underlying structure. tsNET, UMAP miss the global placement of pyramids and MDS stretches them.

Sierpinksi_3d models the Sierpinski pyramid with 2050 vertices and 6144 edges – a finite fractal object with recursively smaller recurring patterns (the pyramid itself is built out of smaller pyramids). These fractal properties are ideal for showcasing the LGS algorithm at work, as small local structures build upon each other to create a global shape; see Fig. 8. We observe that tsNET captures the smallest structures well but places them arbitrarily in the embedding space. UMAP does better at placing the local structures in context but still creates long edges and twists not present in the data. While MDS visually captures the fractal motifs, it 'squishes' local structures. LGS can be used to balance these extremes.

We demonstrate more examples in Table 4, with additional embeddings available in the supplemental material [19]. Note that for lower values of k the embedding obtained by LGS visually resembles the output of tsNET, while for larger values of k the obtained embedding is more similar to the outputs of MDS. We see this reflected numerically in many graphs; see Fig. 6 and Table 1. For intermediate values of k, LGS often outperforms tsNET, UMap and MDS with respect to the intermediate structure preservation, measured by cluster distance (CD).

5 Evaluation

We test LGS on a selection of real-world and synthetic graphs from [4,15,31]; A full list can be found in [19].

Table 1. NE scores on LGS for varying values of k (left) and on competing algorithms (right). The colormap is normalized by row with dark orange representing the lowest score (best) and dark purple representing the highest (worst). Bold text indicates the lowest score in that row.

low middle high

	k=32	k=64	k=85	k=100	k=150	k=200	tsnet	umap	mds
lesmis	0.3761	0.3169	0.3126	0.3174	0.3166	0.3125	**0.3058**	0.3952	0.3141
can_96	0.4393	0.3980	0.4605	0.4650	0.4661	0.4658	0.3523	**0.3417**	0.4653
football	0.4390	0.4469	0.4452	0.4377	0.4371	**0.4369**	0.4636	0.4693	0.4380
rajat11	0.4347	0.4374	0.4377	0.3868	0.3641	0.3615	0.3942	**0.3331**	0.3608
mesh3e1	0.1841	0.1415	0.0933	0.0971	0.1444	**0.0**	0.1210	0.1241	0.0003
connected_watts_300	0.1995	0.3990	0.5311	0.5386	0.5302	0.5365	0.1805	**0.1778**	0.5453
block_model_300	0.6474	0.6767	0.6748	0.6794	0.6874	0.6741	**0.5538**	0.5675	0.6785
powerlaw300	0.5070	0.5075	0.5126	0.5044	0.5233	0.5062	**0.3563**	0.3849	0.5075
netscience	0.4556	0.4597	0.4751	0.4860	0.5017	0.5118	0.4319	**0.3207**	0.5123
dwt_419	0.3016	0.3151	**0.2404**	0.4085	0.2666	0.3203	0.2796	0.2892	0.2421
powerlaw500	0.5942	0.5630	0.5433	0.5712	0.5593	0.5432	**0.4237**	0.4683	0.5604
block_model_500	0.5396	0.6551	0.6832	0.6945	0.7024	0.7115	**0.4377**	0.4430	0.7151
connected_watts_500	0.5788	0.5731	**0.5676**	0.5852	0.6547	0.6513	0.5789	0.5764	0.6808
grid_cluster	0.3767	0.3831	0.3767	**0.3239**	0.3456	0.3863	0.8487	0.3491	0.4742
price_1000	0.7103	0.7108	0.7228	0.7313	0.7400	0.7468	0.6015	**0.5297**	0.7943
connected_watts_1000	0.6217	0.6712	0.7870	0.8199	0.8563	0.8472	0.6283	**0.6090**	0.8390
powerlaw1000	0.6555	0.6362	0.6409	0.6382	0.6241	0.6532	**0.3956**	0.4206	0.6384
block_model_1000	0.5753	0.5552	0.5593	0.6179	0.7549	0.7777	**0.4703**	0.4880	0.7822
dwt_1005	0.4999	0.4517	0.4417	0.4586	0.4499	0.4826	**0.3692**	0.4581	0.4372
btree9	0.7337	0.8082	0.8321	0.8625	0.8812	0.8811	0.6960	**0.5876**	0.9241
CSphd	0.5816	0.5797	0.5711	0.5783	0.6015	0.6128	0.6510	**0.4378**	0.6051
fpga	0.5199	0.5875	0.6043	0.6521	0.6884	0.6998	**0.4136**	0.5179	0.6818
sierpinski3d	0.6225	0.5729	0.5313	0.5439	0.5192	**0.4252**	0.4504	0.4554	0.4600
EVA	0.4544	0.4690	0.4822	0.4797	0.5070	0.5123	0.8763	**0.2295**	0.5629

We compare LGS against state-of-the-art techniques for local and global embeddings: tsNET from the repository linked in [15], UMAP from the umap python library written by the authors of [18], and MDS via the python bindings from [30] with default parameters. Our implementation of LGS is available online. The experiments were performed on an Intel® Core™ i7-3770 machine (CPU @ 3.40 GHz × 8 with 32 GB of RAM) running Ubuntu 20.04.3 LTS.

NE, CD and Stress Values and Trends. To evaluate how well LGS preserves local neighborhoods, we compare the average NE scores over several runs and present our results in Table 1. We can see a general trend: although, tsNET performs better with respect to NE values, LGS has consistently lower NE values than MDS. Additionally, as we increase the size of the neighborhood parameter, the NE values tend to increase by bringing the layouts closer to those of MDS. Interestingly, UMAP also tends to fall somewhere between tsNET and MDS on this metric. As expected, in many cases, LGS 'transitions' from tsNET to UMAP then finally to MDS as one goes from left to right, increasing k.

Next, we report the average CD scores for the graphs in our benchmark in Table 2. Unlike the NE values which increase as we increase k and the stress values which decrease as we increase k, the best CD values are obtained for intermediate values of k. This confirms that a balance between local and global optimization is needed to capture intermediate structures.

Table 2. CD scores following the same scheme as Table 1. NA indicates no clusters present in the data.

low middle high

	k=32	k=64	k=85	k=100	k=150	k=200	tsnet	umap	mds
lesmis	0.7308	**0.6745**	0.7116	0.7210	0.7135	0.7260	0.6949	0.6904	0.7195
can_96	0.7088	0.6701	0.6669	0.6666	0.6666	0.6666	**0.5174**	0.6747	0.6666
football	0.7747	0.7580	0.7911	0.7933	0.7983	0.8002	0.7804	**0.7090**	0.7979
rajat11	0.4092	0.4045	**0.3672**	0.3862	0.4083	0.4075	0.4607	0.5093	0.4080
mesh3e1	0.4087	0.3951	0.3824	0.3879	0.3865	0.3674	0.3948	0.3739	**0.3673**
connected_watts_300	0.3044	0.2445	0.2098	0.1989	0.1966	0.1968	0.2853	0.3028	**0.1929**
block_model_300	0.7454	0.7726	0.7994	0.7646	0.7727	0.7980	0.7767	**0.7307**	0.7901
powerlaw300	N/A	N/A	N/A	N/A	N/A	N/A	N/A	N/A	N/A
netscience	0.2716	0.2720	0.2717	0.2698	0.2649	0.2689	0.3222	0.3775	**0.2439**
dwt_419	0.1938	0.1899	0.1898	0.1962	0.1925	0.1943	**0.1657**	0.2173	0.1880
powerlaw500	0.4923	0.4332	0.4174	0.4336	0.4066	0.4249	**0.3607**	0.4261	0.4271
block_model_500	**0.6414**	0.7065	0.6626	0.6837	0.6819	0.6855	0.6551	0.7569	0.6975
connected_watts_500	0.2480	0.2268	0.2321	0.2378	0.2235	0.2191	0.2417	0.2294	**0.2120**
grid_cluster	0.9131	0.7905	0.8347	0.8040	0.7488	0.7415	0.7786	0.7454	**0.7335**
price_1000	N/A	N/A	N/A	N/A	N/A	N/A	N/A	N/A	N/A
connected_watts_1000	0.2734	0.2304	0.2946	0.2802	0.3648	0.2894	0.2333	0.2323	**0.2184**
powerlaw1000	0.3917	0.3671	0.3754	0.3640	**0.3590**	0.3775	0.4362	0.4210	0.3683
block_model_1000	0.6308	**0.5358**	0.5447	0.5435	0.5853	0.5419	0.5632	0.5977	0.5991
dwt_1005	0.1597	0.1414	0.1413	0.1466	0.1414	0.1415	0.1438	0.1456	**0.1386**
btree9	N/A	N/A	N/A	N/A	N/A	N/A	N/A	N/A	N/A
CSphd	N/A	N/A	N/A	N/A	N/A	N/A	N/A	N/A	N/A
fpga	0.1692	0.1511	0.1493	0.1601	0.1497	0.1506	0.1770	0.1966	**0.1305**
sierpinski3d	0.1252	0.1156	0.1050	0.1038	0.1021	0.1016	0.1244	0.1321	**0.0974**
EVA	N/A	N/A	N/A	N/A	N/A	N/A	N/A	N/A	N/A

We compute and report the averaged stress scores in Table 3. MDS is consistently good at minimizing the stress, but we see a salient trade-off between the stress scores of LGS's tsNET-like embeddings with low k values and LGS's MDS-like embeddings with high k values. When we look at small neighborhoods such as $k = 16$, we tend to see high stress values, however, the values decrease as we expand the neighborhoods. UMAP does not seem to capture global structure well for these graphs, often having the highest stress values.

Table 3. Stress scores following the same scheme as Table 1

	k=32	k=64	k=85	k=100	k=150	k=200		tsnet	umap	mds
lesmis	0.1988	0.1702	0.1673	0.1670	0.1665	**0.1656**		0.2290	0.4058	0.1661
can_96	0.1782	0.1526	0.1400	**0.1390**	0.1391	0.1391		0.2315	0.2237	0.1393
football	0.2719	0.2622	0.2578	0.2549	0.2544	**0.2544**		0.3521	0.3888	0.2548
rajat11	0.1636	0.1685	0.1774	0.1474	0.1250	**0.1246**		0.3167	0.4076	0.1251
mesh3e1	0.1125	0.0655	0.0328	0.0482	0.0613	0.0050		0.0858	0.0251	**0.0049**
connected_watts_300	0.4984	0.2359	0.2353	0.2267	0.2008	0.2081		0.4218	0.4702	**0.1896**
block_model_300	0.3592	0.3132	0.3087	0.3051	0.3052	0.2929		0.3889	0.4437	**0.2863**
powerlaw300	0.1975	0.1848	0.1741	0.1668	0.1644	0.1532		0.2619	0.3731	**0.1505**
netscience	0.1914	0.1539	0.1604	0.1627	0.1493	0.1472		0.3296	0.5198	**0.1131**
dwt_419	0.0688	0.0638	0.0360	0.1206	0.0521	0.0759		0.1417	0.1646	**0.0312**
powerlaw500	0.5465	0.2172	0.2023	0.2133	0.2022	0.1893		0.2643	0.3413	**0.1772**
block_model_500	0.3360	0.3224	0.3160	0.3108	0.3047	0.3016		0.4047	0.4510	**0.2855**
connected_watts_500	0.4816	0.3380	0.3367	0.3279	0.3055	0.3019		0.3884	0.4332	**0.2826**
grid_cluster	0.5494	0.5335	0.5814	0.3049	0.2553	0.2478		0.4195	0.3760	**0.2371**
price_1000	0.3543	0.2428	0.2320	0.2347	0.2127	0.2011		0.3654	0.5721	**0.1848**
connected_watts_1000	0.5578	0.3867	1.0963	0.9221	1.6998	0.8549		0.4169	0.4460	**0.3166**
powerlaw1000	0.3012	0.2418	0.2344	0.2410	0.2129	0.2196		0.2907	0.3615	**0.1815**
block_model_1000	0.4389	0.3456	0.3444	0.3409	0.3292	0.3212		0.4078	0.4379	**0.2921**
dwt_1005	0.1858	0.1206	0.0939	0.1165	0.0891	0.1355		0.3556	0.0669	**0.0424**
btree9	0.4973	0.3090	0.3510	0.2740	0.2818	0.2563		0.3539	0.3603	**0.2307**
CSphd	0.2642	0.2483	0.2218	0.1925	0.2051	0.1883		0.2943	0.4243	**0.1446**
fpga	0.4414	0.3139	0.2888	0.2498	0.2157	0.2113		0.5018	0.4575	**0.1679**
sierpinski3d	0.5063	0.3015	0.2097	0.2304	0.1769	0.1298		0.5194	0.2525	**0.1252**
EVA	0.6843	1.0466	0.7862	0.6387	0.4662	0.6394		0.3535	0.4913	**0.1907**

Effect of k on Evaluation Metrics: To visually explain LGS behavior, we plot examples NE, CD, and stress with respect to k. In Fig. 6, we demonstrate two separate plots for each graph. It can be seen that we often fall in between the values of NE and stress that tsNET and MDS reach. These plots show what we expect to see: as k increases NE increases and stress decreases. In Fig. 7(a-b), we plot the CD values of our layout with tsNET, UMAP, MDS for comparison. In many layouts LGS indeed has the lowest CD score. Values of k were chosen to be representative of the local-global tradeoff.

6 Discussion and Limitations

We described LGS: an adaptable algorithmic framework for embeddings that can prioritize local neighborhoods, global structure, or a balance between the two. LGS provides flexible structure preservation choices with comparable embedding quality to previous single-purpose methods (local or global), while also outperforming state-of-the-art methods in preserving intermediate structures.

There are several limitations: Our results are based on a small number of graphs. Additional systematic experimentation would further support the usefulness of LGS Some experiments for high-dimensional datasets are in [19]. LGS modifies MDS's objective function to accommodate varying neighborhood sizes, similarly, one could adapt the KL divergence cost function of t-SNE. Note that t-SNE's perplexity parameter ostensibly controls the size of a neighborhood, but high perplexity values do not result in global structure preservation [29].

LGS algorithm has several hyperparameters, including c, α, and k. We provide default values for c and α based on experiments, and leave k as a true hyperparameter. Our intention is for a visualization designer to adjust k as needed; to generate a spectrum of embeddings to get a sense of both local and global properties of a dataset. An interactive LGS version is not yet available. While LGS runs in seconds for graphs with a few thousand vertices, the running time can become untenable for larger instances, due to the $O(|V|^2)$ optimization per epoch, and pre-processing with APSP. While LGS's runtime is comparable with those of tsNET and MDS (see Fig. 7(c)) both can be sped up through the use of approximations [9,23,31], Speeding up LGS is a potential future work.

References

1. Börsig, K., Brandes, U., Pasztor, B.: Stochastic gradient descent works really well for stress minimization. In: GD 2020. LNCS, vol. 12590, pp. 18–25. Springer, Cham (2020). https://doi.org/10.1007/978-3-030-68766-3_2
2. Cai, H., Zheng, V.W., Chang, K.C.: A comprehensive survey of graph embedding: problems, techniques, and applications. IEEE Trans. Knowl. Data Eng. **30**(9), 1616–1637 (2018)
3. Chen, L., Buja, A.: Local multidimensional scaling for nonlinear dimension reduction, graph drawing, and proximity analysis. J. Am. Stat. Assoc. **104**(485), 209–219 (2009)
4. Davis, T.A., Hu, Y.: The University of Florida sparse matrix collection. ACM Trans. Math. Softw. **38**(1), 1:1–1:25 (2011)
5. De Leeuw, J.: Applications of convex analysis to multidimensional scaling (2005)
6. Ertoz, L., Steinbach, M., Kumar, V.: A new shared nearest neighbor clustering algorithm and its applications. In: Workshop on Clustering High Dimensional Data and Its Applications at 2nd SIAM International Conference on Data Mining, vol. 8 (2002)
7. Espadoto, M., Martins, R.M., Kerren, A., Hirata, N.S.T., Telea, A.C.: Toward a quantitative survey of dimension reduction techniques. IEEE Trans. Vis. Comput. Graph. **27**(3), 2153–2173 (2021)
8. Frey, D., Pimentel, R.: Principal component analysis and factor analysis (1978)
9. Fu, C., Zhang, Y., Cai, D., Ren, X.: AtSNE: efficient and robust visualization on GPU through hierarchical optimization. In: Proceedings of the 25th ACM SIGKDD International Conference on Knowledge Discovery & Data Mining, pp. 176–186 (2019)
10. Gansner, E.R., Hu, Y., North, S.C.: A maxent-stress model for graph layout. IEEE Trans. Vis. Comput. Graph. **19**(6), 927–940 (2013)
11. Gansner, Emden R.., Koren, Yehuda, North, Stephen: Graph drawing by stress majorization. In: Pach, János. (ed.) GD 2004. LNCS, vol. 3383, pp. 239–250. Springer, Heidelberg (2005). https://doi.org/10.1007/978-3-540-31843-9_25
12. Ghojogh, B., Ghodsi, A., Karray, F., Crowley, M.: Uniform manifold approximation and projection (UMAP) and its variants: Tutorial and survey. CoRR abs/2109.02508 (2021)
13. Jolliffe, I.T.: Principal Component Analysis. Springer Series in Statistics, Springer (1986)
14. Kamada, T., Kawai, S.: An algorithm for drawing general undirected graphs. Inf. Process. Lett. **31**(1), 7–15 (1989)
15. Kruiger, J.F., Rauber, P.E., Martins, R.M., Kerren, A., Kobourov, S.G., Telea, A.C.: Graph layouts by t-SNE. Comput. Graph. Forum **36**(3), 283–294 (2017)

16. Kruskal, J.B.: Multidimensional scaling by optimizing goodness of fit to a non-metric hypothesis. Psychometrika **29**(1), 1–27 (1964)
17. Van der Maaten, L., Hinton, G.: Visualizing data using t-SNE. J. Mach. Learn. Res. **9**(11) (2008)
18. McInnes, L., Healy, J.: UMAP: uniform manifold approximation and projection for dimension reduction. CoRR abs/1802.03426 (2018)
19. Miller, J., Huroyan, V., Kobourov, S.: Balancing between the local and global structures (LGS) in graph embedding (2023). https://arxiv.org/abs/2308.16403
20. Munzner, T.: Visualization Analysis and Design. CRC Press, Boca Raton (2014)
21. Nguyen, Q.H., Eades, P., Hong, S.: On the faithfulness of graph visualizations. In: Carpendale, S., Chen, W., Hong, S. (eds.) IEEE Pacific Visualization Symposium, PacificVis 2013, 27 February 2013 – 1 March 2013, Sydney, NSW, Australia, pp. 209–216. IEEE Computer Society (2013)
22. Nocaj, A., Ortmann, M., Brandes, U.: Untangling the hairballs of multi-centered, small-world online social media networks. J. Graph Algorithms Appl. **19**(2), 595–618 (2015)
23. Ortmann, M., Klimenta, M., Brandes, U.: A sparse stress model. J. Graph Algorithms Appl. **21**(5), 791–821 (2017)
24. Purchase, H.C.: Metrics for graph drawing aesthetics. J. Vis. Lang. Comput. **13**(5), 501–516 (2002)
25. Roweis, S.T., Saul, L.K.: Nonlinear dimensionality reduction by locally linear embedding. Science **290**(5500), 2323–2326 (2000)
26. Shepard, R.N.: The analysis of proximities: multidimensional scaling with an unknown distance function. I. Psychometrika **27**(2), 125–140 (1962)
27. Tamassia, R.: Handbook of Graph Drawing and Visualization. CRC Press, Boca Raton (2013)
28. Tenenbaum, J.B., Silva, V.D., Langford, J.C.: A global geometric framework for nonlinear dimensionality reduction. Science **290**(5500), 2319–2323 (2000)
29. Wattenberg, M., Viégas, F., Johnson, I.: How to use t-SNE effectively. Distill (2016). https://doi.org/10.23915/distill.00002
30. Zheng, J.X., Pawar, S., Goodman, D.F.M.: Graph drawing by stochastic gradient descent. IEEE Trans. Vis. Comput. Graph. **25**(9), 2738–2748 (2019)
31. Zhu, M., Chen, W., Hu, Y., Hou, Y., Liu, L., Zhang, K.: DRGraph: an efficient graph layout algorithm for large-scale graphs by dimensionality reduction. IEEE Trans. Vis. Comput. Graph. **27**(2), 1666–1676 (2021)

Graph Representations

Boxicity and Interval-Orders: Petersen and the Complements of Line Graphs

Marco Caoduro[1](✉) and András Sebő[2]

[1] Sauder School of Business, The University of British Columbia, Vancouver, Canada
marco.caoduro@ubc.ca
[2] CNRS, Laboratoire G-SCOP, Univ. Grenoble Alpes, Grenoble, France
andras.sebo@cnrs.fr

Abstract. The boxicity of a graph is the smallest dimension d allowing a representation of it as the intersection graph of a set of d-dimensional axis-parallel boxes. We present a simple general approach to determining the boxicity of a graph based on studying its "interval-order subgraphs".

The power of the method is first tested on the boxicity of some popular graphs that have resisted previous attempts: the boxicity of the Petersen graph is 3, and more generally, that of the Kneser-graphs $K(n, 2)$ is $n - 2$ if $n \geq 5$, confirming a conjecture of Caoduro and Lichev [Discrete Mathematics, Vol. 346, 5, 2023].

Since every line graph is an induced subgraph of the complement of $K(n, 2)$, the developed tools show furthermore that line graphs have only a polynomial number of edge-maximal interval-order subgraphs. This opens the way to polynomial-time algorithms for problems that are in general $\mathcal{NP}$-hard: for the existence and optimization of interval-order subgraphs of line-graphs, or of interval-completions of their complement.

Keywords: Boxicity · Interval-orders · Interval-completion · Kneser-graphs · Line Graphs

1 Introduction

The *intersection graph* of a finite family of sets $\mathcal{F}$ is the graph $G(\mathcal{F})$ having vertex-set $\{v_A : A \in \mathcal{F}\}$ and edge-set $\{v_A v_B : A, B \in \mathcal{F},\ A \neq B, \text{ and } A \cap B \neq \emptyset\}$, where multiple occurrences of a set is allowed, and different occurrences are considered as different sets.x The *boxicity* of $G = (V, E)$, denoted by $\text{box}(G)$, is the minimum dimension d such that $G = G(\mathcal{B})$, where $\mathcal{B}$ is a family of axis-parallel boxes in $\mathbb{R}^d$. A graph G has $\text{box}(G) = 0$ if and only if G is a complete graph, $\text{box}(G) \leq 1$ if and only if it is an *interval graph*, and $\text{box}(G) \leq k$ if it is the intersection of k interval graphs [8].

Complements of interval graphs will be called here *interval-order graphs*, referring to the natural order of disjoint intervals. Interval graphs, and therefore also interval-order graphs, can be recognized in linear time [2] while determining whether a graph has boxicity at most 2 is $\mathcal{NP}$-complete [14]. In the language of

M. A. Bekos and M. Chimani (Eds.): GD 2023, LNCS 14465, pp. 283–295, 2023.
https://doi.org/10.1007/978-3-031-49272-3_19

parameterized complexity, the computation of boxicity is not in the class *XP* of problems solvable in polynomial time when the parameter boxicity is bounded by a constant.

Boxicity was introduced by Roberts [18] in 1969, and has been a well-studied graph parameter. Roberts [18] proved that any graph G on n vertices has boxicity at most $\left\lfloor \frac{n}{2} \right\rfloor$. Esperet [9] showed that the boxicity of graphs with m edges is $\mathcal{O}(\sqrt{m \log m})$, while Adiga, Bhowmick, and Chandran [1] proved that the boxicity is $\mathcal{O}(\Delta \log^2 \Delta)$ for graphs with maximum degree Δ. In [6], this latter bound was improved to $\mathcal{O}(\Delta \log \log \Delta)$ in the particular case of line graphs. Further relevant results have been proved by Scheinerman [19], Thomassen [22], Chandran and Sivadasan [7], Esperet [11], and Esperet, Joret [10].

We present now an important background of our results. Let k and n be two positive integers such that $n \geq 2k+1$. The *Kneser-graph* $\mathrm{K}(n,k)$ is the graph with vertex-set given by all subsets of $[n] := \{1, 2, \ldots, n\}$ of size k where two vertices are adjacent if their corresponding k-sets are disjoint. Kneser-graphs stimulated deep and fruitful graph theory, "building bridges" with other parts of mathematics. For instance, Lovász's proof [15] of Kneser's conjecture [13] is the source of the celebrated "topological method." This method has proven to be a powerful approach for a range of challenging combinatorial problems [16].

The study of the boxicity of Kneser-graphs $\mathrm{K}(n,k)$ was initiated by Matěj Stehlík as a question [21]. Caoduro and Lichev [4] established a general upper bound of $n-2$, a lower bound of $n - \frac{13k^2 - 11k + 16}{2}$ for $n \geq 2k^3 - 2k^2 + 1$, and a lower bound of $n-3$ for $k=2$ that nearly matches the upper bound. They also conjectured that $\mathrm{box}(\mathrm{K}(n,2)) = n-2$ for any $n \geq 5$. We establish here this conjecture. Essentially new ideas are needed already for the case $n=5$; to the best of our knowledge, all declared solutions for this particular graph finish with a computer-based case-checking (cf. [4, Section 6]).

Theorem 1. *The boxicity of the Kneser-graph* $\mathrm{K}(n,2)$ *with* $n \geq 5$ *is* $n-2$. *In particular, the boxicity of the Petersen graph* $\mathrm{K}(5,2)$ *is* 3.

The proof of this theorem starts with a well-known rephrasing of boxicity in terms of "interval completion" (Sect. 2.1), allowing a graph theory perspective. This approach reveals that each interval completion of $\mathrm{K}(n,2)$ is uniquely determined by the choice of at most 5 of its vertices. Thus, the number of interval-completions is polynomial (at most n^5) in general. Even though the list is non-trivial already for small Kneser graphs, there are only four essentially different interval completions of $\mathrm{K}(n,2)$.

We introduce some additional notation and terminology. Given a graph G, a graph H is a subgraph of G if $V(H) \subseteq V(G)$ and $E(H) \subseteq E(G)$. An *interval-completion* of G is an interval graph containing G as a subgraph, and the *line graph* of G, denoted by $L(G)$, is the intersection graph of the edge-set of G as a family of sets of two vertices. The complement of a graph G is denoted by $\overline{G}$, and the complete graph on n vertices or on the set V is denoted by K_n, and K_V, respectively. Note that $\mathrm{K}(n,2) = \overline{L(K_n)}$.

Leveraging the fact that the complement of every line graph $\overline{L(G)}$ is the induced subgraph of $\overline{L(K_{V(G)})}$ we get that the complement of *any line graph has only a polynomial number of interval completions:*

Lemma 1. *Let $G = (V, E)$ be a graph on n vertices. Then $\overline{L(G)}$ has at most n^5 inclusion-wise minimal interval-completions. These can be listed in $\mathcal{O}(n^7)$ time.*

Computing the minimum number of edges of an interval-completion for any graph is an $\mathcal{NP}$-hard problem [12], called *Interval Graph Completion (IGC)*. However, by Lemma 1 we have:

Theorem 2. *IGC is polynomial-time solvable for complements of line graphs.*

Polynomial complexity follows for detecting bounded boxicity:

Theorem 3. *Let $G = (V, E)$ be a graph on n vertices and k a positive integer. Then it can be decided in $\mathcal{O}(n^{5k+3})$-time whether* $\text{box}(\overline{L(G)}) \leq k$.

Summarizing, our study of Kneser graphs $\text{K}(n, 2)$ has uncovered fundamental structural properties of complements of line graphs. This allows us to compute in polynomial time *all* minimal interval completions of complements of line graphs, that is, to list all "interval-disjointness subgraphs" of "edge-disjointness graphs". This leads to new polynomial algorithms, for instance, the weighted version of the interval completion problem for complements of line graphs.

The paper is structured as follows. Section 2 introduces the first, well-known, classical properties that play a fundamental role in the study of boxicity. It also characterizes interval-order graphs using the orderings of their vertices. Section 3 focuses on the interval-order subgraphs of line graphs and shows that all interval-order subgraphs of the line graph of K_n can be easily described using only five vertices of K_n. Section 4 contains the proofs of the main results. It determines the boxicity of the Petersen graph, which had been considered to be a difficult open problem. More generally, the boxicity of the Kneser-graphs $\text{K}(n, 2)$ for any integer $n \geq 5$ is established. The proof is based on elementary results and can be generalized to solve the interval-completion problem for complements of line graphs in polynomial time. It also leads to detecting constant boxicity, placing boxicity in the class XP for complements of line graphs. It is yet to be proved whether this problem is $\mathcal{NP}$-hard, or can be solved in polynomial time. The last section (Conclusion, Sect. 5) offers applications, possible extensions of the methods, and open questions.

2 Preliminaries

We use standard graph theory notation (mostly following [20]): given a graph $G = (V, E)$ and a vertex $v \in V$, $\delta(v)$ denotes the edges incident to v and $N(v)$ the vertices adjacent to v. The *degree* of a vertex v is $d(v) := |\delta(v)| = |N(v)|$. For $V' \subseteq V$, $G[V']$ is the induced subgraph of G with vertex-set V', and edge-set $\{uv \ : \ uv \in E \text{ and } u, v, \in V'\}$.

Graphs are *simple* in this paper, that is, they do not have loops or parallel edges. (These are either senseless or irrelevant to the results.)

2.1 Defining the Boxicity Using Interval Graphs

An *axis-parallel box* in $\mathbb{R}^d$ is a Cartesian product $I_1 \times I_2 \times \cdots \times I_d$ where each I_i is a closed interval in the real line. Axis-parallel boxes B, B' intersect if and only if for all $1 \leq i \leq d$ the intervals I_i and I'_i intersect. Roberts [18] formalized a set of simple but important statements assuring starting tools for the study of the boxicity: the boxicity of a graph is at most k if and only if it is the intersection of k interval graphs; adding a vertex or even two non-adjacent vertices the boxicity increases by at most one, so the boxicity of a graph is at most $\left\lfloor \frac{n}{2} \right\rfloor$. We will extensively use the former definition of boxicity in the following more comfortable form for us (obtained by complementation using "de Morgan's law"):

We say that the family of edge-sets of a graph $G = (V, E)$, $\mathcal{C} = \{C_1, ..C_k\}$ ($C_i \subseteq E$) is a *interval-order-cover*, or a k*-interval-order-cover* of E (or G) if $\bigcup_{i=1}^{k} C_i = E$, and each (V, C_i) is an interval-order graph. The C_i will mostly be called colors and an edge in C_i will be said to have *color* i. Some edges will have several colors, which may be necessary to make each color an interval-order graph.

Lemma 2 (Cozzens and Roberts [8]**, 1983).** *Let* G *be a graph. Then* $\text{box}(G) \leq k$ *if and only if* $\overline{G}$ *has a* k*-interval-order-cover.*

2.2 Defining Interval-Order Graphs Using Orderings on Their Vertices

Given a family $\mathcal{I}$ of n intervals in $\mathbb{R}$, we order them in a non-decreasing order $\sigma = I_1 \dots I_n$ of their right end-points, and consider the interval-order graph $G = (V, E)$, and $V = \{v_1, \dots, v_n\}$, where v_i corresponds to I_i ($i = 1, \dots n$) and two vertices are joined if and only if the corresponding intervals are disjoint (the complement of their intersection graph).

Orienting the edges of G from the vertices of smaller index towards those of larger index, we have:

$$\textit{if } i > j, \textit{ then } N^+(v_i) \subseteq N^+(v_j), \tag{1}$$

where the *out-neighborhood* of a vertex $v_l \in V(G)$ is the set $N^+(v_l) := \{v_m : v_l v_m \in E(G), l < m\}$.

Clearly, the series of the sizes of $N^+(v_i)$ for $i = 1, \dots, n$ is (not necessarily strictly) monotone decreasing from $d(v_1)$ to 0. In addition, *in the chain of (*1*), only the first at most* D *sets are not empty, where* D *is the maximum degree of* G. Indeed, suppose $N^+(v_i) \neq \emptyset$ for $i \geq D + 1$, and let $x \in N^+(v_i)$. Then by (1), x is also in the neighborhood of all previous vertices, that is, $N_G(x) \supseteq \{v_1, \dots, v_{D+1}\}$, contradicting that D is the maximum degree.

By necessity, one realizes that (1) actually *characterizes* interval-order graphs (the proof is immediate by induction). It is explicitly stated in Olariu's paper [17]:

Lemma 3. *Let G be an undirected graph. Then $G = (V, E)$ is an interval-order graph if and only if V has an ordering $(v_1, \ldots, v_n)$ so that orienting the edges from the vertices of smaller index towards those of larger index (1) holds.*

Inspired by this characterization, we can simply construct interval-order subgraphs of an arbitrary graph, and it turns out that all interval-order subgraphs are of the form determined by Lemma 3.

Let $G = (V, E)$ be a graph and σ an ordering of $V(G)$. We define the graph $G^\sigma = (V, E^\sigma)$ as follows: let $V_0 := V$, $V_i := V_{i-1} \cap N_G(v_i)$ for $1 \le i \le n$, and $E^\sigma := E_1 \cup E_2 \cup ... \cup E_{n-1} \subseteq E$ where E_i is the set of edges from v_i to V_i. (See Fig. 1 for an example.) Note that the set V_i – that is eventually becoming $N^+(v_i)$ and this is the notation we will use for it – depends only on the undirected graph G, and the *i-prefix* $\sigma_i := (v_1, \ldots, v_i)$ of σ. We will also say that σ_i is an i-prefix in V. If $N^+(v_{i+1}) = \emptyset$, v_j for $j > i$ does not bring in any more edges to E^σ, so the *i-suffix* $(v_{i+1}, \ldots, v_n)$ of σ may then be deleted or remain undefined. We saw that $N^+(v_i) = \emptyset$ if $i > D$, so σ can supposed to be an ordered D-tuple.

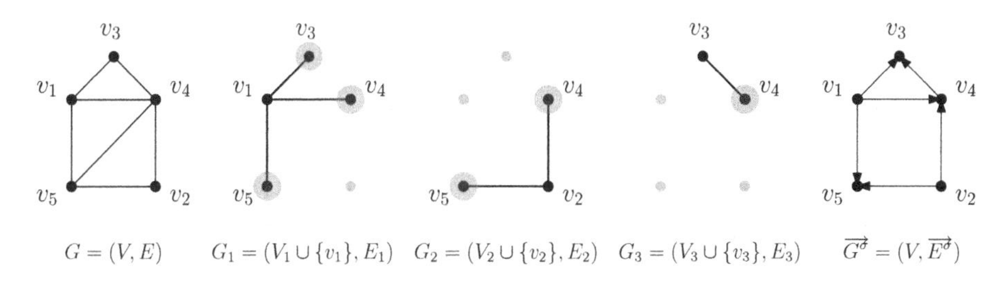

Fig. 1. A graph G with an ordering $\sigma := (v_1, v_2, v_3, v_4, v_5)$ of $V(G)$, the corresponding G_i ($i \in \{1, 2, 3\}$), and $\overrightarrow{G^\sigma}$ (see Lemma 3 and thereafter). For each G_i ($i \in \{1, 2, 3\}$), the vertices of V_i are marked with red disks around them. Note that v_4, the only vertex of V_3, is not contained in $N_G(v_4)$, so $V_4 = \emptyset$, $E_4 = \emptyset$.

Corollary 1. *Let $G = (V, E)$ be a graph. Then for any ordering σ of V, G^σ is an interval-order subgraph of G. Conversely, any inclusion-wise maximal interval-order subgraph of G is G^σ for some ordering σ of V.*

Proof. Denote by $\overrightarrow{G^\sigma} = (V, \overrightarrow{E^\sigma})$ the digraph obtained by orienting each edge of G^σ from its endpoint of smaller index to the one with larger index. Clearly $N^+(v_i) = V_i$ and by construction, (1) is satisfied. Hence, Lemma 3 immediately implies the first part of the corollary. The converse follows by considering the ordering σ satisfying (1) that exists by the reverse implication of Lemma 3. □

3 Interval-Orders in Line Graphs

In this section, we study the inclusion-wise maximal interval-order subgraphs of the line graph of K_n, we can actually list them all (Lemma 5)!

The interval-completion problem for a graph is equivalent, by complementation, to finding interval-order subgraphs in the complementary graph. The following lemma makes it easier to encounter such subgraphs.

Lemma 4. *Let $G = (V, E)$ be a graph, and $\sigma_i = (v_1, v_2, \ldots, v_i)$ an i-prefix in V. If an inclusion-wise maximal interval-order subgraph H belongs to an ordering with prefix σ_i, then*

(i) For $u, v \in V \setminus \{v_1, v_2, \ldots, v_i\}$, $N(u) \supseteq N(v) \cap N^+(v_i)$, there exists an ordering σ with prefix σ_i such that $G^\sigma = H$ and u precedes v in σ.
(ii) In every ordering σ with prefix σ_i and $H = G^\sigma$, σ_i is immediately followed by all $N_i := \{v \in V \setminus \{s_1, s_2, \ldots, s_i\} : N(v) \supseteq N^+(v_i)\}$ in arbitrary order.
(iii) If $|N^+(v_i)| = 2$, then H is one of two possible interval-order graphs.

Note that N_i is exactly the set of elements electable to be added to σ_i as v_{i+1} so that $N^+(v_{i+1}) = N^+(v_i)$.

Proof of Lemma 4. To show (i) just note: if v precedes u, then interchanging u and v, the condition of (i) makes sure that neither $N^+(u)$ nor $N^+(v)$ decrease.

For checking (ii) let σ be an ordering of V with prefix σ_i and such that G^σ is a maximal interval-order subgraph of G. If σ_i is not immediately followed by the vertices of N_i, modify σ by moving all the vertices of N_i immediately after v_i, in arbitrary order. Clearly, each $N^+(v_i)$ is replaced by a superset, and at least one of them by a proper one: E^σ increases, contradicting maximality.

To prove (iii) suppose $N^+(v_i) = \{x, y\}$, and N_x, N_y, $N_{x,y}$ be the vertices in $V \setminus \{v_1, v_2, ..., v_i\}$ adjacent only to x, only to y, or to both respectively. Observe that $N_{x,y} = N_i$, so, by (ii), σ_i is immediately followed by $N_{x,y}$, and then by (i), either an element of N_x or an element of N_y follows, unless both are empty. Applying then (ii) again, the entire N_x or the entire N_y must follow, finishing the proof of the Lemma. □

We switch now to a higher gear, introducing the key tool to compute the boxicity of $\mathrm{K}(n, 2)$ for $n \geq 5$ (Theorem 1), then solve IGC (Lemma 1 and Theorem 2), and finally compute the boxicity of complements of line graphs (Theorem 3).

A solution to these problems necessitates, already for complements of line graphs of small order like the Petersen graph, refined knowledge about their interval completions, that is, about interval-order subgraphs of line graphs. The following lemma establishes that $L(K_n)$ has only four essentially different maximal interval-order subgraphs.

Lemma 5. *For any inclusion-wise maximal interval-order subgraph H of $L(K_n)$ ($n \geq 5$), there exist distinct vertices $a, b, c, d, e \in V(K_n)$, uniquely determining the edge-sets $E_{a,b,c,d,e}, E_{a,b,c,d}, F_{a,b,c,d}, F'_{a,b,c,d} \subseteq E(L(K_n))$ such that $E(H)$ is equal to one of these, where*

$$|E_{a,b,c,d,e}| = \frac{(n+2)(n-1)}{2},\ |E_{a,b,c,d}| = 4(n-1),\ |F_{a,b,c,d}| = |F'_{a,b,c,d}| = 5(n-2).$$

The proof consists in presenting the four orderings of $E(K_n)$ determining these sets as developed in Sect. 2.2. After the proof, these sets can be explicitly given with exact formulas. For these formulas, and already for the proof, we denote by (e, f) the edge of $L(K_n)$ between two incident edges $e, f \in E(K_n)$ and, for simplicity, borrow notation from K_n for some edge-sets in $L(K_n)$:

- for a vertex $v \in V(K_n)$, Q_v denotes the set of $\frac{(n-1)(n-2)}{2}$ edges of the clique formed in $L(K_n)$ by the $n-1$ edges of $\delta(v) \subseteq E(K_n)$ as vertices of $L(K_n)$;
- for an edge $e = uv \in E(K_n)$, δ_{uv} denotes the set of edges $ef \in E(L(K_n))$, where $f \in V(L(K_n))$ is incident to u or v in K_n, that is, δ_{uv} is the star of center $e = uv \in V(L(K_n))$ in $L(K_n)$, $|\delta_{uv}| = 2(n-2)$; the set δ_{uv^-} consists only of the edges ef, where f is incident to u in K_n, $|\delta_{uv^-}| = n - 2$;
- for a set $U \subseteq V(K_n)$, K_U denotes the edge-set of $L(K_n[U])$; $K_{u,v,w^-} := \{(uv, uw), (uv, vw)\}$.

Proof of Lemma 5. We define the orderings of $E(K_n)$ giving rise to the four claimed graphs. Let a, b, c, d, e be five different vertices of K_n:

First, starting with $v_1 = ab$ and $v_2 = ac$ adds already $\delta_{ab} \cup \delta_{ac^-} \cup K_{\{a,b,c\}}$ to E^σ, consisting of $3(n-2)$ edges: $2(n-2)$ edges in $\delta_{ab} \subseteq Q_a \cup Q_b$, $n-3$ edges in $\delta_{ac-} \subseteq Q_a$ (the edge between ab and ac has already been counted) and in addition, the edge between ac and bc, which is neither in Q_a, nor in Q_b, see Fig. 2 (i). Then $v_3, \ldots, v_{n-3}$ is the list of all other edges of K_n incident to a, except ad, ae, and $v_{n-2} = de$; we continue by adding the remaining $n-3$ edges incident to d different from ad, in arbitrary order; and finally, $v_{2(n-2)} := ae$. We added to E^σ in this way all edges of Q_a, and besides that all the $n - 2 + 1$ edges of $\delta_{ad} \cup K_{\{a,d,e\}}$, symmetrically to the effect of the starting two edges, see Fig. 2 (ii). Clearly,

$$|E^\sigma| = \frac{(n-1)(n-2)}{2} + 2(n-1) = \frac{(n+2)(n-1)}{2}.$$

The set E^σ that we get in this way depends only on the ordered set of the chosen five vertices, so we can denote it by $E_{a,b,c,d,e}$.

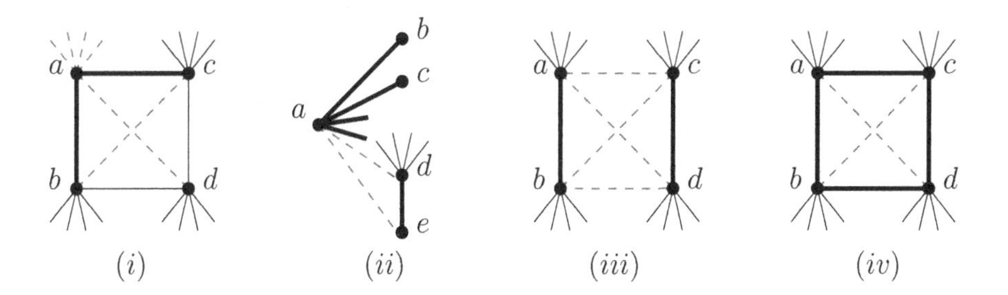

Fig. 2. The edges of $V(K_n)$ in σ; the edges with fat lines correspond to v_1, v_2 in a.left, and in b.; to $v_1, \ldots, v_{n-2}$ in a.right; and to $v_1, \ldots v_4$ in c.; the edges with dashed red lines correspond to the vertices in $N^+(v_i)$ $(i = 2, n-2, 4)$ respectively.

Second, let $v_1 = ab$, $v_2 = cd$, $v_3 = ac$, $v_4 = bd$, and continue to define an ordering σ with the remaining edges of $\delta(a)$ and of $\delta(d)$ as vertices of $L(K_n)$

in arbitrary order, and then E^σ. See Fig. 2 (*iii*) and (*iv*). Besides the 12 edges of $K_{a,b,c,d}$ each of δ_{ab} and δ_{ad} contain $2(n-4)$ edges of $L(K_n)$, so E^σ has now $12+2(n-4)+2(n-4)=4(n-1)$ edges The set E^σ depends only on the 4-tuple a,b,c,d in this order, let us denote it by $E_{a,b,c,d,e}$.

A third kind of interval-order graph arises from the ordering σ with the 4-prefix defined by $v_1 = ab$, $v_2 = ac$, $v_3 = bd$, $v_4 = cd$, leading to $N^+(v_4) = \{ad, bc\}$. See Fig. 2 (*i*) and (*iv*). Then the ordering continues either with the remaining $n-4$ edges of $\delta(a)$ followed by those of $\delta(d)$, or the same for $\delta(b)$ followed by $\delta(c)$. We denote the corresponding E^σ by $F_{a,b,c,d}$, $F'_{a,b,c,d}$ respectively.

We can now calculate $|F_{\{a,b,c,d\}}|$, $|F'_{\{a,b,c,d\}}|$ for instance by counting the size of the out-neighborhood of each v_i: $|N^+(v_1)| + |N^+(v_2)| = 3(n-2)$, as in the beginning of the proof; $|N^+(v_3)| = |N^+(v_4)| = 2$; then we add $2(n-4)$ edges of $L(K_n)$ with out-neighborhoods of size 1, after which the out-neighborhoods are empty, so $|F_{\{a,b,c,d\}}| = |F'_{\{a,b,c,d\}}| = 3(n-2)+4+2(n-4) = 5(n-2)$.

It remains to check that any ordering of the vertices defines a subgraph of one of the listed graphs. Due to space constraints, this part of the proof has been omitted. It can be found in the extended version of the paper [5]. □

The above proof shows how E^σ can be explicitly expressed as a function of the vertices a,b,c,d,e inducing the first edges of σ (Fig. 2). The three formulas are useful to have at hand:

$$E_{a,b,c,d,e} = Q_a \cup \delta_{ab} \cup \delta_{ad} \cup K_{\{a,b,c\}} \cup K_{\{a,d,e\}}, \tag{a}$$

$$E_{a,b,c,d} = \delta_{ab} \cup \delta_{ad} \cup K_{\{a,b,c,d\}}, \tag{b}$$

$$F_{a,b,c,d} = \delta_{ab} \cup \delta_{ad} \cup \delta_{ac^-} \cup K_{\{a,b,c\}} \cup K_{\{a,b,d\}} \cup K_{a,d,c^-} \cup K_{b,c,d^-}. \text{ or } F'_{a,b,c,d}, \text{ where } \delta_{ad} \text{ is replaced by } \delta_{bc}. \tag{c}$$

4 Boxicity of the Petersen Graph and Complements of Line Graphs

In this section, we exploit the description of the inclusion-wise maximal interval-order subgraphs of $L(K_n)$ (Sect. 3), for proving our main results. We say that a subgraph is *of type (a), (b), or (c)* if its edge-set corresponds to that of (a), (b), or (c), respectively; $F_{a,b,c,d}$ and $F'_{a,b,c,d}$ are both considered as of type (c).

In Sect. 4.1 we establish Theorem 1, that is $\text{box}(\overline{L(K_n)}) = n-2$ for every $n \geq 5$. First, we show the easier upper bound (Lemma 6). The lower bound, that is, the tightness of the proven upper bound is then proved for $n = 5, 6$ separately (Lemmas 7 and 8). These three proofs introduce already the general ideas, but with the difference that for $n \leq 6$ the interval-order subgraphs of type (b) and (c) are larger or equal to those of type (a), so they do play a more essential role. Then we take on the challenges of the proof for $n \geq 7$ (Lemma 9). Section 4.2 makes one more small step for generalizing the results to arbitrary complements of line graphs (without the assumption $G = K_n$), finishing the proofs of Theorem 2 and 3. Further possibilities for applying the arguments are discussed in Sect. 5.

4.1 Proof of Theorem 1

Theorem 1.1 in [4] states that the boxicity of the Kneser-graph $\mathrm{K}(n,k)$ is at most $n-2$ for $n \geq 2k+1$. We present a simpler proof for $k=2$ in the extended version of the paper [5].

Lemma 6. $\mathrm{box}(\overline{L(K_n)}) \leq n-2$ *if* $n \geq 5$.

Although the growth of the number of edges is quadratic in (a), while it is only linear in (b) and (c), the three sizes are comparable for small values of n. For this reason, the arguments for proving the lower bound in Theorem 1 for $n \leq 6$ and for $n \geq 7$ are slightly different.

Fact 1. *Let* $\{a,b,c\}$ *and* $\{a',b',c'\}$ *be two sets of three distinct vertices of* $V(K_n)$. *Then* $(\delta_{ab} \cup \delta_{ac}) \cap (\delta_{a'b'} \cup \delta_{a'c'}) = \emptyset$ *if and only if* $\{a,b,c\} \cap \{a',b',c'\} = \emptyset$. □

Lemma 7. $\mathrm{box}(\overline{L(K_5)}) \geq 3$.

Proof. Assume for a contradiction that $\mathrm{box}(\overline{L(K_5)}) \leq 2$. Let $\{E_1, E_2\}$ be a 2-interval-order-cover of $L(K_5)$, and assume that (V, E_1) and (V, E_2) are maximal interval-order subgraphs of $L(K_5)$. By Lemma 5, $|E_i| = 14$ (a), or $|E_i| = 15$ (c), or $|E_i| = 16$ (b), $(i = 1, 2)$. In each of these cases, there are three distinct vertices $a_i, b_i, c_i \in V(K_5)$ such that $\delta_{a_ib_i} \cup \delta_{a_ic_i} \subseteq E_i$ $(i \in \{1,2\})$, (see (a), (b), and (c)). The set $V(K_5)$ has only 5 vertices, so $\{a_1, b_1, c_1\}$ and $\{a_2, b_2, c_2\}$ have to intersect, and then $|E_1 \cap E_2| \geq 1$ by Fact 1.

Since $|E(L(K_5))| = 5\binom{4}{2} = 30$, $\{E_1, E_2\}$ forms an interval-order-cover only if E_1 and E_2 have both at least 15 edges, and at least one of them has 16 edges. Assume that $|E_1| = 16$ (E_1 is of type (b)), and E_2 has either 16 or 15 edges (it is of type (b) or (c)). In both cases, there are two sets of four distinct vertices in $V(K_5)$ defining the edge-set of E_1 and E_2. These two sets have at least three common vertices, say $\{a,b,c\}$, and $K_{a,b,c} \subseteq E_1$ follows, and also $K_{a,b,c} \subseteq E_2$ if it is of type (b), or $|K_{a,b,c} \cap E_2| \geq 2$ if it is of type (c), (see (b), and (c)). Either way, $|E_1| + |E_2| - |E_1 \cap E_2| \leq 16 + 15 - 2 < 30$, contradicting the assumption that $E_1 \cup E_2 = E(L(K_5))$. □

Lemma 8. $\mathrm{box}(\overline{L(K_6)}) \geq 4$.

Proof. Assume for a contradiction that $\mathrm{box}(\overline{L(K_6)}) \leq 3$. Let $\{E_1, E_2, E_3\}$ be a 3-interval-order-cover of $L(K_6)$, and assume that (V, E_1), (V, E_2), and (V, E_3) are maximal interval-order subgraphs of $L(K_5)$.

By Lemma 5, $|E_i| = 20$ (no matter if it is of type (a), (b), or (c)) and in all the three cases there are three distinct vertices $a_i, b_i, c_i \in V(K_6)$ such that $\delta_{a_ib_i} \cup \delta_{a_ic_i} \subseteq E_i$. Since $|E(L(K_6))| = 6\binom{5}{2} = 60$, E_1, E_2 and E_3 are pairwise disjoint. Applying Fact 1 three times, we deduce that the nine vertices of a_i, b_i, c_i for $i \in [3]\}$ are all distinct, contradicting $|V(K_6)| = 6$. □

We now finish the proof of Theorem 1 with the lower bound for $n \geq 7$.

Lemma 9. $\mathrm{box}(\overline{L(K_n)}) \geq n-2$, *for any* $n \geq 7$.

Proof. Assume for a contradiction that $\text{box}(\overline{L(K_n)}) \leq n-3$. Let $\{E_i : i \in [n-3]\}$ be an $(n-3)$-interval-order-cover of $L(K_n)$, and assume that $\{(V, E_i) : i \in [n-3]\}$ are maximal interval-order subgraphs of $L(K_n)$. For each $i \in [n-3]$, there are distinct $a_i, b_i, c_i \in V(K_n)$ such that:

- $E_i \supset Q_{a_i} \cup \delta_{a_i b_i} \cup \delta_{a_i c_i}$, if E_i is of type (a); or
- $E_i \supset \delta_{a_i b_i} \cup \delta_{a_i c_i}$, if E_i is of type (b) or (c).

We show that for any possible assignment of $\{(a_i, b_i, c_i) : i \in [n-3]\}$, the number of edges in $\bigcup_{i \in [n-3]} E_i$ is strictly smaller than $|E(L(K_n))|$.

First, observe that *at least $n-4$ interval-order graphs in this cover are of type (a).* Indeed, if there are at most $n-5$ edge-sets of type (a), then, by Lemma 5, the number of covered edges is at most

$$(n-5)\frac{(n+2)(n-1)}{2} + 10(n-2),$$

quantity that, for $n \geq 7$, is strictly smaller than $|E(L(K_n))| = n\binom{n-1}{2}$. Therefore, the only two cases to consider are: the cover contains $n-3$ edge-sets of type (a), or it contains $n-4$ edge-sets of type (a) and one edge-set of type (b) or (c).

Then, note that $\sum_{i=1}^{n-3} |E_i| = |E(L(K_n))| + \frac{(n-1)(n-6)}{2}$ in the first case, and $\sum_{i=1}^{n-3} |E_i| = |E(L(K_n))| + n - 6$ in the second case. We prove $|\bigcup_{i \in [n-3]} E_i| < |E(L(K_n))|$ by showing that, in both cases, the pairwise intersections of the E_i ($i \in [n-3]$) sum up to more than $\frac{(n-1)(n-6)}{2}$, and $n-6$, respectively. We omit the details in this limited version. □

4.2 Interval-Completion and Boxicity of Complements of Line Graphs

Lemma 5 presents all the maximal interval-order subgraphs of $L(K_n)$. Now, we show how to use this information to generate all the maximal interval-order subgraphs of $L(G)$ for any graph $G = (V, E)$. Since complementing the maximal interval-order subgraphs of $L(G)$, we get the minimal interval-completions of $\overline{L(G)}$, this will prove Lemma 1. We then derive Theorem 2 and 3 as easy applications of this lemma.

Proof of Lemma 1. Complete $G = (V, E)$ by adding all of its *non-edges*, to get the complete graph K_V. Then $L(G) = L(K_V)[E]$ is the subgraph of $L(K_V)$ induced by the vertex-set E of $L(G)$.

Let $\mathcal{A}$ be the family of edge-sets of all the at most $\mathcal{O}(n^5)$ edge-maximal interval-order subgraphs of $L(K_V)$ listed by Lemma 5, and, for each $A \in \mathcal{A}$, denote by G_A the graph $(E(K_V), A)$. Since interval-order graphs are closed under taking induced subgraphs, the graphs in $\mathcal{B} := \{G_A[E] : A \in \mathcal{A}\}$ (induced by the vertices of $L(G)$ corresponding to the edges of G) are also interval-order graphs.

Each interval-order subgraph of $L(G)$ can be completed to an inclusion-wise maximal interval-order subgraph of $L(K_V)$, so the inclusion-wise maximal ones among the edge-sets of the graphs in $\mathcal{B}$ are exactly the edge-sets of the maximal interval-order subgraphs of $L(G)$. The edge-set of each can be computed in $\mathcal{O}(n^2)$ time by following the procedure of Lemma 5 and by working directly in K_V. □

Given Lemma 1 and the notion of k-interval-order-covers, the proofs of Theorem 2 and 3 follow easily:

Proof of Theorem 2. Let G be a graph on n vertices. By Lemma 1 all inclusion-wise minimal interval-completions of $\overline{L(G)}$ can be listed in $\mathcal{O}(n^7)$ time, and we can take the one with a minimum number of edges among them. □

Proof of Theorem 3. We can assume $k \leq n-3$. By Lemma 1, the family $\mathcal{B}$ (defined in the proof of Lemma 1) can be computed in $\mathcal{O}(n^7)$ time and has cardinality $\mathcal{O}(n^5)$. To decide if $\text{box}(\overline{L(G)}) \leq k$, one can simply generate all subsets of k distinct elements from $\mathcal{B}$ and return *True* if at least one of these defines a k-interval-cover of $L(G)$, and *False*, otherwise. This algorithm has a running time bounded by $\mathcal{O}(n^7) + \mathcal{O}(n^{5k})f(n,k)$ where $f(n,k)$ is the time spent to check if a k-set of $\mathcal{B}$ is an interval-order-cover of $L(G)$. Since $L(G)$ has at most n^3 edges and $k < n$, $f(n,k) = \mathcal{O}(n^3)$ using an appropriate data-structure. This concludes the proof of the Theorem. □

5 Conclusion

We have explored in this paper the interval-order subgraphs of line graphs and showed that their number can be bounded by a polynomial of the number of vertices, and the polynomial solvability of related optimization problems follows.

Some other connections extend our arguments concerning the boxicity to other graphs than complements of line graphs, and the new observations have led us to establish the boxicity of some other relevant graphs, about which we provide a short account.

First, of course, the question of determining the boxicity of line graphs comes up. Our meta-method dictates to study the interval-order subgraphs of the complements of line graphs. The inclusion-wise maximal ones among these are in one-to-one correspondence with linear orderings (permutations) of the vertex-set. More concretely, for an arbitrary graph G on n vertices, the boxicity of $L(G)$ turns out to be equivalent to a beautiful extremal problem about the permutations of $V(G)$:

We say that a linear ordering $\prec$ of $V(G)$ *covers* the pair of vertex-disjoint edges $\{ab, cd\}$ of G if $\max\{a,b\} \prec \min\{c,d\}$, or $\max\{c,d\} \prec \min\{a,b\}$, where the max and the min concern the linear ordering $\prec$. We proved that $\text{box}(L(G)) = \mathcal{O}(\log n)$ even if G is the largest possible among line graphs, that is, $G = K_n$. In this case, we also proved $\text{box}(L(K_n)) = \Omega(\log\log n)$. Further investigations, including algorithms and complexity, are under investigation. The only general results about the boxicity of these graphs we found in the literature are in [6].

It is also tempting to apply the meta-method – of coloring the edges of the graph so that each color forms an interval-order subgraph, and an edge may get several colors – to some popular "Mycielsky graphs" and "Ramsey graphs" and follow a program of generalizations similar to what we did in this article. An account of these results can be found in [3, Chapter 4].

We conclude with a few open problems. The first one is from [4].

Problem 1. [4] Determine the boxicity of Kneser-graphs $\mathrm{K}(n,k)$ with $k \geq 3$.

Already establishing $\mathrm{box}(\mathrm{K}(7,3))$ is open. The computation of this value could either support or directly refute the intriguing conjecture $\mathrm{box}(\mathrm{K}(n,k)) = n-k$, for any $n \geq 2k+1$.

Deciding whether the boxicity of the complement of a line graph is at most k can be solved in polynomial time when k is fixed in advance (Theorem 3). We did not manage to show that the dependency on k is really necessary.

Problem 2. Is computing the boxicity of *complements of line graphs* $\mathcal{NP}$-hard?

Assuming a positive answer to Problem 2, we ask if the dependency on k can be essentially improved. For instance, *is it FPT?* Problem 2 arises for line graphs as well, where even the membership to the class XP is open:

Problem 3. Can it be decided in polynomial time whether the boxicity of a *line graph* is 2, or is it $\mathcal{NP}$-complete?

It is well-known, and easy to check, that $\mathrm{box}(L(K_4)) = 3$. Hence, Problem 3 is interesting only for line graphs of graphs with clique number at most three.

Acknowledgment. We thank an anonymous referee for numerous helpful suggestions that significantly improved the structure and quality of our presentation.

References

1. Adiga, A., Bhowmick, D., Chandran, L.S.: Boxicity and poset dimension. In: Thai, M.T., Sahni, S. (eds.) COCOON 2010. LNCS, vol. 6196, pp. 3–12. Springer, Heidelberg (2010). https://doi.org/10.1007/978-3-642-14031-0_3
2. Booth, K.S., Lueker, G.S.: Testing for the consecutive ones property, interval graphs, and graph planarity using PQ-tree algorithms. J. Comput. Syst. Sci. **13**(3), 335–379 (1976)
3. Caoduro, M.: Geometric challenges in combinatorial optimization: packing, hitting, and coloring rectangles. Ph.D. thesis, Laboratoire G-SCOP (Univ. Grenoble Alpes) (2022)
4. Caoduro, M., Lichev, L.: On the boxicity of Kneser graphs and complements of line graphs. Discret. Math. **346**(5), 113333 (2023)
5. Caoduro, M., Sebő, A.: Boxicity and interval-orders: Petersen and the complements of line graphs. arxiv:2309.02062 (2023)
6. Chandran, L.S., Mathew, R., Sivadasan, N.: Boxicity of line graphs. Discret. Math. **311**(21), 2359–2367 (2011)
7. Chandran, L.S., Sivadasan, N.: Boxicity and treewidth. J. Comb. Theory, Ser. B **97**(5), 733–744 (2007)
8. Cozzens, M.B., Roberts, F.S.: Computing the boxicity of a graph by covering its complement by cointerval graphs. Discret. Appl. Math. **6**(3), 217–228 (1983)
9. Esperet, L.: Boxicity and topological invariants. Eur. J. Comb. **51**, 495–499 (2016)

10. Esperet, L.: Box representations of embedded graphs. Discrete Comput. Geom. **57**(3), 590–606 (2017)
11. Esperet, L., Joret, G.: Boxicity of graphs on surfaces. Graphs Comb. **29**, 417–427 (2013)
12. Garey, M.R., Johnson, D.S.: Computers and Intractability: A Guide to the Theory of NP-Completeness (Series of Books in the Mathematical Sciences). Freeman W.H. (ed.) first edition edn. (1979)
13. Kneser, M.: Aufgabe 360. Jahresber. Deutsch. Math.-Verein. **2**, 27 (1956)
14. Kratochvíl, J.: A special planar satisfiability problem and a consequence of its NP-completeness. Discret. Appl. Math. **52**(3), 233–252 (1994)
15. Lovász, L.: Kneser's conjecture, chromatic number, and homotopy. J. Comb. Theory, Ser. A **25**(3), 319–324 (1978)
16. Matoušek, J.: Using the Borsuk-Ulam Theorem, 1st edn. Springer, Berlin (2003)
17. Olariu, S.: An optimal greedy heuristic to color interval graphs. Inf. Process. Lett. **37**(1), 21–25 (1991)
18. Roberts, F.S.: On the boxicity and cubicity of a graph. In: Tutte, W.T. (ed.) Recent Progress in Combinatorics, pp. 301–310. Academic Press, New York (1969)
19. Scheinerman, E.R.: Intersection classes and multiple intersection parameters of graphs. Ph.D. thesis, Princeton University (1984)
20. Schrijver, A.: Combinatorial Optimization. Springer, Berlin (2003)
21. Stehlík, M.: Personal communication
22. Thomassen, C.: Interval representations of planar graphs. J. Comb. Theory, Ser. B **40**(1), 9–20 (1986)

Side-Contact Representations with Convex Polygons in 3D: New Results for Complete Bipartite Graphs

André Schulz(✉)

FernUniversität in Hagen, Universitätsstraße 47, 58097 Hagen, Germany
andre.schulz@fernuni-hagen.de

Abstract. A polyhedral surface $\mathcal{C}$ in $\mathbb{R}^3$ with convex polygons as faces is a *side-contact representation* of a graph G if there is a bijection between the vertices of G and the faces of $\mathcal{C}$ such that the polygons of adjacent vertices are exactly the polygons sharing an entire common side in $\mathcal{C}$.

We show that $K_{3,8}$ has a side-contact representation but $K_{3,250}$ has not. The latter result implies that the number of edges of a graph with side-contact representation and n vertices is bounded by $O(n^{5/3})$.

Keywords: Contact Representations · Polyhedral Surfaces · 3D

1 Introduction

Contact representations are a classical approach to visualize graphs. A graph G has a contact representation if there is a bijection between its vertex set V and a set of interior-disjoint geometric objects from a given class such that two objects touch if and only if the corresponding vertices are adjacent. For a concrete contact representation it has to be specified, which geometric objects are considered (including their embedding space) and what it means for two objects to touch. In this paper we consider convex polygons in 3D as geometric objects. To avoid confusion we call the edges of a polygon its *sides* and the vertices its *corners*. Two polygons touch with a *side-contact* if and only if they have a full side in common. It is not allowed that a side is contained in more than two polygons. Notice that we do not require that all polygon sides are incident to two polygons. For brevity, we call representations of convex polygons in 3D with such side-contacts simply *side-contact representations* throughout the paper and every polygon will be considered as convex.

It is an open question to characterize the graphs that have a side-contact representation. First results were given by Arseneva et al. [2], who introduced this kind of contact representation. We list some of the results from Arseneva et al.: Exactly the planar graphs have a side-contact representation in the plane. The graph K_5 has no side-contact representation in 3D, but $K_{3,5}$ and $K_{4,4}$ have one. Another graph that has no side-contact representation is $K_{5,81}$, which implies by the Kővari–Sós–Turán theorem [5] that graphs with side-contact representation

M. A. Bekos and M. Chimani (Eds.): GD 2023, LNCS 14465, pp. 296–303, 2023.
https://doi.org/10.1007/978-3-031-49272-3_20

have at most $O(n^{9/5})$ edges, for n being the number of vertices. On the other hand all graphs of hypercubes have a side-contact representation and thus there are n-vertex graphs with $\Theta(n \log n)$ edges with side-contact representation.

There exists a large body of literature for other types of contact representations. For a few selected results in 2D we redirect the reader to Arseneva et al. [2]. For 3D we list some selected results here: Due to Tietze [9] every graph has a contact representation with interior-disjoint convex polytopes in $\mathbb{R}^3$ where contacts are given by shared 2-dimensional facets. Evans et al. showed that every graph has a contact representation in 3D where two convex polygons touch if they share a single corner [3]. Every planar graph has a contact representation with axis-parallel cubes in $\mathbb{R}^3$ as shown by Felsner and Francis [4] (two cubes touch if their boundaries intersect), a similar result for boxes was discovered earlier by Thomassen [8]. Kleist and Rahman [6] studied a similar model but required that the intersection has nonzero area. They proved that every subgraph of an Archimedean grid can be represented with unit cubes. Alam et al. [1] showed in the same model with axis-aligned boxes that every 3-connected planar graph and its dual can be represented simultaneously.

Our Contribution. We extend the results of Arseneva et al. [2] for complete bipartite graphs. We construct a side-contact representation of $K_{3,8}$, where previously only a construction for $K_{3,5}$ was known. On the other hand, we prove that $K_{3,250}$ has no side-contact representation. As a consequence the number of edges of an n-vertex graph with a side-contact representation is bounded by $O(n^{5/3})$.

The construction of the side-contact representation of $K_{3,8}$ presented here omits some details; like precise coordinates. For a complete description of the construction we direct the reader to the full version of this article [7].

2 A Side-Contact Representation for $K_{3,8}$

In this section we explain how to construct a side-contact representation of $K_{3,8}$. As an intermediate step we construct a *corner-contact representation.* In contrast to side-contact representations, two polygons touch with corner-contact if they share a single corner. For a polygon p its supporting plane $p^=$ defines two open half-spaces which we label arbitrarily as p^+ and p^-. We say that a representation (either corner-contact or side-contact) is *one-sided,* if for every polygon p its touching polygons lie either all in the closure of p^+, or they lie all in the closure of p^-. Note that the definition of one-sided is slightly stronger than the definition of *one-sided with respect to a set* as used by Arseneva et al. [2].

Lemma 1. *Every one-sided corner-contact representation of* $K_{3,8}$ *can be transformed into a side-contact representation of* $K_{3,8}$.

Proof. We call the polygons from the partition class with eight elements the blue polygons. Every blue polygon b can be trimmed to a triangle touching the (red) polygons r_1, r_2, r_3. We can assume for every b that the red polygons lie in $b^+ \cup b^=$. Consider the plane $b^=$ and the line arrangement $\mathcal{A}$ given by $b^= \cap r_i^=$ for $i \in \{1,2,3\}$. Since the representation is one-sided, $\mathcal{A}$ contains a triangular cell Δ such that every edge of Δ contains exactly one corner of b. Let h be a plane parallel to $b^=$ (inside b^+, very close to $b^=$) such that the line arrangement given by $h \cap r_i^=$ for $i \in \{1,2,3\}$ is combinatorially equivalent to $\mathcal{A}$ and furthermore the cell corresponding to Δ contains on every edge exactly one line segment from $S = \{s_i := h \cap r_i \mid i \in \{1,2,3\}\}$. We can now replace b by the convex hull of S and then restrict the red polygons to $b^= \cup b^+$, w.r.t. the modified b. Since the three segments of S lie on the boundary of Δ, they all appear on its convex hull. Thus we keep all incidences without introducing new ones (see Fig. 1). Also, the one-sidedness property is maintained. Repeating this for every blue polygon yields a side-contact representation of $K_{3,8}$. □

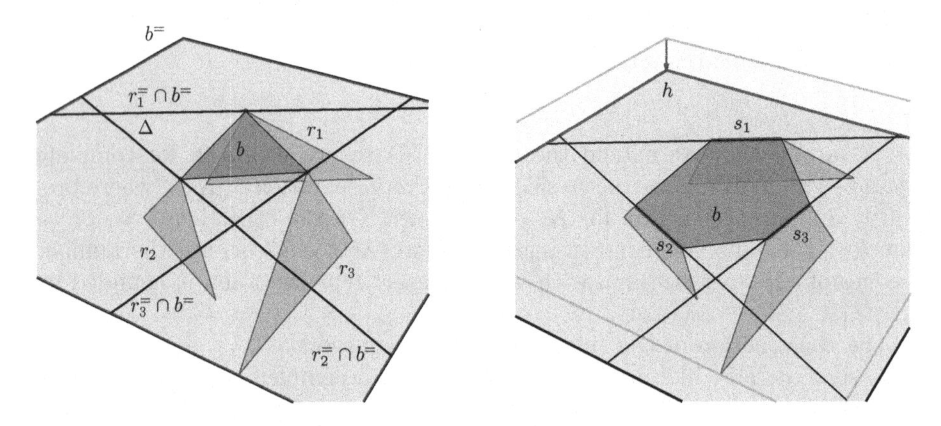

Fig. 1. Offsetting the supporting planes of the blue polygons can transform a one-sided corner-contact representation into a side-contact representation. (Color figure online)

It remains to construct a one-sided corner-contact representation for $K_{3,8}$. We start with a hexagonal prism of height 1. Its base is parallel to the xy-plane and given by a hexagon with alternating side lengths 2 and 14 and interior angles of $2\pi/3$. We name the corners of the bottom base (in cyclic order) $x_0, \ldots, x_5$, and the corners at the top $x'_0, \ldots, x'_5$, such that x_i and x'_i are adjacent. All indices of these points are considered modulo 6. Let ℓ_i be the segment between x_{2i+1} and x'_{2i+4} for $i \in \{1,2,3\}$. For any ℓ_i we define ℓ'_i to be a copy of ℓ_i that is vertically shifted up by 0.2. Now, we subdivide all six segments in the middle and move the subdivision point vertically up by 1.08 in case of the ℓ'_is and vertically down by 1.08 in case of the ℓ_is. We define for all $i \in \{1,2,3\}$ the (red) polygon r_i as the convex hull of ℓ_i and ℓ'_i (including the translated subdivision point). Notice that the polygons are disjoint (see full version [7]). The convex hull of these polygons defines a convex polyhedron $\mathcal{P}$. We observe that $\mathcal{P}$ has eight triangular faces that are incident to all red polygons. These define the blue polygons (see Fig. 2).

Note that at the subdivision points we have one red polygon adjacent to two blue polygons. To resolve this issue we replace all subdivision points by an ε-small side such that the red polygons remain convex and all blue polygons still appear on the convex hull of $\{r_1, r_2, r_3\}$ (details are given in the full version [7]). We then move the corners of the adjacent blue polygons to two distinct endpoints of the new sides. It can be checked (see also full version [7]) that the constructed representation is one-sided (in particular, the red polygons are contained in $\mathcal{P}$) and therefore, by Lemma 1 it can be transformed into a (one-sided) side-contact representation. We summarize our result.

Theorem 1. *The graph $K_{3,8}$ has a side-contact representation with convex polygons in 3D.*

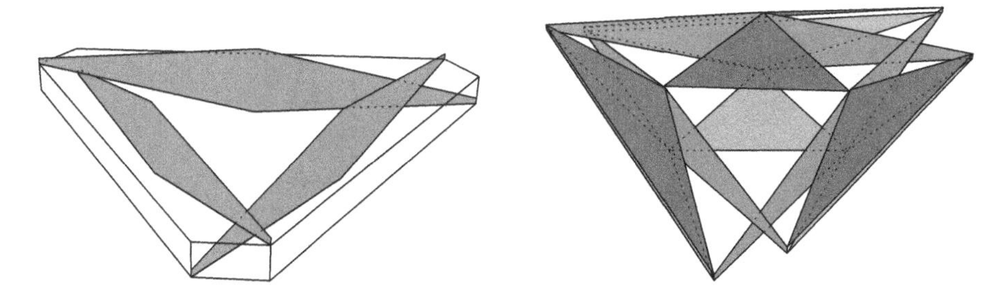

Fig. 2. The configuration of the red polygons and the prism (left). The full configuration (right). (Color figure online)

We remark that the side-contact representation of $K_{3,8}$ is one-sided. As a consequence of a result by Arseneva et al. [2, Lemma 11] no $K_{3,t}$ with $t > 8$ has a one-sided representation with side-contacts.

3 $K_{3,250}$ Has No Side-Contact Representation

In this section we prove the following result:

Theorem 2. *The graph $K_{3,250}$ has no side-contact representation with convex polygons in 3D.*

To prove Theorem 2 we present first some lemmas for configurations of (straight-line) segments in 2D. Thus, until mentioned otherwise, all configurations are considered in 2D from now on. We say that a set of segments $\mathcal{S}$ is convex if every $s \in \mathcal{S}$ lies on the convex hull of $\mathcal{S}$ and no two segments have the same slope. We allow that in a convex set of segments two segments share an endpoint. Assume that the elements of $\mathcal{S}$ are named such that the sequence $s_1, s_2 \ldots, s_m$ lists the segments according to their clockwise appearance on the convex hull. The intersection of the supporting lines of two segments s_i and s_j

is called *support intersection point* (si-point for shorthand notation) of s_i and s_j. If $j = i + 1$, or $i = 1$ and $j = m$, we call the si-point of s_i and s_j a *consecutive support intersection point* (csi-point for shorthand notation). Let $s_i = a_i b_i$ and $s_j = a_j b_j$ such that $a_i b_j$ is a proper segment on the convex hull of $\mathcal{S}$. The csi-point c_{ij} of s_i and s_j is called *flopped* if the line through $a_i b_j$ defines a closed half-space that contains c_{ij} and $\mathcal{S}$. See Fig. 3 for an illustration.

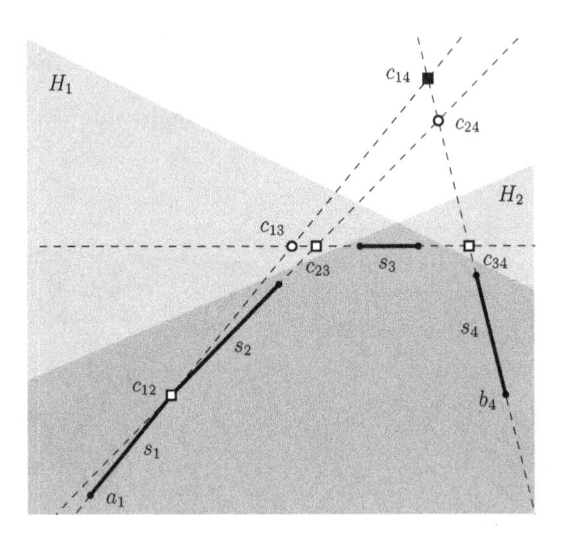

Fig. 3. Four segments in convex position. Csi-points are shown as squares. The only flopped csi-point c_{14} is filled. All other si-points are drawn as (empty) disks.

Lemma 2. *For any set $\mathcal{S}$ of segments in convex positions there is at most one csi-point that is flopped.*

Proof. Let s_i and s_j be two segments with csi-point c_{ij}. We can assume that $j = i + 1$. The point c_{ij} can only be flopped if the clockwise radial sweep of the tangent lines from s_i to s_j requires an angle larger than π, since we need to transition a state in which the tangent line is parallel to s_i. Since in a total angular sweep we rotate by 2π, this can happen only once. □

Lemma 3. *Let $\mathcal{S}$ be a set of at least four segments in convex position. Consider any two closed half-spaces H_1 and H_2 that (i) both contain $\mathcal{S}$, and (ii) no $s \in \mathcal{S}$ is completely part of the boundary of H_1 or H_2. Then at least one csi-point of S lies in the interior of $H_1 \cap H_2$.*

Proof. Assume first that $\mathcal{S}$ contains no flopped csi-point. Then the set of csi-points forms a convex set C. Furthermore, every edge of the convex hull of C contains exactly one segment of $\mathcal{S}$ completely. Consider now a closed half-space H that contains $\mathcal{S}$. If the interior of H misses two points from C, then an edge of the convex hull of C (and therefore a segment of $\mathcal{S}$) lies in the complement of the interior of H. Clearly H violates condition (i) or (ii) from the statement of the lemma in this case.

Now assume that we have a flopped csi-point c. We can augment $\mathcal{S}$ by adding a new segment such that the new set is convex and has no flopped csi-point. All csi-points other than c will remain. Thus, also in this situation, at most one nonflopped csi-point is not in the interior of H if the boundary of H contains no segment from $\mathcal{S}$ completely.

The interior of the intersection of any two closed half-spaces H_1 and H_2 fulfilling (i) and (ii) can therefore miss no more than one nonflopped csi-point per half-space, and possibly a flopped csi-point if it exists. By Lemma 2 there can only be one flopped csi-point. The statement of the lemma follows. □

We remark that the statement of Lemma 3 is "tight" as shown by the configuration in Fig. 3.

Lemma 4. *Let $\mathcal{S} = \{s_1, \ldots, s_m\}$ be a set of segments in convex position indexed in cyclic order. Assume that $s_1 = aa'$ and $s_m = bb'$ define a flopped csi-point c such that the segment ab lies on the convex hull of $\mathcal{S}$. Then all si-points of $\mathcal{S}$ lie inside the triangle Δ spanned by a, b and c.*

Proof. Let ℓ_i be the supporting line of s_i. We denote the intersection of ℓ_i with ℓ_j by x_{ij}. Notice that on ℓ_1 the following points appear in order: $a, x_{12}, x_{13}, \ldots, x_{1m}, c$. On ℓ_m however, we have the order: $b, x_{m(m-1)}, \ldots, x_{m2}, x_{m1}, c$. Both facts can be observed by radially sweeping a tangent line around the convex hull of $\mathcal{S}$.

Now consider any two segments $s_i, s_j \in \mathcal{S}$, with $1 < i < j < m$. Their si-point is denoted as x_{ij}. Since on ac the order of points is a, x_{1i}, x_{1j}, c and on bc the order of points is b, x_{mj}, x_{mi}, c the segments $x_{1i}x_{mi}$ and $x_{1j}x_{mj}$ have to cross inside Δ and hence the si-point defined by s_i and s_j lies in Δ (see also Fig. 4). We have already observed that all other relevant si-points (defined by either s_1 or s_m) lie on the boundary of Δ. □

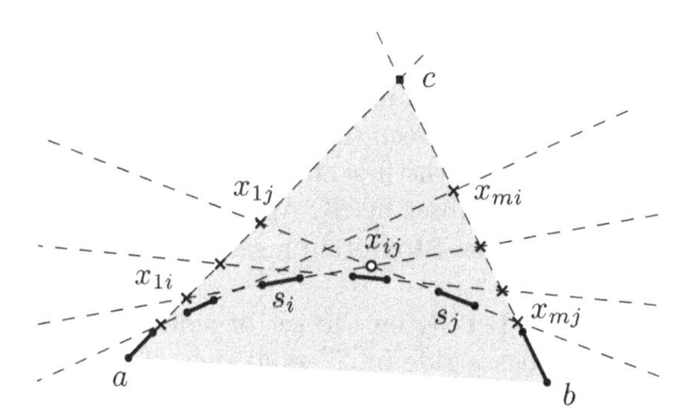

Fig. 4. Illustration of the proof of Lemma 3. The triangle Δ is shaded blue and the si-point induced by s_i and s_j is drawn as empty disk.

We now prove Theroem 2 and go back to 3D.

Proof. (Theorem 2). Assume that we have a side-contact representation of $K_{3,250}$. We call the polygons r_1, r_2, r_3 of the first partition class the *red polygons*. The polygons of the other partition class are called the *blue polygons*. The supporting plane of a polygon r_i is named $r_i^{=}$. Let $\mathcal{A}$ be the arrangement given by $r_1^{=}, r_2^{=}, r_3^{=}$. We can assume that the three planes intersect in a single point r_*, and that no two sides of a polygon are parallel. Otherwise we apply suitable (small) projective transformations to prevent parallel planes and lines without disconnecting the polygons. We call the eight (closed) cells of $\mathcal{A}$ *octants*. Note that every blue polygon has to lie in a single octant, since it has a side-contact with each of the red polygons. A red polygon can only be part of all octants if it contains r_*. Thus, at least two red polygons need to avoid r_* and "miss" at least two octants each, and only one of these octants can be the same. As a consequence there are at most 5 octants that have a piece of every red polygon on the boundary. One of them contains at least $50 = 250/5$ blue polygons. We denote this octant by $\mathcal{C}$.

Let f_i be the bounding face of $\mathcal{C}$ that contains r_i and denote the interior of f_i by $\tilde{f}_i$. Further let ρ_{ij} be $f_i \cap f_j$. Both r_1 and r_2, can have at most one side fully contained in ρ_{12} and no side from r_3 can be completely in ρ_{12} since $\rho_{12} \cap r_3^{=} = \{r_*\}$. Thus, $\rho := \rho_{12} \cup \rho_{23} \cup \rho_{13}$ contains at most six complete sides from blue polygons in $\mathcal{C}$. We ignore any blue polygon with a full side in ρ and remain with a set B of at least 44 blue polygons.

First, we consider the polygon r_1 and select a set $\mathcal{S}_1$ of 44 of its sides that are incident to some polygon in B. As usual, we label the segments $s_1, s_2, \ldots, s_{44}$ cyclically and set $\mathcal{S}_1' = \{s_1, s_{12}, s_{23}, s_{34}\}$. The face f_1 can be obtained by intersecting $r_1^{=}$ with two closed half-spaces. No segments of $\mathcal{S}_1'$ lies completely on the boundary of f_1 and thus, by Lemma 3 at least one csi-point, say c, of $\mathcal{S}_1'$ lies in $\tilde{f}_1$. Take the two segments aa' and bb' (with $a'b'$ on the convex hull on $\mathcal{S}_1'$) defining c and all of the ten segments of $\mathcal{S}$ in between them in the cyclic order. We call this set $\mathcal{S}_1''$. Note that this set has a flopped csi-point, which is c. Since a, b, c lie in $\tilde{f}_1$ we have by Lemma 4 that all si-points of $\mathcal{S}_1''$ lie in $\tilde{f}_1$.

We now deal with polygon r_2. Let $\mathcal{S}_2$ be the set of sides of r_2 that share a side with a blue polygon that has a side in $\mathcal{S}_1''$. We get that $|\mathcal{S}_2| = 12$. We sort the segments in $\mathcal{S}_2$ again by a radial sweep (notice that the order might be different than in $\mathcal{S}_1''$). This time we select the first, fourth, seventh and tenth segment in this order and we denote this subset by $\mathcal{S}_2'$. Again, we apply Lemma 3 to find a csi-point in $\tilde{f}_2$ and then Lemma 4 to obtain a set $\mathcal{S}_2''$ of (this time 4) segments, whose si-points are all in $\tilde{f}_2$.

Finally, we consider r_3. Let $\mathcal{S}_3$ be the set of sides of r_3 that share a side with a blue polygon that has a side in $\mathcal{S}_2''$ (and therefore in $\mathcal{S}_1''$ as well). Since $|\mathcal{S}_2| = 4$ we get by Lemma 3 that one csi-point of $\mathcal{S}_3$ lies in $\tilde{f}_3$. Two segments of $\mathcal{S}_3$ define this point. Call the adjacent blue polygons b_1 and b_2, with supporting planes $b_1^{=}$ and $b_2^{=}$. We denote the restriction of $b_1^{=}/b_2^{=}$ to the boundary of $\mathcal{C}$ by t_1/t_2. By our construction, t_1 and t_2 intersect on $\tilde{f}_3$ in a csi-point. But both blue polygons have also a common side with each of the sets $\mathcal{S}_2''$ and $\mathcal{S}_1''$. As a consequence, t_1 and t_2 intersect in an si-point of $\mathcal{S}_1''$ inside $\tilde{f}_1$ and in an si-point of $\mathcal{S}_2''$ inside $\tilde{f}_2$. The three si-points are distinct and define a plane. We get that $b_{\overline{1}}^{=} = b_{\overline{2}}^{=}$. However, if two blue polygons lie in the same plane, all red polygons and therefore all blue polygons have to lie in this plane as well. Since $K_{3,250}$ is nonplanar, it has no side-contact representation in the plane, and we have obtained the desired contradiction. □

The following is now a simple consequence from the Kővari–Sós–Turán theorem [5], which states that an n-vertex graph that has no $K_{s,t}$ as a subgraph can have at most $O(n^{2-1/s})$ edges.

Corollary 1. *Let G be an n-vertex graph with a side-contact representation of convex polygons in 3D. Then the number of edges in G is bounded by $O(n^{5/3})$.*

References

1. Alam, J., Evans, W., Kobourov, S., Pupyrev, S., Toeniskoetter, J., Ueckerdt, T.: Contact representations of graphs in 3D. In: Dehne, F., Sack, J.-R., Stege, U. (eds.) WADS 2015. LNCS, vol. 9214, pp. 14–27. Springer, Cham (2015). https://doi.org/10.1007/978-3-319-21840-3_2
2. Arseneva, E., et al.: Adjacency graphs of polyhedral surfaces. In: Buchin, K., de Verdière, ÉC. (eds.) Proceedings of the 37th International Symposium on Computational Geometry (SoCG '21), volume 189 of LIPIcs, pp. 1–17. Schloss Dagstuhl - Leibniz-Zentrum für Informatik (2021). https://arxiv.org/abs/2103.09803v2
3. Evans, W., Rzazewski, P., Saeedi, N., Shin, C.-S., Wolff, A.: Representing graphs and hypergraphs by touching polygons in 3D. In: Archambault, D., Tóth, C.D. (eds.) GD 2019. LNCS, vol. 11904, pp. 18–32. Springer, Cham (2019). https://doi.org/10.1007/978-3-030-35802-0_2
4. Felsner, S., Francis, M.C.: Contact representations of planar graphs with cubes. In: Hurtado, F., van Kreveld, M.J. (eds.) Proceedings of the 27th Annual Symposium On Computational Geometry (SoCG'11), pp. 315–320. ACM (2011)
5. Kővari, T., Sós, V.T., Turán, P.: On a problem of K. Zarankiewicz. Coll. Math. **3**(1), 50–57 (1954)
6. Kleist, L., Rahman, B.: Unit contact representations of grid subgraphs with regular polytopes in 2D and 3D. In: Duncan, C., Symvonis, A. (eds.) GD 2014. LNCS, vol. 8871, pp. 137–148. Springer, Heidelberg (2014). https://doi.org/10.1007/978-3-662-45803-7_12
7. Schulz, A.: Side-contact representations with convex polygons in 3D: new results for complete bipartite graphs. CoRR, abs/2308.00380 (2023). http://arxiv.org/abs/2308.00380
8. Thomassen, C.: Interval representations of planar graphs. J. Combin. Theory Ser. B **40**(1), 9–20 (1986)
9. Tietze, H.: Über das Problem der Nachbargebiete im Raum. Monatshefte für Mathematik und Physik **16**(1), 211–216 (1905)

Mutual Witness Proximity Drawings of Isomorphic Trees

Carolina Haase[1(✉)], Philipp Kindermann[1], William J. Lenhart[2], and Giuseppe Liotta[3]

[1] Universität Trier, Trier, Germany
{haasec,kindermann}@uni-trier.de
[2] Williams College, Williamstown, USA
wlenhart@williams.edu
[3] Università degli Studi di Perugia, Perugia, Italy
giuseppe.liotta@unipg.it

Abstract. A pair $\langle G_0, G_1 \rangle$ of graphs admits a mutual witness proximity drawing $\langle \Gamma_0, \Gamma_1 \rangle$ when: (i) Γ_i represents G_i, and (ii) there is an edge (u, v) in Γ_i if and only if there is no vertex w in Γ_{1-i} that is "too close" to both u and v $(i = 0, 1)$. In this paper, we consider infinitely many definitions of closeness by adopting the β-proximity rule for any $\beta \in [1, \infty]$ and study pairs of isomorphic trees that admit a mutual witness β-proximity drawing. Specifically, we show that every two isomorphic trees admit a mutual witness β-proximity drawing for any $\beta \in [1, \infty]$. The constructive technique can be made "robust": For some tree pairs we can suitably prune linearly many leaves from one of the two trees and still retain their mutual witness β-proximity drawability. Notably, in the special case of isomorphic caterpillars and $\beta = 1$, we construct linearly separable mutual witness Gabriel drawings.

Keywords: Mutual witness proximity drawings · β-proximity · Trees

1 Introduction

Proximity drawings are geometric graphs (i.e., straight-line drawings) such that any two vertices are connected by an edge if and only if they are deemed to be close according to some definition of closeness. Therefore, proximity drawings are such that pairs of non-adjacent vertices are relatively far apart while highly connected subgraphs correspond to groups of vertices that can be naturally clustered together in a visual inspection.

In this paper, we investigate *mutual witness proximity drawings*, which employ the concept of closeness to simultaneously represent pairs of graphs.

Research partially supported by: (i) MUR PRIN Proj. 2022TS4Y3N - "EXPAND: scalable algorithms for EXPloratory Analyses of heterogeneous and dynamic Networked Data"; (ii) MUR PRIN Proj. 2022ME9Z78 - "NextGRAAL: Next-generation algorithms for constrained GRAph visuALization".

M. A. Bekos and M. Chimani (Eds.): GD 2023, LNCS 14465, pp. 304–319, 2023.
https://doi.org/10.1007/978-3-031-49272-3_21

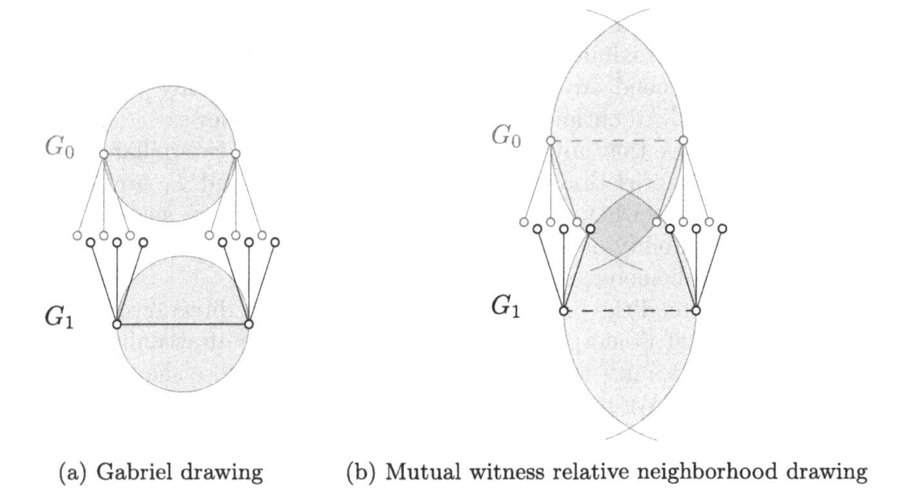

(a) Gabriel drawing (b) Mutual witness relative neighborhood drawing

Fig. 1. Two mutual witness drawings on the same point set.

Specifically, consider a pair of graphs, denoted as $\langle G_0, G_1 \rangle$. The pair admits a mutual witness proximity drawing, denoted as $\langle \Gamma_0, \Gamma_1 \rangle$, under the following conditions: (i) Γ_i represents G_i, and (ii) an edge (u, v) exists in Γ_i if and only if there is no vertex w in Γ_{1-i} that is "too close" to both u and v (where $i = 0, 1$). Vertex w is called a *witness* and its proximity to u and v impedes the presence of the edge. Clearly, by changing the definition of proximity a pair of graphs may or may not admit a mutual witness proximity drawing.

There is general consensus in the literature to define the closeness of w to both u and v by means of a *proximity region* of u and v, which is a convex region in the plane whose area increases when the distance between u and v increases. For example, the *Gabriel region* [8] of u and v is the disk whose diameter is the line segment $\overline{uv}$; the witness w is close to u and v if it is a point of their Gabriel disk. A mutual witness Gabriel drawing of a pair $\langle G_0, G_1 \rangle$ is therefore a pair of drawings Γ_0 of G_0 and Γ_1 of G_1 such that for any two non-adjacent vertices in one drawing their Gabriel disk contains a witness from the other drawing, while for any two adjacent vertices their Gabriel region does not contain any witnesses. Figure 1a shows a mutual witness Gabriel drawing of two caterpillars. As another example, the *relative neighborhood region* [20] of u and v is the intersection of the two disks of radius $d(u, v)$ centered at u and v, respectively. Figure 1b depicts a mutual proximity drawing that adopts the relative neighborhood region: The drawing has the same vertex set but fewer edges than the drawing in Fig. 1a.

We want to understand what families of graph pairs admit a mutual witness proximity drawing for a given definition of proximity. Intuitively, the denser the two graphs are, the more likely they admit such a representation: If the graphs are complete, we can draw them sufficiently far apart so that the proximity regions of their edges do not contain any witnesses. On the other hand, when the graphs are sparse there are many non-adjacent vertices requiring the presence of witnesses in their proximity regions, which makes the geometry of the two drawings strongly depend on one another. We specifically study very sparse graphs, namely trees. An outline of our contribution is as follows.

In Sect. 4, we prove that any pair $\langle G_0, G_1 \rangle$ of isomorphic caterpillars admits a mutual witness Gabriel drawing $\langle \Gamma_0, \Gamma_1 \rangle$ such that Γ_0 and Γ_1 are linearly separable. This is somewhat surprising as caterpillars are very sparse graphs and the linear separability of mutual witness Gabriel drawings was known only for graphs of small diameter, namely at most two [14].

In Sect. 5, we extend the previous result in two different directions: We consider pairs of general isomorphic trees and we study their drawability for an infinite family of proximity regions called *β-regions* [13], whose shape depends on a parameter $\beta \in \mathbb{R}$. We show that any pair $\langle G_0, G_1 \rangle$ of isomorphic trees admits a mutual witness proximity drawing for any β-region such that $\beta \in [1, \infty]$. While the two drawings are no longer linearly separable, they have the property that the coordinates of their vertex sets remain the same for any possible value of β. It is worth recalling that the Gabriel disk is the β-region for $\beta = 1$ and that the relative neighborhood region corresponds to the β-region for $\beta = 2$.

In Sect. 6, we investigate the "robustness" of the construction of Sect. 5: We show that for some tree pairs, this construction can be modified so that the drawing remains valid even after pruning a suitable set of leaves. While it is known that any two star trees admit a mutual witness Gabriel drawing if and only if the cardinalities of their vertex sets differ by at most two [14], we show that there exist tree pairs which can differ by linearly many leaves and still admit a mutual witness proximity drawing for any β-region such that $\beta \in [1, \infty]$.

Results marked with a "$\star$" are proved in the full version of the paper [9].

2 Related Work

Proximity drawings are a classical research topic in graph drawing; they find application in several areas, including pattern recognition, data mining, machine learning, computational biology, and computational morphology. Proximity drawings have also been used to determine the faithfulness of large graph visualizations. A limited list of references includes [7,11,15–17,21].

In the context of designing trained classifiers, mutual witness proximity drawings were first introduced by Ichino and Slansky [10] under the name of *interclass rectangle of influence graphs*. In [10] the proximity region of a pair of vertices, called the *rectangle of influence*, is the smallest axis-aligned rectangle containing the two vertices. This study was then extended to other families of proximity regions, including the Gabriel region, in a sequence of papers by Aronov et al. [1–4]. Notably, in [4] it is said that once the combinatorial properties of those pairs of graphs that admit a mutual witness Gabriel drawing are understood, *"we would have useful tools for the description of the interaction between two point sets"*. Aronov et al. prove in [3] that any pair of complete graphs admits a mutual witness Gabriel drawing where the two drawings are linearly separable. The linear separability property of mutual witness Gabriel drawings is

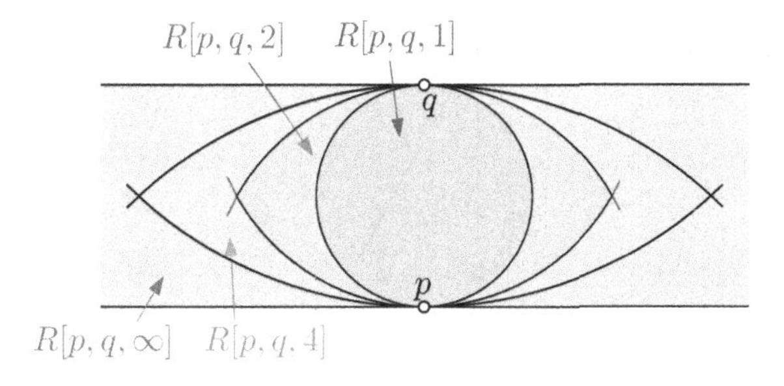

Fig. 2. Examples of β-proximity regions for $\beta \geq 1$.

extended to diameter-2 graphs by Lenhart and Liotta, who also give a complete characterization of those complete bipartite graphs that admit a mutual witness Gabriel drawing [14]. Another related contribution of Aronov et al. [1–4] is to introduce and study *witness proximity drawings*, which can be shortly described as a relaxation of mutual proximity drawings where one of the two drawings has no edges, independently of whether the proximity regions of its vertices do or do not contain any witnesses.

3 Preliminaries

We assume familiarity with basic graph drawing concepts; see e.g. [5,12,18,19].

Let p and q be two distinct points in the plane. We denote by $\overline{pq}$ the straight-line segment having p and q as its extreme points. We define β-regions adopting the notation in [6]. A region in the plane is *open* if it is an open set, that is the points on its boundary are not part of the region, and *closed* if all of the points of the boundary are part of the region. Given a pair p, q of points in the plane and a real number $\beta \in [1, \infty]$, the *open β-region* of p and q, denoted by $R(p, q, \beta)$, is defined as follows. For $1 \leq \beta < \infty$, $R(p, q, \beta)$ is the intersection of the two open disks of radius $\beta d(p, q)/2$ and centered at the points $(1 - \beta/2)p + (\beta/2)q$ and $(\beta/2)p + (1 - \beta/2)q$. $R(p, q, \infty)$ is the open infinite strip perpendicular to the line segment $\overline{pq}$ and for $\beta \in [1, \infty]$, the closed β-region $R[p, q, \beta]$ is simply the open region $R(p, q, \beta)$ along with its boundary; see Fig. 2.

Note that $R[p, q, 1]$ is the Gabriel region of p, q and that $R(p, q, 2)$ is the relative neighborhood region of p, q. We shall denote as a *MW-$[\beta]$ drawing* a mutual witness proximity drawing such that for any two vertices p and q the proximity region is $R[p, q, \beta]$. In particular, an *MW-$[1]$ drawing* is a mutual witness Gabriel drawing. Similarly, a *MW-(β) drawing* is a mutual witness proximity drawing that uses the open β-region.

As we shall see, some of our constructive arguments produce drawings that are simultaneously *MW-(β)* and *MW-$[\beta]$* drawings; in this case we refer to them simply as *MW-β drawings*. Note that for any pair p, q of vertices in an MW-β drawing, $R(p, q, \beta)$ contains a witness if p and q are not adjacent, while $R[p, q, \beta]$ contains no witnesses if p and q are adjacent.

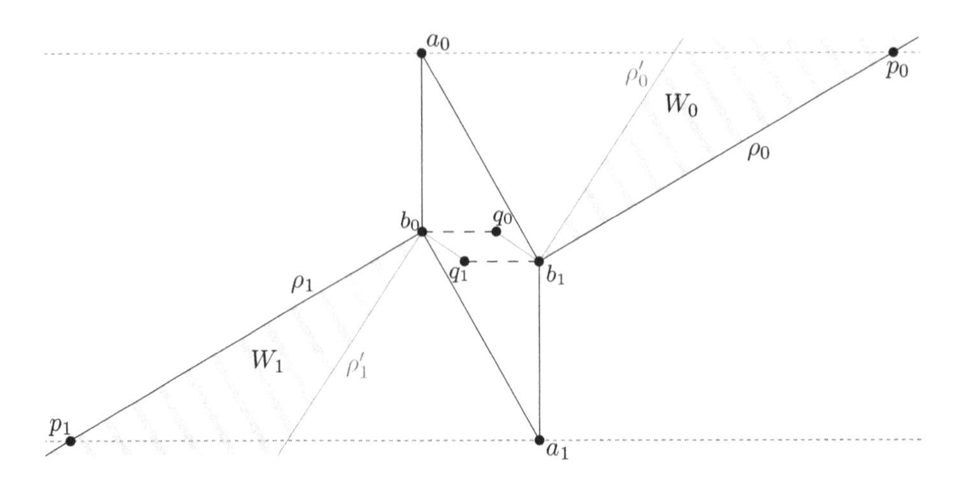

Fig. 3. A winged parallelogram with anchors q_0, q_1, safe wedges W_0, W_1, and ports p_0, p_1.

Let $\langle \Gamma_0, \Gamma_1 \rangle$ be an MW-β drawing of graphs $\langle G_0, G_1 \rangle$ for some value of β. We say that the drawing is *linearly separable* if there exists a line ℓ such that Γ_0 and Γ_1 lie in opposite half-planes with respect to ℓ. The following property rephrases an observation of [14] and will be used in the proof of Theorem 1.

Property 1. Let $\langle \Gamma_0, \Gamma_1 \rangle$ be a linearly separable MW-[1] drawing and let u and v be any two non-adjacent vertices of Γ_i, for $i = 0, 1$. Then any witness in $R[u, v, 1]$ is also a point in $R[u, v, \infty]$

4 MW-[1] Drawings of Isomorphic Caterpillars

A *caterpillar* is a tree T such that, when removing the leaves of T one is left with a non-empty path called *spine* of T. We call the graph $K_{1,n}$ with $n \geq 1$ a *star*; if $n > 1$ the non-leaf vertex of a star is the *center* of the star, otherwise (i.e., when the star is an edge) either vertex can be chosen as the center.

In this section, we prove that any two isomorphic caterpillars admit a linearly separable MW-[1] drawing, that is they admit a linearly separable mutual witness Gabriel drawing. As pointed out both in [3] and in [10], the linear separability of mutual witness proximity drawings is a desirable property because it gives useful information about the inter-class structure of two sets of points.

Let $P = \langle a_0, b_0, a_1, b_1 \rangle$ be a parallelogram such that $y(a_0) > y(b_0) > y(b_1) > y(a_1)$ and $x(a_0) = x(b_0) < x(a_1) = x(b_1)$. Let q_0 and q_1 be two points in the interior of P satisfying $y(b_i) = y(q_i)$, $x(q_1) < x(q_0)$, and $x(q_0) - x(b_0) = x(b_1) - x(q_1)$. Let W_i be the wedge with apex b_i not containing any vertex of P other than b_i and defined by two rays ρ_i, ρ_i' such that ρ_i is perpendicular to $\overline{a_i b_{1-i}}$ and ρ_i' is perpendicular to $\overline{q_i b_{1-i}}$. We call W_i *safe wedges* of P and the q_i *anchors*. We assume W_i to be an open set. Finally, we identify two *ports*, the points

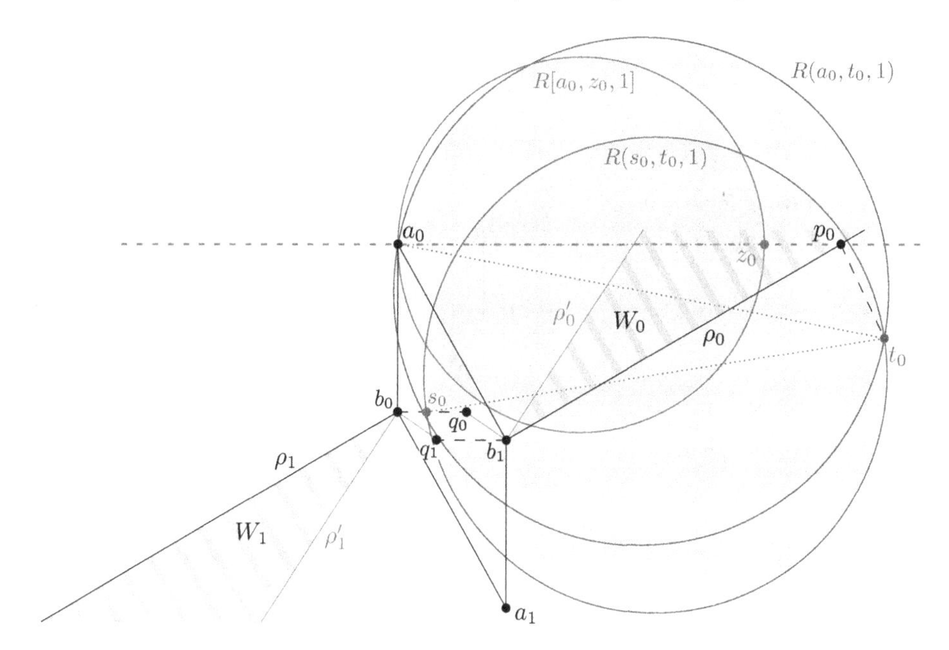

Fig. 4. Illustration for Property 2. (P1) Neither a_1 nor any points of $\overline{q_1 b_1}$ are points of $R[a_0, z_0, 1]$; (P2) $b_1 \in R(s_0, t_0, 1)$; (P3) $b_1 \in R(a_0, t_0, 1)$.

p_i, where p_i is the point along ρ_i such that $y(p_i) = y(a_i)$. The parallelogram P together with its anchors, safe wedges, and ports is called a *winged parallelogram* $WP(P, q_0, q_1, W_0, W_1, p_0, p_1)$. Figure 3 shows an example of a winged parallelogram. The following property is an immediate consequence of the definition of winged parallelogram; see also Fig. 4.

Property 2. Let $WP(P, q_0, q_1, W_0, W_1, p_0, p_1)$ be a winged parallelogram such that the interior angles at points a_i $(i = 0, 1)$ are at most $\frac{\pi}{4}$. Let s_i, t_i, z_i $(i = 0, 1)$ be any three points such that $s_i \in \overline{b_i q_i}$, $t_i \notin W_i$ with $x(t_0) \geq x(p_0)$, $x(t_1) \leq x(p_1)$, and $z_i \in W_i$ with $y(z_i) = y(a_i)$. Then: (P1) neither s_{1-i} nor a_{1-i} are points of $R[a_i, z_i, 1]$; (P2) $b_{1-i} \in R(s_i, t_i, 1)$; (P3) $b_{1-i} \in R[a_i, t_i, 1]$ if t_i on ρ_i and $b_{1-i} \in R(a_i, t_i, 1)$ if t_i is not on ρ_i.

We first show how to draw pairs of isomorphic stars into a winged parallelogram and then generalize the construction to pairs of isomorphic caterpillars.

Lemma 1 ($\star$). *Let $\langle T_0, T_1 \rangle$ be a pair of isomorphic stars such that, for $i = 0, 1$, T_i has root r_i and leaves $v_{i,0}, \ldots, v_{i,k}$. Then $\langle T_0, T_1 \rangle$ admits an MW-[1] drawing $\langle \Gamma_0, \Gamma_1 \rangle$ contained in a winged parallelogram $WP(P, q_0, q_1, W_0, W_1, p_0, p_1)$ such that: (i) r_i is drawn at a_i and the the internal angle of $WP(P, q_0, q_1, W_0, W_1, p_0, p_1)$ at a_i is at most $\frac{\pi}{4}$; (ii) $v_{i,0}$ is drawn at b_i; and (iii) for $0 < j \leq k$, $v_{i,j}$ is drawn at an interior point of the segment $\overline{b_i q_i}$.*

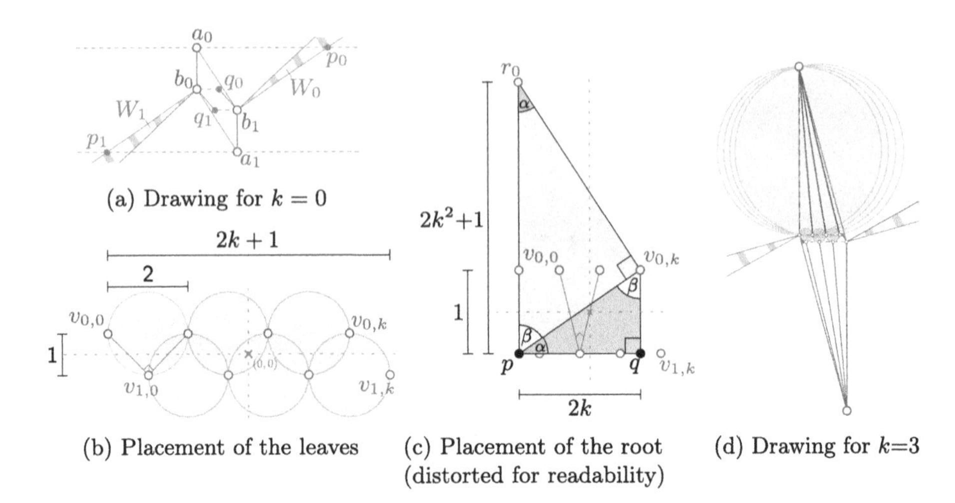

Fig. 5. Illustration for the Proof of Lemma 1.

Proof sketch. For $i = 0, 1$, if T_i has only one leaf, the construction is trivial; see Fig. 5a. Otherwise, we draw the leaves of T_i uniformly spaced along a horizontal segment σ_i and then place σ_0 and σ_1 relative to each other so that for every pair of consecutive leaves of T_i, there is a witness for that pair among the leaves of T_{1-i}; see Fig. 5b.

The horizontal line midway between σ_0 and σ_1 will form a separating line for $\langle \Gamma_0, \Gamma_1 \rangle$ once the centers r_i of T_i are placed. The center r_0 of T_0 is then placed vertically above the leftmost leaf of T_0 and the center r_1 of T_1 is placed vertically below the rightmost leaf of T_1, each center far enough from the separating line so that for $i = 0, 1$ and $0 \leq j \leq k$, no proximity region $R[r_i, v_{i,j}, 1]$ contains any witness from T_{1-i}; see Fig. 5c.

□

In the following we call an MW-[1] drawing $\langle \Gamma_0, \Gamma_1 \rangle$ of two isomorphic stars computed as in the proof of Lemma 1 a *WP-drawing on P* and say that the winged parallelogram *supports* the drawing; see Fig. 6. Note that, by construction, the horizontal line L having $y(L) = (y(b_0) + y(b_1))/2$ is a separating line for the WP-drawing of two isomorphic stars.

Lemma 2. *Let $\langle \Gamma_0, \Gamma_1 \rangle$ be a WP-drawing of two isomorphic stars $\langle T_0, T_1 \rangle$ and let P be the winged parallelogram that supports $\langle \Gamma_0, \Gamma_1 \rangle$. Then, any pair $\langle T_0', T_1' \rangle$ of isomorphic stars with at least one leaf and $T_i' \subset T_i$ has a WP-drawing on P.*

Proof. Let r_i be the root of T_i and $v_{i,0}, \ldots, v_{i,k}$ be the leaves of T_i. Consider the drawing $\langle \Gamma_0, \Gamma_1 \rangle$ computed in Lemma 1; see Fig. 5. We use the same notation as in the proof of Lemma 1. Remove all leaves $v_{i,j}$ that are not in T_i' and reposition the remaining leaves uniformly along σ_i as in the proof of Lemma 1.

By construction, the Gabriel region $R[v_{0,i}, v_{0,j}, 1]$ for every $v_{0,i}, v_{0,j} \in T_i, 1 \leq i < j \leq k$ still contains the vertex $v_{1,i}$, while the Gabriel region $R[v_{1,i}, v_{1,j}, 1]$

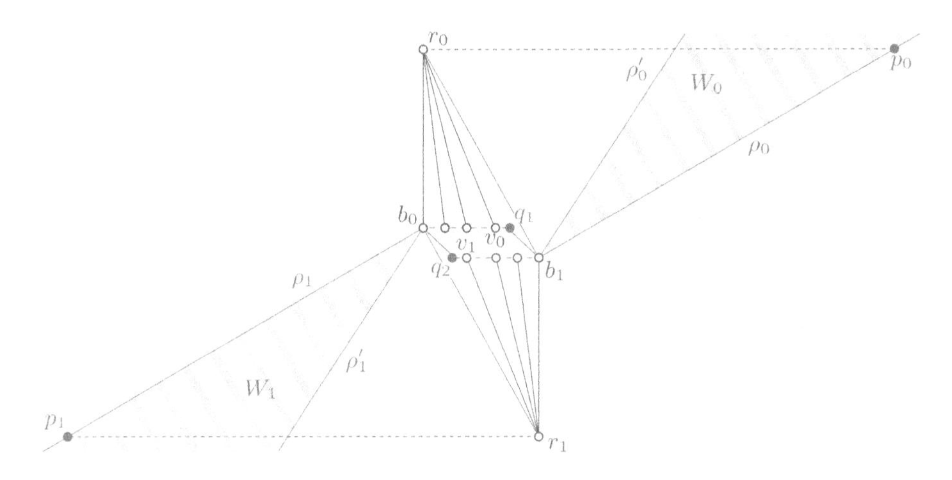

Fig. 6. A WP-drawing of two isomorphic stars on a parallelogram P

for every $v_{1,i}, v_{1,j} \in T_i, 1 \leq i < j \leq k$ still contains the vertex $v_{1,j}$. Otherwise, if $v_{i,0} \notin T'_i$, then take any leaf $v_{i,j} \in T'_i$, switch its position with $v_{i,0}$ in Γ_i, and then proceed as above. □

Theorem 1. *Any pair $\langle T_0, T_1 \rangle$ of isomorphic caterpillars admits a linearly separable MW-[1] drawing.*

Proof. For $i = 0, 1$, if each T_i is a path, the pair can easily be realized by two horizontal paths, such that corresponding vertices of T_0 and T_1 have the same x-coordinates, all edges have the same length and the y-distance between T_0 and T_1 is at most the edge length. So we can assume that the spine of T_i is a path such that at least one spine vertex has degree greater than two.

Let $r_{i,0}, \ldots, r_{i,k}$ be the spine vertices of T_i in the order that they appear along the spine. Decompose T_i into subtrees $T_{i,0}, \ldots T_{i,k}$, having roots $r_{i,0}, \ldots, r_{i,k}$ respectively. Note that each $\langle T_{0,j}, T_{1,j} \rangle$ is either an isomorphic pair of stars with centers $r_{0,j}$ and $r_{1,j}$, respectively, or it is a pair of isolated vertices.

Let h be an index such that $r_{i,h}$ is a vertex of highest degree in T_i. Compute a WP-drawing $\langle \Gamma_{0,h}, \Gamma_{1,h} \rangle$ of $\langle T_{0,h}, T_{1,h} \rangle$ by means of Lemma 1 and let $WP_h = WP(P_h, q_{0,h}, q_{1,h}, W_{0,h}, W_{1,h}, p_{0,h}, p_{1,h})$ be the winged parallelogram that supports the drawing. Let $N = y(r_{0,h})$ and $S = y(r_{1,h})$ and let L_0 and L_1 be the two horizontal lines at heights N and S, respectively. We will construct a MW-[1] drawing of the two caterpillars such that all spine vertices of T_i lie on L_i and such that the horizontal line L at height $(N+S)/2$ separates T_0 from T_1.

For any $0 \leq j \leq k$, $j \neq h$ such that $r_{i,j}$ has at least one leaf, we use Lemma 2 to compute a WP-drawing $\langle \Gamma_{0,j}, \Gamma_{1,j} \rangle$ of $\langle T_{0,j}, T_{1,j} \rangle$ in a winged parallelogram WP_j congruent to WP_h that will be placed so that $r_{i,j}$ lies on L_i. For any $0 \leq j \leq k$, $j \neq h$ such that $r_{i,j}$ has no children, we will place $r_{i,j}$ on L_i so that the line through $r_{0,j}, r_{1,j}$ is perpendicular to the line through $r_{1,h}, p_{0,h}$.

We now describe how to place each pair $\langle \Gamma_{0,j}, \Gamma_{1,j} \rangle$; note that placing $r_{0,j}$ completely determines the placement of $\langle \Gamma_{0,j}, \Gamma_{1,j} \rangle$. Vertex $r_{0,0}$ can be

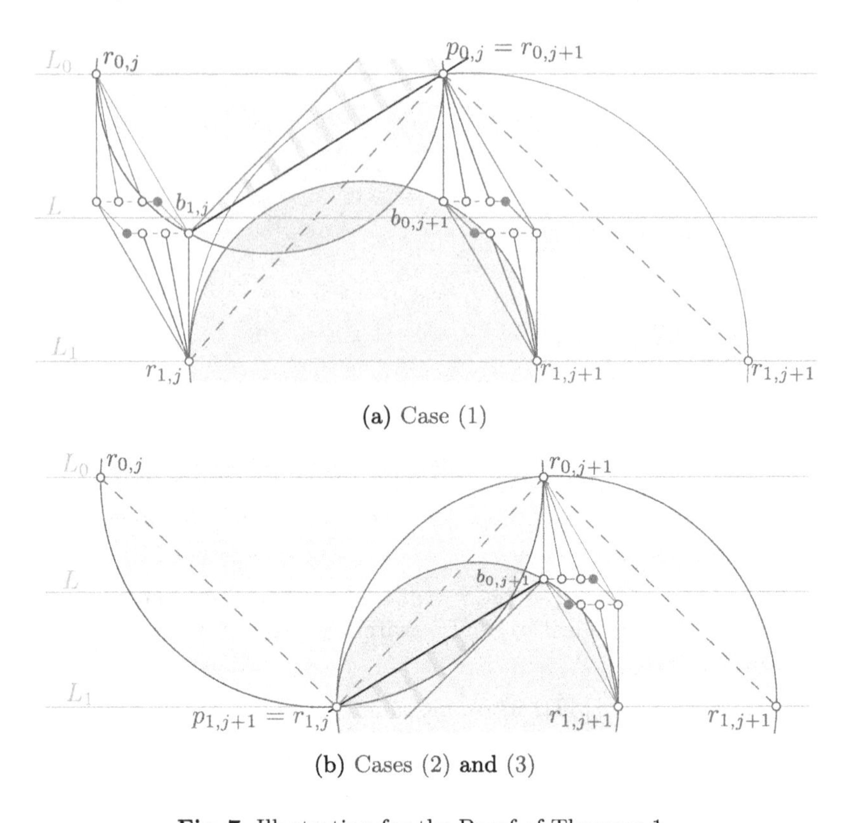

(a) Case (1)

(b) Cases (2) and (3)

Fig. 7. Illustration for the Proof of Theorem 1.

placed arbitrarily along L_0. Assume now that, for some $j \geq 0$, the pairs $\{r_{0,0}, r_{1,0}\}, \ldots \{r_{0,j}, r_{1,j}\}$ have been placed along L_0 and L_1. We describe how to place $r_{0,j+1}$. There are three cases; see Fig. 7: (1) If $r_{0,j}$ has at least one leaf, place $r_{0,j+1}$ at port $p_{0,j}$. (2) If $r_{0,j}$ has no leaves and $r_{0,j+1}$ has at least one leaf, place $r_{0,j+1}$ so that $r_{1,j}$ is at port $p_{1,j+1}$. (3) If both $r_{0,j}$ and $r_{0,j+1}$ have no leaves, place $r_{0,j+1}$ at the intersection of L_0 with the line through $r_{1,j}$ that is perpendicular to $\overline{r_{0,j}r_{1,j}}$.

This construction is *almost* an MW-[1] drawing of $\langle T_0, T_1 \rangle$. Consider the mutual witness Gabriel drawing Γ induced by the placement of the vertices of $\langle T_0, T_1 \rangle$ described above. Note that in our constructed drawing: (i) The pairs $\langle T_{0,j}, T_{1,j} \rangle$ are drawn in vertically disjoint strips and by Property 1 form MW-[1] drawings of those pairs. (ii) For any non-spine vertex $u_{0,j} \in T_{0,j}$, and *any* vertex $u_{0,t} \in T_{0,t}$ $(0 \leq j < t \leq k)$, by Property 2 (P2), $b_{1,j} \in R(u_{0,j}, u_{0,t}, 1)$ and so the pair $\{u_{0,j}, u_{0,t}\}$ is not an edge in Γ. (iii) For any spine vertex $r_{0,j} \in T_{0,j}$, and non-spine vertex $u_{0,t} \in T_{0,t}$ $(0 \leq j < t \leq k)$, either $r_{0,j}$ has a leaf, and so by Property 2 (P3), $b_{1,j} \in R(r_{0,j}, u_{0,t}, 1)$ or $r_{0,j}$ has no leaves and $r_{1,j} \in R[r_{0,j}, u_{0,t}, 1]$ by the construction described above. Similar statements hold for pairs of vertices in T_1 by the symmetry of the construction.

The drawing Γ is not yet an MW-[1] drawing of $\langle T_0, T_1 \rangle$ because there are *no* edges in Γ between *any* pair of consecutive spine vertices of T_i. This problem can be easily rectified, however. Note that in Γ there are only two types of non-adjacent vertex pairs that only have witnesses on the boundaries of their Gabriel regions (that is, that only have witnesses forming right angles), namely, consecutive leaves in an individual subtree $T_{i,j}$, and consecutive spine vertices in T_i. Let $r_{i,j}$ and $r_{i,j+1}$ be any two consecutive spine vertices of T_i. We can always perturb Γ so that by very slightly moving to the left all vertices of $\langle T_{0,j+1}, T_{1,j+1} \rangle$, we have, by Property 2 (P1), that $R[r_{i,j}, r_{i,j+1}, 1]$ contains no witnesses while for every other pair of non-adjacent vertices their Gabriel regions still contain a witness. Once all spine vertices have been properly connected, the resulting drawing is a linearly separable MW-[1] drawing of $\langle T_0, T_1 \rangle$. □

5 MW-β Drawings of Isomorphic Trees

In this section we show that, at the expense of losing linear separability, the result of Theorem 1 can be extended to any two isomorphic trees and to any mutual witness proximity drawing that adopts either the open or the closed β-region for all values of $\beta \in [1, \infty]$. A nice property of our algorithm is that it does not depend on the exact choice of β, i.e., it produces a single drawing that is an MW-β proximity drawing for every $\beta \geq 1$.

Similar to the previous section, we show a construction to recursively draw subtrees inside suitable parallelograms, which are however not winged parallelograms. We start by defining these parallelograms. In the remainder of the section, we shall sometimes assume that our trees are rooted, in which case we denote as (T, r) a tree T with root r.

Let $P = \langle a_0, b_0, a_1, b_1 \rangle$ be a parallelogram where $\overline{a_0 a_1}$ is the longer diagonal and no angle is equal to $\frac{\pi}{2}$. We say that P is *nicely oriented* if $y(a_0) > y(b_1) > y(b_0) > y(a_1)$ and $x(a_0) < x(b_0) < x(b_1) < x(a_1)$; see Fig. 8a.

Let $\langle (T_0, r_0), (T_1, r_1) \rangle$ be a pair of isomorphic rooted trees with n vertices each. An *MW-β parallelogram drawing* of $\langle (T_0, r_0), (T_1, r_1) \rangle$ is an MW-β proximity drawing $\langle \Gamma_0, \Gamma_1 \rangle$ contained in a nicely oriented parallelogram $P = \langle a_0, b_0, a_1, b_1 \rangle$ such that, for $i = 0, 1$, the following holds: (i) point a_i represents the root r_i of T_i; (ii) if $n > 0$, point b_i represents a vertex of T_i adjacent to r_i; (iii) for every other vertex $v_i \in \Gamma_i$ such that v_i is neither the root of T_i nor the vertex at b_i, we have $y(b_1) > y(v_i) > y(b_0)$; (iv) no edge of Γ_i is a vertical segment. Figure 8b shows an example of an MW-1 parallelogram drawing.

Theorem 2. *Any two isomorphic trees $\langle T_0, T_1 \rangle$ admit a parallelogram drawing that is an MW-β-drawing for all $\beta \in [1, \infty]$.*

Proof. Let r_0 be any vertex of T_0 and let $r_1 \in T_1$ be the isomorphic image of r_0. We will show by induction on the depth δ of (T_0, r_0) that $\langle (T_0, r_0), (T_1, r_1) \rangle$ admits an MW-β parallelogram drawing of $\langle T_0, T_1 \rangle$ for any $\beta \geq 1$.

If $\delta = 0$ each T_i consists of only its root r_i. Choosing any nicely oriented parallelogram with $\overline{r_0 r_1}$ as its long diagonal will result in a valid MW-β drawing. Assume the claim holds for $\delta \leq k$ and suppose $\delta = k + 1$.

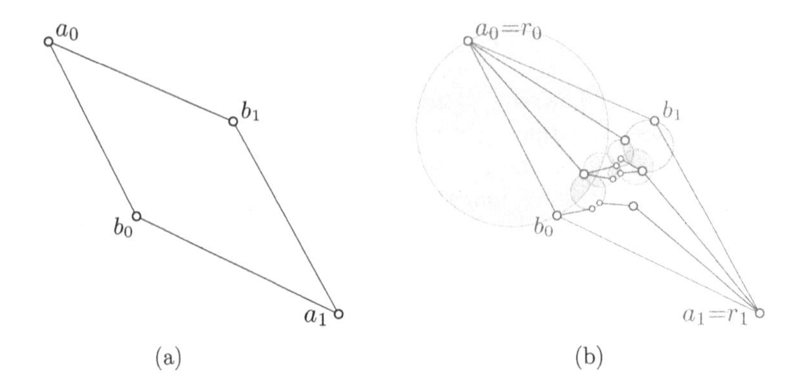

Fig. 8. (a) A nicely oriented parallelogram; (b) an MW-1 parallelogram drawing.

Let $\langle (T_{0,0}, r_{0,1}), (T_{1,0}, r_{1,0}) \rangle, \ldots, \langle (T_{0,m}, r_{0,m}), (T_{1,m}, r_{1,m}) \rangle$ be the pairs of isomorphic rooted trees resulting from deleting r_i from T_i. By induction, each $\langle (T_{0,j}, r_{0,j}), (T_{1,j}, r_{1,j}) \rangle$ with $0 \leq j \leq m$ admits a parallelogram drawing which is an MW-β drawing. Let H be any horizontal strip defined by two parallel lines $y = s$ and $y = t$ such that $s < t$. We uniformly scale and translate the parallelogram drawings of $\langle (T_{0,j}, r_{0,j}), (T_{1,j}, r_{1,j}) \rangle$ such that $y(r_{0,j}) = t$ and $y(r_{1,j}) = s$. Note that this operation does not change any of the β-proximity properties of any of the tree pairs.

Let $P_j = (a_{0,j}, b_{0,j}, a_{1,j}, b_{1,j})$ be the parallelogram that supports the MW-β drawing $\langle (\Gamma_{0,j}), (\Gamma_{1,j}) \rangle$ of $\langle (T_{0,j}, r_{0,j}), (T_{1,j}, r_{1,j}) \rangle$. Let ℓ_j and ℓ'_j be two half-lines such that ℓ_j starts at $r_{0,j}$, is orthogonal to $\overline{r_{0,j}, b_{0,j}}$, and crosses H, and ℓ'_j starts at $r_{1,j}$, is orthogonal to $\overline{r_{1,j}, b_{1,j}}$, and crosses H; see Fig. 9. We position P_{j+1} such that (i) ℓ_{j+1} is to the right of ℓ'_j; (ii) for any edge $e_{1,j} = (u_{1,j}, v_{1,j})$ in $T_{1,j}$, $r_{0,j+1}$ is to the right of the rightmost intersection point between H and $R[u_{1,j}, v_{1,j}, \infty]$ (since by inductive hypothesis no edge of $\Gamma_{1,j}$ is vertical, the coordinates of such points are finite); and (iii) for any edge $e_{0,j+1} = (u_{0,j+1}, v_{0,j+1}) \in T_{0,j+1}$, $r_{1,j}$ is to the left of the leftmost intersection point between H and $R[u_{0,j+1}, v_{0,j+1}, \infty]$ (by inductive hypothesis, the coordinates of such points are finite).

Condition (i) guarantees that for any vertices $v_{1,j+1} \in \Gamma_{1,j+1}$ and $v_{1,j} \in \Gamma_{1,j}$, we have $\angle(v_{1,j+1}, r_{0,j+1}, v_{1,j}) > \frac{\pi}{2}$ and thus $r_{0,j+1} \in R(v_{1,j+1}, v_{1,j}, 1)$ and $r_{0,j+1} \in R(v_{1,j+1}, v_{1,j}, \beta)$ for any $\beta \geq 1$. Similarly, for any vertices $v_{0,j+1} \in \Gamma_{0,j+1}$ and $v_{0,j} \in \Gamma_{0,j}$, we have $r_{1,j} \in R(v_{0,j+1}, v_{0,j}, \beta)$ for any $\beta \geq 1$. Conditions (ii) and (iii) guarantee that for any pair of adjacent $v_{i,j}, u_{i,j}$ in $\Gamma_{i,j}$, there is no witness in $R[v_{i,j}, u_{i,j}, \infty]$ and thus no witness in $R[v_{i,j}, u_{i,j}, \beta]$ for any finite $\beta \geq 1$. We now show how to place the roots $r_0 \in T_0$ and $r_1 \in T_1$ to produce an MW-β parallelogram drawing of $\langle (T_0, r_0), (T_1, r_1) \rangle$ for any $\beta \in [1, \infty]$.

Let L_0 be the vertical line through $r_{0,0}$ and let L_1 be the vertical line through $r_{1,m}$; see Fig. 10a. We show how to place r_i on L_i such that the *closed* β-region $R[r_i, r_{i,j}, \infty]$ does not contain any witness, while for any other vertex $v \in T_i$, the *open* β-region $R(r_i, v, 1)$ contains a witness. This implies that $R[r_i, r_{i,j}, \beta]$ does

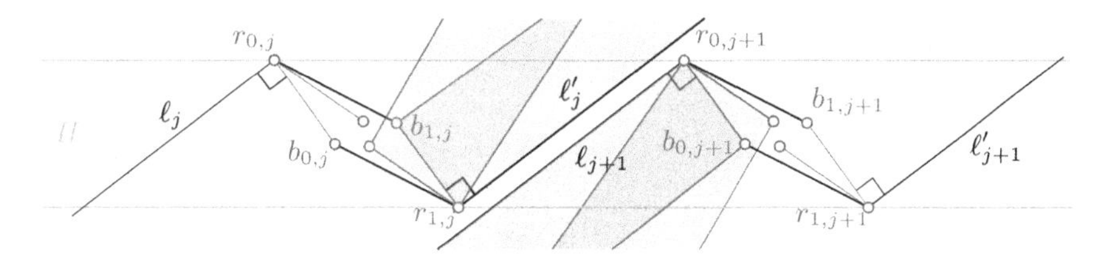

Fig. 9. Parallelograms P_j and P_{j+1} placed inside H.

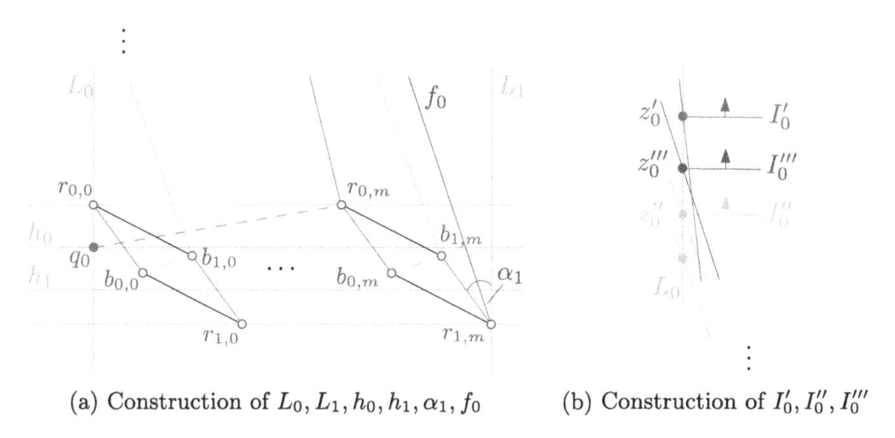

(a) Construction of $L_0, L_1, h_0, h_1, \alpha_1, f_0$ (b) Construction of I'_0, I''_0, I'''_0

Fig. 10. Placing r_0 in the proof of Theorem 2.

not contain any witnesses for all finite values of β and that $R(r_i, v, \beta)$ contains some witnesses for every $\beta \geq 1$.

We proceed in three steps. In the first step, we identify an interval I'_i of L_i such that for any point $p' \in I'_i$ and for any $r_{i,j}$, $R[p', r_{i,j}, \beta]$ contains no witnesses ($i = 0, 1$, $0 \leq j \leq m$). In the second step, we identify an interval I''_i of L_i such that for each vertex $v \neq r_{i,j}$ ($i = 0, 1$, $0 \leq j \leq m$) in Γ_i and for each point $p'' \in I''_i$, $R(p'', v, \beta)$ contains a witness. In the third step, we identify an interval I'''_i of L_i such that for any point $p_0 \in I'''_0$ the segment $\overline{p_0 b_{1,m}}$ does not intersect any parallelogram P_j with $0 \leq j \leq m$. Similarly, for any point $p_1 \in I'''_1$, the segment $\overline{p_1, b_{0,0}}$ does not intersect P_j for $0 \leq j \leq m$. As we will see, $I'_i \cap I''_i \cap I'''_i$ is a half-infinite strip for $i = 0, 1$; see Fig. 10b. We will describe how to obtain the intervals I'_0, I''_0, I'''_0; the intervals I'_1, I''_1, I'''_1 can be constructed symmetrically.

We start by defining I'_0. By construction of the MW-β drawing of the forests $T_{0,0}, \ldots T_{0,m}$ and $T_{1,0} \ldots T_{1,m}$, there exist horizontal lines h_0, h_1 in the interior of H such that h_i separates $r_{i,0}, \ldots r_{i,m}$ from every other vertex in the forest; see Fig. 10a. Let q_0 be the intersection point of h_0 and L_0. Let z'_0 be the intersection point of L_0 with the line through $r_{0,m}$ perpendicular to $\overline{r_{0,m} q_0}$. Let $I'_0 = \{z \in L_0 : y(z) \geq y(z'_0)\}$. Observe that for any $p_0 \in I'_0$ and any $r_{0,j}$, $R[p_0, r_{0,j}, \infty]$ contains no witnesses.

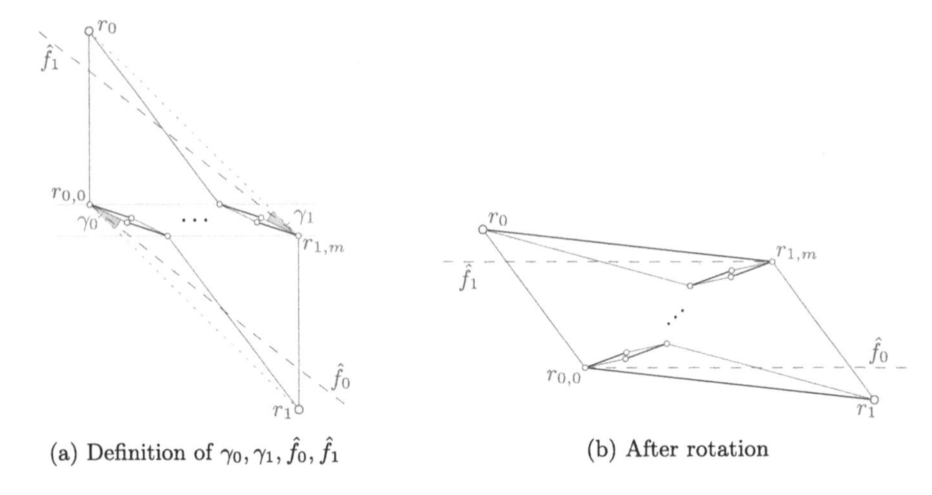

(a) Definition of $\gamma_0, \gamma_1, \hat{f}_0, \hat{f}_1$ (b) After rotation

Fig. 11. Rotating the drawing to obtain an MW-β parallelogram drawing.

We now define I_0''. For any parallelogram P_j and any vertex $v_{0,j} \in T_{0,j} \setminus \{r_{0,j}\}$, let $z_{0,j}$ be the intersection of L_0 with the line through $b_{1,j}$ perpendicular to $\overline{b_{1,j}v_{0,j}}$. Let z_0'' be the $z_{0,j}$ of maximum y-value over all $z_{0,j}$ $(0 \leq j \leq m)$. Let $I_0'' = \{z \in L_0 : y(z) \geq y(z_0'')\}$. Observe that for any point $p_0 \in I_0''$ and for any $v_{0,j} \in T_{0,j} \setminus \{r_{0,j}\}$, we have that $\angle(v_{0,j}, b_{1,j}, p_0) \geq \frac{\pi}{2}$ and thus $b_{1,j} \in R[p_0, v_{0,j}, 1]$.

We now define I_0'''. Let α_0 be the acute angle formed by L_0 and the segment $\overline{r_{0,0}b_{0,0}}$. Let α_1 be the acute angle formed by L_1 and the segment $\overline{r_{1,m}b_{1,m}}$ and let $\alpha = \min\{\alpha_0, \alpha_1\}$. Let f_0 be a half-line starting at $r_{1,m}$, having negative slope, and forming an acute angle of $\alpha/2$ with L_1. Let z_0''' be $f_0 \cap L_0$ and let $I_0''' = \{z \in L_0 : y(z) \geq y(z_0''')\}$.

Let $I_i = I_i' \cap I_i'' \cap I_i'''$ and let $p_i \in I_i$ be such that $\overline{p_0 r_{1,m}}$ is parallel to $\overline{p_1 r_{0,0}}$. We draw r_i at p_i, which produces an MW-β drawing of $\langle T_0, T_1 \rangle$ in a parallelogram $P = \langle a_0, b_0, a_1, b_1 \rangle = \langle r_0, b_{1,m}, r_1, b_{0,0} \rangle$. This is however not yet a parallelogram drawing, as $y(b_0) = y(b_{0,0}) > y(b_{1,m}) = y(b_1)$ and some edges are vertical.

To complete the proof, we thus show how to rotate P to produce an MW-β parallelogram drawing. Refer to Fig. 11a. Let γ_0 be the angle between $\overline{r_{0,0}r_1}$ and $\overline{r_{0,0}b_{0,0}}$ and let γ_1 be the angle between $\overline{r_{1,m}r_0}$ and $\overline{r_{1,m}b_{1,m}}$; Let $\gamma = \min\{\gamma_0, \gamma_1\}$. Let $\hat{f}_0$ be the ray originating at $r_{0,0}$, forming an angle $\gamma' < \gamma$ with, and lying above, segment $\overline{r_{0,0}r_1}$, so that no edge of the drawing is perpendicular to $\hat{f}_0$. Let $\hat{f}_1$ be the ray originating at $r_{1,m}$ having opposite direction to $\hat{f}_0$. Observe that $\hat{f}_0$ and $\hat{f}_1$ are parallel and that any vertex of T_i except r_i is in the strip between $\hat{f}_0$ and $\hat{f}_1$. We now rotate P counterclockwise until $\hat{f}_0$ and $\hat{f}_1$ become horizontal; see Fig. 11b. This produces a parallelogram drawing of $\langle T_0, T_1 \rangle$, since no edge is vertical, $y(r_0) > y(r_{1,m}) > y(r_{0,0}) > y(r_1)$, and $x(r_0) < x(r_{0,0}) < x(r_{1,m}) < x(r_1)$. $\square$

6 Pruning Leaves from MW-β Drawings of Isomorphic Trees

In this section, we explore the question of how far from isomorphic two trees might be while still allowing an MW-β drawing. We consider the MW-β drawing $\langle \Gamma_0, \Gamma_1 \rangle$ constructed in the proof of Theorem 2 and ask whether it is possible to prune some leaves from Γ_1 and still have an MW-β drawing of the resulting trees. Precisely, we show that there are cases when we can remove linearly many leaves from Γ_1 and still obtain an MW-β drawing of the resulting tree for any $\beta \in [1, \infty]$. It may be worth recalling that Lenhart and Liotta proved that two stars admit an MW-1 drawing if and only if the cardinalities of their vertex sets differ by at most two [14].

Let (T, r) be a rooted tree and let $\mathcal{L}$ be a set of leaves of T. The vertex v is a *cousin* of a vertex v' if v and v' have a common grandparent but no common parent, i.e., there is a vertex w such that a length-2 directed path w, p, v and a length-2 directed path w, p', v' with $p \neq p'$ exist. We say that $\mathcal{L} \neq \emptyset$ is *sparse* if, for every $v \in \mathcal{L}$, (i) v has at least one sibling, (ii) every sibling v' of v is a leaf with $v' \notin \mathcal{L}$, and (iii) v has a cousin w such that $w \notin \mathcal{L}$ and, for all siblings w' of w, $w' \notin \mathcal{L}$. Note that the existence of a sparse set implies that (T, r) has height at least 2, otherwise there is no vertex that has a cousin.

Theorem 3 ($\star$). *Let (T, r) be a rooted tree and let $\mathcal{L}$ be a sparse set of leaves of T. Then the pair $\langle T, T \setminus \mathcal{L} \rangle$ of trees admits an MW-β drawing for all $\beta \in [1, \infty]$.*

Corollary 1 ($\star$). *For any $m \geq 1$ and $n = 7m+1$, there exist tree pairs $\langle T_0, T_1 \rangle$ with $|V(T_1)| \leq 1 + \frac{5}{6}(|V(T_0)| - 1)$ that admit an MW-β drawing for all $\beta \in [1, \infty]$.*

7 Concluding Remarks

In this paper, we studied the mutual witness proximity drawability of pairs of isomorphic trees. We adopted the well-known concept of open/closed β-proximity regions and considered any value of the parameter β such that $\beta \geq 1$. For the special case of $\beta = 1$, the definition of closed β-proximity region coincides with the definition of Gabriel proximity region. We showed in Theorem 1 that any pair of isomorphic caterpillars admits a linearly separable mutual witness Gabriel drawing. We then extended this result in Theorem 2 to any value of $\beta \geq 1$ and to any pair of isomorphic trees, but at the cost of losing linear separability.

It would be interesting to establish whether any two isomorphic trees admit a linearly separable MW-β drawing for $\beta \geq 1$. Also, even for the special case of caterpillars, extending the result of Theorem 1 to values of $\beta > 1$ does not seem immediate. Finally, a characterization of those non-isomorphic pairs of trees that admit a mutual witness β-drawing continues to be elusive. Theorem 3 shows that the trees in the pair may differ by linearly many vertices.

Acknowledgements. We thank Stefan Näher for many helpful discussions, for implementing the caterpillar algorithm, and for creating a program to edit and verify MW-[1] drawings that was very helpful in verifying our constructions.

References

1. Aronov, B., Dulieu, M., Hurtado, F.: Witness (Delaunay) graphs. Comput. Geom. **44**(6–7), 329–344 (2011). https://doi.org/10.1016/j.comgeo.2011.01.001
2. Aronov, B., Dulieu, M., Hurtado, F.: Witness Gabriel graphs. Comput. Geom. **46**(7), 894–908 (2013). https://doi.org/10.1016/j.comgeo.2011.06.004
3. Aronov, B., Dulieu, M., Hurtado, F.: Mutual witness proximity graphs. Inf. Process. Lett. **114**(10), 519–523 (2014). https://doi.org/10.1016/j.ipl.2014.04.001
4. Aronov, B., Dulieu, M., Hurtado, F.: Witness rectangle graphs. Graphs Comb. **30**(4), 827–846 (2014). https://doi.org/10.1007/s00373-013-1316-x
5. Battista, G.D., Eades, P., Tamassia, R., Tollis, I.G.: Graph Drawing: Algorithms for the Visualization of Graphs. Prentice-Hall, Hoboken (1999)
6. Battista, G.D., Liotta, G., Whitesides, S.: The strength of weak proximity. J. Discrete Algorithms **4**(3), 384–400 (2006). https://doi.org/10.1016/j.jda.2005.12.004
7. Eades, P., Hong, S., Nguyen, A., Klein, K.: Shape-based quality metrics for large graph visualization. J. Graph Algorithms Appl. **21**(1), 29–53 (2017). https://doi.org/10.7155/jgaa.00405
8. Gabriel, K.R., Sokal, R.R.: A new statistical approach to geographic variation analysis. Syst. Zool. **18**, 259–278 (1969). https://doi.org/10.2307/2412323
9. Haase, C., Kindermann, P., Lenhart, W.J., Liotta, G.: Mutual witness proximity drawings of isomorphic trees (2023). https://arxiv.org/abs/2309.01463
10. Ichino, M., Sklansky, J.: The relative neighborhood graph for mixed feature variables. Pattern Recognit. **18**(2), 161–167 (1985). https://doi.org/10.1016/0031-3203(85)90040-8
11. Jaromczyk, J.W., Toussaint, G.T.: Relative neighborhood graphs and their relatives. Proc. IEEE **80**(9), 1502–1517 (1992). https://doi.org/10.1109/5.163414
12. Kaufmann, M., Wagner, D. (eds.): Drawing Graphs. LNCS, vol. 2025. Springer, Heidelberg (2001). https://doi.org/10.1007/3-540-44969-8
13. Kirkpatrick, D.G., Radke, J.D.: A framework for computational morphology. Mach. Intelligence Pattern Recogn. **2**, 217–248 (1985). https://doi.org/10.1016/B978-0-444-87806-9.50013-X
14. Lenhart, W.J., Liotta, G.: Mutual witness Gabriel drawings of complete bipartite graphs. In: Angelini, P., von Hanxleden, R. (eds.) Graph Drawing and Network Visualization - 30th International Symposium. Lecture Notes in Computer Science, vol. 13764, pp. 25–39. Springer, Cham (2022). https://doi.org/10.1007/978-3-031-22203-0_3
15. Liotta, G.: Proximity drawings. In: Tamassia, R. (ed.) Handbook on Graph Drawing and Visualization, pp. 115–154. Chapman and Hall/CRC, Boca Raton (2013). https://cs.brown.edu/people/rtamassi/gdhandbook/chapters/proximity.pdf
16. Okabe, A., Boots, B., Sugihara, K., Chiu, S.N., Kendall, D.G.: Spatial Tessellations: Concepts and Applications of Voronoi Diagrams, Second Edition. Wiley Series in Probability and Mathematical Statistics, Wiley, Hoboken (2000). https://doi.org/10.1002/9780470317013
17. O'Rourke, J., Toussaint, G.T.: Pattern recognition. In: Goodman, J.E., O'Rourke, J., Toth, C. (eds.) Handbook of Discrete and Computational Geometry, Third Edition. Chapman and Hall/CRC, Boca Raton (2017). http://www.csun.edu/ctoth/Handbook/chap54.pdf
18. Tamassia, R. (ed.): Handbook on Graph Drawing and Visualization. Chapman and Hall/CRC, Boca Raton (2013). https://www.crcpress.com/Handbook-of-Graph-Drawing-and-Visualization/Tamassia/9781584884125

19. Tamassia, R., Liotta, G.: Graph drawing. In: Goodman, J.E., O'Rourke, J. (eds.) Handbook of Discrete and Computational Geometry, Second Edition, pp. 1163–1185. Chapman and Hall/CRC, Boca Raton (2004). https://doi.org/10.1201/9781420035315.ch52
20. Toussaint, G.T.: The relative neighbourhood graph of a finite planar set. Pattern Recognit. **12**(4), 261–268 (1980). https://doi.org/10.1016/0031-3203(80)90066-7
21. Toussaint, G.T., Berzan, C.: Proximity-graph instance-based learning, support vector machines, and high dimensionality: an empirical comparison. In: Perner, P. (ed.) MLDM 2012. LNCS (LNAI), vol. 7376, pp. 222–236. Springer, Heidelberg (2012). https://doi.org/10.1007/978-3-642-31537-4_18

Graph Decompositions

Three Edge-Disjoint Plane Spanning Paths in a Point Set

P. Kindermann[1](✉), J. Kratochvíl[2], G. Liotta[3], and P. Valtr[2]

[1] Universität Trier, Trier, Germany
kindermann@uni-trier.de
[2] Department of Applied Mathematics, Faculty of Mathematics and Physics, Charles University, Prague, Czech Republic
{honza,valtr}@kam.mff.cuni.cz
[3] University of Perugia, Perugia, Italy
giuseppe.liotta@unipg.it

Abstract. We study the following problem: Given a set S of n distinct points in the plane, how many edge-disjoint plane straight-line spanning paths of S can one draw? While each spanning path is crossing-free, the edges of distinct paths may cross each other (i.e., they may intersect at points that are not elements of S). A well-known result is that when the n points are in convex position, $\lfloor n/2 \rfloor$ such paths always exist, but when the points of S are in general position the only known construction gives rise to two edge-disjoint plane straight-line spanning paths. In this paper, we show that for any set S of at least ten points in the plane, no three of which are collinear, one can draw at least three edge-disjoint plane straight-line spanning paths of S. Our proof is based on a structural theorem on halving lines of point configurations and a strengthening of the theorem about two spanning paths, which we find interesting in its own right: if S has at least six points, and we prescribe any two points on the boundary of its convex hull, then the set contains two edge-disjoint plane spanning paths starting at the prescribed points.

Keywords: Plane Spanning Paths · Point Sets · Geometric Graph Theory

1 Introduction

Let S be a set of distinct points (locations) in the plane. We want to compute three *edge-disjoint* spanning paths of S. Note that the edges of each path are straight-line segments and that no two edges of a same path can cross while the edges of distinct spanning paths may cross (i.e. share points that are not elements of S).

At a first glance, one would be tempted to generalize the question and study the existence of k such edge-disjoint spanning paths with $k \geq 2$. Namely, the proof of Bernhart and Kainen about the book thickness of a complete graph (Theorem 3.4 of [3]) already gives a partial answer: if the n points are in convex

M. A. Bekos and M. Chimani (Eds.): GD 2023, LNCS 14465, pp. 323–338, 2023.
https://doi.org/10.1007/978-3-031-49272-3_22

position, then it is possible to draw $\lfloor \frac{n}{2} \rfloor$ edge-disjoint plane straight-line spanning paths of the point set which is also a tight upper bound for even values of n (the complete graph has $\frac{n(n-1)}{2}$ edges). However, little is known when the n points are not in convex position: The only result we are aware of is by Aichholzer et al. [2], who show the existence of two edge-disjoint plane straight-line spanning paths for any set of $n \geq 4$ points in general position (no three collinear). Aichholzer et al. leave open the problem of proving whether three or more paths always exist. Our main result is as follows.

Theorem 1. *Let S be any set of at least ten points in general position in the plane. There are three edge-disjoint plane straight-line spanning paths of S.*

Besides addressing an open problem by Aichholzer et al. [2], Theorem 1 relates with some classical topics in the graph drawing literature. Among them, the *graph packing problem* asks whether it is possible to map a set of smaller graphs into a larger graph, called the *host graph*, without using the same edge of the host graph twice. A rich body of literature is devoted to this problem, both when the host graph is the complete graph and when the smaller graph is either planar or near-planar (see, e.g. [4,6–8,11,13]). While most papers devoted to graph packing do not assume that a drawing of the host graph is given as part of the input, our study considers a *geometric graph packing problem*, as we want to map three plane geometric paths with n vertices into a complete geometric graph K_n. Bose et al. [5] give a characterization of those plane trees that can be packed in a complete geometric graph K_n in the special case that the vertices of K_n are in convex position. Aichholzer et al. [2] show that $\Omega(\sqrt{n})$ edge-disjoint plane trees can be packed into a complete geometric graph with n vertices, but it is not known whether this lower bound extends to paths. From the perspective of geometric graph packing problems, Theorem 1 directly implies the following.

Corollary 1. *Three edge-disjoint plane Hamiltonian paths can be packed into every complete geometric graph with at least ten vertices.*

Paper Organization. The rest of the paper is organized as follows. In Sect. 2, we introduce technical notions and briefly recall the concept of zig-zag paths of [1], and then in Sect. 3, we present an overview of the proof of our main theorem. Section 4 is devoted to a strengthening of the result of [2] on the existence of two paths even when the starting points are prescribed, which we believe is also interesting on its own. Section 5 then presents the proof of our main theorem in detail. For space reasons, the proofs of some statements (marked by a ($\star$)) can only be found in the full version of the paper [10], but we include their figures for illustration.

2 Preliminaries

We denote a path in a graph by its sequence of vertices and edges. For a path P starting at v_1, ending at v_h, and visiting its vertices v_i in increasing order of the subscript i from 1 to h, we represent the sequence as $P = v_1 \circ v_1v_2 \circ v_2 \circ \cdots \circ v_{h-1}v_h \circ v_h$. Thus, the concatenation of vertex-disjoint paths P (ending with vertex x) and Q (starting with vertex y adjacent to x) is the path $P \circ xy \circ Q$. By P^{-1} we denote the path P traversed in the reversed order.

Let S be a set of points on general position (i.e., no three collinear) in the plane and let p, q be two points of S. We denote by $\overline{pq}$ the line passing through them, and by pq its line segment with endpoints p and q. The *geometric graph* $G(S)$ determined by S is the straight-line drawing of the complete graph with vertex set S. A *path on* S is a straight-line drawing of a path in $G(S)$. For the sake of brevity we shall omit the term "straight-line". A path on S is *spanning* if its set of vertices is the entire set S. We shall say *plane (spanning) path* to mean that the (spanning) path is crossing-free.

We denote by $\mathrm{CH}(S)$ the convex hull of S and by $\partial\mathrm{CH}(S)$ the boundary of its convex hull. Points of $S \cap \partial\mathrm{CH}(S)$ are the *extreme points* of S. We will call a partition $S = S_1 \cup S_2$ a *balanced separated partition* if the two sets are almost equal in sizes (i.e., $||S_1| - |S_2|| \leq 1$) and $\mathrm{CH}(S_1) \cap \mathrm{CH}(S_2) = \emptyset$. In such a case we denote the partition as (S_1, S_2). The boundary of $\mathrm{CH}(S)$ contains two edges with one end-point in S_1 and the other one in S_2; such edges are called *bridges* of the partition, and each vertex incident with a bridge is called *bridged*. A line ℓ is a *balancing line* for a set S if the intersections of S with the open half-planes determined by ℓ form a balanced separated partition of $S \setminus \ell$. Note that a balancing line for S may contain 0, 1 or 2 points of S. Every point of S belongs to at least one balancing line passing through this point, but not every two points of S belong to the same balancing line. However, every set (of size at least 2) contains two points which determine a balancing line.

If (S_1, S_2) is a balanced separated partition of S, a *zig-zag* (S_1, S_2)*-path* is a plane spanning path in S on which the points from S_1 and S_2 alternate. When S_1 and S_2 are clear from the context, we just call it a zig-zag path. It is well known that a zig-zag path exists for every balanced separated partition of S [1,9].

Lemma 1 ([1]). *Every balanced separated partition admits a zig-zag path.*

The algorithm by Abellanas et al. [1] works roughly as follows. Assume that (S_1, S_2) is a partition of S, S_1 and S_2 are separated by a horizontal line and that $|S_1| \geq |S_2|$. Let (p_1, q_1) be the *left bridge* of S, i.e., the left edge of $\mathrm{CH}(S)$ crossing the separating line, with $p_1 \in S_1$ and $q_1 \in S_2$. Starting with $P = p_1$, compute inductively the left bridge pq of $S \setminus V(P)$, and set $P = P \circ p$ if the last point of P was in S_2, or set $P = P \circ q$ if the last point of P was in S_1. Continue this process until all vertices of S are added to P. Then P is a zig-zag path. Note here that if $|S_1| = |S_2|$, we may choose if the zig-zag path starts in S_1 or in S_2, but when the sets S_1 and S_2 are not equal in sizes, every zig-zag path must start in the bigger one of them. And that on a zig-zag path constructed in the above sketched way, the crossing points of the edges of the path with the separating horizontal line have the same linear order along the path and along the separating line.

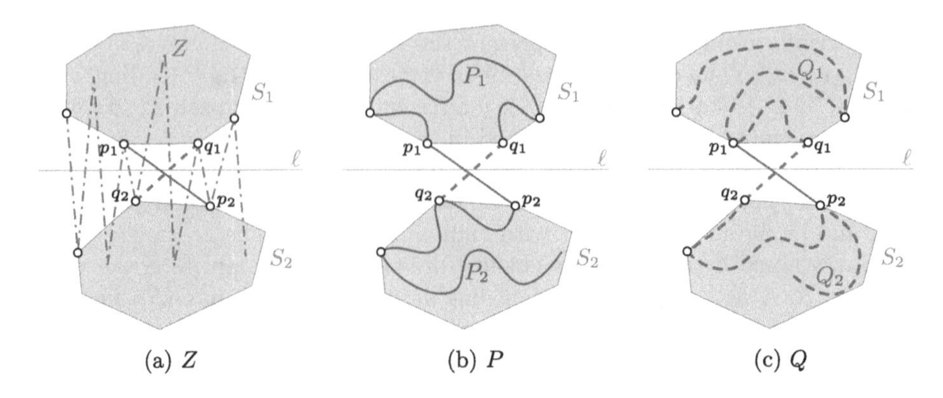

Fig. 1. Schematic illustration of the approach behind the proof of Theorem 1.

3 Approach Overview

The main idea of our approach is to construct three edge-disjoint plane spanning paths Z, P, Q on a point set S with $|S| \geq 10$ as follows. We find a suitable balanced separated partition (S_1, S_2) of S. The first path Z will be the zig-zag path obtained by Lemma 1. For P and Q, we seek to find two edge-disjoint plane spanning paths P_1, Q_1 (ending in p_1 and q_1, respectively) in S_1 and two edge-disjoint plane spanning paths P_2, Q_2 (starting in p_2 and q_2, respectively) in S_2; these are obviously edge-disjoint with Z. If p_1p_2 and q_1q_2 do not belong to Z and their interiors are disjoint with $\mathrm{CH}(S_1) \cup \mathrm{CH}(S_2)$, we combine these four paths to two edge-disjoint plane spanning paths $P = P_1 \circ p_1p_2 \circ P_2$ and $Q_1 \circ q_1q_2 \circ Q_2$ in S. To this end, we reverse the strategy and try to find two pairs of vertices on $\mathrm{CH}(S_1)$ and $\mathrm{CH}(S_2)$ that see each other (i.e., their connections do not go through $\mathrm{CH}(S_1) \cup \mathrm{CH}(S_2)$) and that are not connected by an edge in Z. See also Fig. 1 for a schematic description of our approach.

There are several difficulties: First, we cannot use the algorithm by Aichholzer et al. [2] to find the two paths in S_1 and S_2, as that does not control the starting and ending points of the paths. Hence, we strengthen their theorem and prove that one can always find two edge-disjoint plane spanning paths even if their starting points are prescribed (Theorem 2). Secondly, it might not be possible to find two pairs of vertices on the convex hulls with our desired properties. However, we prove that in most of such situations, the zig-zag path can be slightly modified so that the connection is possible. Lastly, we show that all of the previously described moves fail in one and only one very specific configuration, which allows three edge-disjoint plane spanning paths to be constructed easily in an ad hoc way, thus establishing Theorem 1.

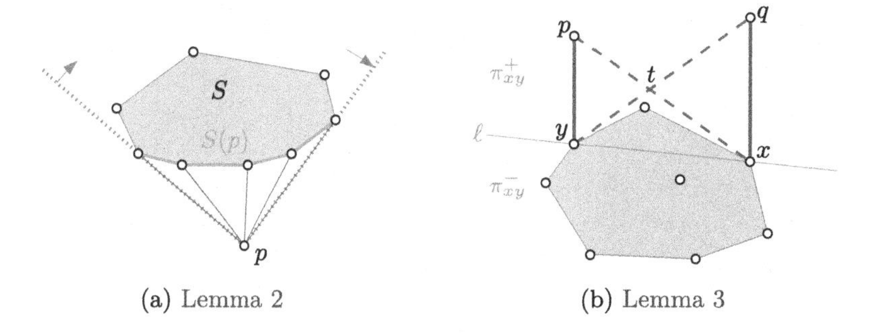

(a) Lemma 2

(b) Lemma 3

Fig. 2. Illustration for the proof of (a) Lemma 2 and (b) Lemma 3.

4 Two Edge-Disjoint Plane Spanning Paths with Prescribed Starting Points

Let S be a set of at least five distinct points in the plane in general position and let s and t be two distinguished elements of S, possibly coincident. In this section we show that there exist two edge-disjoint plane spanning paths of S, one starting at s and the other one starting at t such that st is not an edge of either path. To this aim, we start with some basic properties of planar point sets.

Let p be a point outside $\mathrm{CH}(S)$. We say that p *sees* a point $q \in S$ if $pq \cap \mathrm{CH}(S) = \{q\}$. We denote by $S(p)$ the set of (extreme) points of S that are seen from p.

Lemma 2 (⋆ – Fig. 2a). *Let $|S| \geq 3$ and let p be a point outside $\mathrm{CH}(S)$. Then p sees at least two points of S. Moreover, $S(p)$ forms a continuous interval on $S \cap \partial\mathrm{CH}(S)$ (along $\partial\mathrm{CH}(S)$).*

Lemma 3 (⋆ – Fig. 2b). *Let $|S| \geq 3$ and let p, q be 2 distinct points outside $\mathrm{CH}(S)$ such that $S \cup \{p, q\}$ is in general position; let x and y be two extreme points of S. If $x \in S(p), y \in S(q)$ and the (visibility) segments px and qy cross in an interior point, then $\{x, y\} \subseteq S(p) \cap S(q)$ and $py \cap qx = \emptyset$.*

Lemma 4 (⋆ – Fig. 3). *Let $|S| \geq 3$ and let p, q be 2 distinct points outside $\mathrm{CH}(S)$ such that $S \cup \{p, q\}$ is in general position. Assume $|S(p) \cup S(q)| \geq 3$. Then for any point $c \in S(q)$, there exist points $a \in S(p)$ and $b \in S(q) \setminus \{c\}$ such that $ap \cap bq = \emptyset$.*

Lemma 5. *Let s and t be two distinct points of S. Then S contains a plane spanning path which starts at s and ends at t.*

Proof. Suppose first that both s and t lie on $\partial\mathrm{CH}(S)$; see Fig. 4a. Let ℓ_s (ℓ_t) be a supporting line of S passing through s (through t) and let x be the crossing point of ℓ_s and ℓ_t (since the points of S are in general position, we may assume

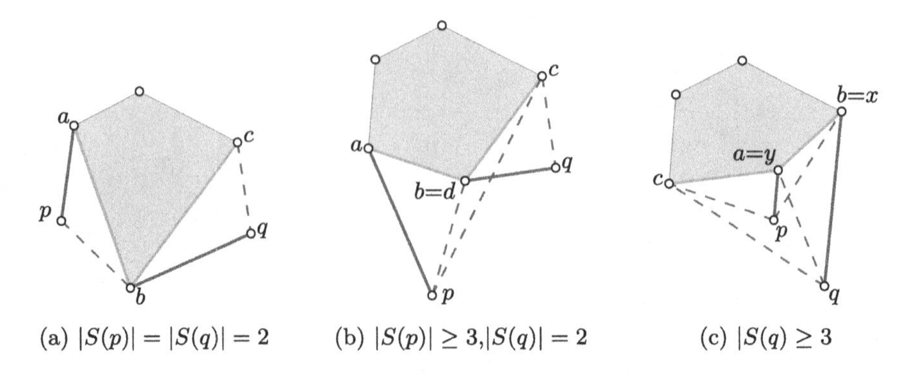

(a) $|S(p)| = |S(q)| = 2$ (b) $|S(p)| \geq 3, |S(q)| = 2$ (c) $|S(q) \geq 3$

Fig. 3. Illustration the proof of Lemma 4.

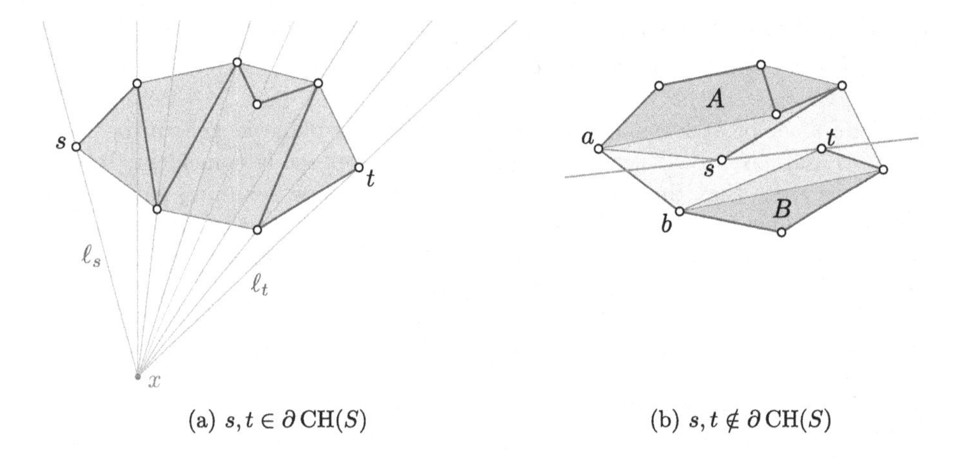

(a) $s, t \in \partial\,\mathrm{CH}(S)$ (b) $s, t \notin \partial\,\mathrm{CH}(S)$

Fig. 4. Illustration for the proof of Lemma 5.

without loss of generality that ℓ_s and ℓ_t are not parallel and that $S \cup \{x\}$ is in general position). Consider the lines $\overline{xy}$, for $y \in S$, and order them $\ell_1, \ell_2, \ldots, \ell_{|S|}$ as they form a rotation scheme around x from $\ell_1 = \ell_s$ to $\ell_{|S|} = \ell_t$. Rename the points of S as $y_i \in \ell_i$, $i = 1, 2, \ldots, |S|$. Then $s{=}y_1 \circ y_1y_2 \circ y_2 \circ \ldots \circ y_{|S|}{=}t$ is a plane spanning path starting at s and ending at t.

Now assume that at least one of s, t is an interior point of $\mathrm{CH}(S)$, say s; see Fig. 4b. The line $\overline{st}$ separates $S \setminus \{s, t\}$ into two disjoint nonempty sets A, B. Also, this line intersects the relative interior of an edge of $\partial\mathrm{CH}(S)$, say ab with $a \in A$ and $b \in B$. Now both s and a lie on $\partial\mathrm{CH}(A \cup \{s\})$, and the previously proven case implies existence of a plane spanning path P_A in $A \cup \{s\}$ which starts in s and ends in a. Similarly, $B \cup \{t\}$ contains a plane spanning path P_B which starts in b and ends in t. Then $P_A \circ ab \circ P_B$ is the desired path. □

The following result will be used in the proof of Theorem 1, but it also provides a strengthening of the result by Aichholzer et al. on the existence of two edge-disjoint plane straight-line spanning paths in a point set [2].

Theorem 2. (⋆). *Let $|S| \geq 5$ and let s and t be two (not necessarily distinct) points of $\partial\mathrm{CH}(S)$. Then S contains two edge-disjoint plane spanning paths, one starting at s and the other one at t. Moreover, if the points s and t are distinct, then the paths can be chosen so that none of them contains the edge st.*

Proof sketch. We first deal with the case $s \neq t$; see Fig. 5. If it is possible to choose a balancing line ℓ passing through the point s in such a way that t belongs to a smaller part of the balanced separated partition of $S \setminus \{s\}$ determined by this line, we connect s to a bridged vertex from the other part and continue on the zig-zag path starting in this point; see Fig. 5a. The second path is constructed by taking a path starting in t, traversing all vertices of its part, connecting via s to the other part and traversing that one. The paths in the parts exist by Lemma 5, suitable neighbors of s are guaranteed by Lemma 2.

If $|S|$ is even and the line $\overline{st}$ is a balancing line, the above described choice of ℓ is not possible. This case is more tricky. Again, one of the paths starts in s and continues on a zig-zag path for the balanced separated partition of $S \setminus \{s, t\}$ determined by the line $\overline{st}$; see Fig. 5b. The second path is constructed with additional help of Lemma 4 if s and t together see at least 3 points of at least one part of the partition, while an ad hoc construction is needed in the last case, when s and t together see only 2 vertices in each part of the partition.

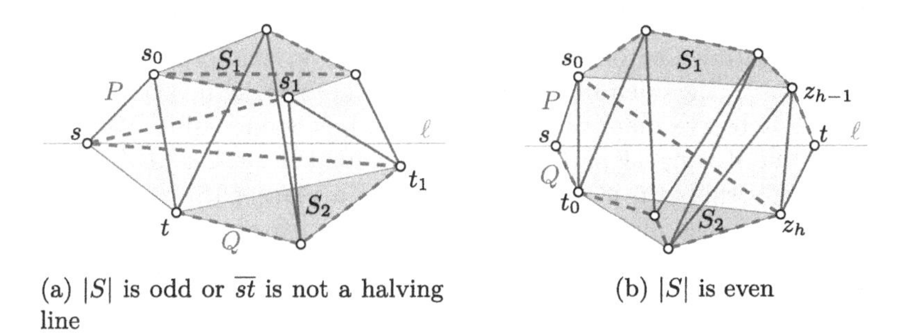

(a) $|S|$ is odd or $\overline{st}$ is not a halving line

(b) $|S|$ is even

Fig. 5. Illustration for the case that $s \neq t$ of the proof of Theorem 2.

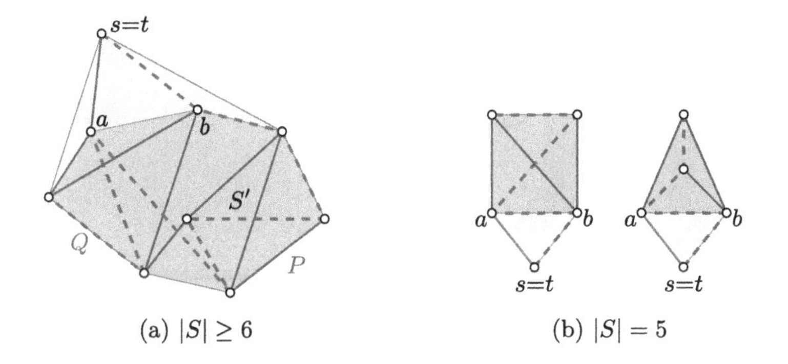

(a) $|S| \geq 6$

(b) $|S| = 5$

Fig. 6. Illustration to the case that $s = t$ of the proof of Theorem 2.

If the points s and t coincide, we consider two consecutive extreme points of $\mathrm{CH}(S \setminus \{s\})$ seen by s and obtain the desired paths starting in s from spanning paths of $S \setminus \{s\}$ starting in these points; see Fig. 6. The existence of such paths was proven in the preceding paragraph if S is large enough ($|S| \geq 6$), and can be easily shown by an ad hoc construction for small S ($|S| = 5$). □

5 Three Edge-Disjoint Plane Spanning Paths

We first introduce a few technical notions. For two points u, v in the plane, we denote by $(uv)^+$ the open halfplane to the right of the line $\overline{uv}$, if the line is traversed in the way that u precedes v. The opposite open halfplane is denoted by $(uv)^-$. Note that $(uv)^- = (vu)^+$. Let Q be a convex polygon and let u, v be two adjacent vertices of Q such that $Q \subseteq (uv)^+$. Then we say that v is the *clockwise neighbor* of u along (the boundary of) Q and u is the *counterclockwise neighbor* of v along (the boundary of) Q. We shall omit the words "the boundary of" when talking about the (counter-)clockwise neighbor along Q. Let (S_1, S_2) be a balanced separated partition of S. The *visibility graph* of the partition is

$$\mathcal{V}(S_1, S_2) = (S, \{ab : a \in S_1, b \in S_2, ab \cap (\mathrm{CH}(S_1) \cup \mathrm{CH}(S_2)) = \{a, b\}\}),$$

i.e., $ab \in E(\mathcal{V}(S_1, S_2))$ if and only if $a \in S_1(b)$ and $b \in S_2(a)$; see Fig. 7. A path $v_1 \circ v_1v_2 \circ \ldots \circ v_k$ of $\mathcal{V}(S_1, S_2)$ is called *switchable* if its edges $v_1v_2, v_2v_3, \ldots, v_{k-1}v_k$ cross the separating line of the partition (S_1, S_2) in this order and for each $i = 1, 2, \ldots, k-2$, the open region bounded by the triangle $v_iv_{i+1}v_{i+2}$ contains no point of S. Observe that every switchable path is non-crossing. We will show in Lemma 7 that if a zig-zag path Z contains a switchable path $a \circ ab \circ b \circ bc \circ c \circ cd \circ d$ of length 3 as a subpath, it can be modified to a plane path $\ldots \circ a \circ ac \circ c \circ cb \circ b \circ bd \circ d \ldots$ which allows spanning paths in the two classes of the balanced separated partition to be concatenated via the edges ab and cd (the proof in full details will appear in the journal version of the paper, an illustrative sketch is in Fig. 13).

The *n-wheel configuration* W_n of points (in general position) is a set of $n-1$ points in convex position, augmented with one point lying inside the convex hull

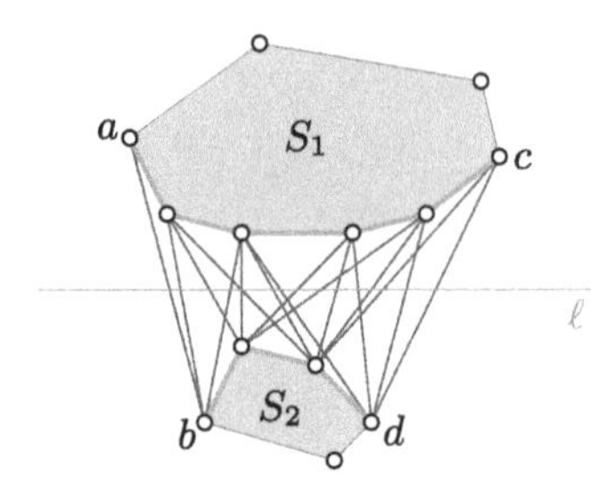

Fig. 7. Illustration for the definition of the visibility graph of a balanced separated partition. The edges of the visibility graph are drawn purple.

of these $n-1$ points in such a position that every line that passes through the augmenting point and any other point is a balancing line of W_n. This point configuration plays an important role in the proof below, and we need to show that it contains three edge-disjoint plane spanning paths by an ad hoc construction, at least for the case of n even. This has already been sketched by Aichholzer et al [2].

Proposition 1. **($\star$ – Fig. 8).** *For even $n \geq 6$, the maximum number of edge-disjoint plane spanning paths in the wheel configuration W_n is $\frac{n}{2}-1$.*

Our later proof of Theorem 1 is based on the following structural result.

Theorem 3. *Let S be a set of $n \geq 5$ points in general position in the plane. Then at least one of the following holds true*

1. *S has a balanced separated partition (S_1, S_2) such that $\mathcal{V}(S_1, S_2)$ contains two crossing edges, or*
2. *S has a balanced separated partition (S_1, S_2) such that $\mathcal{V}(S_1, S_2)$ contains a switchable path of length 3 and a bridged vertex not included in the path which is incident with at least 2 edges of $\mathcal{V}(S_1, S_2)$, or*
3. *n is even and S is the wheel configuration W_n.*

Proof. **Case 1: n is odd**. A line $\overline{xy}$ passing through two points $x, y \in S$ is an *almost-balancing line* of S, if exactly $(n-1)/2$ points of $S \setminus \{x, y\}$ lie on one of the sides of $\overline{xy}$ and the remaining $(n-3)/2$ points of $S \setminus \{x, y\}$ lie on the other side of $\overline{xy}$. We fix an extreme point u of S. Let $\overline{ua}$, $\overline{ub}$ be the two almost-balancing lines passing through u; see Fig. 9a. Suppose $b \in (ua)^+$. Note that the interior of the convex wedge bounded by the two rays emanating from u, one passing through a and the other passing through b, contains no point of S. Each of the lines $\overline{ua}$ and $\overline{ub}$ partitions the set $S \setminus \{u, a, b\}$ into two sets A and B of equal size $(n-3)/2$, such that A lies to the left of the lines $\overline{ua}$ and $\overline{ub}$ and B lies to the right of them. The line $\overline{ab}$ partitions A into A_1 and A_2, and B into B_1 and B_2, such that A_1 and B_1 lie in $(ab)^+$, and A_2 and B_2 lie in $(ab)^-$.

Suppose first that $A_1 \neq \emptyset$, and let $a_{\circlearrowright}$ be the clockwise neighbor of a along $\text{CH}(A \cup \{a\})$; see Fig. 9b. Consider the balanced separated partition (S_1, S_2) with

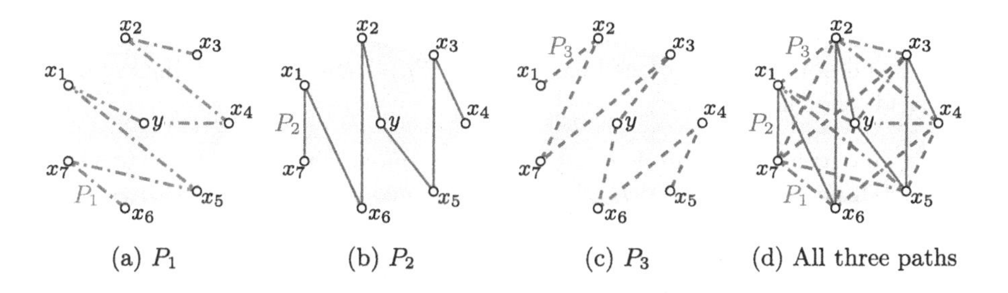

Fig. 8. Illustration for the proof of Proposition 1 for W_8.

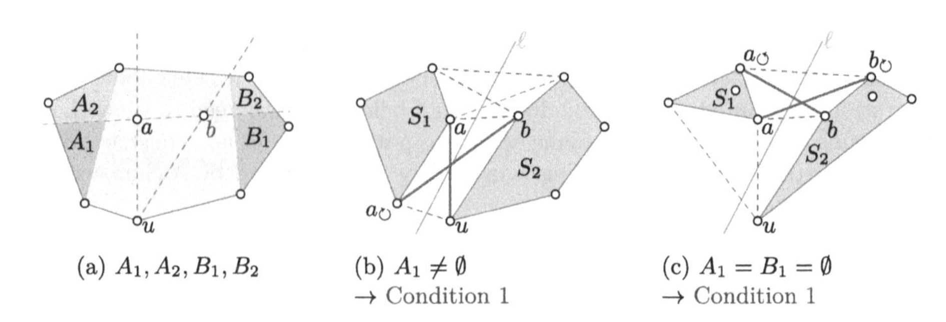

(a) A_1, A_2, B_1, B_2 (b) $A_1 \neq \emptyset$ $\rightarrow$ Condition 1 (c) $A_1 = B_1 = \emptyset$ $\rightarrow$ Condition 1

Fig. 9. Illustration for Case 1 of the proof of Theorem 3.

$S_1 = A \cup \{a\}$ and $S_2 = B \cup \{b, u\}$) of S. Its visibility graph $\mathcal{V}(S_1, S_2)$ contains the crossing edges au and $a_{\circlearrowleft}b$, which proves the result in this case.

If $B_1 \neq \emptyset$, then we can analogously find a crossing pair of edges in $\mathcal{V}(A \cup \{a, u\}, B \cup \{b\})$. Thus, we may further assume that $A_1 = B_1 = \emptyset$. Then we have $|A_2| = |B_2| = (n-3)/2 > 0$. Let $a_{\circlearrowleft} \in A_2$ be the counterclockwise neighbor of a along $\text{CH}(A \cup \{a\})$, and let $b_{\circlearrowright}$ be the clockwise neighbor of b along $\text{CH}(B \cup \{u, b\})$. Then $a_{\circlearrowleft}b$, $ab_{\circlearrowright}$ is a crossing pair of edges of $\mathcal{V}(A \cup \{a\}, B \cup \{b, u\})$.

Case 2: n is even. A line passing through two points $x, y \in S$ is a *halving line* of S if exactly $\frac{n-2}{2}$ points of S lie on each of its two sides. If $\overline{xy}$ is a halving line of S, where $x, y \in S$, then the segment xy is called a *halving segment* of S.

Claim 1. Let uv be a halving segment of a set S of n points in general position in the plane such that $u \in \partial\text{CH}(S)$. Then there is another halving segment pq of S such that the following three conditions hold: (1) an unbounded part of the ray emanating from p and passing through q lies in $(uv)^+$; (2) no point of $S \setminus \{u, v, p, q\}$ lies in the double-wedge $((uv)^+ \cap (pq)^-) \cup ((uv)^- \cap (pq)^+)$; (3) $p = v$, or the two open segments uv and pq cross.

Proof. Suppose without loss of generality that uv is a vertical line and u lies below v; see Fig. 10. Let $X := \text{CH}(S \cap (uv)^-)$ and $Y := \text{CH}(S \cap (uv)^+)$. Let $x \in X$ and $y \in Y$ be the extreme points of X and Y, respectively, such that $\overline{xy}$ avoids the interiors of X and Y, the set X lies above $\overline{xy}$, and Y lies below $\overline{xy}$.

If the open segments uv and xy cross, then $p = x$ and $q = y$ yields the claim; see Fig. 10a. Otherwise, xy crosses the line $\overline{uv}$ above v; see Fig. 10b. Let $p = v$ and $q \in Y$ be the extreme point of Y seen from p as the highest point of Y. (That is, all points of $Y \setminus \{q\}$ lie below the line $\overline{pq}$.) Then pq is a halving segment and yields the claim. □

Fix an extreme point u of S. Let uv be the unique halving segment incident to u. Let $A := S \cap (uv)^-$ and $B := S \cap (uv)^+$. Since uv is a halving segment, we have $|A| = |B| = \frac{n-2}{2}$. Let pq be the halving segment guaranteed by Claim 1.

If the halving segments uv and pq cross each other, then they both belong to the visibility graph $\mathcal{V}(A', B')$, where $A' := A \cup \{v\}$ and $B' := B \cup \{u\}$, and Condition 1 applies; see Fig. 11a. Thus we may assume that $p = v$ and there are

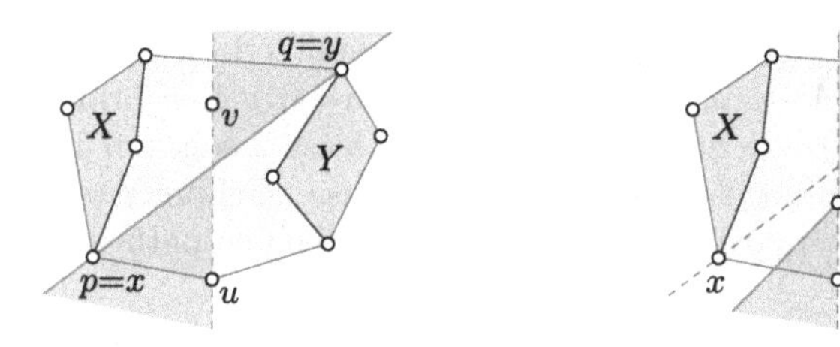

(a) Segments uv and xy cross (b) Segments uv and xy do not cross

Fig. 10. Illustration for Claim 1.

no points of S inside the double wedge $((uv)^+ \cap (vq)^-) \cup ((uv)^- \cap (vq)^+)$. By a mirror argument, we can also assume that there is a point $r \in A$ such that there are no points of S inside the double wedge $((uv)^- \cap (vr)^+) \cup ((uv)^+ \cap (vr)^-)$.

The line $\overline{qr}$ partitions $A \setminus \{r\}$ ($B \setminus \{q\}$, respectively) into two sets A_1 and A_2 (B_1 and B_2, respectively) such that A_1, B_1 lie in $(qr)^-$ and A_2, B_2 lie in $(qr)^+$.

We now distinguish three cases. Consider first the case $A_2 \neq \emptyset$ and $B_2 \neq \emptyset$; see Fig. 11b. Then the counterclockwise neighbor $r_{\circlearrowleft}$ of r along $\mathrm{CH}(A \cup \{u\})$ lies in A_2. Similarly, the clockwise neighbor $q_{\circlearrowright}$ of q along $\mathrm{CH}(B \cup \{v\})$ lies in B_2. It follows that the edges $qr_{\circlearrowleft}$ and $rq_{\circlearrowright}$ form a crossing pair in $\mathcal{V}((A \cup \{u\}), (B \cup \{v\}))$, and Condition 1 applies..

Consider now that one of the sets A_2 and B_2 is empty and the other one is non-empty; see Fig. 11c. By symmetry, we may assume that $A_2 \neq \emptyset$ and $B_2 = \emptyset$. Then $B_1 \neq \emptyset$. We consider the balanced separated partition (A', B'), where

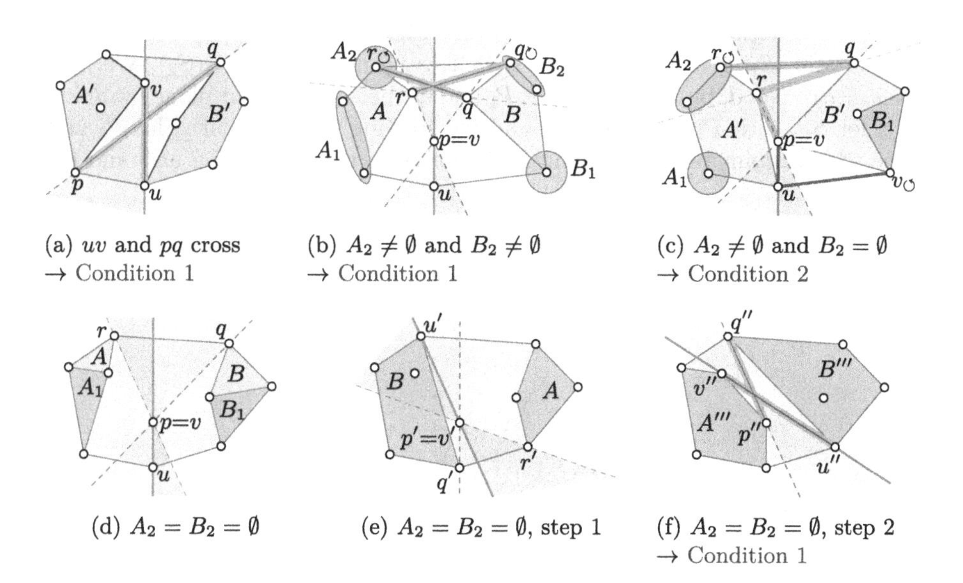

(a) uv and pq cross → Condition 1 (b) $A_2 \neq \emptyset$ and $B_2 \neq \emptyset$ → Condition 1 (c) $A_2 \neq \emptyset$ and $B_2 = \emptyset$ → Condition 2

(d) $A_2 = B_2 = \emptyset$ (e) $A_2 = B_2 = \emptyset$, step 1 (f) $A_2 = B_2 = \emptyset$, step 2 → Condition 1

Fig. 11. Illustration for Case 2 of Theorem 3.

$A' := A \cup \{u\}$ and $B' := B \cup \{v\}$. Let $r_{\circlearrowleft}$ be the counterclockwise neighbor of r along CH(A'). Since $A_2 \neq \emptyset$, $r_{\circlearrowleft}$ lies in A_2. Then (1) u is a bridged vertex for the partition (A', B'), (2) u is incident with at least 2 edges of $\mathcal{V}(A', B')$ – the edge uv and the edge $uv_{\circlearrowleft}$ for the counter-clockwise neighbor $v_{\circlearrowleft}s$ of v along CH(B'), and (3) $v \circ vr \circ r \circ rq \circ q \circ qr_{\circlearrowleft} \circ r_{\circlearrowleft}$ is a switchable path in $\mathcal{V}(A', B')$, and Condition 2 applies.

Finally, consider that $A_2 = \emptyset$ and $B_2 = \emptyset$; see Fig. 11d. Then q and r are neighbors along CH(S), and we again consider the whole analysis which started with fixing an extreme point of S but now we fix the point $u' := r$ instead of u; see Fig. 11e. Either we find a balanced separated partition satisfying Condition 1 or Condition 2, or we find two neighbors q' and r' along CH(S). In the first case we are done. In the latter case, the point q' is actually equal to u and it is clockwise of r' along CH(S). We then again consider the analysis which started with fixing an extreme point S but now we fix the point $u'' := r'$ instead of u; see Fig. 11f. Continuing this process, at some point we find a balanced separated partition satisfying condition 1) or 2) of the theorem, or otherwise after $n/2$ repetitions of the analysis starting with fixing an extreme point of S we would conclude that the set S is necessarily a wheel on n vertices. □

Let Z be a zig-zag path for a partition (S_1, S_2) of S. An edge ab of S is called *free* (with respect to Z) if $ab \in E(\mathcal{V}(S_1, S_2))$ and $ab \notin E(Z)$.

Lemma 6. *Let $|S| \geq 10$ and let Z be a zig-zag path for a balanced separated partition (S_1, S_2) of S which leaves at least two free edges. Then S allows three edge-disjoint plane spanning paths.*

Proof. Let ab and cd be two free edges with respect to a zig-zag path Z and a balanced separated partition (S_1, S_2); see Fig. 12a. Since $|S| \geq 10$, we have $|S_1| \geq 5$ and $|S_2| \geq 5$. Suppose $a, c \in \partial\text{CH}(S_1)$ and $b, d \in \partial\text{CH}(S_2)$. We may have $a = c$ or $b = d$, but not both. Let P_1 and Q_1 be edge-disjoint plane spanning paths for S_1, with P_1 starting at a and Q_1 starting at c. Similarly, let P_2 and Q_2 be edge-disjoint plane spanning paths for S_2, with P_2 starting at b and Q_2

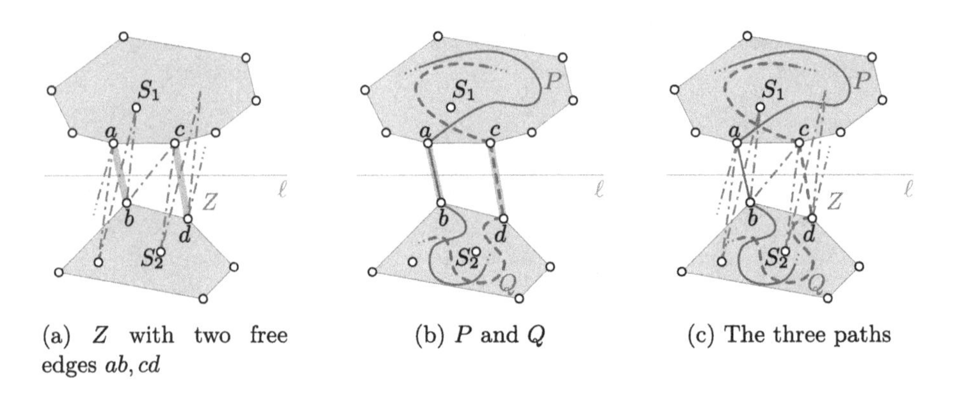

(a) Z with two free edges ab, cd (b) P and Q (c) The three paths

Fig. 12. Illustration for the proof of Lemma 6

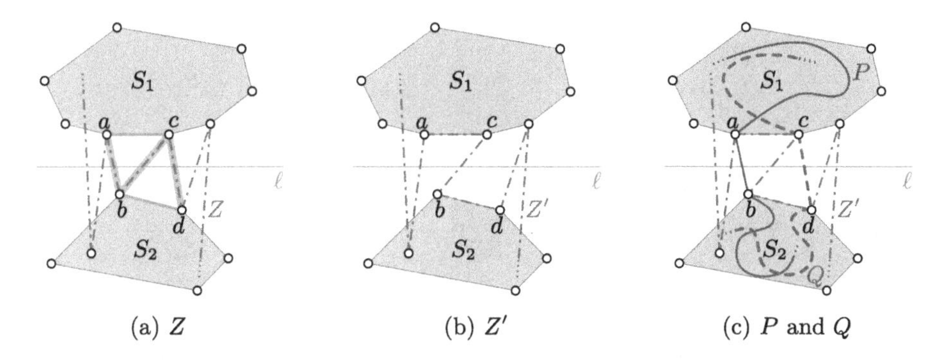

(a) Z (b) Z' (c) P and Q

Fig. 13. Illustration for the proof of Lemma 7

starting at d. The existence of such paths is guaranteed by Theorem 2. Then $P = P_2^{-1} \circ ba \circ P_1$ and $Q = Q_2^{-1} \circ dc \circ Q_1$ are edge-disjoint plane spanning paths for S; see Fig. 12b. Each of them is plane because the edge ab (cd, respectively) contains no point in the interior of $\mathrm{CH}(S_1)$ (of $\mathrm{CH}(S_2)$, respectively). Both of them are edge-disjoint with Z, because the only two edges of them that are incident with vertices from both S_1 and S_2 are ab and cd, and these are by assumption free w.r.t. Z; see Fig. 12c. □

The following lemmas can be proved similarly with Lemmas 2 and 3.

Lemma 7 ($\star$ – **Fig.** *13*). *Let $|S| \geq 10$ and let Z be a zig-zag path for a balanced separated partition (S_1, S_2) of S which contains all three edges of a switchable path of length 3 in $\mathcal{V}(S_1, S_2)$. Then S contains three edge-disjoint plane spanning paths.*

Lemma 8 ($\star$ – **Fig.** *14*). *Let $|S| \geq 10$ and let (S_1, S_2) be a balanced separated partition of S such that $\mathcal{V}(S_1, S_2)$ contains two crossing edges. Then S contains three edge-disjoint plane spanning paths.*

Lemma 9 ($\star$ – **Fig.** *15*). *Let $|S| \geq 10$ and let (S_1, S_2) be a balanced separated partition of S with $|S_1| \geq |S_2|$ such that $\mathcal{V}(S_1, S_2)$ contains a switchable path of length 3 and a bridged vertex in S_1 which does not belong to the switchable path and which is incident with at least two edges of $\mathcal{V}(S_1, S_2)$. Then S contains three edge-disjoint plane spanning paths.*

We are now ready to prove our main result Theorem 1.

Theorem 1. *Let S be any set of at least ten points in general position in the plane. There are three edge-disjoint plane straight-line spanning paths of S.*

Proof. Given a set of points S, apply Theorem 3. If S allows a balanced separated partition with two crossing edges in its visibility graph, S has three edge-disjoint

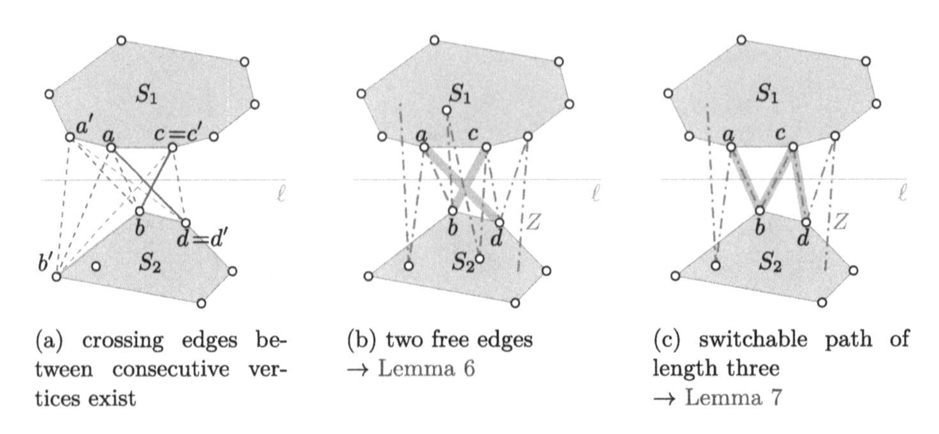

(a) crossing edges between consecutive vertices exist

(b) two free edges → Lemma 6

(c) switchable path of length three → Lemma 7

Fig. 14. Illustration for the proof of Lemma 8

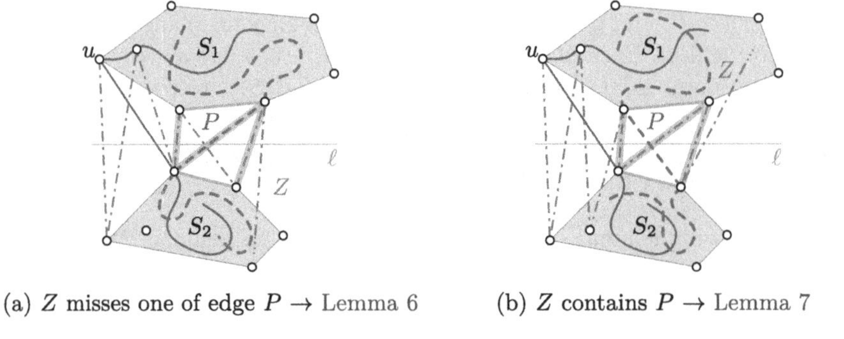

(a) Z misses one of edge P → Lemma 6

(b) Z contains P → Lemma 7

Fig. 15. Illustration for the proof of Lemma 9

plane spanning paths according to Lemma 8. If S allows a balanced separated partition whose visibility graph contains a switchable path of length three and a bridged vertex, then S has three edge-disjoint plane spanning paths according to Lemma 9. If none of these cases apply, then by Theorem 3 S is the wheel configuration W_n and n is even, in which case S has $n/2 - 1 \geq 4$ edge-disjoint plane spanning paths according to Proposition 1. □

6 Conclusion

In this paper, we showed that every set of at least 10 points in general position admits three edge-disjoint plane spanning paths. While we mostly focused on the combinatorial part, it is easy to see that our constructive arguments give rise to a polynomial time algorithm. We note that it's a simple exercise to verify that the 6-wheel configuration does not contain three edge-disjoint spanning paths. On the other hand, it was verified by a computer program that all sets of 7, 8, or 9 points contain three edge-disjoint spanning paths [12].

Can Theorem 2 be strengthened? Does any set of n points (for large enough n) in general position contain, for any choice of two distinct points s, t (not necessarily lying on the boundary of the convex hull of the set), edge-disjoint plane spanning paths starting in these points and not containing the edge st?

Let us mention in this connection that Theorem 1 cannot be strengthened in the way of Theorem 2. If the points of S are in convex position and the starting points of the three paths are prescribed to be the same point of S, then three edge-disjoint plane spanning paths do not exist (for a convex position, every path must start with an edge of $\partial \mathrm{CH}(S)$, and for a single point, there are only two such edges). The question is currently open to us if the starting points are required to be distinct.

Acknowledgments. The second and fourth authors gratefully acknowledge the support of Czech Science Foundation through research grant GAČR 23-04949X. The work of the third author is partially supported by " (i) MUR PRIN Proj. 2022TS4Y3N - "EXPAND: scalable algorithms for EXPloratory Analyses of heterogeneous and dynamic Networked Data"; (ii) MUR PRIN Proj. 2022ME9Z78 - "NextGRAAL: Next-generation algorithms for constrained GRAph visuALization". All authors acknowledge the working atmosphere of Homonolo meetings where the research was initiated and part of the results were obtained, as well as of Bertinoro Workshops on Graph Drawing, during which we could meet and informally work on the project. Our special thanks go to Manfred Scheucher whose experimental results encouraged us to keep working on the problem in the time when all hopes for a solution seemed far out of sight.

References

1. Abellanas, M., Garcia-Lopez, J., Hernández-Peñalver, G., Noy, M., Ramos, P.A.: Bipartite embeddings of trees in the plane. Discret. Appl. Math. **93**(2–3), 141–148 (1999). https://doi.org/10.1016/S0166-218X(99)00042-6
2. Aichholzer, O., et al.: Packing plane spanning trees and paths in complete geometric graphs. Inf. Process. Lett. **124**, 35–41 (2017). https://doi.org/10.1016/j.ipl.2017.04.006
3. Bernhart, F., Kainen, P.C.: The book thickness of a graph. J. Comb. Theory, Ser. B **27**(3), 320–331 (1979). https://doi.org/10.1016/0095-8956(79)90021-2
4. Bollobás, B., Eldridge, S.E.: Packings of graphs and applications to computational complexity. J. Comb. Theory Ser. B **25**(2), 105–124 (1978). https://doi.org/10.1016/0095-8956(78)90030-8
5. Bose, P., Hurtado, F., Rivera-Campo, E., Wood, D.R.: Partitions of complete geometric graphs into plane trees. Comput. Geom. **34**(2), 116–125 (2006). https://doi.org/10.1016/j.comgeo.2005.08.006
6. Geyer, M., Hoffmann, M., Kaufmann, M., Kusters, V., Tóth, C.D.: The planar tree packing theorem. J. Comput. Geom. **8**(2), 109–177 (2017). https://doi.org/10.20382/jocg.v8i2a6 https://doi.org/10.20382/jocg.v8i2a6 https://doi.org/10.20382/jocg.v8i2a6
7. Haler, S.P., Wang, H.: Packing four copies of a tree into a complete graph. Australas. J. Comb. **59**, 323–332 (2014). http://ajc.maths.uq.edu.au/pdf/59/ajc_v59_p323.pdf

8. Hedetniemi, S.M., Hedetniemi, S.T., Slater, P.J.: A note on packing two trees into k_n. Ars Combin. **11**, 149–153 (1981)
9. Hershberger, J., Suri, S.: Applications of a semi-dynamic convex hull algorithm. BIT **32**(2), 249–267 (1992). https://doi.org/10.1007/BF01994880
10. Kindermann, P., Kratochvíl, J., Liotta, G., Valtr, P.: Three edge-disjoint plane spanning paths in a point set. Arxiv report 2306.07237 (2023). https://doi.org/10.48550/arXiv.2306.07237
11. Luca, F.D., et al.: Packing trees into 1-planar graphs. J. Graph Algorithms Appl. **25**(2), 605–624 (2021). https://doi.org/10.7155/jgaa.00574
12. Scheucher, M.: Personal communication
13. Teo, S.K., Yap, H.P.: Packing two graphs of order n having total size at most 2n–2. Graphs Comb. **6**(2), 197–205 (1990). https://doi.org/10.1007/BF01787731

Decomposition of Geometric Graphs into Star-Forests

János Pach[1,2], Morteza Saghafian[2], and Patrick Schnider[3](✉)

[1] Rényi Institute of Mathematics, Budapest, Hungary
pach@cims.nyu.edu
[2] ISTA (Institute of Science and Technology Austria), Klosterneuburg, Austria
morteza.saghafian@ist.ac.at
[3] Department of Computer Science, ETH Zürich, Zürich, Switzerland
patrick.schnider@inf.ethz.ch

Abstract. We solve a problem of Dujmović and Wood (2007) by showing that a complete convex geometric graph on n vertices cannot be decomposed into fewer than $n-1$ star-forests, each consisting of non-crossing edges. This bound is clearly tight. We also discuss similar questions for abstract graphs.

Keywords: Geometric graphs · Graph Decomposition · Graph Thickness · Star Forests

1 Introduction

To determine the smallest number of subgraphs of some special kind that a graph G can be partitioned into is a large and classical theme in graph theory. In particular, the parts may be required to be matchings (as in Vizing's theorem [12]), complete bipartite graphs (as in the Graham-Pollak theorem [7]), paths and cycles (as in Lovász' theorem [9]), forests (as in the Nash-Williams theorem [10]), etc.

Most likely, it was Erdős who first realized that one can ask many interesting new extremal questions for graphs drawn in the plane or in some other surface, if we replace the purely combinatorial conditions by geometric ones; see [11]. For instance, we may require that the edges participating in a matching or a path do not cross each other [4,8]. In the 80s and 90s, the emergence of Graph Drawing as a separate discipline gave fresh impetus to this line of research.

János Pach's Research partially supported by European Research Council (ERC), grant "GeoScape" No. 882971 and by the Hungarian Science Foundation (NKFIH), grant K-131529. Work by Morteza Saghafian is partially supported by the European Research Council (ERC), grant No. 788183, and by the Wittgenstein Prize, Austrian Science Fund (FWF), grant No. Z 342-N31.

M. A. Bekos and M. Chimani (Eds.): GD 2023, LNCS 14465, pp. 339–346, 2023.
https://doi.org/10.1007/978-3-031-49272-3_23

A *geometric graph* G is a graph whose vertex set is a set of points in the plane, no 3 of which are collinear, and whose edges are (possibly crossing) line segments connecting certain pairs of vertices. If the vertices of G are in *convex position*, that is, they form the vertex set of a convex polygon, then G is called a *convex geometric graph.* In the sequel, whenever we say that a graph or a geometric graph G *can be decomposed* into certain parts, we mean that its *edge set*, $E(G)$, can be partitioned into such parts. Each part can be regarded as a different *color class* in the corresponding coloring.

A *star* is a graph consisting of a vertex together with some edges incident to it. In particular, a single vertex is counted as a star. A graph whose every connected component is a star is called a *star-forest.* The (edge set of a) complete graph K_n with n vertices can be decomposed into $n-1$ stars. Akiyama and Kano [2] proved that fewer stars do not suffice. (This also follows from the Graham-Pollak theorem [7], mentioned above.) However, it was also shown in [2] that one can decompose K_n into much fewer *star-forests*: one needs only $\lceil n/2 \rceil + 1$ of them. Can one also decompose a *complete convex geometric graph* on n vertices into fewer than $n-1$ star-forests, if we insist that each star-forest is a *plane graph*, that is, its edges do not cross each other? This question was raised by Dujmović and Wood [6] (Section 10).

The aim of this note is to answer this question in the negative.

Theorem 1. *Let $n \geq 1$. The complete convex geometric graph with n vertices cannot be decomposed into fewer than $n-1$ plane star-forests.*

On the other hand, there are complete geometric graphs where fewer than $n-1$ plane star-forests suffice: consider $P = A_1 \cup A_2 \cup A_3 \cup A_4$ a point set consisting of four pairwise disjoint sets $A_1, \ldots, A_4$, each of size k, such that for every choice $P_1 \in A_1, \ldots, P_4 \in A_4$ we have that P_4 lies inside the convex hull of P_1, P_2 and P_3. Then, it can be seen that the complete geometric graph on P can be decomposed into $3k = 3n/4$ plane star-forests, which come in three families: the first family consists of stars emanating from points in A_1 connecting to all points in A_1 and A_2 together with stars emanating from points in A_3 connecting to all points in A_3 and A_4. Similarly, we draw stars emanating from points in A_2 connecting to all points in A_2 and A_3 and from points in A_4 connecting to all points in A_4 and A_1, and for the last family stars from points in A_1 connecting to all points in A_1 and A_3 and from points in A_2 connecting to all points in A_2 and A_4.

The most important unsolved question in this direction is, how much the bound in Theorem 1 can be improved if we drop the assumption that the vertices are in convex position. We conjecture that the above example is optimal.

Conjecture 2. *Let $n \geq 1$. There is no complete geometric graph with n vertices that be decomposed into fewer than $\lceil 3n/4 \rceil$ plane star-forests.*

Note that in the example above, all star-forests had exactly two components. A star-forest consisting of at most k connected components (stars) is said to be a *k-star-forest.*

It is also an interesting open problem to determine the minimum number of plane k-star-forests that a complete (convex) geometric graph of n vertices can be decomposed into. We do not even know the answer to the analogous question for abstract graphs.

Problem 3. *Let k and n be fixed positive integers. What is the minimum number of k-star-forests that a complete graph K_n of n vertices can be decomposed into?*

As was mentioned earlier, for $k = 1$, the minimum is $n - 1$. The following result settles the first nontrivial case.

Theorem 4. *The complete graph with $n > 3$ vertices can be decomposed into $\lceil 3n/4 \rceil$ 2-star-forests. This bound cannot be improved.*

In particular, this shows that any counterexample to Conjecture 2 would require the use of star-forests with more than 2 components.

Many other variants of decomposing complete geometric graphs have been studied in the literature, including decompositions into plane spanning trees. The conjecture that every complete geometric graph on $2m$ vertices can be decomposed into m plane spanning trees has been recently disproved in [1]. Several notions of thickness studied in [6] are concerned with decompositions of graphs into plane substructures. For many other interesting questions on abstract and geometric graph parameters, consult [3] and [5].

In Sects. 2 and 3, we prove Theorems 1 and 4, respectively.

2 Covering with Plane Star-Forests–Proof of Theorem 1

Recall that a *plane star-forest* is a star-forest which is a plane graph, i.e., its edges do not cross each other. In this section, in a slight abuse of notations, we will denote the complete convex geometric graph on n points as K_n. Instead of *decompositions* of K_n into plane star-forests, it will be more convenient to consider *coverings*, that is, to allow an edge to belong to more than one star-forest (to have more than one "color"). This does not change the problem, because by keeping just one color for each edge, we turn any covering of the edge set of K_n into a decomposition.

Definition 5. *A collection of plane star-forests, $F_1, F_2, \dots, F_t$ forms a* covering *of K_n if every edge of K_n belongs to* at least one *F_i.*

For the proof, we need to introduce some simple terminology. The graphs consisting of just one vertex or a single edge are also regarded as stars. Every star S has a *center*. If S is a vertex, then it is its own center. If S is a single edge, we arbitrarily fix one of its endpoints and call it the center of S. The center of a star S is also said to be the *center of any edge* of S. Accordingly, if F is a (plane) star-forest, we always assume that each of its components is a star with a *fixed center*.

Proof (Proof of Theorem 1). For $n = 1, 2$, the statement is trivial. Assume for contradiction and let $n \geq 3$ be the smallest number for which the statement is not true. Let K_n be a complete convex geometric graph, and denote its vertices by $P_1, P_2, \ldots, P_n$, in clockwise order. The indices are taken modulo n, so that $P_{n+1} = P_1, P_{n+2} = P_2$, etc.

Suppose that K_n is covered by t plane star-forests, $F_1, F_2, \ldots, F_t$, for some $t < n - 1$. Our goal is to move some edges from one star-forest to another (i.e., to "recolor" them) in order to turn at least one F_i into a single star. We make sure that after each step of this process, we obtain a covering of K_n with plane star-forests. As soon as one of the F_is becomes a single star, we remove its center from K_n, and contradict with n being the smallest number for which we have a covering of K_n with fewer than $n - 1$ plane star-forests.

For every a, $1 \leq a \leq n$, and for every k, $1 < k < n$, we call the edge $P_a P_{a+k}$ a k-edge. Note that every k-edge is also a $(n - k)$-edge.

Definition 6. *A k-edge $P_a P_{a+k}$ is called* supported *if there exists F_i such that $P_a P_{a+k}$ belongs to F_i, and*

(i) either all edges $P_a P_{a+1}, P_a P_{a+2}, \ldots, P_a P_{a+k-1}$ belong to F_i,
(ii) or all edges $P_{a+1} P_{a+k}, P_{a+2} P_{a+k}, \ldots, P_{a+k-1} P_{a+k}$ belong to F_i.

Otherwise, we call it unsupported.

The goal is to recolor the edges step by step in order to make all the edges supported. For this purpose, the following observation is useful for the recoloring process.

Observation 7. *Suppose that the complete geometric graph K_n is covered by t plane star-forests, $F_1, F_2, \ldots F_t$. Let S be a connected component of F_i (that is, a star) where $1 \leq i \leq t$. Assume that no edge of S crosses an edge of F_j where $1 \leq j \leq t$, $j \neq i$. Remove the edges in S from F_i and add them to F_j. Then any edge that was supported before is still supported.*

Lemma 8. *Suppose that the complete geometric graph K_n can be covered by t plane star-forests, for some positive integer t.*

Then, for every k, $1 < k < n$, there exists a covering of K_n by t plane star-forests $F_1, F_2 \ldots, F_t$ such that every k'-edge with $1 < k' \leq k$ is supported.

Proof. We prove the lemma by induction on k.

Suppose that $k = 2$. By symmetry, it is sufficient to consider the 2-edge $P_1 P_3$ (that is, $a = 1$). We can assume without loss of generality that $P_1 P_3$ belongs to F_i, for some i, and its center is P_1 (which implies that $P_2 P_3$ is not in F_i). If $P_1 P_2$ belongs to F_i, condition (i) in Definition 6 is satisfied, and we are done. If $P_1 P_2$ does not belong to F_i, then add it to F_i. Obviously, it cannot cross any other edge in F_i. The only problem that may occur is that until now P_2 was a single vertex star in F_i, and now F_i has two stars that have a point in common. In this case, simply erase the single vertex star P_2 from F_i. Thus, the lemma is true for $k = 2$.

Suppose next that $k > 2$ and the statement has already been verified for $k-1$. We want to prove it for k.

By symmetry, it is enough to consider the k-edge P_1P_{k+1} and make it supported without making the already supported edges unsupported. Suppose without loss of generality that P_1P_{k+1} belongs to a star in F_i and the center of this star is P_1. The edges in F_i are marked *blue*.

Let $l < k+1$ be the *largest* index such that P_1P_l does *not* belong to F_i. Then the edges $P_1P_{l+1}, \dots, P_1P_{k+1}$ are all blue. If there is no such index l, then we are done, because F_i satisfies condition (i) in Definition 6.

By the induction hypothesis the edge P_1P_l is supported, so there exists a star-forest F_j, $j \neq i$, which contains P_1P_l along with all the edges $P_1P_2, P_1P_3, \dots, P_1P_{l-1}$ or along with all the edges P_2P_l, $P_3P_l, \dots, P_{l-1}P_l$. The edges of F_j are marked *red.* We distinguish two cases depending on these two possibilities.

Case 1: *The edges $P_1P_2, P_1P_3, \dots, P_1P_l$ belong to F_j.*

We make two changes. See Fig. 1.

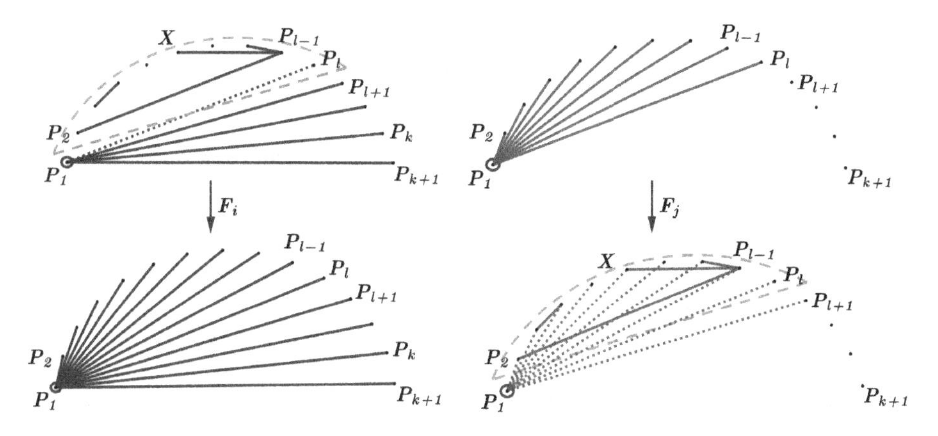

Fig. 1. In Case 1, recolor $P_1P_2, \dots, P_1P_l$ from red to blue, and all blue stars spanned by $\{P_2, \dots, P_l\}$ to red. A dotted line marks the absence of an edge. (Color figure online)

STEP 1: Remove the edges $P_1P_2, P_1P_3, \dots, P_1P_l$ from F_j and add all of them to F_i (unless they were already in F_i).

Then P_1P_{k+1} will satisfy condition (i) of definition 6 in F_i (with $a = 1$). However, in the process, we may have created some crossings within F_i, and F_i may also cease to be a star-forest. Both of these problems can be avoided by performing

STEP 2: Remove from F_i all (blue) edges connecting two elements of $\{P_2, P_3, \dots, P_l\}$ and add them to F_j.

Note that by recoloring the blue edges within $\{P_2, P_3, \dots, P_l\}$ to red, we do not violate the condition that F_j is a plane star-forest. Indeed, unless $l = 2$, originally, no element of $\{P_2, P_3, \ldots, P_l\}$ was connected by a red edge to any vertex other than P_1. Also by Observation 7, neither of the two steps results in any previously supported edge becoming unsupported.

Case 2: *The edges $P_2P_l, P_3P_l, \ldots, P_{l-1}P_l$ belong to F_j.*

First, we will modify F_i by including the edge P_1P_l. This will require some care, to make sure that the new covering does not violate the conditions. See Fig. 2.

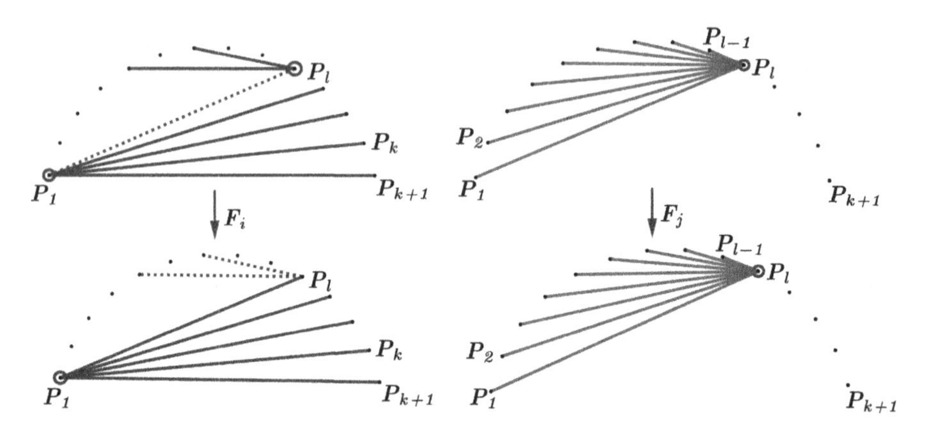

Fig. 2. In Case 2, P_1P_l will have two colors: red and blue. Remove the color blue from all previously blue edges incident to P_l. (Color figure online)

STEP 1: Add the edge P_1P_l to F_i, but also keep it in F_j. Remove from F_i all other edges incident to P_l.

Notice that after performing this step, we still have a covering of K_n by plane star-forests. It is a *covering*, because all edges deleted from F_i also belonged, and continue to belong, to F_j. Obviously, F_i remains a *star-forest*: its component containing P_1 remains a star, because we removed from F_i any other edge incident to P_l. Finally, F_i remains a *plane graph*, because its newly added edge, P_1P_l cannot cross any other blue edge. Indeed, such an edge should be incident to P_{l+1}, contradicting our assumption that P_1P_{k+1} originally belonged to a star in F_i, whose center is P_1. Also note that edges incident to P_l in F_i form a connected component which is already in F_j. So removing them is equivalent to recoloring them as red, which, by Observation 7, does not make any already supported edge unsupported.

Now we go back to the beginning of the proof, and again find the largest index l' such that $P_1P_{l'}$ does not belong to F_i. Obviously, we have $l' < l$. As before, we distinguish two cases. In Case 1, we conclude that P_1P_{k+1} satisfies condition (i) of Definition 6 in F_i (with $a = 1$), and we are done with the induction step. In Case 2, we can include the edge $P_1P_{l'}$ in F_i. Continuing like this, in fewer than k steps, we arrive at a situation where either P_1P_{k+1} satisfies condition (i) of definition 6 in F_i, or one by one, we manage to include all of the edges

$P_1P_{k+1}, P_1P_k, \ldots, P_1P_3, P_1P_2$ in F_i, which again means that P_1P_{k+1} satisfies condition (i) of definition 6 in F_i. This completes the proof of Lemma 8. □

Applying the lemma with $k = n - 1$ and $a = 1$, we can construct a covering of K_n by fewer than $n - 1$ plane star-forests such that one of them, again denoted by F_i, has the property that either $P_1P_2, P_1P_3, \ldots, P_1P_n$ belong to F_i, or $P_1P_n, P_2P_n, \ldots, P_{n-1}P_n$ belong to F_i. That is, F_i is a single star of degree $n - 1$, centered at P_1 or P_n. Deleting P_1 or P_n, resp., from K_n, we obtain a covering of K_{n-1} with fewer than $n - 2$ plane star-forests, which contradicts our assumption that Theorem 1 is true for decompositions and, hence, for coverings of the complete convex geometric graph K_{n-1}. This completes the proof of Theorem 1. □

3 2-Star-Forests–Proof of Theorem 4

Proof. Let V be an n-element set, and let $V = V_1 \cup V_2 \cup V_3 \cup V_4$ be a partition of V into 4 subsets as equal as possible. Suppose without loss of generality that

$$\lfloor n/4 \rfloor \leq |V_1| \leq |V_2| \leq |V_3| \leq |V_4| \leq \lceil n/4 \rceil.$$

Let $f : V_2 \to V_1$ be a surjection (onto mapping). For every $u \in V_2$, consider the two-star-forest F_u consisting of all edges connecting u to a every vertex in $V_2 \cup V_3$, and connecting $f(u)$ to every vertex in $V_1 \cup V_4$. These two-star-forests completely cover all edges within V_2 and V_1, and all edges in $V_1 \times V_4$ and in $V_2 \times V_3$. In a similar manner, we can construct $|V_4|$ two-star-forests that cover all edges within V_4 and V_3, and all edges in $V_4 \times V_2$ and $V_3 \times V_1$. Finally, with $|V_3|$ two-star-forests (with one center in V_3 and one in V_1), we can cover all edges in $V_3 \times V_4$ and $V_1 \times V_2$. Thus, we covered K_n with $|V_2| + |V_3| + |V_4| = \lceil 3n/4 \rceil$ two-star-forests, as required.

Next, we show that K_n cannot be covered by fewer than $\lceil 3n/4 \rceil$ two-star-forests, for any $n \geq 4$. The case $n = 4$ is easy. The proof is by contradiction. Let n be the smallest value greater than 4 for which there exists a covering of K_n by $t \leq \lceil 3n/4 \rceil - 1$ two-star-forests. Denote the two-star-forests participating in such a covering by $F_1, \ldots, F_t$. If any F_i has only one center, then deleting it from K_n, together with all edges incident to it, we reduce the number of vertices by 1 and the number of two-star-forests by 1. This would contradict the minimal choice of n. Thus, we can and will assume that every F_i, $1 \leq i \leq t$, has two centers.

Now consider a graph G with the same set of vertices as K_n, and for every 2-star-forest F_i, draw an edge in G between the two centers of stars in F_i. The resulting graph G has at most $\lceil 3n/4 \rceil - 1$ edges and, therefore, at least $n - \lceil 3n/4 \rceil + 1$ connected components. Note that $3(n - \lceil 3n/4 \rceil + 1) > \lceil 3n/4 \rceil - 1$, so there exists a connected component C in G with fewer than 3 edges.

If C is a single vertex u, then by construction it cannot be the center of any two-star-forest. Thus, we would need at least $n - 1$ two-star-forests just to cover the edges incident to u in K_n. If C consists of only one edge u_1u_2, then neither of these vertices can be the center of any other two-star-forest. Thus, the edge u_1u_2 was not covered by any two-star-forest F_j, which is a contradiction.

Finally, if C consists of two edges, u_1u_2 and u_1u_3, say, then it is not difficult to see that at least one of the edges between u_1, u_2, u_3 in K_n is not covered by any two-star-forest F_j. In each of the above cases, we obtained a contradiction. This completes the proof of Theorem 4. □

In view of Theorem 4, we state the following conjecture.

Conjecture 9. *For any $n \geq k \geq 2$, the number of k-star-forests needed to cover the complete graph K_n is at least $\lceil \frac{(k+1)n}{2k} \rceil$.*

For $k = 2$, the conjecture is true, by Theorem 4. We construct an example inspired by the construction in [2], showing that Conjecture 9, if true, is best possible. For simplicity, we describe it only for the case where n is divisible by 2. Assuming $n = 2t$, and labeling the vertices by $\{v_1, v_2, \cdots, v_n\}$, we create t 2-star-forests $F_1, F_2, \cdots F_t$ by picking vertices v_i and v_{i+t} as centers of F_i, $1 \leq i \leq t$ and connecting v_i to all vertices v_j, $i < j < i+t$, and connecting v_{i+t} to all vertices v_{j+t}, $i < j < i+t$ (the indices are taken modulo n). The introduced 2-star-forests cover all edges of K_n, except the set of edges v_iv_{i+t} which can be simply decomposed into $\lceil \frac{n}{2k} \rceil$ k-star-forests. Altogether, K_n can be covered by $\frac{n}{2} + \lceil \frac{n}{2k} \rceil$ k-star-forests.

References

1. Aichholzer, O., et al: Edge partitions of complete geometric graphs. In: 38th International Symposium on Computational Geometry (SoCG 2022) (2022)
2. Akiyama, J., Kano, M.: Path factors of a graph. In: Graphs and Applications (Boulder, Colo., 1982), pp. 1–21. Wiley, New York (1985)
3. Araujo, G., Dumitrescu, A., Hurtado, F., Noy, M., Urrutia, J.: On the chromatic number of some geometric type Kneser graphs. Comput. Geom. **32**(1), 59–69 (2005)
4. Avital, S., Hanani, H.: Graphs, continuation (in Hebrew). Gilyonot Le'matematika **3**(2), 2–8 (1966)
5. Bose, P., Hurtado, F., Rivera-Campo, E., Wood, D.R.: Partitions of complete geometric graphs into plane trees. Comput. Geom. **34**(2), 116–125 (2006)
6. Dujmović, V., Wood, D.R.: Graph treewidth and geometric thickness parameters. Discrete Comput. Geom. **37**(4), 641–670 (2007)
7. Graham, R.L., Pollak, H.O.: On the addressing problem for loop switching. Bell Syst. Tech. J. **50**, 2495–2519 (1971)
8. Kupitz, Y.S.: Extremal problems of combinatorial geometry. In: Lecture Notes Series, vol. 53. Aarhus University, Denmark (1979)
9. Lovász, L.: On covering of graphs. In: Theory of Graphs (Proc. Colloq., Tihany, 1966), pp. 231–236. Academic Press, New York (1968)
10. Nash-Williams, C.S.J.: Decomposition of finite graphs into forests. J. London Math. Soc. **39**, 12 (1964)
11. Pach, J.: The beginnings of geometric graph theory. In: Erdős centennial, Bolyai Soc. Math. Stud. János Bolyai Math. Soc., Budapest **25**, 465–484 (2013)
12. Vizing, V.G.: On an estimate of the chromatic class of a p-graph (in Russian). Diskret. Analiz (3), 25–30 (1964)

Graph Drawing Contest Report

Graph Drawing Contest Report

Philipp Kindermann[1], Fabian Klute[2], Tamara Mchedlidze[3], Wouter Meulemans[4(✉)], and Debajyoti Mondal[5]

[1] Universität Trier, Trier, Germany
kindermann@uni-trier.de
[2] Polytechnic University of Catalunya, Barcelona, Spain
fabian.klute@upc.edu
[3] Utrecht University, Utrecht, The Netherlands
t.mtsentlintze@uu.nl
[4] TU Eindhoven, Eindhoven, The Netherlands
w.meulemans@tue.nl
[5] University of Saskatchewan, Saskatoon, Canada
dmondal@cs.usask.ca

Abstract. This report describes the 30th Annual Graph Drawing Contest, held in conjunction with the 31st International Symposium on Graph Drawing and Network Visualization (GD'23) in Isola delle Femmine (Palermo), Italy. The contest still allowed for remote participation, running in a hybrid format, with both on-site and online participants. Nearly all participants of the manual live contest though were onsight. The mission of the Graph Drawing Contest is to monitor and challenge the current state of the art in graph-drawing technology.

1 Introduction

Following the tradition of the past years, the Graph Drawing Contest was divided into two parts: the *creative topic* and the *live challenge*.

Changing the format slightly, there was only one data set in the creative category. The set was the *Boardgame Recommendations*: The data represent the collection of the top 100 boardgames as ranked on http://www.boardgamegeek.com and the recommendations between the games by users of the platform. The data set was published about half a year in advance, and contestants submitted their visualizations before the conference started.

The live challenge took place during the conference in a format similar to a typical programming contest. Teams were presented with a collection of *challenge graphs* and had one hour to submit their highest scoring drawings. This year's topic was new: given an undirected simple graph and a point set, find straight-line drawing of the graph with the vertices drawn ontop of the points such that the number of crossing edges is minimized.

Overall, we received 41 submissions: 15 submissions for the creative topics and 26 submissions for the live challenge (19 manual and 7 automatic).

M. A. Bekos and M. Chimani (Eds.): GD 2023, LNCS 14465, pp. 349–357, 2023.
https://doi.org/10.1007/978-3-031-49272-3

2 Creative Topic

The dataset for the Creative Topic represents a collection of the top 100 boardgames on http://www.boardgamegeek.com. Of these 100 games we required the contestants to visualize the top 40. Most teams opted for focusing on this reduced dataset. Each entry corresponds to a boardgame and contains the following information:

- The id of the game (`id`)
- The name of the boardgame (`title`)
- The publishing year (`year`)
- The rank of the game in the top 100 (`rank`)
- Minimum/maximum number of players (`minplayers/maxplayers`)
- Minimum/maximum playtime in minutes (`minplaytime/maxplaytime`)
- Minimum age to play the game (`minage`)
- Record with the rating and number of reviews (`rating`)
- An entry containing a list with the ids of games that fans of this game recommend (`recommendations`)
- Record containing the categories and mechanics of the game (`types`)
- Record containing the designers of the game (`credit`)

The data was extracted from the http://www.boardgamegeek.com website via a Python script. The network to visualize was the graph given by the recommendations, i.e., the graph in which every boardgame is a vertex and two vertices u and v are connected by a directed edge if the boardgame corresponding to vertex u contains the one corresponding to v in its recommendations.

The general goal of the creative topics was visualize the dataset with complete artistic freedom, and with the aim of communicating as much information as possible from the provided data in the most readable and clear way. We received 15 submissions for the creative challenge. Submissions were evaluated according to four criteria:

(i) readability and clarity of the visualization,
(ii) aesthetic quality,
(iii) novelty of the visualization concept, and
(iv) design quality.

We noticed overall that it is a complex combination of several aspects that make a submission stand out. These aspects include but are not limited to the understanding of the structure of the data, investigation of the additional data sources, applying intuitive and powerful data visual metaphors, careful design choices, combining automatically created visualizations with post-processing by hand, as well as keeping the visualization, especially the text labels, readable. We selected the top six submissions before the conference, which were printed on large poster boards and presented at the Graph Drawing Symposium. We also made all the submissions available on the contest website in the form of a virtual poster exhibition. During the conference, we presented these submissions and announced the winners. For a complete list of submissions, refer to https://www.graphdrawing.org/gdcontest/2023/results/. Eight of the submissions were accompanied by an online tool, which are linked on the web page.

3rd Place: Florentina Voboril, Felicia Schmidt and Viktoria Pogrzebacz (TU Wien).

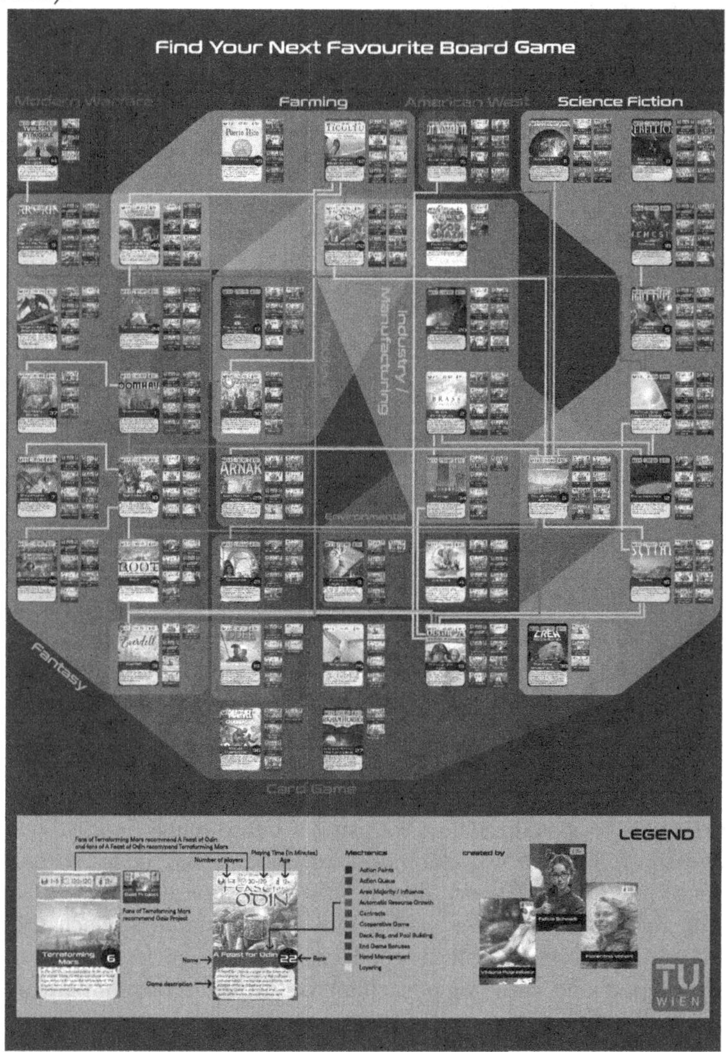

This submission was valued for its great ideas and clever choices. Representing the games as cards resembling popular trading card games is appealing at least to other players of such games. The choice to only use edges for bidirectional relations and visualize unidirectional ones via a tabulation is clever and reliefs some clutter.

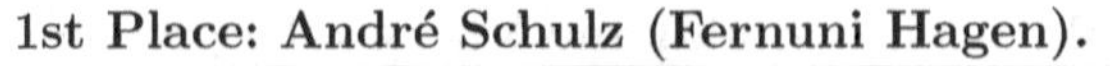

1st Place: André Schulz (Fernuni Hagen).

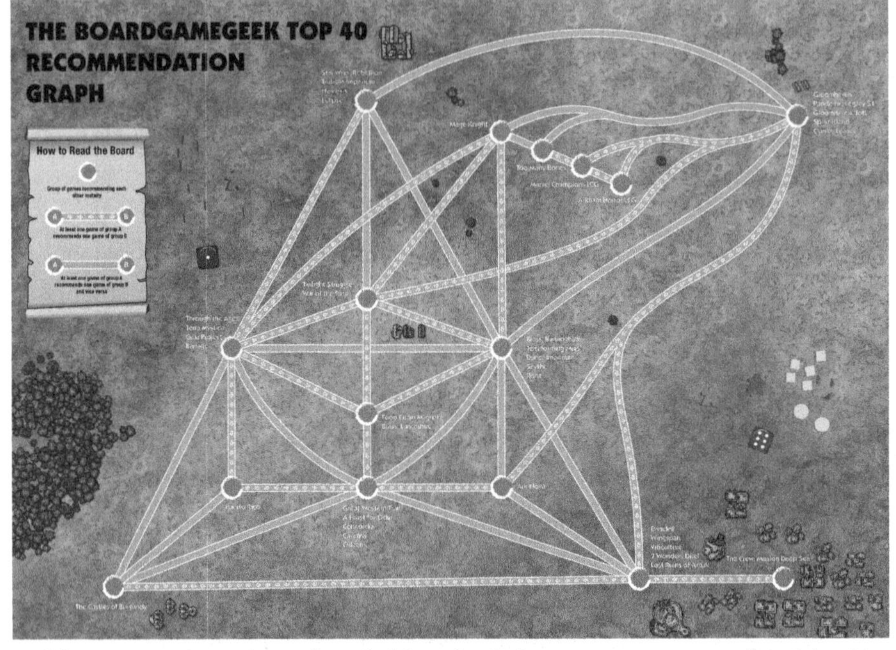

The contest committee found this submission very easy to read and inviting for further investigation of the poster. The design quality is high, both in the visual qualities and for the graph layout. Especially, the usage of clustering enabled a very nice vertex placement and edge layout. The choice to solve some crossings via confluency was appreciated.

> “I had the idea to visualize the data using a board game board metaphor. Showing the full graph was not possible, since its 337 edges surpass the number of edges of a network drawn on a game board. I decided to save edges by representing games, which mutually recommend each other as a single cluster-node. Furthermore, I only added an edge between two cluster-nodes if one game in one cluster-node recommends at least one game in another cluster-node. By this some information got lost but many edges were saved. The best distribution of clusters was found by exhaustive search. I managed to reduce the graph down to 16 cluster-nodes and 57 directed edges. For the layout I started with a forced-directed layout, which was manually adjusted to reflect the symmetry of the graph better. Arrows drawn on the directed edges were used to indicate the direction. I curved some of the edges to make room for the labels, which were placed next to the vertices. Also some edges were bundled if there were to many incident edges at a node.”
>
> *André Schulz*

1st Place: Christoph Kern, Manuel Oberbacher, and Horst Zahradnik (TU Wien).

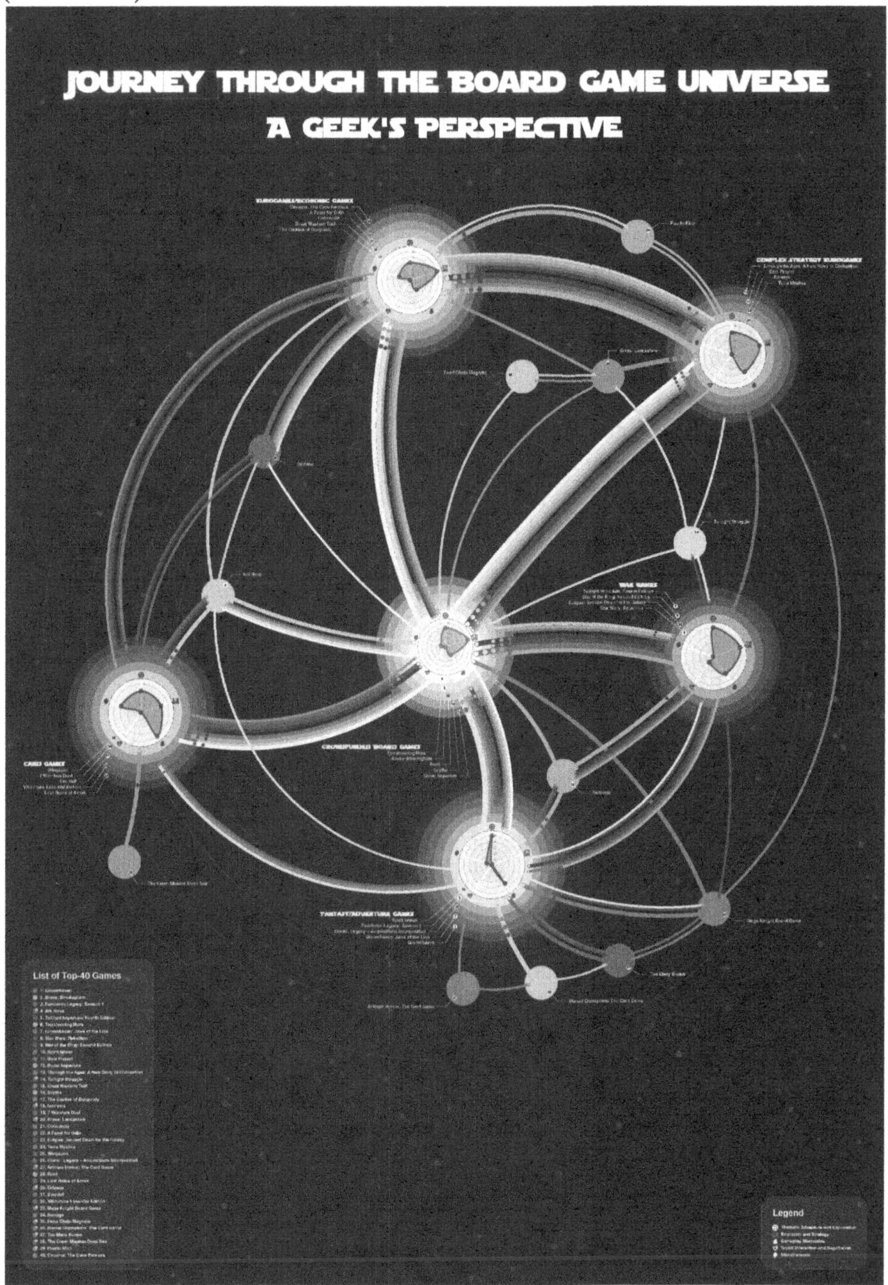

The committee found this visualization to look "quite amazing". While at first maybe overwhelming with visual details, the choices and execution make sense and is of high quality. The layout of the vertices and edges make sense throughout and meta-information is chosen appropriately without trying to fit all the information possible. The usage of clustering also makes a lot of sense, and the rings fit the galactic theme.

"For our poster, our goal was to reduce the impact of the numerous edges within the dataset. Therefore, we needed to group games effectively. We explored various approaches and ultimately adopted an ILP-based clustering technique aimed at minimizing the number of cliques, each comprising at least 3 games, based on the "fans_liked" relationship. This strategy significantly reduced the influence of edges in the graph, allowing us to organize games into cliques stacked upon one another, leaving only the remaining edges to be drawn. Furthermore, we implemented edge bundling, adjusting the edge widths based on the number of games liked within other cliques by a particular game, thereby adding an extra dimension to the graph. Finally, the positioning of nodes and edges was done manually. The central clique, connected to all other cliques (akin to the Sun), was placed at the center, while the remaining cliques (representing planets) were positioned on outer spheres, replicating a Solar System-like structure. The remaining games not included in cliques were positioned as moons around the planets."
Christoph Kern, Manuel Oberbacher and Horst Zahradnik

3 Live Challenge

The live challenge took place during the conference and lasted exactly one hour. During this hour, local participants of the conference could take part in the manual category (in which they could attempt to draw the graphs using a supplied tool: http://graphdrawing.org/gdcontest/tool/), or in the automatic category (in which they could use their own software to draw the graphs). Because of the global COVID-19 pandemic, we allowed everybody in both categories to participate remotely. To coordinate the contest, give a brief introduction, answer questions, and give participants the possibility to form teams, we were kindly provided with both a room in the conference building, and a Zoom stream for the conference; furthermore, participants could also meet and follow the contest via a dedicated room in gather.town.

The challenge focused on placing the vertices of an undirected simple graph on a given point set with the goal to minimize the edge crossings in the resulting straight-line drawing. We allowed for points of the point set to be collinear and for vertices to lie ontop of edges. For each proper crossings we added one to the quality measure and for each vertex-edge overlap we added n to the quality measure where n was the number of vertices. Embedding vertices at fixed or constraint locations is a researched topic in information visualization and graph drawing often with a focus on achieving plane drawings. With this challenge we hope to point to the possibility in this topic to also look at classic quality measures, such as edge crossings.

3.1 The Graphs

In the manual category, participants were presented with seven graphs. These were arranged from small to large with the exception of the last graph and chosen to contain different types of graph structures. In the automatic category, participants had to draw the same seven graphs as in the manual category, and in addition another seven larger graphs. Again, the graphs were constructed to have different structures.

For illustration, we include below the fourth graph, where the contestants were given a bipartite graph on a classic pointset. The best manual solution (by team *RoMaMa*) managed to find a drawing with 5 crossings while the best automatic solution achieved 4 crossings (by team *Baseline*).

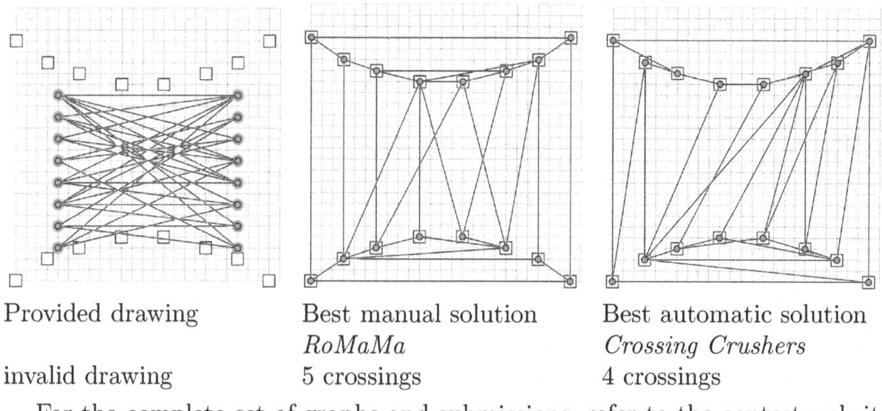

Provided drawing	Best manual solution	Best automatic solution
	RoMaMa	*Crossing Crushers*
invalid drawing	5 crossings	4 crossings

For the complete set of graphs and submissions, refer to the contest website at https://www.graphdrawing.org/gdcontest/2023/results/. The graphs are still available for exploration and solving Graph Drawing Contest Submission System: https://www.graphdrawing.org/gdcontest/tool/.

Similarly to the past years, the committee observed that manual (human) drawings of graphs often display a deeper understanding of the underlying graph structure than automatic and therefore gain in readability. Moreover, on all but two graphs the humans were able to find a solution with the same number of crossings (presumably the best possible) as the automatic solutions. For the larger graphs, the automatic solutions had problems with instance number eleven. It is not yet clear if this is due to an error in their implementation or in the data. Since no team managed to submit a valid solution, this was without consequence to the contest.

3.2 Results: Manual Category

Below we present the full list of scores for all teams. The numbers listed are the edge-length ratios of the drawings; the horizontal bars visualize the corresponding scores.

graph	1	2	3	4	5	6	7
Phoenicopteridae	3	25	1	22	10	587	81
Hello KITty	4	33	5	8	4	531	78
Goose	9	41	18	48	18	581	29
TU RACs	4	29	9	31	347	622	28
yPeoplez	8	25	25	56	149	622	40
Power Graph Girls	5	29	15	16	4	502	95
We have GD at home	3	33	0	9	4	1361	19
Beppos	5	29	0	17	110	622	29
Esri	5	38	17	20	160	689	40
We found a new Martin Gronemann!	5	27	0	10	9	593	33
HliKra	3	27	0	19	10	590	19
yWorks Team Yellow	3	25	6	9	16	622	25
Hebonk	3	29	0	14	4	583	25
???????????? keyboard, who dis?	6	27	0	17	379	1415	19
Grand Theft Edge Crossing	5	35	18	21	1602	1565	76
The Backseats	3	29	20	59	482	567	34
RoMaMa	3	25	5	5	4	681	29
Duck	4	29	0	32	106	2038	25
Konstanz	5	29	2	12	575	1005	33

Third place: **We have GD at home**, consisting of Alexander Dobler, Jules Wulms.
Second place: **RoMaMa**, consisting of Robert Ganian, Maarten Löffler, Martin Nöllenburg
Winner: **Hebonk**, consisting of Tim Hegemann, Florentina Voboril, Johannes Zink.

“ GD'23 conference day one, Sicily 17:45. A tentative team has assembled, but the final lineup has not yet been decided. 17:50. It's time to stick with it. An hour earlier, nothing of it was foreseeable, but it was a stroke of luck, as it would later turn out. 17:55. Intense discussions about the team name are taking place. Combining parts of our names? Vohezi? Flotijo? No! 18:00. The focus is now for one hour on vertices on points, on crossings, and on overlaps. Our strategy is, of course, a meticulous inspection of graph properties, e.g., which vertices can and cannot be placed onto the middle points if we want to avoid overlaps, and luck. 19:00. That's it. First place and the clear lead on the frozen scoreboard, but previous contests have taught us that this does not need to mean anything. And, indeed, it turns out to be again a very close race of many great teams, but it is enough. ”
Tim Hegemann, Florentina Voboril, Johannes Zink

3.3 Results: Automatic Category

In the following we present the full list of scores for all teams that participated in the automatic category. The numbers listed are the edge-length ratios of the drawings; the horizontal bars visualize the corresponding scores.

graph	1	2	3	4	5	6	7	8	9	10	11	12	13	14
Crossing Crushers	3	25	0	4	4	409	19	5.3E6	7.6E4	6.2E5	-2	4.6E7	8.2E5	3.9E8
Baseline	3	25	37	4	1566	2546	199	5.8E6	5.3E5	7.2E7	-2			
GKLAJER	3	25	5	5	4		20		1.8E6					
Minamo	3	25	7	4	4	443	19	6.7E6	3E5	4.1E6	-2	1.5E4	1.9E4	3.9E8
OMeGA	3	25	0	4	4	395	19	4.7E6	7.4E4	3994	-2	1.4E4	9.1E5	2.1E8
12345	6	25	0	14	14	1216	26	5.3E6	2E5					
LookMumNoCrossing	3	25	423	12	1448	574	387	5E6	5.1E4	7.7E7	-2	4.1E7	9E5	3.6E8

Third place: **Crossing Crushers**, consisting of Martin Siebenhaller and Benjamin Niedermann.
Second place: **Minamo**, consisting of Saman Miran, Jakub Nawrocki, Hesham Morgan.
Winner: **OMeGA**, consisting of Laurent Moalic, Dominique Schmitt, and Julien Bianchetti.

“Our algorithm is based on a simulated annealing approach (SA). In order to get an initial solution, we consider three embeddings. The first is the one given by the input file, in case the nodes are already assigned to points. The second one is randomly generated. To get the third one, we apply the Fast Multiple Multilevel Embedder (FMME) algorithm, implemented in the OGDF library. This layout is then scaled to match the axes-parallel bounding box of the given points. Afterwards, every node is assigned to its closest available point. Notice that graphs 11 and 14 had issues with the FMME algorithm. Out of these three potential layouts, we keep the one with the lowest number of crossings. A finely tuned simulated annealing is then used to iteratively select one node at random and move it to a random point. In case another node had already been assigned to that point, the two nodes are swapped. According to the SA principle, the move is always accepted if it decreases the current number of intersections. To avoid falling in a local optimum, the move is also accepted with a certain probability when it degrades the current solution. The probability is high at the beginning of the algorithm and is low at the end. We ran this algorithm on the 14 given graphs simultaneously on a 12-core CPU but stopped the computation early for the smaller graphs.”
Laurent Moalic, Dominique Schmitt, and Julien Bianchetti

Acknowledgments. The contest committee would like to thank the organizing and program committee of the conference; the organizers who provided us with a room with hardware for the live challenge and monetary prizes; the generous sponsors of the symposium; and all the contestants for their participation. Further details including all submitted drawings and challenge graphs can be found on the contest website: https://www.graphdrawing.org/gdcontest/2023/results/.

Author Index

M. A. Bekos and M. Chimani (Eds.): GD 2023, LNCS 14465, pp. 359–361, 2023.
https://doi.org/10.1007/978-3-031-49272-3

Printed in the United States
by Baker & Taylor Publisher Services